Handbook of Experimental Pharmacology

Volume 251

More information about this series at http://www.springer.com/series/164

Alexander Pfeifer • Martin Klingenspor •
Stephan Herzig
Editors

Brown Adipose Tissue

Springer

Editors
Alexander Pfeifer
Institute of Pharmacology & Toxicology
University Hospital Bonn
University of Bonn
Bonn, Germany

Martin Klingenspor
Institute of Molecular Nutritional Medicine
TU München
Freising, Bayern, Germany

Stephan Herzig
Institute for Diabetes and Cancer IDC
Helmholtz Center München
Neuherberg, Germany

ISSN 0171-2004 ISSN 1865-0325 (electronic)
Handbook of Experimental Pharmacology
ISBN 978-3-030-10512-9 ISBN 978-3-030-10513-6 (eBook)
https://doi.org/10.1007/978-3-030-10513-6

Library of Congress Control Number: 2019933559

This Springer imprint is published by the registered company Springer Nature Switzerland AG.
The registered company address is: Gewerbestrasse 11, 6330 Cham, Switzerland

Preface

Spotlight on Brown Adipose Tissue: Four Short Questions

What Is Brown Adipose Tissue?

This simple question is not so easy to answer, because brown fat is a special type of fat that has not much in common with the adipose tissue that we usually associate with "fat," i.e., the white adipose tissue. White fat is our largest store of energy, whereas brown fat generates heat – this process is also called non-shivering thermogenesis, because it does not rely on muscle contraction and shivering. Energy expenditure by brown fat is physiologically induced by cold stress but can also be induced pharmacologically by the stress hormone norepinephrine. Brown adipose tissue is special because it expresses the unique uncoupling protein 1 (UCP1) that uncouples mitochondrial respiration from ATP production, thus generating heat. The location and quantity of brown fat vary with age and sex; it can be found mainly between the shoulder blades, in the neck, deep within the chest around the great vessels, and around the kidney.

What Is Beige Fat?

Adipose tissue exhibits a substantial degree of plasticity and depots can change their phenotype/color. White fat – especially the subcutaneous depot – can take on the appearance of brown fat. This process can be observed after cold exposure and is also known as "browning." These cells share many characteristics with brown adipocytes and are therefore called brown-in-white (thus the abbreviation brite) or beige cells. Importantly, beige fat can contribute to whole body energy expenditure, albeit to a lesser extent than brown fat.

Why Is BAT so Interesting at Present or Why Do We Need a Compendium of Reviews on BAT Now?

Obesity has reached pandemic dimensions, and there is a lack of specific and efficient pharmacological treatment of overweight and obesity. Thus, there is high medical need for novel treatment strategies, and increasing energy expenditure has been suggested to be a potential strategy to fight obesity. But how can energy expenditure be achieved? Telling obese people to exercise is obviously not enough. The publications in 2007 and 2009 showing that adult humans possess metabolically active brown fat and that its activity correlates with leanness sparked off new interest in this special type of fat. Many labs shifted their attention to brown and beige fat, and since the last decade a wealth of new studies have been published on this topic.

This review is especially dedicated to those scientists newly intrigued by the metabolic power of brown adipose tissue.

How Is This Handbook Structured?

The 21 articles of the handbook are arranged into four parts:

- Part I focuses on the differences in the development of brown versus beige adipocytes and how brown adipocytes can be cultured in vitro. The focus of the articles on adipocyte models lies on human brown adipocytes. In addition to technical aspects of lineage tracing in vivo, aspects of brown adipocyte aging are covered in this chapter.
- Part II centers on molecular mechanisms of BAT function, especially on UCP1, and signaling mechanisms. The latter encompass papers on novel lipid signals that control BAT and the second messengers cAMP that plays a major role in BAT activation and its "smaller sister/brother" cGMP, which is getting more attention as an important enhancer of brown adipocyte differentiation. This chapter also deals with the expanding field of noncoding RNAs (microRNAs and long-noncoding RNAs) of BAT and beige fat. It is well established that white fat secretes a broad spectrum of hormones and endocrine factors (e.g., leptin and adiponectin), and recent studies suggest that BAT also has endocrine functions.
- The function of BAT in human adults is still not completely understood. A major hurdle is the lack of efficient and cheap diagnostic markers that do not expose subjects to radiation (X-ray and radioactive tracer). This is the topic of Part III.
- Finally, Part IV deals with BAT activation in humans by foods and drugs. The handbook closes with a detailed review of the potential of BAT as a pharmacological target.

Bonn, Germany — Alexander Pfeifer
Freising, Germany — Martin Klingenspor
Neuherberg, Germany — Stephan Herzig

Contents

Part I Origin of Brown and Beige Adipocytes

Brown Adipose Tissue Development and Metabolism 3
Su Myung Jung, Joan Sanchez-Gurmaches, and David A. Guertin

Lessons from Cre-Mice and Indicator Mice 37
Christian Wolfrum and Leon Gabriel Straub

Aging of Brown and Beige/Brite Adipose Tissue 55
Antonia Graja, Sabrina Gohlke, and Tim J. Schulz

Adipogenesis in Primary Cell Culture 73
Therese Juhlin Larsen, Naja Zenius Jespersen, and Camilla Scheele

In Vitro Models for Study of Brown Adipocyte Biology 85
Mark Christian

Brown-Like Adipocyte Progenitors Derived from Human iPS Cells: A New Tool for Anti-obesity Drug Discovery and Cell-Based Therapy? 97
Xi Yao, Barbara Salingova, and Christian Dani

Brown Adipose Tissue in Human Infants 107
Martin E. Lidell

Part II Molecular Mechanisms of BAT Function and Signaling

Evolution of UCP1 127
Michael J. Gaudry, Kevin L. Campbell, and Martin Jastroch

The Mechanism FA-Dependent H^+ Transport by UCP1 143
Ambre M. Bertholet and Yuriy Kirichok

Role of cAMP and cGMP Signaling in Brown Fat 161
Laia Reverte-Salisa, Abhishek Sanyal, and Alexander Pfeifer

Fatty Acid Metabolites as Novel Regulators of Non-shivering Thermogenesis 183
Stefanie F. Maurer, Sebastian Dieckmann, Karin Kleigrewe, Cécilia Colson, Ez-Zoubir Amri, and Martin Klingenspor

Regulatory Small and Long Noncoding RNAs in Brite/Brown Adipose Tissue 215
Marcel Scheideler

Brown Adipokines 239
Francesc Villarroya, Aleix Gavaldà-Navarro, Marion Peyrou, Joan Villarroya, and Marta Giralt

Part III Detection of BAT In Vivo

Infrared Thermography 259
James Law, David E. Morris, Helen Budge, and Michael E. Symonds

In Vivo Detection of Human Brown Adipose Tissue During Cold and Exercise by PET/CT 283
Emmani B. M. Nascimento and Wouter D. van Marken Lichtenbelt

Techniques and Applications of Magnetic Resonance Imaging for Studying Brown Adipose Tissue Morphometry and Function 299
Dimitrios C. Karampinos, Dominik Weidlich, Mingming Wu, Houchun H. Hu, and Daniela Franz

Multispectral Optoacoustic Tomography of Brown Adipose Tissue . . . 325
Angelos Karlas, Josefine Reber, Evangelos Liapis, Korbinian Paul-Yuan, and Vasilis Ntziachristos

BAT Exosomes: Metabolic Crosstalk with Other Organs and Biomarkers for BAT Activity 337
Deborah Goody and Alexander Pfeifer

Part IV Recruitment and Activation of Human BAT

Activation of Human Brown Adipose Tissue (BAT): Focus on Nutrition and Eating 349
Kirsi A. Virtanen

Translational Aspects of Brown Fat Activation by Food-Derived Stimulants 359
Takeshi Yoneshiro, Mami Matsushita, and Masayuki Saito

Translational Pharmacology and Physiology of Brown Adipose Tissue in Human Disease and Treatment 381
Christopher J. Larson

Part I

Origin of Brown and Beige Adipocytes

Brown Adipose Tissue Development and Metabolism

Su Myung Jung, Joan Sanchez-Gurmaches, and David A. Guertin

Contents

1 Introduction and Background 4
1.1 Overview 4
1.2 Basics of Non-shivering Thermogenesis 5
1.3 Brown Fat Anatomy and Morphology 6
1.4 BAT Vascularization and Innervation 10
1.5 Transcriptional Control of Brown Adipocyte Differentiation 11
2 Brown Fat Growth 12
2.1 Techniques for Studying BAT Development 12
2.2 Brown Adipocyte Origins 14
2.3 Postnatal and Adult Brown Fat Growth and Metabolism 18
3 Other Thermogenic Adipocytes 19
3.1 Brite/Beige Adipocytes 19
3.2 Brite/Beige Adipocyte Origins 20
4 Going Forward 23

S. M. Jung
Program in Molecular Medicine, University of Massachusetts Medical School, Worcester, MA, USA

J. Sanchez-Gurmaches (✉)
Division of Endocrinology, Division of Developmental Biology, Cincinnati Children's Hospital Research Foundation, Cincinnati, OH, USA

Department of Pediatrics, University of Cincinnati College of Medicine, Cincinnati, OH, USA
e-mail: juan.sanchezgurmaches@cchmc.org

D. A. Guertin (✉)
Program in Molecular Medicine, University of Massachusetts Medical School, Worcester, MA, USA

Molecular, Cell and Cancer Biology Program, University of Massachusetts Medical School, Worcester, MA, USA

Lei Weibo Institute for Rare Diseases, University of Massachusetts Medical School, Worcester, MA, USA
e-mail: David.guertin@umassmed.edu

A. Pfeifer et al. (eds.), *Brown Adipose Tissue*,
Handbook of Experimental Pharmacology 251, https://doi.org/10.1007/164_2018_168

4.1 Unanswered Questions and Future Goals 23
4.2 Prospects for BAT-Based Therapeutics 25
References 26

Abstract

Brown adipose tissue is well known to be a thermoregulatory organ particularly important in small rodents and human infants, but it was only recently that its existence and significance to metabolic fitness in adult humans have been widely realized. The ability of active brown fat to expend high amounts of energy has raised interest in stimulating thermogenesis therapeutically to treat metabolic diseases related to obesity and type 2 diabetes. In parallel, there has been a surge of research aimed at understanding the biology of rodent and human brown fat development, its remarkable metabolic properties, and the phenomenon of white fat browning, in which white adipocytes can be converted into brown like adipocytes with similar thermogenic properties. Here, we review the current understanding of the developmental and metabolic pathways involved in forming thermogenic adipocytes, and highlight some of the many unknown functions of brown fat that make its study a rich and exciting area for future research.

Keywords

Adipogenesis · Beige adipocyte · Brite adipocyte · Brown adipose tissue · Development · Glucose and lipid metabolism · Lineage tracing · Progenitor cells · Thermogenesis · Ucp1

1 Introduction and Background

1.1 Overview

Brown adipocytes, which reside in specific depots called brown adipose tissues (BAT), produce heat in a process called non-shivering thermogenesis. Thermogenesis in BAT is stimulated mainly by the sympathetic nervous system in response to cold exposure, and it helps maintain body temperature (euthermia) in placental mammals. The acquisition of BAT in early mammalian evolution is considered one key evolutionary advantage that allowed for the successful expansion of mammals, and its functional importance in newborn humans and small rodents has long been appreciated. More recently, it has become apparent that adult humans also have functionally relevant BAT and possibly the additional capacity to induce the formation of brown-like adipocytes within white adipose tissues (WAT) (called brite or beige adipocytes) under certain conditions. Because these thermogenic cells, when active, have a high rate of nutrient consumption and energy expenditure, their existence in adult humans not only correlates with improved metabolic profiles (Betz and Enerback 2018) but has stimulated interest in targeting them therapeutically to fight obesity and improve glycemic control (Hanssen et al. 2015; Ouellet et al. 2012; Yoneshiro et al. 2011b, c). This has gone hand-in-hand with renewed interest in understanding the basic biological mechanisms of brown fat development and metabolic regulation, which includes understanding the cellular lineages and precursor cell pools that give rise to brown and brite/beige adipocytes, and the signals that govern their fuel selection and unique metabolism. Identifying brown adipocyte stem and progenitor cells, and elucidating the mechanisms that stimulate their differentiation into mature thermogenic adipocytes, could have important implications in developing brown fat-based therapeutics. Here, we will discuss our present understanding of brown adipocyte development and function, the related topic of brite/beige adipocytes, and key future goals and unanswered questions especially as they relate to potential therapies.

1.2 Basics of Non-shivering Thermogenesis

Cold-stimulated non-shivering thermogenesis (NST) in the brown adipocyte is dependent upon the intrinsic expression and function of uncoupling protein 1 (UCP1), an inner mitochondrial membrane transporter that dissipates the energy stored in the mitochondrial electrochemical gradient as heat, "uncoupled" from ATP synthesis (Betz and Enerback 2018). In the absence of thermal stress, brown adipocyte UCP1 is thought to be inhibited by purine nucleotides (Nicholls 2006; Sluse et al. 2006). During cold stress, brown fat thermogenesis is classically stimulated by norepinephrine released from the sympathetic nervous system (SNS), which activates β3-adrenergic receptors on brown adipocytes to stimulate intracellular synthesis of the second messenger cyclic AMP (cAMP), leading to cAMP-driven protein kinase A (PKA) signaling activation. This stimulates lipid catabolism processes such as lipolysis which liberates free fatty acids from triacylglycerol lipid storage droplets and increases expression of a thermogenic gene expression program that includes UCP1 mRNA (Nicholls 2006; Sluse et al. 2006; Fedorenko et al. 2012; Lehr et al. 2006).

Exactly how brown adipocytes choose and utilize fuel remains an important and open question. Recent studies suggest that active lipolysis in brown adipocytes may not be required for sustaining thermogenesis so long as exogenous lipids are available; nevertheless, cellular free fatty acids reportedly directly activate UCP1 (Fedorenko et al. 2012; Shin et al. 2017; Schreiber et al. 2017). Active brown adipocytes also take up glucose from circulation and synthesize free fatty acids de novo from glucose and possibly other lipogenic precursors (i.e., the process of de novo lipogenesis) to continuously fuel NST or to provide other yet-to-be-appreciated metabolic advantages (Sanchez-Gurmaches et al. 2018; McCormack and Denton 1977; Mottillo et al. 2014; Shimazu and Takahashi 1980; Trayhurn 1979; Yu et al. 2002). In addition, BAT thermogenesis is fueled by liver-derived plasma-lipid metabolites (acyl-carnitines), the release of which is stimulated by cold-induced lipolysis in the WAT (Simcox et al. 2017). It has also been suggested recently that UCP1-independent mechanisms of thermogenesis might exist under certain circumstances (Bertholet et al. 2017; Ikeda et al. 2018; Kazak et al. 2015). Brown adipocytes might also have key metabolic functions in addition to thermogenesis, such as secreting special adipokines (called BATokines) and exosomes containing miRNAs that might have both autocrine function and paracrine functions on nearby immune cells, as well as endocrine functions related to glucose homeostasis and cardiovascular health (Thomou et al. 2017; Villarroya et al. 2013; Hansen et al. 2014; Svensson et al. 2016; Long et al. 2016; Wang et al. 2014a; Villarroya and Giralt 2015).

1.3 Brown Fat Anatomy and Morphology

The color distinction between a "brown" and a "white" adipocyte largely reflects the many more mitochondria (which are high in iron) in brown adipocytes compared to white adipocytes (Fig. 1). A stimulated brown adipocyte actively generating heat also contains many small lipid droplets and is referred to as being multilocular, while white adipocytes, such as those in subcutaneous and visceral depots (sWAT and vWAT, respectively), typically have a single large unilocular lipid droplet (Fig. 1). Having many small lipid droplets increases lipid droplet surface area and presumably promotes metabolite exchange with mitochondria (Blanchette-Mackie and Scow 1983; Benador et al. 2018). A less active brown adipocyte that is not engaged in thermogenesis (e.g., after acclimation to thermoneutrality) adopts a morphology more similar to a white adipocyte although it retains an epigenetic cellular identity that differentiates it from a white adipocyte (Roh et al. 2018).

As indicated above, brown adipocytes exist in defined BAT depots in the mouse, which is the main model organism used to study brown fat. Notably, the size and composition of each BAT depot differs with age, gender, and mouse strain background (Frontini and Cinti 2010; Murano et al. 2009). The largest BAT depots are clustered in the dorsal anterior regions of the mouse body and include the interscapular (iBAT), subscapular (sBAT), and cervical depots (cBAT) (Frontini and Cinti 2010; de Jong et al. 2015; Walden et al. 2012; Cinti 2005) (Fig. 2). In addition, there are several small BAT depots proximal to major blood vessels and specific organs, such as the periaortic BAT depot (paBAT) that aligns to aortic vessels and the perirenal BAT depot (prBAT) that localizes in a fibrous capsule of the kidney (Frontini and Cinti 2010) (Fig. 2). Recent studies using ^{18}F-FDG PET-CT or ($^{123/125}$I)-β-methyl-p-iodophenyl-pentadecanoic acid with SPECT/CT imaging, which traces glucose and lipid uptake, respectively, suggest additional small pockets of cold responsive fat depots exist in suprascapular, supraspinal, infrascapular, and ventral spinal regions (Zhang et al. 2018; Mo et al. 2017).

Similar to the mouse, newborn humans have active brown adipocytes present at birth in large interscapular BAT depots and perirenal depots (Fig. 3a), which presumably help maintain core body temperature though could also have other neonatal functions not yet appreciated. Until recently, it was widely believed that after neonatal BAT recedes, adult humans lacked brown fat. However, about a decade ago, the widespread existence of active BAT in adults was revealed by retrospective analyses of ^{18}F-fluodeoxyglucose (FDG) uptake assays, which use positron emission tomography-computed tomography (PET-CT) to measure glucose uptake into organs (Yoneshiro et al. 2011a, 2013; van der Lans et al. 2013; Ouellet

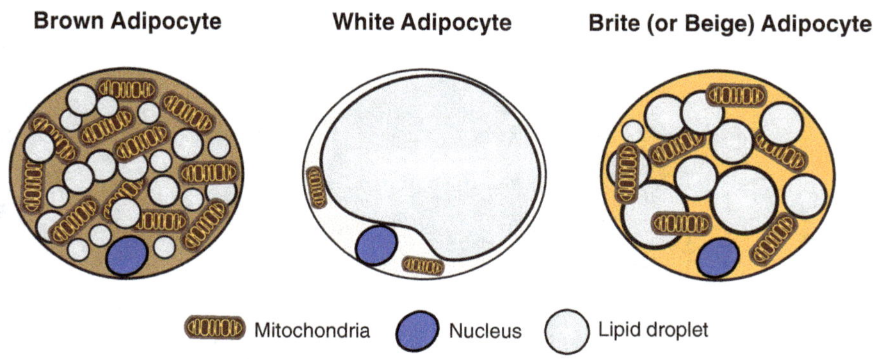

	Brown	***White***	***Brite/beige***
UCP1 Expression	Positive	Negative	Positive
Mitochondrial Density	High	Low	Medium
LD Morphology	Multi-locular	Uni-locular	Multi-locular
Primary Function	Thermogenesis Endocrine	Energy storage Endocrine	Thermogenesis? Endocrine?

Fig. 1 General characteristics of brown, white, and brite/beige adipocytes. A stimulated brown adipocyte (left) contains numerous small lipid droplets and many mitochondria and expresses high levels of uncoupling protein 1 (UCP1), which is embedded in the inner mitochondrial membrane and required for thermogenesis. The color of brown fat reflects the high iron content of mitochondria. A white adipocyte (middle) in contrast contains a single large lipid droplet and fewer mitochondria, and does not express UCP1. A brite/beige adipocyte (right) is characteristically intermediate between brown and white adipocyte, having multiple lipid droplets (though often larger than those seen in a brown adipocyte), more mitochondria than a white adipocyte, and it expresses UCP1

et al. 2012; Hanssen et al. 2015; Nedergaard et al. 2007; Cypess et al. 2009; van Marken Lichtenbelt et al. 2009; Saito et al. 2009; Virtanen et al. 2009; Kortelainen et al. 1993). These studies also revealed a correlation between BAT activity/amount and metabolic fitness. More recent studies show that BAT depots in adult humans exist in the supraclavicular, axillar, and paravertebral regions, though the variability across individuals and populations is still being worked out (Zhang et al. 2018; Nedergaard et al. 2007; Cypess et al. 2009; van Marken Lichtenbelt et al. 2009; Virtanen et al. 2009; Ouellet et al. 2012) (Fig. 3b). There are also small BAT depots in perivascular regions (aorta, common carotid artery), and near the heart wall (epicardium), lung bronchia, and some solid organs (hilum of kidney and spleen, adrenal, pancreas, liver) (Sacks and Symonds 2013) (Fig. 3b).

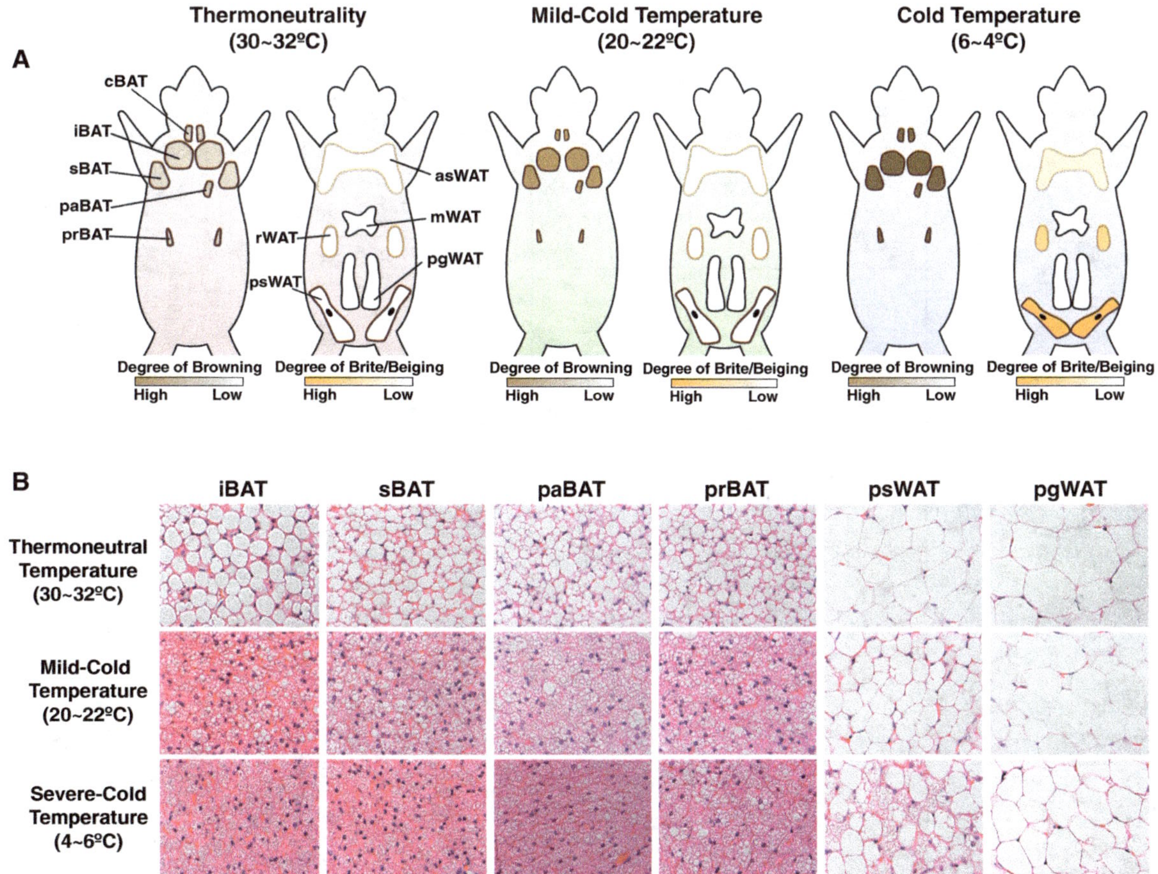

Fig. 2 Adipose tissue anatomy and plasticity. (**a**) Cartoons showing brown and white fat depots in mice that are acclimated to thermoneutrality (30–32°C), mild cold (20–22°C), and severe cold (4–6°C). The color and size of each depot is modeled such that it reflects the observed differences in mice acclimated to each

Fig. 2 (continued) temperature. A key showing the gradient of "browning" or "britening/beiging" is provided below each model. (**b**) Hematoxylin and eosin staining of the indicated brown and white fat depots at each temperature. Note that at thermoneutrality, brown adipocytes contain larger single lipid droplets. At 20–22°C, the standard mouse facility temperature, brown adipocytes exhibit their stimulated morphology of being multilocular (see Fig. 1), while white adipocytes remain unilocular though SWAT adipocyte size is reduced likely reflecting in part a higher level of lipolysis that is necessary to fuel the active brown fat depots. At severe cold temperatures (6–10°C), additional morphological changes can been seen in BAT (i.e., lipid droplets become more uniform), and under these conditions, brite/beige adipocytes also form in the subcutaneous WAT. Of note, the browning capacity of WAT depots is not dependent on a depot being subcutaneous or visceral because, for example, the retroperitoneal visceral WAT depot has high britening/beiging capacity (not shown) while the perigonadal visceral WAT (shown) does not. *iBAT* interscapular BAT, *sBAT* subscapular BAT, *cBAT* cervical BAT, *paBAT* periaortic BAT, *prBAT* perirenal BAT, *asWAT* anterior subcutaneous WAT, *psWAT* posterior subcutaneous WAT, *mWAT* mesenteric WAT, *rWAT* retroperitoneal WAT, *pgWAT* perigonadal WAT. The images in this figure are based primarily on experiments with C57Bl/6 mice

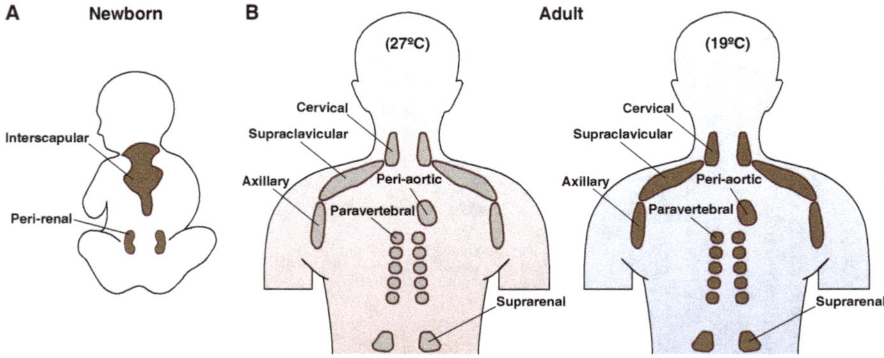

Fig. 3 Brown fat locations in humans. (**a**) Newborn infants have large interscapular and perirenal BAT depots. (**b**) In adults, smaller BAT depots are located in the cervical, supraclavicular, axillary, periaortic, paravertebral, and suprarenal regions. The mapping of these depots in adults is largely based on glucose uptake measurements by ^{18}F-FDG-PET-CT imaging, which shows increased glucose flux at colder temperatures (shown in figure) and on postmortem resections. The molecular and functional nature of individual (putative) BAT depots remains unclear in humans. Also note that the amount of BAT is highly variable between individuals, but when active BAT is present, it has been shown to correlate with improved metabolism (not shown, discussed in text). Emerging advances in BAT imaging will likely confirm additional depots

1.4 BAT Vascularization and Innervation

Brown fat depots are also highly vascularized, which facilitates the exchange of oxygen and nutrients and the dissipation of heat and release of BATokines into circulation (Bartelt et al. 2011; Labbe et al. 2015; Sacks and Symonds 2013). In fact, BAT requires increased blood infusion rate during BAT recruitment (i.e., cold stimulation) to obtain sufficient metabolic substrates and oxygen. Brown adipocytes also generate vascular endothelial growth factor-A (VEGF-A) and nitric oxide (NO), which facilitates BAT angiogenesis and vascularization (Xue et al. 2009; Sun et al. 2014; Nisoli et al. 1998; Mahdaviani et al. 2016), a process that is reduced in obese mice resulting in loss of thermogenic activity (Shimizu et al. 2014). Other recent work suggests that brown adipocytes may have a vasoprotective role that might be mediated by the secretion of hydrogen peroxide (H_2O_2), which inhibits vessel contractions in nearby vascular cells (Friederich-Persson et al. 2017).

In addition to being highly vascularized, BAT is extensively innervated allowing for its rapid stimulation by the sympathetic nervous system (SNS). The SNS releases catecholamines such as norepinephrine that activate G-protein-coupled β3-adrenergic receptors that are highly expressed on mature brown adipocytes and β1-adrenergic receptors on brown adipocyte precursors (Cannon and Nedergaard 2004; Morrison et al. 2012; Bukowiecki et al. 1986; Bronnikov et al. 1992). While β3-adrenergic receptor signaling stimulates mature brown adipocyte lipid catabolic activity and thermogenesis, β1-adrenergic receptor signaling stimulates brown fat adipogenesis upon prolonged cold challenge (Bronnikov et al. 1992). Classic denervation studies reveal the indispensability of the SNS connections for thermogenesis (Silva and Larsen 1983; Rothwell and Stock 1984; Takahashi et al. 1992; Labbe et al. 2015). Emerging research also suggests that innervation may also be critical for BAT to communicate directly with other non-SNS tissues, such as the WAT (Schulz et al. 2013; Garretson et al. 2016; Nguyen et al. 2018).

In summary, BAT is a dynamic and heterogeneous tissue, and the extensive networks of vessels and nerves found in BAT suggest that during brown fat development, there is tight coordination between brown adipocyte precursors (discussed below), endothelial lineages, and nerve cell lineages and likely immune cells too (Lumeng and Saltiel 2011; Olefsky and Glass 2010; Villarroya et al. 2018). The signaling and metabolic interactions between different cell lineages during brown fat development have not yet been extensively studied by system-based approaches.

1.5 Transcriptional Control of Brown Adipocyte Differentiation

Much of the general transcriptional cascade that promotes adipogenesis is shared between brown and white adipocytes and has been studied at length using in vitro models (e.g., 3T3-L1 cells). The master regulator of adipogenesis, PPARγ, is both necessary and sufficient for adipogenesis (Rosen et al. 1999; Tontonoz et al. 1994; Wang et al. 2013a). Other key components of the general adipogenesis transcriptional cascade also important in brown and brite/beige adipocyte differentiation include the members of the C/EBP family (C/EBPα, C/EBPβ, C/EBPδ) (Farmer 2006). While PPARγ is the dominant factor, overexpression of all C/EBP family members induces adipocyte formation. In culture, C/EBPβ and C/EBPδ function in the first wave of adipogenic transcription factors (hours after adipogenic induction) that eventually triggers a second wave (days after adipogenic induction) that includes C/EBPα and PPARγ, which feed-forward activate themselves (Farmer 2006).

More recently, efforts to identify brown adipocyte lineage-specific transcription factors have identified new additional components that may contribute to the brown (or brite/beige) adipocyte fate. PRDM16 (PRD1-BF1-RIZ1 homologous domain containing 16) was originally described as a BAT transcriptional determination factor that induces a robust thermogenic adipocyte phenotype in white adipocytes both in vitro and in vivo and can direct muscle precursors to differentiate into brown adipocytes in vitro (Seale et al. 2007, 2008, 2011). In vivo, other PRDM family members can compensate for the loss of PRDM16 in BAT precursors to maintain normal BAT formation (Harms et al. 2014). In addition, the EBF2 (early B-cell factor 2) transcription factor is selectively expressed in both BAT and brite/beige precursors, and it is required for BAT identity and efficient brite/beige cell formation (Rajakumari et al. 2013; Stine et al. 2016; Wang et al. 2014b). Recent studies also identified zinc-finger protein 516 (Zfp516), whose expression in brown fat is markedly increased in response to cold exposure or β-adrenergic stimulation via β-AR-cAMP pathway, and it directly interacts with PRDM16 to promote BAT development and WAT browning while suppressing myogenesis (Dempersmier et al. 2015; Sambeat et al. 2016). Whether there are brite-/beige-specific transcription factors that do not function in brown adipocyte lineages remains an important area of investigation.

In contrast to pro-thermogenic transcription factors, less is known about the transcriptional machinery that promotes and/or maintains the white adipocyte phenotype. One interesting candidate is Zfp423, which has recently emerged as a critical brake that prevents white adipocytes from converting to thermogenic adipocytes. Zfp423 is expressed in white adipocyte precursor cells and functions to block the brite/beige thermogenic program by inhibiting the EBF2 and PRDM16 (Gupta et al. 2010, 2012; Shao and Gupta 2018; Shao et al. 2016). While these studies are opening the door to our understanding of adipocyte fate determination at the level of gene expression, there is still much to be learned especially if this information is to be harnessed for therapeutic opportunities. Moreover, other key gene expression factors that contribute to fate decisions, such as epigenetic marks and higher-order chromatin regulation, are just beginning to be explored (Roh et al. 2017, 2018; Zhao et al. 2016; Carrer et al. 2017) making this an important area of investigation for many years to come.

2 Brown Fat Growth

2.1 Techniques for Studying BAT Development

Understanding how brown fat grows begins with understanding its developmental origins. We begin this section with a brief commentary on the two main methods that have been instrumental in beginning to elucidate the developmental origins of both brown and white adipocytes: (1) *fluorescence-activated cell sorting (FACS)*, in which stem and progenitor cells are isolated based on their expression of cell surface markers or engineered genetic labels and then tested for their ability to function as adipocyte precursors, and (2) *lineage tracing*, in which stem and progenitor cells are indelibly labeled with a genetic mark that can be followed, or traced, throughout development in all descendant cells.

2.1.1 FACS

Adipocyte precursors reside within whole adipose tissue depots in a highly heterogeneous non-adipocyte cell population commonly referred to as the stromal vascular fraction or "SVF." In addition to adipocyte stem and progenitor cells, the SVF contains endothelial, immune, nerve, and other cells that support tissue function. Adipocyte precursors are necessary not only for establishing fat depots but also for expanding and regenerating adipocytes. Starting with only the SVF population from white adipocytes, several studies have used FACS technology with cell surface markers thought to label the adipocyte precursor population to enrich for pools of adipocyte stem and progenitor cells (ASPCs) (Berry et al. 2014). Although a single marker for prospective isolation of adipocyte precursors has not been found, combinations of surface markers have been use in this regard to isolate white ASPCs (Berry and Rodeheffer 2013; Rodeheffer et al. 2008). One common example in mice is the $CD31^{neg}$, $CD45^{neg}$, $Ter119^{neg}$, $CD29^{pos}$, $Sca1^{pos}$, $CD34^{pos}$, and $CD24^{pos}$ population, which has enhanced adipogenic potential compared to the total SVF. Although brown and white adipocytes have many functional, anatomical, and morphological differences, a similar population of ASPCs can be isolated from BAT depots (Sanchez-Gurmaches et al. 2012; Wang et al. 2014b).

Recently, PDGFRα was also reported to be a marker for ASPCs. PDGFRα can be used to isolate ASPCs using flow cytometry from the $CD31^{neg}$ and $CD45^{neg}$ population within the SVF of all WAT and BAT (Church et al. 2014; Berry and Rodeheffer 2013). These findings have been further validated using lineage-tracing approaches (discussed below), which confirm that adipocyte lineages express Cre recombinase driven by the PDGFRα promoter (Berry and Rodeheffer 2013; Vishvanath et al. 2016; Lee et al. 2012, 2015). From a technical perspective, this finding is important because it simplifies the enrichment protocol for ASPCs. Interestingly, PDGFRα also labels a fibro/adipogenic precursor cell population within skeletal muscles and skin (Joe et al. 2010; Rivera-Gonzalez et al. 2016) suggesting PDGFRα may be a broadly relevant marker of ASPCs, and recent studies further conclude that PDGFRα signaling may functionally contribute to ASPCs fate and adipose tissue organogenesis (Rivera-Gonzalez et al. 2016; Sun et al. 2017). However, PDGFRα also expresses in many non-adipocyte cells, and it will be important to delineate its different roles within the heterogeneous SVF population of adipose tissues.

A current key challenge of using FACS-isolated adipocyte precursors is that the ASPCs, although enriched for adipogenic precursors, are still a heterogeneous population containing subpopulations of cells that remain largely undefined by molecular approaches, and whether a true adipocyte stem cell can be purified is still an open question. Recent studies using single-cell RNA-seq are beginning to provide key insights into this problem (discussed below). Other studies have identified markers of differentiated brown or beige adipocytes (Ussar et al. 2014). However, highly specific and reliable surface markers that can differentiate between brown, beige, or white adipocyte progenitors have not yet been identified. On the other hand, the prospective nature of using FACS to isolated adipocyte precursors may facilitate the isolation and application of human ASPCs for use in cell-based therapies. Several different protocols for the isolation of human adipocyte progenitors are being developed (van Harmelen et al. 2005; Baglioni et al. 2009, 2012; Perrini et al. 2013).

2.1.2 Lineage Tracing

Lineage tracing is a classic developmental biology technique that has been used to study adipose tissue development mainly in mouse models due to its genetic nature. In classic lineage-tracing experiments, an indelible mark, often a fluorescent reporter, is expressed in a specific population of precursor cells by homologous recombination or by a transgenic approach, and the permanently modified cells then transmit the reporter to all of their descendant cells or lineages. A common strategy for studying adipocyte lineages in mice is to use a cell-specific Cre recombinase that activates the reporter's expression. Cre drivers can be always on in the specific population being studied, called constitutive, or stimulated to be active only transiently in a cell population, called inducible. The latter requires the administration of a stimulus to turn on Cre activity.

Both methods have advantages and disadvantages that need to be considered for data interpretation of adipocyte lineages. For starters, there is currently no known Cre driver that only expresses in ASPCs. Inducible Cre recombinases have the advantage that they allow the timing of activation to be regulated such that cells are only labeled for a brief moment, and then those specific cells can be followed. This is not achievable with constitutive Cre drivers, making it difficult to determine precisely which cells first express the Cre in a particular lineage using these drivers. However, the inducers used to turn on the inducible Cre drivers, typically tamoxifen or doxycycline, can have unintended toxic effects on cells (Ye et al. 2015; Moullan et al. 2015). Even when pools of cells are inducibly labeled, it is difficult to distinguish whether two descendant labeled cells originate from common or distinct Cre-expressing precursors. Another consideration is whether a particular Cre driver reflects the expression of the actual endogenous gene/protein whose promoter is used to drive the Cre or whether it only reflects the promoter activity uncoupled from the normal expression of the associated gene and/or protein. The use of knock-in Cre drivers, which are expressed from endogenous promoters, can help mitigate against this concern. Related to this point, caution should be taken in inferring whether the activity of a specific Cre (i.e., promoter) reflects a functional role for the associated gene in lineage specification.

The choice of a reporter (often a fluorescent reporter) is also important when performing lineage tracing in adipocytes. One issue with adipocytes relative to non-adipocytes is the small amount of cytoplasm and large quantity of lipid droplets, which both make the use of cytoplasmic fluorescent reporters challenging to detect and make it difficult to obtain high-quality frozen sections. Thus, the reporter of choice for adipocytes is typically a membrane-targeted reporter, such as the dual-fluorescent membrane-targeted Tomato, membrane-targeted GFP, or mTmG reporter (Muzumdar et al. 2007). This reporter has two major advantages: (1) all cells are labeled, the mGFP reporter only being activated in Cre-positive lineages, and (2) both fluorescent reporters are membrane targeted. Its utility in adipose tissue both for lineage tracing as well as for use in FACS-based studies has been demonstrated in many reports (Berry and Rodeheffer 2013; Sanchez-Gurmaches and Guertin 2014; Shao et al. 2016; Wang et al. 2014b). Related to lineage tracing is cell-labeling or cell-marking, which is a common technique to study mature adipocyte dynamics. By this strategy, only mature adipocytes are labeled (rather than precursors), which allows single mature adipocytes to be followed over time especially when combined with inducible Cre drivers of reporter expression.

2.2 Brown Adipocyte Origins

Brown adipocytes are thought to originate from the mesoderm during embryonic development and thus share a very early developmental origin with skeletal muscle,

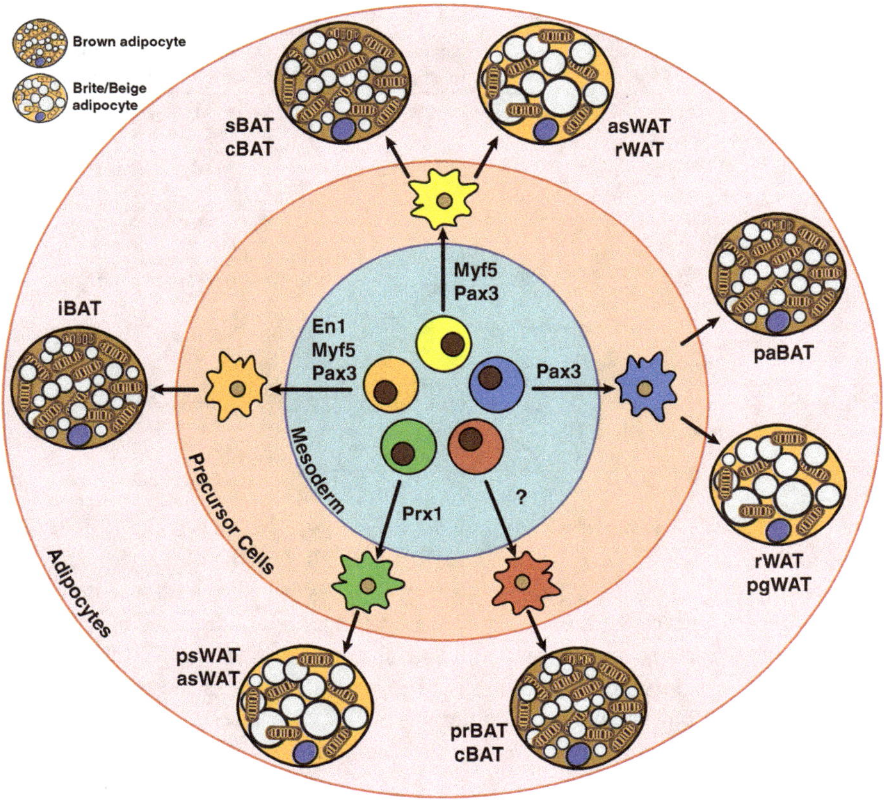

Fig. 4 Model of the heterogeneity and complexity in brown and brite/beige adipocyte development. Several multipotent cell populations that are mainly mesodermal and express specific transcription factors (e.g., En1, Myf5, Pax3, Prx1) appear to give rise heterogeneously to thermogenic adipocytes in different depots. Note that there is overlap shared with some markers but not with others. For example, Pax3 and Myf5 together may mark a pool of early precursors that give rise to iBAT, but only Pax3 marks a precursor pool that gives rise to some visceral pgWAT adipocytes (discussed in text). The significance of this heterogeneity is not understood. Additionally, there are several populations of brown and brite/beige adipocytes for which potential lineage markers remain unidentified. Also note that the brown and brite/beige adipocytes shown in this figure are depicted in their active state (i.e., upon β-adrenergic stimulation), but in vivo brown and brite/beige adipocytes are not necessarily present at the same time, such as in mild cold conditions (see Fig. 2)

bone, white adipocytes, and connective tissues (Wang et al. 2014b; Atit et al. 2006; Seale et al. 2008; Lepper and Fan 2010; Sanchez-Gurmaches et al. 2012). However, the pathways that specify the brown adipocyte developmental lineage is not fully clear. In accordance with a mesodermal origin, a population of cells within the central dermomyotome that is labeled at E9.5 by expression of the homeobox transcription factor Engrailed 1 (En1) gives rise to iBAT, dermis, and epaxial muscles (Figs. 2a, 4 and Table 1) (Atit et al. 2006). However, these E9.5 En1^{+} progenitors do not appear to give rise to sBAT or any of the major white fat depots

Table 1 Depot-specific developmental origins of BAT

Tissue type	Anatomical locations (Human)	Anatomical locations (Mouse)	Developmental origins (Lineage tracing study)
Brown adipocyte	Supraclavicular/ paravertebral Perivascular: (aorta, artery) Periviscus: (heart, lung bronchia) Solid organs: Kidney/spleen hilum, pancreas, liver	Interscapular	$En1^{+}$, $Myf5^{+}$, $Pax7^{+}$, $Pax3^{+}$, $Prx1^{-}$
		Subscapular	$En1^{-}$, $Myf5^{+}$, $Pax7^{+}$, $Pax3^{+}$, $Prx1^{-}$
		Cervical	$En1^{?}$, $Myf5^{\pm}$, $Pax7^{+}$, $Pax3^{\pm}$, $Prx1^{-}$
		Perirenal	$En1^{?}$, $Myf5^{-}$, $Pax7^{+}$, $Pax3^{\pm}$, $Prx1^{-}$
		Periaortic	$En1^{?}$, $Myf5^{-}$, $Pax7^{+}$, $Pax3^{-}$, $Prx1^{\pm}$
Brite/beige adipocyte	Supraclavicular? Subcutaneous	Posterior subcutaneous (Inguinal)	$En1^{?}$, $Myf5^{-}$, $Pax3^{-}$, $Prx1^{+}$

(Atit et al. 2006) (Atit personal communication) suggesting that some brown and white adipocyte origins may differ and that not all brown adipocytes share a common origin (see below). This concept of adipocyte heterogeneity within and between depots, as we will discuss, is now a central tenet of adipocyte biology.

The model tilted toward brown fat and skeletal muscle sharing a common developmental origin with the finding that brown adipocytes in the iBAT and the skeletal muscles, but not certain populations of WAT, share a common cellular origin in the dermomyotome defined by the expression of Myf5-Cre (Seale et al. 2008). Using the constitutively expressing Myf5-Cre knock-in driver with a cytoplasmic reporter, this study found that Cre recombinase activity labels mature brown adipocytes in the iBAT in addition to skeletal muscles (Seale et al. 2008). Myf5 is a classic myogenic determination factor from the basic helix loop helix (bHLH) family, and thus the labeling of brown adipocytes with Myf5-Cre (Tallquist et al. 2000) was predicated to explain the metabolic similarities between brown fat and skeletal muscle with respect to high oxygen consumption and fuel usage and conversely the metabolic difference between BAT and the less metabolically active and energy-storing WAT depots (Harms and Seale 2013). Notably, at the time, most studies used mice that were mildly cold stressed in which the BAT is hyperactive, rather than mice living at thermoneutrality, when brown adipocytes are more similar morphologically and metabolically to white adipocytes. Nevertheless, in support of this model, an inducible Cre driver under control of the Pax7 promoter (the PAX transcription factor family member 7 collaborates with Myf5 and other myogenic factors during skeletal myogenesis) showed that Pax7+ progenitors that arise between E9.5 and E10.5 (but not later in development) also give rise to interscapular brown adipocytes (Lepper and Fan 2010). This also suggested an early divergence between BAT and muscle lineages.

While the Myf5-lineage model of BAT specification was elegant in its simplicity, studies challenging its uniformity soon after revealed that the brown adipocyte developmental landscape is more complicated. Similar fate-mapping experiments using the same Myf5-Cre driver, but more broadly examining brown and white fat depots, and using the mTmG reporter, showed that many white adipocytes are also Myf5-Cre lineage positive and unexpectedly that many brown adipocytes are Myf5-Cre lineage negative (Sanchez-Gurmaches and Guertin 2014). For example, Myf5-Cre labeled precursors appear to give rise to nearly all brown adipocytes in iBAT and sBAT depots, but only about half of the brown adipocytes in the cervical BAT, and none of the brown adipocytes in prBAT or paBAT. Moreover, Myf5-Cre-positive adipocytes populate the asWAT and rWAT depots (Fig. 2a), indicating that Myf5-Cre neither uniformly nor specifically labels brown adipocytes. Other studies have replicated these findings confirming the heterogeneous labeling of adipocytes with Myf5-Cre (Sanchez-Gurmaches and Guertin 2014; Sanchez-Gurmaches et al. 2012; Shan et al. 2013; Wang et al. 2014b).

Interestingly, lineage tracing using a Pax3-Cre driver, (Pax3 is another myogenic Pax family transcription factor that expresses just prior to Myf5) labels similar populations of cells with a few key differences. Notably, Pax3-Cre cells give rise to most of the brown adipocyte in iBAT, sBAT, cBAT, and prBAT, but none of the brown adipocytes in the paBAT (Sanchez-Gurmaches and Guertin 2014; Liu et al. 2013), and also to nearly 50% of the white adipocytes in the large visceral pgWAT depot. For comparison, MyoD-Cre (another classic myogenic transcription factor) does not label any brown or white adipocytes but importantly does label skeletal muscles (Sanchez-Gurmaches and Guertin 2014). Thus, there may be specificity within skeletal muscle lineages in which some precursors (i.e., Pax3/Myf5/Pax7^{positive}) can also become adipocytes while others (i.e., MyoD^{positive}) cannot or rather that some adipocyte and muscle precursors can independently express Pax3/Myf5/Pax7-Cre (see discussion above on the challenges of lineage-tracing studies) (Sanchez-Gurmaches and Guertin 2014; Haldar et al. 2008; Gensch et al. 2008). The most interesting possibility is that there is a temporal or spatial separation between certain lineages, and understanding this may help in understanding the commitment phase to brown adipocytes. Regardless, these studies conclusively revealed an unanticipated heterogeneity in both brown and white adipocyte development that suggests brown adipocytes residing in different depots could have different embryonic origins.

The developmental heterogeneity observed between brown adipocyte lineages is not likely due to low efficiency or specificity of the Cre drivers because independent experiments with Myf5-Cre, Pax3-Cre, and Pax7-CreER lines are remarkably similar (Lepper and Fan 2010; Sanchez-Gurmaches and Guertin 2014; Liu et al. 2013; Sanchez-Gurmaches et al. 2012; Seale et al. 2008; Shan et al. 2013; Wang et al. 2014b). Moreover, heterogeneous Myf5 labeling is also observed in skeletal muscle lineages in which Myf5 only labels around 50% of the satellite cells in the limb muscles but around 80% in epaxial muscles (Haldar et al. 2008; Gensch et al. 2008). An unanswered question is whether developmentally distinct brown (or white) adipocytes differ only in their anatomical location or whether they have unique functions (e.g., metabolic efficiency, BATokine production, exosome secretion, etc.) that might be specific by their developmental origins. Answering these questions will require an improved ability to isolate and study single brown adipocytes, a deeper understanding of the regulatory mechanisms of BAT development, and markers that label the unidentified (Myf5-Cre; Pax3-Cre^{negative}) brown adipocyte lineages.

2.3 Postnatal and Adult Brown Fat Growth and Metabolism

In older laboratory mice (i.e., juveniles and adults), individual BAT depots can expand their mass by either increasing brown adipocyte number (hyperplasia) or by increasing individual cell size (hypertrophy) depending upon their initial housing temperature and the duration and degree of cold exposure. For example, hypertrophic growth of brown adipocytes is observed when mice living in standard housing conditions (22°C) are acclimated to their thermoneutral zone (e.g., 30–32°C). Under these conditions, the sympathetic tone is reduced by removing thermal stress, and the brown adipocytes decrease their thermogenic activity. This results in lipids accumulating and coalescing into a single large unilocular lipid droplet, thereby increasing individual cell size. Notably, while thermoneutral BAT displays a WAT-like morphology and gene expression signature, it maintains its BAT epigenetic signature (Hung et al. 2014; Veniant et al. 2015; Roh et al. 2018). Nevertheless, the net result of increasing cell size is in an increase in total depot size compared to mice living in the mild cold temperatures of most mouse facilities (Fig. 2).

Conversely, if mice living at thermoneutrality are moved to the mild cold (20–22°C) and BAT thermogenesis is activated, the mobilization and metabolism of lipids and other metabolites reduce individual adipocyte cell size and thereby overall BAT depot size (Fig. 2). However, if these mice are then further adapted to more severely cold temperatures (e.g., in 4–6°C range), additional new active brown adipocytes are recruited into the BAT depots (presumably from the brown ASPC pool described above), which increases BAT mass but by hyperplastic growth (Bukowiecki et al. 1982; Rehnmark and Nedergaard 1989; Geloen et al. 1992; Lee et al. 2015; Razzoli et al. 2018). Indeed, de novo adipogenesis of brown adipocyte precursor cells occurs in response to chronic cold (Rosenwald et al. 2013; Lee et al. 2015). Brown adipocyte size also increases by denervation, during extended high caloric (fat) feeding, or with aging (Hung et al. 2014; Roberts-Toler et al. 2015). Thus, while there is an underlying natural turnover of brown adipocytes (Sakaguchi et al. 2017), the iBAT depots in laboratory mice are smallest when mice are acclimated to standard lab conditions (mild cold), and it grows with increased or decreased temperature mainly by hypertrophic or hyperplasic growth, respectively.

Gene expression profiling of BAT tissue reveals greater differences between mice acclimated to thermoneutrality (30–32°C) and mild cold (20–22°C) than between mice acclimated to mild cold (20–22°C) and severe cold (6°C) (Sanchez-Gurmaches et al. 2018). This is consistent with brown fat morphology at these temperatures, which shows individual brown adipocytes in an "off" state (unilocular) in thermoneutrality and an "on" state (multilocular lipid droplets) at 22°C. Further reductions in temperature (e.g., to 6°C) increase the magnitude of thermogenesis and many genes associated with thermogenesis, and this is additionally reflected by morphological "ordering" of the lipid droplets (Fig. 2). A survey of metabolic genes upregulated in the mild cold indicates that genes encoding regulators fatty acid oxidation and de novo lipogenesis are both upregulated as are genes whose products function in respiratory metabolism (e.g., the tricarboxylic (TCA) cycle, electron transport chain) and thermogenesis (Sanchez-Gurmaches et al. 2018; McCormack and Denton 1977; Mottillo et al. 2014; Shimazu and Takahashi 1980; Townsend and Tseng 2015; Trayhurn 1979; Yu et al. 2002). This emphasizes an interesting metabolic paradox of brown adipocytes that increasing BAT catabolic activity by cold is also associated with induction of anabolic lipid synthesis pathways. This may be another BAT characteristic futile cycling mechanism, or alternatively, the stimulation of de novo lipogenesis may have other metabolic implications since many intermediates in the de novo lipogenesis pathway, such as acetyl-coA, also function as second messengers (Pietrocola et al. 2015).

3 Other Thermogenic Adipocytes

3.1 Brite/Beige Adipocytes

A second type of UCP1-expressing adipocyte called a brite (brown-like in white) adipocyte, also known as a beige adipocyte, is also attracting interest as a potential therapeutic target in obesity and metabolic disease. As the synonymous names imply, brite/beige adipocytes appear within specific WAT depots under certain stresses, and their morphology (lipid droplet size and mitochondria content) is intermediary between that of classic brown and white adipocytes (Fig. 1). There have been two experimental methods used to drive brite/beige cell formation in sWAT. Acclimation to severe cold temperatures (4–6°C) may be the most physiological approach, which strongly induces BAT thermogenesis along with the formation of brite/beige adipocytes that express UCP1 in sWAT. Similar to how brown adipocytes change their appearance between thermoneutrality and mild cold, brite/beige adipocytes undergo morphological changes between mild cold and severe cold that include the typical multilocular morphology in severe cold (Fig. 2b). A second common method to induce brite/beige cell formation is to treat mice with the β3-adrenergic agonists CL-316243, which resembles the effects of cold exposure on mature adipocytes. Many other stresses can also lead to the formation of brite/beige adipocytes including exercise, cancer cachexia, and peripheral tissue injury (Ikeda et al. 2018; Singh and Dalton 2018) suggesting brite/beige adipocyte formation may reflect a general stress response rather than specifically the response to cold. Whether these alternative modes of browning indicate a physiologically relevant role for thermogenesis or reflect a secondary consequence of altered adipocyte state is not yet clear. Nevertheless, increasing brite/beige adipocyte number could also be a strategy to fight obesity, and thus there is strong interest in understanding the biology of how brite/beige adipocytes develop.

The location and number of brite/beige adipocytes in adult humans are less clear. Studies suggest that UCP1-positive brown-like adipocytes purified from human supraclavicular BAT depots have a similar gene expression pattern to murine brite/beige adipocytes (Lidell et al. 2013; Wu et al. 2012; Sharp et al. 2012; Shinoda et al. 2015), whereas brown-like adipocytes isolated from other human BAT depots (neck, cervical, perirenal) appear to more closely resemble classic brown adipocytes in mice (Cypess et al. 2013; Xue et al. 2015). More recent studies using ^{18}F-fluodeoxyglucose (FDG) PET-CT imaging noticed that several additional metabolically active adipocytes reside in the abdominal and subcutaneous areas of adult humans, which could be brown or brite/beige adipocytes (Leitner et al. 2017). Other studies of BAT in the supraclavicular region of adult humans have shown more mixed transcriptional profiling representative of both brown and brite/beige adipocytes (Leitner et al. 2017). One caveat of comparing human studies to mouse studies is that often the comparisons are made between thermoneutral humans and cold stressed mice, and it is possible that some of the human brite/beige adipocytes could be less-stimulated brown adipocytes. Another open question is whether brite/beige adipocytes make significant contributions to overall thermogenesis (Singh and Dalton 2018; Kajimura et al. 2015: Nedergaard and Cannon 2014). Nevertheless, these studies support the idea that stimulating brite/beige adipocyte formation in humans could be another way to improve glucose homeostasis.

3.2 Brite/Beige Adipocyte Origins

Understanding brite/beige adipocyte origins is important because it may provide insight into therapeutic strategies to induce their formation. Currently, there are two main competing theories to explain where brite/beige adipocytes originate from that are not necessarily mutually exclusive. The first theory posits that brite/beige adipocytes form de novo upon stimulation from a precursor cell pool (Wang et al. 2013b); the second argues that they intercovert from existing adipocytes between a dormant and active state depending upon the presence of stimulus (Lee et al. 2015; Barbatelli et al. 2010). A third likely possibility is that both mechanisms occur, perhaps in a context-dependent manner dependent upon many factors including type of stimulation, its duration, the depot analyzed, and proximity to the sympathetic nervous system input.

Using strategies to fluorescently mark individual adipocytes, it has been shown that around 60% of the total UCP1+ adipocytes that form in the sWAT after cold acclimation (7 days) are generated de novo by the process of adipogenesis (Wang et al. 2013b; Berry et al. 2016). These new brite/beige adipocytes originate from smooth muscle actin (SMA)-positive progenitors and require β1-adrenergic receptor signaling similar to how nascent brown adipocytes form upon cold exposure (Berry et al. 2016; Jiang et al. 2017; Bukowiecki et al. 1982, 1986; Bronnikov et al. 1992; Rehnmark and Nedergaard 1989; Geloen et al. 1992; Razzoli et al. 2018). However, SMA^+ progenitors also give rise to all white adipocytes in both subcutaneous and visceral fats (Jiang et al. 2014). Thus, whether these reflect two distinct sub-pools of SMA^+ progenitors for white and brite/beige adipocytes, or a common precursor pool, is unclear. The other implication of these data is that the other 40% of the UCP1+ adipocytes that formed originate, or interconvert, from preexisting white adipocytes, which have also been referred to as transdifferentiation to reflect the fundamental changes in gene expression and morphology (Cinti 2002). However, there is inconsistency between these and other studies that may be related to the lack of a standard experimental approach across studies or differences in strain background, age, and/or previous exposure to environmental or dietary variables (Lee et al. 2015).

Administering the β3-adrenergic agonist CL-316243, which is widely used to induce brite/beige cell formation, appears to induce brite/beige adipocyte formation from preexisting mature white adipocytes (Jiang et al. 2017; de Jong et al. 2017). However, because CL-316423 acts only on the mature cells, a systemically derived signal that might act on the precursors may be absent. Notably, lack of β3-adrenergic receptor activity does not prevent the "browning" capacity of sWAT by cold further suggesting that multiple pathways to brite/beige adipocyte formation exist that could have compensatory capability (Jiang et al. 2017). Again, these results must be interpreted carefully because distinct responses to β3-adrenergic receptor inactivation are observed depending on mouse background (Barbatelli et al. 2010). An additional confusing factor is that the browning capacity of the psWAT depends on the genetic background. For instance, the A/J mouse strain shows higher UCP1 induction upon β3-adrenergic stimulation (in the psWAT) than the more commonly used C57Bl6 mice (Chabowska-Kita and Kozak 2016; Collins et al. 1997). More research is clearly needed to fully understand which modality of brite/beige adipocyte formation is most tractable for therapeutic targeting.

An interesting question is whether all white adipocytes can become brite/beige under certain conditions or whether there is a fundamental cell intrinsic feature of like the sWAT adipocytes that give them their "britening" capacity. A related question is whether there is a specific brite/beige adipocyte cell lineage that is different from the lineages that give rise to the white adipocytes that do not become brite/beige and the brown adipocytes (Nedergaard and Cannon 2014; Kozak 2011). Originally it was suggested that Myf5 expression could delineate between the brown adipocyte lineage and the brite/beige adipocyte lineage, which was thought to be Myf5-negative. However, more comprehensive Myf5-lineage-tracing studies later showed that the brite/beige adipocytes that form in the Myf5-positive asWAT and retroperitoneal WAT depots are also Myf5-positive (Shan et al. 2013; Sanchez-Gurmaches and Guertin 2014) (Fig. 2a). In contrast, none of the brite/beige adipocytes that form in psWAT are Myf5-positive, suggesting Myf5 expression likely delineates between anatomical positioning rather than function; however, this has not yet been fully resolved, and it is unknown if Myf5-positive brite/beige adipocytes are functionally identical to Myf5-negative brite/beige adipocytes (Berry et al. 2016; Sanchez-Gurmaches and Guertin 2014; Sanchez-Gurmaches et al. 2012; Seale et al. 2008).

One study searching for markers of a brite/beige adipocyte lineage found that CD137-positive precursors isolated from psWAT have a greater propensity to induce UCP1 mRNA in culture compared to CD137-negative precursors, suggesting an intrinsic heterogeneity in the capacity to adopt different metabolic profiles (Wu et al. 2012). However, it has not yet been shown that this population is specific to a brite/beige adipocyte lineage in vivo such as by lineage-tracing studies.

Ribosome-profiling studies have also shown that brite/beige adipocytes in the psWAT possess a gene expression signature that has similarity to smooth muscle-like cells, which is not observed in brown adipocytes, and it is independent of anatomical position (Long et al. 2014). However, this appears to only represent a subset of the total brite/beige adipocyte population because lineage-tracing experiments with a Cre recombinase driven by the *Myh11* promoter, which is a marker of smooth muscle cells, only label ~10% of the UCP1+ brite/beige adipocytes following prolonged cold acclimation (Long et al. 2014; Berry et al. 2016), and deleting *PPARγ* in this lineage does prevent not WAT browning (Berry et al. 2016). These data may reflect the inherent heterogeneity in adipocytes, and thus, a specific brite/beige adipocyte lineage marker remains elusive. High expression of Ebf2 is found in precursor cells with high thermogenic capacity in psWAT (Wang et al. 2014b; Stine et al. 2016). It will be interesting to see whether Ebf2 functionally commits progenitors to a thermogenic lineage.

In contrast to the intra-depot heterogeneity seen with other brite/beige cell markers, all psWAT and bone marrow adipocytes are homogeneously labeled by a Cre recombinase driven from the paired related homeobox transcription factor 1 (Prx1) promoter, which expresses in mesenchymal precursors during development in what appears to be a multipotent precursor population that also gives rise to limb and head tissues (Sanchez-Gurmaches et al. 2015; Krueger et al. 2014; Ambrosi et al. 2017). Interestingly, the *Prx1* transcription factor itself has been linked to cell-fate decisions including adipocyte specification (Logan et al. 2002: Cserjesi et al. 1992; Du et al. 2013; Hu et al. 1998; Lu et al. 1999: Peterson et al. 2005; ten Berge et al. 1998). What is noteworthy about Prx1-Cre is that it does not significantly label any other WAT or BAT depots, allowing for some degree of depot specificity when used for targeting WAT that cannot be achieved with Adiponectin-Cre (Eguchi et al. 2011). Because Prx1-Cre does not label the brite/beige adipocytes that form in other depots, such as the rWAT, it is not a universal brite/beige marker, and it likely expresses very early before adipocyte specification. However, its labeling pattern could be a clue to understanding inter-depot heterogeneity, but this remains to be seen.

One factor to consider in studying brite/beige lineages is not only the potential heterogeneity of the ASPC pool or the individual adipocyte functional identities (which is still mysterious) but also the morphological heterogeneity across the depot with respect to where these cells form. For example, cold can induce browning in an irregular "patchy" pattern in sWAT such that distinct islands of brite/beige adipocytes can sometimes be seen. It is possible that there are differences in the local concentration of adrenaline/noradrenaline that is dependent upon proximity to nerves or neurite density and could explain the erratic patterning. However, recent discoveries suggest that almost all sWAT adipocytes are in direct contact with sympathetic innervation (Chi et al. 2018; Jiang et al. 2017) suggesting the alternative possibilities that the pattering is cell autonomously regulated or could reflect different yet-to-be-defined niches within the depot. In sum, a distinct lineage or ASPC population that exclusively gives rise to the brite/beige cells is still lacking, suggesting their formation is likely more complex and multifactorial.

4 Going Forward

4.1 Unanswered Questions and Future Goals

Understanding the developmental origins of brown and brite/beige adipocytes, and the cell intrinsic and extrinsic signals that specify their fate and metabolic properties, is not only of biological interest but critical to advancing potential therapies that target thermogenesis as a means to increase energy expenditure. One of the major themes in adipose tissue biology that has emerged in recent years, driven by both developmental and metabolic studies, is that adipose tissues are highly heterogeneous. Developmental studies suggest brown and white adipocytes in different depots, as well as brown and white adipocytes within the same depot, may have different embryonic origins. Metabolic studies indicate that different white fat depots have different metabolic properties; for example, excess vWAT is metabolically unhealthy, while excess sWAT can be protective against metabolic disease (Reaven 1988; Snijder et al. 2003, 2004; Van Pelt et al. 2005). Even within WAT depots, some neighboring adipocytes may have different metabolic activity (Lee et al. 2017). Perhaps an interesting comparison is to skeletal muscle, which can have both fast and slow twitch fibers. Whether different BAT depots or brown adipocytes within single BAT depots have different metabolic properties or other functions is less clear. Understanding the functional significance of BAT heterogeneity, both at the mature adipocyte level and within the ASPC pool, is one important future goal.

While adipose tissue heterogeneity can be visualized by imaging studies, understanding the biochemical significance of BAT heterogeneity, and adipose tissue heterogeneity in general, has been more complicated by the fact that whole depots (as well as FACS-isolated ASPC pools) contain many non-adipocyte cells that can "contaminate" experiments that are focused on the adipocyte linages. Thus, key unanswered questions include whether there are genetic or epigenetic differences between lineages; whether there are lineage-specific transcription factors, receptors, or other factors; and whether different lineages produce different amounts of adipokines/BATokines or other transmissible signals. Exciting technological advances in single-cell RNA/DNA sequencing, metabolomics, and proteomics, combined with emerging tools that can purify organelles (including nuclei) and translating RNA away from non-adipocyte cells (Roh et al. 2017; Chen et al. 2016; Abu-Remaileh et al. 2017), are opening the door to a much higher-resolution view of adipocyte heterogeneity.

Single-cell profiling, for example, will allow us to define the cellular heterogeneity of ASPCs (and mature adipocyte populations) based on their genetic and epigenetic expression profiles. Single-cell analysis on cells captured at different differentiation stages has been useful in other systems to understand the differentiation path of a particular cell type (following what has been called pseudotime). In one application for lineage-tracing purposes, mutations or single-nucleotide polymorphism can be introduced in precursors in a manner that will accumulate over time such that unique individual cell sequences can be followed cumulatively in descendent cells (McKenna et al. 2016). Additionally, algorithms capable of deciphering spatial differences in gene expression are being developed to help understand region-specific functions within a tissue (Potter 2018; Griffiths et al. 2018; Kumar et al. 2017; McKenna et al. 2016).

A recent study focusing on adipocytes that will undoubtedly usher more, single-cell RNA sequencing of the ASPCs population in WAT identified at least three different cell populations involved in adipocyte regulation, one of which surprisingly secretes an unknown signal that inhibits adipogenesis (Schwalie et al. 2018). Another recent study performed single-cell transcriptomic analysis with human sWAT precursor cells and demonstrated that ASPCs are largely clustered in a single population (Acosta et al. 2017). Combining such studies with new tools that can isolate adipocytes away from the many non-adipocytes in a whole depot will be powerful. For example, a novel mouse model called NuTRAP, when combined with a fat-specific Cre driver, enables transcriptional and epigenomic profiling of only the Cre-marked adipocytes isolated from whole-fat tissues (Roh et al. 2017, 2018). Advances such as these will undoubtedly refine our understanding of the adipocyte lineages and will be a major focus area for the near future. As such, it will be important to standardize strains, diets, temperature, and sex across experiments as well as consider each depot as separate and functionally distinct entities so that results are comparable across laboratories.

Another important goal is to understand BAT fuel utilization and whether developmental origins have any role in specifying metabolic activities. Since thermogenesis requires free fatty acid exchange with mitochondria, it will be important to understand how BAT handles lipids. For example, why does BAT have both catabolic and anabolic lipid pathways working simultaneously (Sanchez-Gurmaches et al. 2018; McCormack and Denton 1977; Mottillo et al. 2014; Shimazu and Takahashi 1980; Townsend and Tseng 2015; Trayhurn 1979; Yu et al. 2002)? What is the significance of BAT lipolysis (Schreiber et al. 2017; Shin et al. 2017)? Recent studies also suggest that mitochondria proximal to lipid droplets, called peridroplet mitochondria, are functionally different from cytoplasmic mitochondria not associated with lipid droplets (Benador et al. 2018; Rambold et al. 2015; Nguyen et al. 2017; Stone et al. 2009; Wang et al. 2011). The prospect of mitochondria heterogeneity within a single cell opens up a whole new avenue of interest in understanding how organelles communicate with each other and the genome to control BAT metabolism. For example, although brown adipocytes have lower endoplasmic reticulum compared to other cell types, a recent study shows that brown adipocyte thermogenesis is regulated by an ER-membrane-embedded transcription factor (Bartelt et al. 2018). Sorting out intracellular BAT metabolism, and the influence of anatomical positioning, developmental patterning, innervation/vascularization, and immune cell communication on these processes will be critical in guiding the development of better therapeutic models.

While the role of the SNS in stimulating brown fat activity has long been understood (Kawate et al. 1994; Muzik et al. 2017; Owen et al. 2014), there are many interesting future questions about the role of the SNS in brite/beige adipocyte formation, as well as in the ability of BAT to communicate back to the brain and to other WAT depots. For example, it was recently proposed that iBAT cross talks to the sWAT through an "sWAT sensory neuron—Brain—iBAT" SNS connection (Garretson et al. 2016; Nguyen et al. 2018). According to this model, cold-induced sWAT lipolysis activates local afferent neurons triggering a neuronal circuit from sWAT to iBAT that controls iBAT thermogenesis, and this effect is abolished when the sWAT is denervated (Garretson et al. 2016; Nguyen et al. 2018). These findings emphasize that BAT development is likely tightly coordinated with nerve development and the concept of "neurometabolism" remains an understudied aspect of BAT growth and overall metabolic homeostasis.

Regarding potential connections between brown (and white) adipocyte origins and human fat disorders, a curious observation is that many lipodystrophy disorders present as selective adipose tissue atrophy, in which some depots shrink or disappear while others expand possibly as a compensatory response (Garg 2011). A similar type of fat body redistribution is observed when Myf5-Cre is used to genetically ablate regulators of the insulin signaling pathway in mice (Sanchez-Gurmaches et al. 2012; Sanchez-Gurmaches and Guertin 2014). For example, deleting *PTEN* (a negative regulator of insulin signaling) in the Myf5 lineage expands Myf5-positive brown *and* white fat, converts the brown fat into a white fat-like tissue, and causes the non-Myf5-lineage-positive adipocytes to disappear (Sanchez-Gurmaches et al. 2012); in contrast, deleting insulin receptor-beta (IR-β) with Myf5-Cre redistributes body fat in the other direction and reduces individual adipocyte size (Sanchez-Gurmaches and Guertin 2014; Gesta et al. 2007). The former model is strikingly similar to a rare fat disorder called multiple symmetric lipomatosis or Madelung's disease (Guastella et al. 2002; Ramos et al. 2010; Herbst 2012). Thus, another key question is whether differences in body fat distribution, whether pathological—such as in lipodystrophy or obesity—or even normal fat distribution across the population, may have some link to the developmental heterogeneityof fat.

4.2 Prospects for BAT-Based Therapeutics

There are a number of key issues that if resolved could help inform the development of BAT-based therapies. First, the commitment of progenitor cells to the thermogenic lineage is not understood. Knowing the mechanisms of brown fat specification could greatly aid in promoting the conversion of non-thermogenic cells to brown adipocytes. Whether better to focus on BAT or brite/beige adipocytes as a target for increasing energy expenditure remains unknown. Brite/beige fat may be promising because many obese or overweight adults seem to have a low abundance of BAT, at least based on classic BAT descriptions; however, better detection methods are needed (Hanssen et al. 2015; Ouellet et al. 2012; van der Lans et al. 2013; Yoneshiro et al. 2011a, b; Betz and Enerback 2018). Moreover, fully activated individual brite/beige adipocytes seem to have the same amount of UCP1 protein as an individual brown adipocyte even though total depot levels are quite different (Shabalina et al. 2013) and perhaps even slight increases in energy expenditure could have large effects over time. On the other hand, it may be possible to "train" adults to increase BAT activity (van der Lans et al. 2013; Hanssen et al. 2015). Understanding the development of thermogenic adipocytes is also relevant to stem cell-based models of thermogenesis such as in isogenic cell therapy programs (Singh and Dalton 2018). For example, it may be possible to generate human-induced pluripotent stem cells from patient-derived somatic cells that are reprogrammed to have a thermogenic adipocyte fate when transplanted into recipients (Ahfeldt et al. 2012; Kishida et al. 2015; Guenantin et al. 2017; Pisani et al. 2011). Chemical/hormonal induction protocols to generate such cells have not yet been described but would be of interest. Directly transplanting patient-derived brown adipocytes into obese individuals to improve metabolism may also be possible and has been demonstrated in rodent models (Min et al. 2016). At this point, there is not a clear consensus as to the best strategy for increasing brown fat activity to fight obesity, and both classic brown and brite/beige adipocytes should be considered until we know more about BAT and brite/beige adipocyte development and function.

It is remarkable that it was only about a decade ago that it became widely appreciated that adults have brown and brite/beige adipocytes, and thus while excitement about the therapeutic potential of targeting these amazing cells to increase energy expenditure is high, there is much research to be done to better understand their biology and, in particular, understanding their development and metabolic control, which are major challenge areas ahead.

Acknowledgments SMJ is supported by a postdoctoral fellowship award from the American Diabetes Association (1-18-PDF-128). JSG is supported by an American Heart Association Career Development award (18CDA34080527). DAG is supported by grants from the NIH (R01DK094004 and R01CA196986) and a Leukemia and Lymphoma Society Career Development Award.

References

Abu-Remaileh M, Wyant GA, Kim C, Laqtom NN, Abbasi M, Chan SH, Freinkman E, Sabatini DM (2017) Lysosomal metabolomics reveals V-ATPase- and mTOR-dependent regulation of amino acid efflux from lysosomes. Science 358:807–813

Acosta JR, Joost S, Karlsson K, Ehrlund A, Li X, Aouadi M, Kasper M, Arner P, Ryden M, Laurencikiene J (2017) Single cell transcriptomics suggest that human adipocyte progenitor cells constitute a homogeneous cell population. Stem Cell Res Ther 8:250

Ahfeldt T, Schinzel RT, Lee Y-K, Hendrickson D, Kaplan A, Lum DH, Camahort R, Xia F, Shay J, Rhee EP, Clish CB, Deo RC, Shen T, Lau FH, Cowley A, Mowrer G, Al-Siddiqi H, Nahrendorf M, Musunuru K, Gerszten RE, Rinn JL, Cowan CA (2012) Programming human pluripotent stem cells into white and brown adipocytes. Nat Cell Biol 14:209–219

Ambrosi TH, Scialdone A, Graja A, Gohlke S, Jank AM, Bocian C, Woelk L, Fan H, Logan DW, Schurmann A, Saraiva LR, Schulz TJ (2017) Adipocyte accumulation in the bone marrow during obesity and aging impairs stem cell-based hematopoietic and bone regeneration. Cell Stem Cell 20:771–784.e6

Atit R, Sgaier SK, Mohamed OA, Taketo MM, Dufort D, Joyner AL, Niswander L, Conlon RA (2006) beta-Catenin activation is necessary and sufficient to specify the dorsal dermal fate in the mouse. Dev Biol 296:164–176

Baglioni S, Francalanci M, Squecco R, Lombardi A, Cantini G, Angeli R, Gelmini S, Guasti D, Benvenuti S, Annunziato F, Bani D, Liotta F, Francini F, Perigli G, Serio M, Luconi M (2009) Characterization of human adult stem-cell populations isolated from visceral and subcutaneous adipose tissue. FASEB J 23:3494–3505

Baglioni S, Cantini G, Poli G, Francalanci M, Squecco R, Di Franco A, Borgogni E, Frontera S, Nesi G, Liotta F, Lucchese M, Perigli G, Francini F, Forti G, Serio M, Luconi M (2012) Functional differences in visceral and subcutaneous fat pads originate from differences in the adipose stem cell. PLoS One 7:e36569

Barbatelli G, Murano I, Madsen L, Hao Q, Jimenez M, Kristiansen K, Giacobino JP, De Matteis R, Cinti S (2010) The emergence of cold-induced brown adipocytes in mouse white fat depots is determined predominantly by white to brown adipocyte transdifferentiation. Am J Physiol Endocrinol Metab 298:E1244–E1253

Bartelt A, Bruns OT, Reimer R, Hohenberg H, Ittrich H, Peldschus K, Kaul MG, Tromsdorf UI, Weller H, Waurisch C, Eychmuller A, Gordts PL, Rinninger F, Bruegelmann K, Freund B, Nielsen P, Merkel M, Heeren J (2011) Brown adipose tissue activity controls triglyceride clearance. Nat Med 17:200–205

Bartelt A, Widenmaier SB, Schlein C, Johann K, Goncalves RLS, Eguchi K, Fischer AW, Parlakgul G, Snyder NA, Nguyen TB, Bruns OT, Franke D, Bawendi MG, Lynes MD, Leiria LO, Tseng YH, Inouye KE, Arruda AP, Hotamisligil GS (2018) Brown adipose tissue thermogenic adaptation requires Nrf1-mediated proteasomal activity. Nat Med 24:292–303

Benador IY, Veliova M, Mahdaviani K, Petcherski A, Wikstrom JD, Assali EA, Acin-Perez R, Shum M, Oliveira MF, Cinti S, Sztalryd C, Barshop WD, Wohlschlegel JA, Corkey BE, Liesa M, Shirihai OS (2018) Mitochondria bound to lipid droplets have unique bioenergetics, composition, and dynamics that support lipid droplet expansion. Cell Metab 27:869–885 e6

Berry R, Rodeheffer MS (2013) Characterization of the adipocyte cellular lineage in vivo. Nat Cell Biol 15:302–308

Berry R, Jeffery E, Rodeheffer MS (2014) Weighing in on adipocyte precursors. Cell Metab 19:8–20

Berry DC, Jiang Y, Graff JM (2016) Mouse strains to study cold-inducible beige progenitors and beige adipocyte formation and function. Nat Commun 7:10184

Bertholet AM, Kazak L, Chouchani ET, Bogaczynska MG, Paranjpe I, Wainwright GL, Betourne A, Kajimura S, Spiegelman BM, Kirichok Y (2017) Mitochondrial patch clamp of beige adipocytes reveals UCP1-positive and UCP1-negative cells both exhibiting futile creatine cycling. Cell Metab 25:811–822 e4

Betz MJ, Enerback S (2018) Targeting thermogenesis in brown fat and muscle to treat obesity and metabolic disease. Nat Rev Endocrinol 14:77–87

Blanchette-Mackie EJ, Scow RO (1983) Movement of lipolytic products to mitochondria in brown adipose tissue of young rats: an electron microscope study. J Lipid Res 24:229–244

Bronnikov G, Houstek J, Nedergaard J (1992) Beta-adrenergic, cAMP-mediated stimulation of proliferation of brown fat cells in primary culture. Mediation via beta 1 but not via beta 3 adrenoceptors. J Biol Chem 267:2006–2013

Bukowiecki L, Collet AJ, Follea N, Guay G, Jahjah L (1982) Brown adipose tissue hyperplasia: a fundamental mechanism of adaptation to cold and hyperphagia. Am J Physiol 242:E353–E359

Bukowiecki LJ, Geloen A, Collet AJ (1986) Proliferation and differentiation of brown adipocytes from interstitial cells during cold acclimation. Am J Physiol 250:C880–C887

Cannon B, Nedergaard J (2004) Brown adipose tissue: function and physiological significance. Physiol Rev 84:277–359

Carrer A, Parris JL, Trefely S, Henry RA, Montgomery DC, Torres A, Viola JM, Kuo YM, Blair IA, Meier JL, Andrews AJ, Snyder NW, Wellen KE (2017) Impact of a high-fat diet on tissue acyl-CoA and histone acetylation levels. J Biol Chem 292:3312–3322

Chabowska-Kita A, Kozak LP (2016) The critical period for brown adipocyte development: genetic and environmental influences. Obesity (Silver Spring) 24:283–290

Chen WW, Freinkman E, Wang T, Birsoy K, Sabatini DM (2016) Absolute quantification of matrix metabolites reveals the dynamics of mitochondrial metabolism. Cell 166:1324–1337 e11

Chi J, Wu Z, Choi CHJ, Nguyen L, Tegegne S, Ackerman SE, Crane A, Marchildon F, Tessier-Lavigne M, Cohen P (2018) Three-dimensional adipose tissue imaging reveals regional variation in beige fat biogenesis and PRDM16-dependent sympathetic neurite density. Cell Metab 27:226–236.e3

Church CD, Berry R, Rodeheffer MS (2014) Isolation and study of adipocyte precursors. Methods Enzymol 537:31–46

Cinti S (2002) Adipocyte differentiation and transdifferentiation: plasticity of the adipose organ. J Endocrinol Investig 25:823–835

Cinti S (2005) The adipose organ. Prostaglandins Leukot Essent Fatty Acids 73:9–15

Collins S, Daniel KW, Petro AE, Surwit RS (1997) Strain-specific response to beta 3-adrenergic receptor agonist treatment of diet-induced obesity in mice. Endocrinology 138:405–413

Cserjesi P, Lilly B, Bryson L, Wang Y, Sassoon DA, Olson EN (1992) MHox: a mesodermally restricted homeodomain protein that binds an essential site in the muscle creatine kinase enhancer. Development 115:1087–1101

Cypess AM, Lehman S, Williams G, Tal I, Rodman D, Goldfine AB, Kuo FC, Palmer EL, Tseng YH, Doria A, Kolodny GM, Kahn CR (2009) Identification and importance of brown adipose tissue in adult humans. N Engl J Med 360:1509–1517

Cypess AM, White AP, Vernochet C, Schulz TJ, Xue R, Sass CA, Huang TL, Roberts-Toler C, Weiner LS, Sze C, Chacko AT, Deschamps LN, Herder LM, Truchan N, Glasgow AL, Holman AR, Gavrila A, Hasselgren PO, Mori MA, Molla M, Tseng YH (2013) Anatomical localization, gene expression profiling and functional characterization of adult human neck brown fat. Nat Med 19:635–639

de Jong JM, Larsson O, Cannon B, Nedergaard J (2015) A stringent validation of mouse adipose tissue identity markers. Am J Physiol Endocrinol Metab 308:E1085–E1105

de Jong JMA, Wouters RTF, Boulet N, Cannon B, Nedergaard J, Petrovic N (2017) The beta3-adrenergic receptor is dispensable for browning of adipose tissues. Am J Physiol Endocrinol Metab 312:E508–e518

Dempersmier J, Sambeat A, Gulyaeva O, Paul SM, Hudak CS, Raposo HF, Kwan HY, Kang C, Wong RH, Sul HS (2015) Cold-inducible Zfp516 activates UCP1 transcription to promote browning of white fat and development of brown fat. Mol Cell 57:235–246

Du B, Cawthorn WP, Su A, Doucette CR, Yao Y, Hemati N, Kampert S, Mccoin C, Broome DT, Rosen CJ, Yang G, Macdougald OA (2013) The transcription factor paired-related homeobox 1 (Prrx1) inhibits adipogenesis by activating transforming growth factor-beta (TGFbeta) signaling. J Biol Chem 288:3036–3047

Eguchi J, Wang X, Yu S, Kershaw EE, Chiu PC, Dushay J, Estall JL, Klein U, Maratos-Flier E, Rosen ED (2011) Transcriptional control of adipose lipid handling by IRF4. Cell Metab 13:249–259

Farmer SR (2006) Transcriptional control of adipocyte formation. Cell Metab 4:263–273

Fedorenko A, Lishko PV, Kirichok Y (2012) Mechanism of fatty-acid-dependent UCP1 uncoupling in brown fat mitochondria. Cell 151:400–413

Friederich-Persson M, Nguyen Dinh Cat A, Persson P, Montezano AC, Touyz RM (2017) Brown adipose tissue regulates small artery function through NADPH oxidase 4-derived hydrogen peroxide and redox-sensitive protein kinase G-1alpha. Arterioscler Thromb Vasc Biol 37:455–465

Frontini A, Cinti S (2010) Distribution and development of brown adipocytes in the murine and human adipose organ. Cell Metab 11:253–256

Garg A (2011) Clinical review#: lipodystrophies: genetic and acquired body fat disorders. J Clin Endocrinol Metab 96:3313–3325

Garretson JT, Szymanski LA, Schwartz GJ, Xue B, Ryu V, Bartness TJ (2016) Lipolysis sensation by white fat afferent nerves triggers brown fat thermogenesis. Mol Metab 5:626–634

Geloen A, Collet AJ, Bukowiecki LJ (1992) Role of sympathetic innervation in brown adipocyte proliferation. Am J Physiol 263:R1176–R1181

Gensch N, Borchardt T, Schneider A, Riethmacher D, Braun T (2008) Different autonomous myogenic cell populations revealed by ablation of Myf5-expressing cells during mouse embryogenesis. Development 135:1597–1604

Gesta S, Tseng YH, Kahn CR (2007) Developmental origin of fat: tracking obesity to its source. Cell 131:242–256

Griffiths JA, Scialdone A, Marioni JC (2018) Using single-cell genomics to understand developmental processes and cell fate decisions. Mol Syst Biol 14:e8046

Guastella C, Borsi C, Gibelli S, Della Berta LG (2002) Madelung's lipomatosis associated with head and neck malignant neoplasia: a study of 2 cases. Otolaryngol Head Neck Surg 126:191–192

Guenantin AC, Briand N, Capel E, Dumont F, Morichon R, Provost C, Stillitano F, Jeziorowska D, Siffroi JP, Hajjar RJ, Feve B, Hulot JS, Collas P, Capeau J, Vigouroux C (2017) Functional human beige adipocytes from induced pluripotent stem cells. Diabetes 66:1470–1478

Gupta RK, Arany Z, Seale P, Mepani RJ, Ye L, Conroe HM, Roby YA, Kulaga H, Reed RR, Spiegelman BM (2010) Transcriptional control of preadipocyte determination by Zfp423. Nature 464:619–623

Gupta RK, Mepani RJ, Kleiner S, Lo JC, Khandekar MJ, Cohen P, Frontini A, Bhowmick DC, Ye L, Cinti S, Spiegelman BM (2012) Zfp423 expression identifies committed preadipocytes and localizes to adipose endothelial and perivascular cells. Cell Metab 15:230–239

Haldar M, Karan G, Tvrdik P, Capecchi MR (2008) Two cell lineages, myf5 and myf5-independent, participate in mouse skeletal myogenesis. Dev Cell 14:437–445

Hansen IR, Jansson KM, Cannon B, Nedergaard J (2014) Contrasting effects of cold acclimation versus obesogenic diets on chemerin gene expression in brown and brite adipose tissues. Biochim Biophys Acta 1841:1691–1699

Hanssen MJ, Hoeks J, Brans B, Van Der Lans AA, Schaart G, Van Den Driessche JJ, Jorgensen JA, Boekschoten MV, Hesselink MK, Havekes B, Kersten S, Mottaghy FM, Van Marken Lichtenbelt WD, Schrauwen P (2015) Short-term cold acclimation improves insulin sensitivity in patients with type 2 diabetes mellitus. Nat Med 21:863–865

Harms M, Seale P (2013) Brown and beige fat: development, function and therapeutic potential. Nat Med 19:1252–1263

Harms MJ, Ishibashi J, Wang W, Lim HW, Goyama S, Sato T, Kurokawa M, Won KJ, Seale P (2014) Prdm16 is required for the maintenance of brown adipocyte identity and function in adult mice. Cell Metab 19:593–604

Herbst KL (2012) Rare adipose disorders (RADs) masquerading as obesity. Acta Pharmacol Sin 33:155–172

Hu YS, Zhou H, Kartsogiannis V, Eisman JA, Martin TJ, Ng KW (1998) Expression of rat homeobox gene, rHOX, in developing and adult tissues in mice and regulation of its mRNA expression in osteoblasts by bone morphogenetic protein 2 and parathyroid hormone-related protein. Mol Endocrinol 12:1721–1732

Hung CM, Calejman CM, Sanchez-Gurmaches J, Li H, Clish CB, Hettmer S, Wagers AJ, Guertin DA (2014) Rictor/mTORC2 loss in the Myf5 lineage reprograms brown fat metabolism and protects mice against obesity and metabolic disease. Cell Rep 8:256–271

Ikeda K, Maretich P, Kajimura S (2018) The common and distinct features of brown and beige adipocytes. Trends Endocrinol Metab 29:191–200

Jiang Y, Berry DC, Tang W, Graff JM (2014) Independent stem cell lineages regulate adipose organogenesis and adipose homeostasis. Cell Rep 9:1007–1022

Jiang Y, Berry DC, Graff JM (2017) Distinct cellular and molecular mechanisms for beta3 adrenergic receptor-induced beige adipocyte formation. Elife 6:e30329

Joe AW, Yi L, Natarajan A, Le Grand F, So L, Wang J, Rudnicki MA, Rossi FM (2010) Muscle injury activates resident fibro/adipogenic progenitors that facilitate myogenesis. Nat Cell Biol 12:153–163

Kajimura S, Spiegelman BM, Seale P (2015) Brown and beige fat: physiological roles beyond heat generation. Cell Metab 22:546–559

Kawate R, Talan MI, Engel BT (1994) Sympathetic nervous activity to brown adipose tissue increases in cold-tolerant mice. Physiol Behav 55:921–925

Kazak L, Chouchani ET, Jedrychowski MP, Erickson BK, Shinoda K, Cohen P, Vetrivelan R, Lu GZ, Laznik-Bogoslavski D, Hasenfuss SC, Kajimura S, Gygi SP, Spiegelman BM (2015) A creatine-driven substrate cycle enhances energy expenditure and thermogenesis in beige fat. Cell 163:643–655

Kishida T, Ejima A, Yamamoto K, Tanaka S, Yamamoto T, Mazda O (2015) Reprogrammed functional brown adipocytes ameliorate insulin resistance and dyslipidemia in diet-induced obesity and type 2 diabetes. Stem Cell Reports 5:569–581

Kortelainen ML, Pelletier G, Ricquier D, Bukowiecki LJ (1993) Immunohistochemical detection of human brown adipose tissue uncoupling protein in an autopsy series. J Histochem Cytochem 41:759–764

Kozak LP (2011) The genetics of brown adipocyte induction in white fat depots. Front Endocrinol (Lausanne) 2:64

Krueger KC, Costa MJ, Du H, Feldman BJ (2014) Characterization of Cre recombinase activity for in vivo targeting of adipocyte precursor cells. Stem Cell Reports 3:1147–1158

Kumar P, Tan Y, Cahan P (2017) Understanding development and stem cells using single cell-based analyses of gene expression. Development 144:17–32

Labbe SM, Caron A, Bakan I, Laplante M, Carpentier AC, Lecomte R, Richard D (2015) In vivo measurement of energy substrate contribution to cold-induced brown adipose tissue thermogenesis. FASEB J 29:2046–2058

Lee YH, Petkova AP, Mottillo EP, Granneman JG (2012) In vivo identification of bipotential adipocyte progenitors recruited by beta3-adrenoceptor activation and high-fat feeding. Cell Metab 15:480–491

Lee YH, Petkova AP, Konkar AA, Granneman JG (2015) Cellular origins of cold-induced brown adipocytes in adult mice. FASEB J 29:286–299

Lee KY, Sharma R, Gase G, Ussar S, Li Y, Welch L, Berryman DE, Kispert A, Bluher M, Kahn CR (2017) Tbx15 defines a glycolytic subpopulation and white adipocyte heterogeneity. Diabetes 66:2822–2829

Lehr L, Canola K, Asensio C, Jimenez M, Kuehne F, Giacobino JP, Muzzin P (2006) The control of UCP1 is dissociated from that of PGC-1alpha or of mitochondriogenesis as revealed by a study using beta-less mouse brown adipocytes in culture. FEBS Lett 580:4661–4666

Leitner BP, Huang S, Brychta RJ, Duckworth CJ, Baskin AS, Mcgehee S, Tal I, Dieckmann W, Gupta G, Kolodny GM, Pacak K, Herscovitch P, Cypess AM, Chen KY (2017) Mapping of human brown adipose tissue in lean and obese young men. Proc Natl Acad Sci U S A 114:8649–8654

Lepper C, Fan C-M (2010) Inducible lineage tracing of Pax7-descendant cells reveals embryonic origin of adult satellite cells. Genesis 48:424–436

Lidell ME, Betz MJ, Leinhard OD, Heglind M, Elander L, Slawik M, Mussack T, Nilsson D, Romu T, Nuutila P, Virtanen KA, Beuschlein F, Persson A, Borga M, Enerback S (2013) Evidence for two types of brown adipose tissue in humans. Nat Med 19:631

Liu W, Shan T, Yang X, Liang S, Zhang P, Liu Y, Liu X, Kuang S (2013) A heterogeneous lineage origin underlies the phenotypic and molecular differences of white and beige adipocytes. J Cell Sci 126:3527–3532

Logan M, Martin JF, Nagy A, Lobe C, Olson EN, Tabin CJ (2002) Expression of Cre recombinase in the developing mouse limb bud driven by a Prxl enhancer. Genesis 33:77–80

Long JZ, Svensson KJ, Tsai L, Zeng X, Roh HC, Kong X, Rao RR, Lou J, Lokurkar I, Baur W, Castellot JJ Jr, Rosen ED, Spiegelman BM (2014) A smooth muscle-like origin for beige adipocytes. Cell Metab 19:810–820

Long JZ, Svensson KJ, Bateman LA, Lin H, Kamenecka T, Lokurkar IA, Lou J, Rao RR, Chang MR, Jedrychowski MP, Paulo JA, Gygi SP, Griffin PR, Nomura DK, Spiegelman BM

(2016) The secreted enzyme PM20D1 regulates lipidated amino acid uncouplers of mitochondria. Cell 166:424–435
Lu MF, Cheng HT, Kern MJ, Potter SS, Tran B, Diekwisch TG, Martin JF (1999) prx-1 functions cooperatively with another paired-related homeobox gene, prx-2, to maintain cell fates within the craniofacial mesenchyme. Development 126:495–504
Lumeng CN, Saltiel AR (2011) Inflammatory links between obesity and metabolic disease. J Clin Invest 121:2111–2117
Mahdaviani K, Chess D, Wu Y, Shirihai O, Aprahamian TR (2016) Autocrine effect of vascular endothelial growth factor-A is essential for mitochondrial function in brown adipocytes. Metabolism 65:26–35
McCormack JG, Denton RM (1977) Evidence that fatty acid synthesis in the interscapular brown adipose tissue of cold-adapted rats is increased in vivo by insulin by mechanisms involving parallel activation of pyruvate dehydrogenase and acetyl-coenzyme A carboxylase. Biochem J 166:627–630
McKenna A, Findlay GM, Gagnon JA, Horwitz MS, Schier AF, Shendure J (2016) Whole-organism lineage tracing by combinatorial and cumulative genome editing. Science 353:aaf7907
Min SY, Kady J, Nam M, Rojas-Rodriguez R, Berkenwald A, Kim JH, Noh HL, Kim JK, Cooper MP, Fitzgibbons T, Brehm MA, Corvera S (2016) Human 'brite/beige' adipocytes develop from capillary networks, and their implantation improves metabolic homeostasis in mice. Nat Med 22:312–318
Mo Q, Salley J, Roshan T, Baer LA, May FJ, Jaehnig EJ, Lehnig AC, Guo X, Tong Q, Nuotio-Antar AM, Shamsi F, Tseng YH, Stanford KI, Chen MH (2017) Identification and characterization of a supraclavicular brown adipose tissue in mice. JCI Insight 2
Morrison SF, Madden CJ, Tupone D (2012) Central control of brown adipose tissue thermogenesis. Front Endocrinol (Lausanne) 3
Mottillo EP, Balasubramanian P, Lee YH, Weng C, Kershaw EE, Granneman JG (2014) Coupling of lipolysis and de novo lipogenesis in brown, beige, and white adipose tissues during chronic beta3-adrenergic receptor activation. J Lipid Res 55:2276–2286
Moullan N, Mouchiroud L, Wang X, Ryu D, Williams EG, Mottis A, Jovaisaite V, Frochaux MV, Quiros PM, Deplancke B, Houtkooper RH, Auwerx J (2015) Tetracyclines disturb mitochondrial function across eukaryotic models: a call for caution in biomedical research. Cell Rep 10:1681–1691
Murano I, Barbatelli G, Giordano A, Cinti S (2009) Noradrenergic parenchymal nerve fiber branching after cold acclimatisation correlates with brown adipocyte density in mouse adipose organ. J Anat 214:171–178
Muzik O, Mangner TJ, Leonard WR, Kumar A, Granneman JG (2017) Sympathetic innervation of cold-activated brown and white fat in lean young adults. J Nucl Med 58:799–806
Muzumdar MD, Tasic B, Miyamichi K, Li L, Luo L (2007) A global double-fluorescent Cre reporter mouse. Genesis 45:593–605
Nedergaard J, Cannon B (2014) The browning of white adipose tissue: some burning issues. Cell Metab 20:396–407
Nedergaard J, Bengtsson T, Cannon B (2007) Unexpected evidence for active brown adipose tissue in adult humans. Am J Physiol Endocrinol Metab 293:E444–E452
Nguyen TB, Louie SM, Daniele JR, Tran Q, Dillin A, Zoncu R, Nomura DK, Olzmann JA (2017) DGAT1-dependent lipid droplet biogenesis protects mitochondrial function during starvation-induced autophagy. Dev Cell 42:9–21 e5
Nguyen NLT, Xue B, Bartness TJ (2018) Sensory denervation of inguinal white fat modifies sympathetic outflow to white and brown fat in Siberian hamsters. Physiol Behav 190:28–33
Nicholls DG (2006) The physiological regulation of uncoupling proteins. Biochim Biophys Acta 1757:459–466

Nisoli E, Clementi E, Tonello C, Sciorati C, Briscini L, Carruba MO (1998) Effects of nitric oxide on proliferation and differentiation of rat brown adipocytes in primary cultures. Br J Pharmacol 125:888–894

Olefsky JM, Glass CK (2010) Macrophages, inflammation, and insulin resistance. Annu Rev Physiol 72:219–246

Ouellet V, Labbe SM, Blondin DP, Phoenix S, Guerin B, Haman F, Turcotte EE, Richard D, Carpentier AC (2012) Brown adipose tissue oxidative metabolism contributes to energy expenditure during acute cold exposure in humans. J Clin Investig 122:545–552

Owen BM, Ding X, Morgan DA, Coate KC, Bookout AL, Rahmouni K, Kliewer SA, Mangelsdorf DJ (2014) FGF21 acts centrally to induce sympathetic nerve activity, energy expenditure, and weight loss. Cell Metab 20:670–677

Perrini S, Ficarella R, Picardi E, Cignarelli A, Barbaro M, Nigro P, Peschechera A, Palumbo O, Carella M, De Fazio M, Natalicchio A, Laviola L, Pesole G, Giorgino F (2013) Differences in gene expression and cytokine release profiles highlight the heterogeneity of distinct subsets of adipose tissue-derived stem cells in the subcutaneous and visceral adipose tissue in humans. PLoS One 8:e57892

Peterson RE, Hoffman S, Kern MJ (2005) Opposing roles of two isoforms of the Prx1 homeobox gene in chondrogenesis. Dev Dyn 233:811–821

Pietrocola F, Galluzzi L, Bravo-San Pedro JM, Madeo F, Kroemer G (2015) Acetyl coenzyme A: a central metabolite and second messenger. Cell Metab 21:805–821

Pisani DF, Djedaini M, Beranger GE, Elabd C, Scheideler M, Ailhaud G, Amri EZ (2011) Differentiation of human adipose-derived stem cells into "brite" (brown-in-white) adipocytes. Front Endocrinol (Lausanne) 2:87

Potter SS (2018) Single-cell RNA sequencing for the study of development, physiology and disease. Nat Rev Nephrol 14:479–492

Rajakumari S, Wu J, Ishibashi J, Lim HW, Giang AH, Won KJ, Reed RR, Seale P (2013) EBF2 determines and maintains brown adipocyte identity. Cell Metab 17:562–574

Rambold AS, Cohen S, Lippincott-Schwartz J (2015) Fatty acid trafficking in starved cells: regulation by lipid droplet lipolysis, autophagy, and mitochondrial fusion dynamics. Dev Cell 32:678–692

Ramos S, Pinheiro S, Diogo C, Cabral L, Cruzeiro C (2010) Madelung disease: a not-so-rare disorder. Ann Plast Surg 64:122–124

Razzoli M, Emmett MJ, Lazar MA, Bartolomucci A (2018) beta-Adrenergic receptors control brown adipose UCP-1 tone and cold response without affecting its circadian rhythmicity. FASEB J:fj201800452R

Reaven GM (1988) Banting lecture 1988. Role of insulin resistance in human disease. Diabetes 37:1595–1607

Rehnmark S, Nedergaard J (1989) DNA synthesis in mouse brown adipose tissue is under beta-adrenergic control. Exp Cell Res 180:574–579

Rivera-Gonzalez GC, Shook BA, Andrae J, Holtrup B, Bollag K, Betsholtz C, Rodeheffer MS, Horsley V (2016) Skin adipocyte stem cell self-renewal is regulated by a PDGFA/AKT-signaling axis. Cell Stem Cell 19:738–751

Roberts-Toler C, O'neill BT, Cypess AM (2015) Diet-induced obesity causes insulin resistance in mouse brown adipose tissue. Obesity (Silver Spring) 23:1765–1770

Rodeheffer MS, Birsoy K, Friedman JM (2008) Identification of white adipocyte progenitor cells in vivo. Cell 135:240–249

Roh HC, Tsai LT, Lyubetskaya A, Tenen D, Kumari M, Rosen ED (2017) Simultaneous transcriptional and epigenomic profiling from specific cell types within heterogeneous tissues in vivo. Cell Rep 18:1048–1061

Roh HC, Tsai LTY, Shao M, Tenen D, Shen Y, Kumari M, Lyubetskaya A, Jacobs C, Dawes B, Gupta RK, Rosen ED (2018) Warming induces significant reprogramming of beige, but not brown, adipocyte cellular identity. Cell Metab 27:1121–1137.e5

Rosen ED, Sarraf P, Troy AE, Bradwin G, Moore K, Milstone DS, Spiegelman BM, Mortensen RM (1999) PPAR gamma is required for the differentiation of adipose tissue in vivo and in vitro. Mol Cell 4:611–617

Rosenwald M, Perdikari A, Rulicke T, Wolfrum C (2013) Bi-directional interconversion of brite and white adipocytes. Nat Cell Biol 15:659–667

Rothwell NJ, Stock MJ (1984) Effects of denervating brown adipose tissue on the responses to cold, hyperphagia and noradrenaline treatment in the rat. J Physiol 355:457–463

Sacks H, Symonds ME (2013) Anatomical locations of human brown adipose tissue: functional relevance and implications in obesity and type 2 diabetes. Diabetes 62:1783–1790

Saito M, Okamatsu-Ogura Y, Matsushita M, Watanabe K, Yoneshiro T, Nio-Kobayashi J, Iwanaga T, Miyagawa M, Kameya T, Nakada K, Kawai Y, Tsujisaki M (2009) High incidence of metabolically active brown adipose tissue in healthy adult humans: effects of cold exposure and adiposity. Diabetes 58:1526–1531

Sakaguchi M, Fujisaka S, Cai W, Winnay JN, Konishi M, O'neill BT, Li M, Garcia-Martin R, Takahashi H, Hu J, Kulkarni RN, Kahn CR (2017) Adipocyte dynamics and reversible metabolic syndrome in mice with an inducible adipocyte-specific deletion of the insulin receptor. Cell Metab 25:448–462

Sambeat A, Gulyaeva O, Dempersmier J, Tharp KM, Stahl A, Paul SM, Sul HS (2016) LSD1 interacts with Zfp516 to promote UCP1 transcription and brown fat program. Cell Rep 15:2536–2549

Sanchez-Gurmaches J, Guertin DA (2014) Adipocytes arise from multiple lineages that are heterogeneously and dynamically distributed. Nat Commun 5:4099

Sanchez-Gurmaches J, Hung CM, Sparks CA, Tang Y, Li H, Guertin DA (2012) PTEN loss in the Myf5 lineage redistributes body fat and reveals subsets of white adipocytes that arise from Myf5 precursors. Cell Metab 16:348–362

Sanchez-Gurmaches J, Hsiao WY, Guertin DA (2015) Highly selective in vivo labeling of subcutaneous white adipocyte precursors with Prx1-Cre. Stem Cell Reports 4:541–550

Sanchez-Gurmaches J, Tang Y, Jespersen NZ, Wallace M, Martinez Calejman C, Gujja S, Li H, Edwards YJK, Wolfrum C, Metallo CM, Nielsen S, Scheele C, Guertin DA (2018) Brown fat AKT2 is a cold-induced kinase that stimulates ChREBP-mediated de novo lipogenesis to optimize fuel storage and thermogenesis. Cell Metab 27:195–209.e6

Schreiber R, Diwoky C, Schoiswohl G, Feiler U, Wongsiriroj N, Abdellatif M, Kolb D, Hoeks J, Kershaw EE, Sedej S, Schrauwen P, Haemmerle G, Zechner R (2017) Cold-induced thermogenesis depends on ATGL-mediated lipolysis in cardiac muscle, but not brown adipose tissue. Cell Metab 26:753–763.e7

Schulz TJ, Huang P, Huang TL, Xue R, McDougall LE, Townsend KL, Cypess AM, Mishina Y, Gussoni E, Tseng YH (2013) Brown-fat paucity due to impaired BMP signalling induces compensatory browning of white fat. Nature 495:379–383

Schwalie PC, Dong H, Zachara M, Russeil J, Alpern D, Akchiche N, Caprara C, Sun W, Schlaudraff K-U, Soldati G, Wolfrum C, Deplancke B (2018) A stromal cell population that inhibits adipogenesis in mammalian fat depots. Nature 559:103–108

Seale P, Kajimura S, Yang W, Chin S, Rohas LM, Uldry M, Tavernier G, Langin D, Spiegelman BM (2007) Transcriptional control of brown fat determination by PRDM16. Cell Metab 6:38–54

Seale P, Bjork B, Yang W, Kajimura S, Chin S, Kuang S, Scime A, Devarakonda S, Conroe HM, Erdjument-Bromage H, Tempst P, Rudnicki MA, Beier DR, Spiegelman BM (2008) PRDM16 controls a brown fat/skeletal muscle switch. Nature 454:961–967

Seale P, Conroe HM, Estall J, Kajimura S, Frontini A, Ishibashi J, Cohen P, Cinti S, Spiegelman BM (2011) Prdm16 determines the thermogenic program of subcutaneous white adipose tissue in mice. J Clin Invest 121:96–105

Shabalina IG, Petrovic N, de Jong JM, Kalinovich AV, Cannon B, Nedergaard J (2013) UCP1 in brite/beige adipose tissue mitochondria is functionally thermogenic. Cell Rep 5:1196–1203

Shan T, Liang X, Bi P, Zhang P, Liu W, Kuang S (2013) Distinct populations of adipogenic and myogenic Myf5-lineage progenitors in white adipose tissues. J Lipid Res 54:2214–2224
Shao M, Gupta RK (2018) Transcriptional brakes on the road to adipocyte thermogenesis. Biochim Biophys Acta Mol Cell Biol Lipids. https://doi.org/10.1016/j.bbalip.2018.05.010
Shao M, Ishibashi J, Kusminski CM, Wang QA, Hepler C, Vishvanath L, Macpherson KA, Spurgin SB, Sun K, Holland WL, Seale P, Gupta RK (2016) Zfp423 maintains white adipocyte identity through suppression of the beige cell thermogenic gene program. Cell Metab 23:1167–1184
Sharp LZ, Shinoda K, Ohno H, Scheel DW, Tomoda E, Ruiz L, Hu H, Wang L, Pavlova Z, Gilsanz V, Kajimura S (2012) Human BAT possesses molecular signatures that resemble beige/brite cells. PLoS One 7:e49452
Shimazu T, Takahashi A (1980) Stimulation of hypothalamic nuclei has differential effects on lipid synthesis in brown and white adipose tissue. Nature 284:62–63
Shimizu I, Aprahamian T, Kikuchi R, Shimizu A, Papanicolaou KN, MacLauchlan S, Maruyama S, Walsh K (2014) Vascular rarefaction mediates whitening of brown fat in obesity. J Clin Invest 124:2099–2112
Shin H, Ma Y, Chanturiya T, Cao Q, Wang Y, Kadegowda AKG, Jackson R, Rumore D, Xue B, Shi H, Gavrilova O, Yu L (2017) Lipolysis in brown adipocytes is not essential for cold-induced thermogenesis in mice. Cell Metab 26:764–777.e5
Shinoda K, Luijten IH, Hasegawa Y, Hong H, Sonne SB, Kim M, Xue R, Chondronikola M, Cypess AM, Tseng YH, Nedergaard J, Sidossis LS, Kajimura S (2015) Genetic and functional characterization of clonally derived adult human brown adipocytes. Nat Med 21:389–394
Silva JE, Larsen PR (1983) Adrenergic activation of triiodothyronine production in brown adipose tissue. Nature 305:712–713
Simcox J, Geoghegan G, Maschek JA, Bensard CL, Pasquali M, Miao R, Lee S, Jiang L, Huck I, Kershaw EE, Donato AJ, Apte U, Longo N, Rutter J, Schreiber R, Zechner R, Cox J, Villanueva CJ (2017) Global analysis of plasma lipids identifies liver-derived acylcarnitines as a fuel source for brown fat thermogenesis. Cell Metab 26:509–522 e6
Singh AM, Dalton S (2018) What can 'brown-ing' do for you? Trends Endocrinol Metab 29:349–359
Sluse FE, Jarmuszkiewicz W, Navet R, Douette P, Mathy G, Sluse-Goffart CM (2006) Mitochondrial UCPs: new insights into regulation and impact. Biochim Biophys Acta 1757:480–485
Snijder MB, Dekker JM, Visser M, Bouter LM, Stehouwer CD, Kostense PJ, Yudkin JS, Heine RJ, Nijpels G, Seidell JC (2003) Associations of hip and thigh circumferences independent of waist circumference with the incidence of type 2 diabetes: the Hoorn study. Am J Clin Nutr 77:1192–1197
Snijder MB, Dekker JM, Visser M, Bouter LM, Stehouwer CD, Yudkin JS, Heine RJ, Nijpels G, Seidell JC, Hoorn S (2004) Trunk fat and leg fat have independent and opposite associations with fasting and postload glucose levels: the Hoorn study. Diabetes Care 27:372–377
Stine RR, Shapira SN, Lim HW, Ishibashi J, Harms M, Won KJ, Seale P (2016) EBF2 promotes the recruitment of beige adipocytes in white adipose tissue. Mol Metab 5:57–65
Stone SJ, Levin MC, Zhou P, Han J, Walther TC, Farese RV Jr (2009) The endoplasmic reticulum enzyme DGAT2 is found in mitochondria-associated membranes and has a mitochondrial targeting signal that promotes its association with mitochondria. J Biol Chem 284:5352–5361
Sun K, Kusminski CM, Luby-Phelps K, Spurgin SB, An YA, Wang QA, Holland WL, Scherer PE (2014) Brown adipose tissue derived VEGF-A modulates cold tolerance and energy expenditure. Mol Metab 3:474–483
Sun C, Berry WL, Olson LE (2017) PDGFRalpha controls the balance of stromal and adipogenic cells during adipose tissue organogenesis. Development 144:83–94
Svensson KJ, Long JZ, Jedrychowski MP, Cohen P, Lo JC, Serag S, Kir S, Shinoda K, Tartaglia JA, Rao RR, Chedotal A, Kajimura S, Gygi SP, Spiegelman BM (2016) A secreted Slit2 fragment regulates adipose tissue thermogenesis and metabolic function. Cell Metab 23:454–466

Takahashi A, Shimazu T, Maruyama Y (1992) Importance of sympathetic nerves for the stimulatory effect of cold exposure on glucose utilization in brown adipose tissue. Jpn J Physiol 42:653–664

Tallquist MD, Weismann KE, Hellstrom M, Soriano P (2000) Early myotome specification regulates PDGFA expression and axial skeleton development. Development 127:5059–5070

ten Berge D, Brouwer A, Korving J, Martin JF, Meijlink F (1998) Prx1 and Prx2 in skeletogenesis: roles in the craniofacial region, inner ear and limbs. Development 125:3831–3842

Thomou T, Mori MA, Dreyfuss JM, Konishi M, Sakaguchi M, Wolfrum C, Rao TN, Winnay JN, Garcia-Martin R, Grinspoon SK, Gorden P, Kahn CR (2017) Adipose-derived circulating miRNAs regulate gene expression in other tissues. Nature 542:450–455

Tontonoz P, Hu E, Spiegelman BM (1994) Stimulation of adipogenesis in fibroblasts by PPAR gamma 2, a lipid-activated transcription factor. Cell 79:1147–1156

Townsend KL, Tseng YH (2015) Of mice and men: novel insights regarding constitutive and recruitable brown adipocytes. Int J Obes Suppl 5:S15–S20

Trayhurn P (1979) Fatty acid synthesis in vivo in brown adipose tissue, liver and white adipose tissue of the cold-acclimated rat. FEBS Lett 104:13–16

Ussar S, Lee KY, Dankel SN, Boucher J, Haering MF, Kleinridders A, Thomou T, Xue R, Macotela Y, Cypess AM, Tseng YH, Mellgren G, Kahn CR (2014) ASC-1, PAT2, and P2RX5 are cell surface markers for white, beige, and brown adipocytes. Sci Transl Med 6:247ra103

van der Lans AA, Hoeks J, Brans B, Vijgen GH, Visser MG, Vosselman MJ, Hansen J, Jorgensen JA, Wu J, Mottaghy FM, Schrauwen P, van Marken Lichtenbelt WD (2013) Cold acclimation recruits human brown fat and increases nonshivering thermogenesis. J Clin Invest 123:3395–3403

van Harmelen V, Skurk T, Hauner H (2005) Primary culture and differentiation of human adipocyte precursor cells. Methods Mol Med 107:125–135

van Marken Lichtenbelt WD, Vanhommerig JW, Smulders NM, Drossaerts JM, Kemerink GJ, Bouvy ND, Schrauwen P, Teule GJ (2009) Cold-activated brown adipose tissue in healthy men. N Engl J Med 360:1500–1508

Van Pelt RE, Jankowski CM, Gozansky WS, Schwartz RS, Kohrt WM (2005) Lower-body adiposity and metabolic protection in postmenopausal women. J Clin Endocrinol Metab 90:4573–4578

Veniant MM, Sivits G, Helmering J, Komorowski R, Lee J, Fan W, Moyer C, Lloyd DJ (2015) Pharmacologic effects of FGF21 are independent of the "browning" of white adipose tissue. Cell Metab 21:731–738

Villarroya F, Giralt M (2015) The beneficial effects of brown fat transplantation: further evidence of an endocrine role of brown adipose tissue. Endocrinology 156:2368–2370

Villarroya J, Cereijo R, Villarroya F (2013) An endocrine role for brown adipose tissue? Am J Physiol Endocrinol Metab 305:E567–E572

Villarroya F, Cereijo R, Villarroya J, Gavalda-Navarro A, Giralt M (2018) Toward an understanding of how immune cells control brown and beige adipobiology. Cell Metab 27:954–961

Virtanen KA, Lidell ME, Orava J, Heglind M, Westergren R, Niemi T, Taittonen M, Laine J, Savisto NJ, Enerback S, Nuutila P (2009) Functional brown adipose tissue in healthy adults. N Engl J Med 360:1518–1525

Vishvanath L, Macpherson KA, Hepler C, Wang QA, Shao M, Spurgin SB, Wang MY, Kusminski CM, Morley TS, Gupta RK (2016) Pdgfrbeta+ mural preadipocytes contribute to adipocyte hyperplasia induced by high-fat-diet feeding and prolonged cold exposure in adult mice. Cell Metab 23:350–359

Walden TB, Hansen IR, Timmons JA, Cannon B, Nedergaard J (2012) Recruited vs. nonrecruited molecular signatures of brown, "brite," and white adipose tissues. Am J Physiol Endocrinol Metab 302:E19–E31

Wang H, Sreenivasan U, Hu H, Saladino A, Polster BM, Lund LM, Gong DW, Stanley WC, Sztalryd C (2011) Perilipin 5, a lipid droplet-associated protein, provides physical and metabolic linkage to mitochondria. J Lipid Res 52:2159–2168

Wang F, Mullican SE, Dispirito JR, Peed LC, Lazar MA (2013a) Lipoatrophy and severe metabolic disturbance in mice with fat-specific deletion of PPARgamma. Proc Natl Acad Sci U S A 110:18656–18661

Wang QA, Tao C, Gupta RK, Scherer PE (2013b) Tracking adipogenesis during white adipose tissue development, expansion and regeneration. Nat Med 19:1338–1344

Wang GX, Zhao XY, Meng ZX, Kern M, Dietrich A, Chen Z, Cozacov Z, Zhou D, Okunade AL, Su X, Li S, Bluher M, Lin JD (2014a) The brown fat-enriched secreted factor Nrg4 preserves metabolic homeostasis through attenuation of hepatic lipogenesis. Nat Med 20:1436–1443

Wang W, Kissig M, Rajakumari S, Huang L, Lim HW, Won KJ, Seale P (2014b) Ebf2 is a selective marker of brown and beige adipogenic precursor cells. Proc Natl Acad Sci U S A 111:14466–14471

Wu J, Bostrom P, Sparks LM, Ye L, Choi JH, Giang AH, Khandekar M, Virtanen KA, Nuutila P, Schaart G, Huang K, Tu H, van Marken Lichtenbelt WD, Hoeks J, Enerback S, Schrauwen P, Spiegelman BM (2012) Beige adipocytes are a distinct type of thermogenic fat cell in mouse and human. Cell 150:366–376

Xue Y, Petrovic N, Cao R, Larsson O, Lim S, Chen S, Feldmann HM, Liang Z, Zhu Z, Nedergaard J, Cannon B, Cao Y (2009) Hypoxia-independent angiogenesis in adipose tissues during cold acclimation. Cell Metab 9:99–109

Xue R, Lynes MD, Dreyfuss JM, Shamsi F, Schulz TJ, Zhang H, Huang TL, Townsend KL, Li Y, Takahashi H, Weiner LS, White AP, Lynes MS, Rubin LL, Goodyear LJ, Cypess AM, Tseng YH (2015) Clonal analyses and gene profiling identify genetic biomarkers of the thermogenic potential of human brown and white preadipocytes. Nat Med 21:760–768

Ye R, Wang QA, Tao C, Vishvanath L, Shao M, McDonald JG, Gupta RK, Scherer PE (2015) Impact of tamoxifen on adipocyte lineage tracing: inducer of adipogenesis and prolonged nuclear translocation of Cre recombinase. Mol Metab 4:771–778

Yoneshiro T, Aita S, Mastushita M, Ogawa T, Okamatsu-Ogura Y, Kawai Y, Saito M (2011a) Age-related decrease in brown adipose tissue and obesity in humans. Obesity 19:S79–S79

Yoneshiro T, Aita S, Matsushita M, Kameya T, Nakada K, Kawai Y, Saito M (2011b) Brown adipose tissue, whole-body energy expenditure, and thermogenesis in healthy adult men. Obesity 19:13–16

Yoneshiro T, Aita S, Matsushita M, Okamatsu-Ogura Y, Kameya T, Kawai Y, Miyagawa M, Tsujisaki M, Saito M (2011c) Age-related decrease in cold-activated brown adipose tissue and accumulation of body fat in healthy humans. Obesity (Silver Spring) 19:1755–1760

Yoneshiro T, Aita S, Matsushita M, Kayahara T, Kameya T, Kawai Y, Iwanaga T, Saito M (2013) Recruited brown adipose tissue as an antiobesity agent in humans. J Clin Invest 123:3404–3408

Yu XX, Lewin DA, Forrest W, Adams SH (2002) Cold elicits the simultaneous induction of fatty acid synthesis and beta-oxidation in murine brown adipose tissue: prediction from differential gene expression and confirmation in vivo. FASEB J 16:155–168

Zhang F, Hao G, Shao M, Nham K, An Y, Wang Q, Zhu Y, Kusminski CM, Hassan G, Gupta RK, Zhai Q, Sun X, Scherer PE, Oz OK (2018) An adipose tissue atlas: an image-guided identification of human-like BAT and beige depots in rodents. Cell Metab 27:252–262.e3

Zhao S, Torres A, Henry RA, Trefely S, Wallace M, Lee JV, Carrer A, Sengupta A, Campbell SL, Kuo YM, Frey AJ, Meurs N, Viola JM, Blair IA, Weljie AM, Metallo CM, Snyder NW, Andrews AJ, Wellen KE (2016) ATP-citrate lyase controls a glucose-to-acetate metabolic switch. Cell Rep 17:1037–1052

Lessons from Cre-Mice and Indicator Mice

Christian Wolfrum and Leon Gabriel Straub

Contents

1 Introduction 38
2 Techniques to Create Transgenic Mouse Lines 40
3 Tamoxifen Inducible vs. Dox-Inducible Recombinase Systems 41
4 The ap2 Promoter 42
5 The Adiponectin Promoter 43
6 The Ucp1 Promoter 44
7 Genetic Mouse Models to Disrupt Mature Brown Adipocyte Function 44
8 Lessons from Fat but Non-brown Fat-Specific Gene Targeting and Tracing 45
9 Indicator Mice of Brown Fat Function 47
10 White/Brown/Brite Adipocyte-Specific Labeling 48
References 50

Abstract

The adult human adipose tissue is predominantly composed of white adipocytes. However, within certain depots, adipose tissue contains thermogenically active brown-like adipocytes, which have been evolutionarily conserved in mammals. This chapter will give a brief overview on the methods used to genetically target and trace both white and brown adipocytes using techniques such as bacterial artificial chromosome (BAC) cloning to create transgenic mouse models and the tools with which genetic recombination is mediated in vivo (e.g., Cre-loxP, CreERT, and Tet-On). The chapter furthermore critically discusses the strength and limitation of the various systems used to target mature white and brown adipocytes (ap2-Cre, Adipoq-Cre, and Ucp1-Cre). Based on these systems, it is evident that our knowledge of mature adipocyte categorization into brown, white, brite, or beige adipocytes is strongly influenced by the use of the various genetic

C. Wolfrum · L. G. Straub (✉)
Institute of Food, Nutrition, and Health, ETH Zurich, Zürich, Switzerland
e-mail: leon-straub@ethz.ch

A. Pfeifer et al. (eds.), *Brown Adipose Tissue*,
Handbook of Experimental Pharmacology 251, https://doi.org/10.1007/164_2018_146

mouse models described in this chapter. Our evaluation of different studies using the aforementioned systems focuses on key genes, which have been reported to maintain adipocyte's function (insulin receptor, Raptor, or Atgl).

Keywords
AdipoChaser · Adipocytes · Adiponectin · aP2 · Atgl · BAT · Beige · Brite · Brown · Cre · Doxycycline · FABP4 · Indicator mouse · Interconversion · IR · mTORC1 · Raptor · Rptor · Tamoxifen · Tet-On · ThermoMouse · Transdifferentiation · Ucp1 · Ucp1-Cre · Ucp1-CreERT · White

1 Introduction

Adipose tissue is the main energy storage organ of the body. Energy is deposited in the form of fat molecules (triacylglycerol and cholesterol esters) in membrane-covered lipid droplets in mature adipocytes, which gives the name to this tissue. The fat tissue's energy storage function is relevant during times of starvation and has been shown to be sex-dependent. Thus, it was reported that during chronic starvation of male rats, 33% of the total energy expended was derived from protein and 67% from fat. In contrast, in females, which have a larger adipose tissue storage, only 8% were derived from protein and 92% from fat (Widdowson 1976). Because of its name, it is generally believed that adipocytes are the main cell type in the adipose tissue. This is clearly correct when comparing the volume contribution of different cells, due to the fact that adipocytes are substantially larger than the average cell and furthermore can change in size very rapidly depending on storage needs (Jo et al. 2009). In line with this, the adipose tissue enlargement is mainly due to an adipocyte volume increase (Johnson and Hirsch 1972). It was reported that the adipocyte volume is 400 pL in a healthy human female with 20 kg adipose tissue mass and that it can reach a volume of more than 1,000 pL in humans with 100 kg adipose tissue mass. Besides, adipocyte size adipose tissue cell number is the most important factor determining adipose tissue mass, as has been suggested by the nonlinear relationship between adipocyte cell volume and adipocyte mass (Spalding et al. 2008). Compared to the large adipocyte, all other cells of the adipose tissue are dwarfish. Lymphocytes, neutrophils, fibroblasts, and alveolar macrophages range around volumes of 0.13, 0.3, 2, and 5 pL, respectively (Krombach et al. 1997; Mitsui and Schneider 1976; Rosengren et al. 1994; Schmid-Schonbein et al. 1980). Therefore, it is generally believed that an increase in the numbers of non-adipocyte cells does not affect adipose tissue volume directly. Taken together these points have led to the hypothesis that adipocytes account for only a minority of the cells found within adipose tissue (Müller et al. 2015; Rosenwald et al. 2013). Recent work from our lab suggests that for the inguinal adipose tissue, the percentage of mature adipocytes is between 10 and 20%. Therefore, any study of the adipose tissue, which focuses on mature adipocytes, has to take into account this complication as it can lead to wrong conclusions on specific marker gene expression, since only a fraction of the tissue is composed of mature adipocytes.

Around 150 years ago, adipose tissue morphology was already discussed by Flemming who described adipocytes as "fixed branched connective tissue cells" that are filled with lipids (Flemming 1870; Shaw 1901). With the introduction of density centrifugation, cells of the adipose tissue could be divided based on their physical properties into the mature adipose tissue fraction with a density below one, which mainly contain mature adipocytes. The rest of the adipose tissue's cells, which exhibit a density of greater than one, were assigned to the stromal vascular fraction (SVF). The SVF contains a population of fibroblast-like cells including adipocyte precursor cells, immune cells, as well as endothelial and smooth muscle cells. Given the interest in adipose tissue development, the last 10 years have seen a surge of papers on the composition of the SVF fraction and the definition of the adipocyte progenitor population, which have been extensively reviewed (Hepler et al. 2017; Sanchez-Gurmaches and Guertin 2014).

This book chapter focuses on the mature adipocyte as the key functional cell regulating lipid homeostasis through storage and release of lipids to fuel whole-body energy expenditure. For this reason, it is important to discuss the different types of mature adipocytes. The most important adipocyte is the white adipocyte. It is defined by its mainly unilocular appearance, and its function is mainly lipid storage. In adult humans, the adipose tissue is predominantly composed of white adipocytes.

In his monumental work, *Historiae Animalium*, between 1551 and 1558, the Swiss natural scientist Conrad Gessner did not only describe the unicorn but also described the anatomy of a glandular tissue, which is located in between the shoulder plates of marmots (Gessner 1551). This tissue was later termed brown adipose tissue. Several hundred years later, Hatai described a similar tissue in humans (Hatai 1902); the unicorn until today remains to be found. Brown adipose tissue is conserved within mammals; for this reason most of the experimental work was conducted using the mouse model system (Zhang et al. 2018). Similar to the white adipocyte, a significant part of the brown adipocyte is composed of lipid droplets; however, in the case of the brown adipocyte, these lipid droplets present a multilocular appearance. The third cell type, which has been termed "inducible brown adipocyte," "beige adipocyte," or "brite adipocyte" (for brown in white adipocyte), is a multilocular adipocyte that appears interspersed within what could be considered white adipose tissue depots. In contrast to the white adipocyte, the brown and brite adipocytes are so-called thermogenic cells which convert chemical energy (mainly glucose and lipids) into heat to aid temperature control in response to acute cold bouts (Cannon and Nedergaard 2004). Even though brown and white cells have opposing functions, they share a significant number of genes, which complicate adipocyte-specific targeting as shall be discussed later in this chapter.

In recent years the brown and brite adipocytes have received tremendous interest as a target for energy expenditure control (Bartelt and Heeren 2012). This has been due to the pandemic of obesity, which shows no sign of diminishing, and the large number of associated comorbidities that pose a serious problem for life quality but also have a severe impact on health and economy. As obesity is per definition a pathological storage of excess energy, it is reasonable to propose to ameliorate this problem by increasing energy expenditure, while keeping energy intake constant.

This has been supported by multiple findings that show a tight inverse correlation between obesity development and the presence of brown and brite adipocytes in multiple species, including humans (Bartelt and Heeren 2014). In addition, our knowledge of brown adipose tissue function, which is mainly derived from rodent research, has demonstrated that targeting brown adipose tissue can not only be used to induce weight loss. Furthermore, brown adipose tissue can be quite effective in reducing circulating metabolites such as lipid and glucose and thereby preventing the development of the obesity-associated comorbidities such as type 2 diabetes and cardiovascular complications (Bartelt et al. 2011). For these reasons, it is important to generate animal model systems, which can be used to target brown adipose tissue to study specific genes in the context of whole-body metabolism. The individual contribution of brown vs. brite adipocytes to energy homeostasis is still part of an ongoing debate. Similarly, it remains unclear whether thermogenic adipocytes found in humans should be considered brown or brite adipocytes (see chapter "Lineage Tracing In Vivo" of David Guertin).

To target brown adipocytes, an overview of the specific marker gene expression is required. Brown adipocytes produce heat via the uncoupling protein 1 (*Ucp1*), which can dissipate the proton gradient in the inner mitochondrial membrane (Cannon and Nedergaard 2004). Given the fact that uncoupling of the proton gradient is a unique cellular feature, from an evolutionary perspective, such a process would require careful restriction, because during times of energy scarcity, it might threaten survival. The *Ucp1* gene thus is one of the best defining markers of brown adipose tissue. Since *Ucp1* expression depends on the activity of the brown adipocyte, it is mainly expressed in thermogenically active brown adipocytes. Therefore, other markers have been proposed which are specific for the adipocyte cell (even if they are not exclusive to brown adipocytes). Given the fact that adipocytes per se have a unique set of expressed genes and given the long-standing interest in white adipose tissue research, specific markers have been validated for adipose tissue in general.

2 Techniques to Create Transgenic Mouse Lines

The study of brown fat biology in vivo required the establishment of new genetic mouse models. There are two common methods to integrate transgenes into the genome by targeted or nontargeted strategies. In nontargeted transgenesis, cDNA or DNA from artificial chromosomes is randomly integrated into the mouse genome after the injection into the one-cell embryo. To ensure specificity of transgene expression in nontargeted transgenesis, gene regulatory elements need to be present in the transgene sequence. A widely used technique to supply transgenic DNA with gene-specific regulatory sequences (includes promoter, enhancer, and silencer) is the cloning of a coding sequence into mouse genomic DNA regions that are stored in bacterial artificial chromosomes (BAC) (Osoegawa et al. 2000). BAC cloning is most widely used to create large transgenes that contain the complete regulatory sequence of a certain gene. BAC cloning bases upon principles of a particular form of homologous recombination that is termed recombineering (Copeland and Jenkins 2001). The common BAC length ranges from 150 k base pairs (kb) to 350 kb, and the linearized BAC will be integrated randomly into the mouse genome. A comprehensive protocol was developed for the rapid generation of BAC transgenic mice (Johansson et al. 2010). Another technique that is distinct from BAC recombineering is the knock-in method. The knock-in method is a targeted transgenesis, which allows the capture of a gene's promoter and its gene regulatory sequences at the correct genomic location (Rickert et al. 1997). The disadvantage of the knock-in method is that it usually creates a null allele where the transgene was inserted. Its advantage is that one restricts the transgene expression to half of the endogenous expression level of the captured gene. While the copy number of the transgene inserted with the BAC can be quite high due to formation of concatemers, the knock-in method limits the reading frame to one copy.

3 Tamoxifen Inducible vs. Dox-Inducible Recombinase Systems

The genetic toolbox of the Cre-loxP system has led geneticists to conditionally recombine the mouse genome and thereby create disease models of great value (Orban et al. 1992; Sternberg and Hamilton 1981). LoxP sites (locus of X-over P1) were derived from the bacteriophage P1 and consist of 34 base pairs (bp). An asymmetric 8 bp sequence is positioned in between two 13 bp symmetric sequence elements. Cre (cyclic recombinase) is a site-specific recombinase that very efficiently mediates deletion of a sequence flanked by two loxP sites in vivo (Rajewsky et al. 1996). This deletion of a DNA sequence is mediated by cleavage and ligation of two loxP sequences. When a tissue-specific promoter regulates the Cre recombinase expression, conditional tissue-specific DNA recombination can be used to achieve spatial resolution. Inducible DNA recombination is a more complex version of a conditional DNA recombination. The inducible Cre-loxP system adds the temporal control to the spatial resolution of the recombination process. This development of the Cre-loxP system allows to knockout a gene in a specific tissue at a specific time. The two most commonly used inducible Cre-loxP systems are the doxycycline-inducible Tet-On system (Gossen et al. 1995) and the tamoxifen-inducible Cre recombinase (CreERT) (Feil et al. 1997).

The Tet-On system requires the conditional expression of the transcription factor reverse tetracycline transactivator (rtTA). The rtTA drives Cre recombinase expression by binding to the Tet response element (TRE) in the promoter region of the transgene that encodes the Cre recombinase. The third transgene is the loxP-containing transgene, which will be recombined by Cre recombinase. Distinct from the components of the Tet-On system, the tamoxifen-inducible Cre recombinase adds the inducibility function to the Cre recombinase itself. The tamoxifen-inducible Cre recombinase, or CreERT, is a fusion protein of the Cre recombinase and a mutated version of the estrogen receptor (ERT). Tamoxifen is the ligand of the ERT part of the fusion protein. While being excluded from the nucleus in the unbound state, the estrogen receptor shuttles into the nucleus upon ligand binding. Conformational changes of the mutated estrogen receptor upon ligand binding expose a nuclear localization protein motif. Mutations were directed at creating an estrogen receptor that exclusively binds tamoxifen, and physiological concentrations of endogenous estrogen do not lead to nuclear localization of CreERT.

Currently an ongoing debate exists about which one of the inducible systems has greater limitations. The toxicity of supraphysiological tamoxifen doses questioned the feasibility of using the tamoxifen-inducible CreERT system for genetic tracing studies of the adipose tissue (Ye et al. 2015). In addition, tamoxifen was developed as a selective estrogen receptor modulator and might disrupt certain estrogen receptor pathways in the adipocyte (Deroo and Korach 2006). On the other hand, doxycycline has been criticized for its antibiotic effects (Moullan et al. 2015). While the CreERT requires two transgenes (CreERT and loxP flanked target-coding sequence), the Tet-On system requires three transgenes. The increased breeding time due to an additional transgene in the Tet-On system increases cost and reduces speed of the development of a new mouse model. In cell lineage tracing experiments, the doxycycline-inducible system can be considered superior due to the fact that tamoxifen is a strongly hydrophobic molecule, and it takes a longer time until tamoxifen is cleared from circulation. Doxycycline in contrast is water soluble, which accelerates its clearance time to less than a day (Wang and Scherer 2014). To genetically target brown adipocytes, transgenes with different promoters have been utilized. In the following, we will discuss the brown fat-specific *Ucp1* promoter as well as the adipocyte-specific *adiponectin* and *ap2* promoters, which have both been used in recent years to study the brown and white adipose tissue.

4 The ap2 Promoter

A widely used marker, which has been employed to define adipocytes, is the adipocyte fatty acid-binding protein (*A-Fabp*, *Fabp4*, or *ap2*). The vast majority of attempts to genetically manipulate the adipose tissue were performed with Cre recombinase expression under the *ap2* gene regulatory sequences. Recent work, however, has demonstrated that *ap2* is not specific for adipocytes. Therefore, several conclusions, which have been drawn based on these mouse model systems, will have to be reevaluated. Interestingly, expression of *ap2* gene by mouse macrophages has already been described in 2001 (Fu et al. 2000; Furuhashi et al. 2007; Makowski et al. 2001). Recombination of macrophage genomes in models using the ap2-Cre system could therefore have contributed to observed metabolic effects through modulation of inflammatory processes, which is in line with the hypothesis that adipose tissue inflammation per se can affect insulin resistance (Hotamisligil et al. 1993; Kahn et al. 2006). In 2009 it was published that mouse endothelial cells express *ap2* (Elmasri et al. 2009), a finding that was already suggested based on the use of a porcine model in 2007 (Lee et al. 2007). Furthermore, in both brown and white adipose tissue, *ap2* marks a distinct population of adipocyte progenitors (Shan et al. 2013). Lastly, transgene expression under the *ap2* promoter was shown to localize within the embryo from mid-gestation, which is 9.5 days after fertilization (Urs et al. 2006). At 15.5 days after fertilization, β-gal activity was detected primarily in the brown adipose tissue, cartilage primordia, dorsal root ganglia, trigeminal ganglia, and vertebrae. For this reason, the *ap2* gene promoter should only be used in well-controlled circumstances when targeting expression of genes that are specific for mature adipocytes.

5 The Adiponectin Promoter

One explanation for the widespread use of the ap2-Cre model clearly was the lack of a better driver line for genetic recombination experiments. This changed with the publication of the Adiponectin-Cre transgenic mouse line by Wang and colleagues from the Scherer Lab in 2010 (Wang et al. 2010). This mouse model system is based on the fat-specific adipokine, adiponectin, which is expressed both in white and brown mature adipocytes. Adiponectin was demonstrated to improve systemic insulin sensitivity and to ameliorate several obesity-associated comorbidities (Holland et al. 2017; Xia et al. 2017). Although the expression levels of adiponectin are negatively correlated with adipocyte size, even low expression levels in large adipocytes are enough to drive recombination. In the adiponectin-Cre mouse line, adipocyte-specific expression is mediated by a 5.4 kb adiponectin gene fragment containing key gene regulatory elements. The adiponectin-Cre mouse line published in 2011 by Eguchi and colleagues was based on a different technique of transgene creation. Eguchi and colleagues used the BAC recombineering technique to ensure adipocyte-specific expression of Cre recombinase (Eguchi et al. 2011). Until today, adiponectin expression is considered to be well confined to the adipocyte, and its expression can be found in both white and brown adipocytes.

Das and colleagues systematically studied the regulatory elements of the mouse adiponectin gene, also called *Acrp30*, in cell culture systems (Das et al. 2001). In subsequent in vivo studies, it was shown that adipocyte specificity can be achieved when a 5.4 kb regulatory fragment is used to drive Cre recombinase expression (Wang et al. 2010). 5.4 kb is too large to fit into conventional adeno-associated virus (AAV) gene vectors. The adiponectin regulatory fragment of choice contains 4.9 kb of the 5'regulatory region, 100 bp of exon 1, 200 bp of the 5'end of intron 1, and 200 bp of the 3'end of intron 1. Exon 1 of the adiponectin gene encodes the 5'end of the untranslated region. The starting methionine codon is located in exon 2. The premise for including regions of intron 1 was to preserve the endogenous splicing information. Adiponectin expression is transcriptionally regulated by activating transcription factor 3 (ATF3), CCAAT/enhancer-binding protein α (Cebpα), and peroxisome proliferator-activated receptor γ (Pparγ) (Iwaki et al. 2003; Kim et al. 2006; Park et al. 2004).

Another systematic investigation of the adiponectin promoter was performed by Segawa et al. (2009) and focused on the human gene. O'Neill and colleagues chose human adiponectin regulatory regions according to detailed studies of the 5'distal regulatory region of the human adiponectin gene (O'Neill et al. 2014). One proof of principle study delivered the leptin gene regulated by the 5'distal regulatory region of adiponectin through an AAV vector into leptin-deficient mice. By using the reduced regulatory regions of the human adiponectin gene, O'Neill and colleagues attempted to regulate transgene expression from the AAV vector only in adipocytes. The results of this study suggest that viral systems might also be used to target adipocytes specifically (O'Neill et al. 2014). However, a systemic injection of a high virus titer requires a supplementation of the AAV viral genome with a miRNA system that prevents the transgene translation in the liver. Further efforts need to be made to design a shorter adiponectin promoter with the key gene regulatory sequences that allow faithful use of AAV for adipocyte-specific targeting.

6 The Ucp1 Promoter

The deletion analysis of the uncoupling protein 1 (*Ucp1*) gene's 5'regulatory regions in transfected mammalian cells (Cassard-Doulcier et al. 1993) uncovered its cis-regulatory elements. The minimal region of promoter activity was found to be positioned between −157 bp and −57 bp (1 bp = A of methionine-encoding ATG). The location of a repressor region was found to be between −400 bp and −157 bp. One enhancer element of the *Ucp1* promoter in its 5'flanking region was found to be 211 bp long and its detailed location defined as between −2,494 and −2,283 bp. In subsequent in vivo studies, this enhancer element was shown to be sufficient to regulate brown adipocyte-specific expression (Cassard-Doulcier et al. 1998). The AAV serotypes 8 and 9 mediate efficient and long-term transduction of adipocytes (Jimenez et al. 2013). Although further independent studies are needed to validate the brown adipocyte-specific expression, the supplementation of AAV vector with the 211 bp long enhancer in addition to the *Ucp1* minimal promoter seems to have ensured brown adipocyte-selective expression in the mouse.

7 Genetic Mouse Models to Disrupt Mature Brown Adipocyte Function

The first two separately developed Ucp1-Cre lines were published by Moulin and colleagues and Guerra and colleagues (Guerra et al. 2001; Moulin et al. 2001). The brown fat-specific insulin receptor knockout (BATIRKO) mouse created by Guerra and colleagues was the first brown fat-specific knockout model ever published (Guerra et al. 2001). The Ucp1-Cre transgene was around 15 kb long and consisted of 8.4 kb of the 5'untranslated region (UTR) of the *Ucp1* gene, 2.4 kb of Cre cDNA including a nuclear localization sequence, and 4.3 kb of a *Ucp1* gene region of exons 3–6. The transgenic DNA was integrated randomly into the mouse genome. After that it took 11 years until a study reported the use of this same Ucp1-Cre mouse line to knockout hypoxia-inducible factor 1 alpha subunit (*Hif1α*) and create the BATcKO mouse (Krishnan et al. 2012). In 2014, Kong and colleagues reported the following (Kong et al. 2014): "Unfortunately, the available

Ucp1-promoter driven Cre line (Guerra et al. 2001) has problems with both specificity and recombination efficiency." It is unknown what caused the problems in the Ucp1-Cre line and if they arose from continued breeding of the line. To solve the problem, a new Ucp1-Cre mouse line was created using the BAC recombineering approach to induce overexpression of an IRF4 transgene. In the same year, Turpin and colleagues published the brown fat-specific Ceramidase 6 knockout with a Ucp1-Cre (Turpin et al. 2014). In the following years, the Ucp1-Cre got much more widely distributed, and many genes were targeted using this line.

A physiological effect of a conditional knockout in Ucp1-expressing cells might be caused by developmental effects, because most adipocytes express Ucp1 in early phases of adipose tissue development. To rule out any contribution of developmental effects, one can use an inducible Cre-loxP system. Our lab was the first to publish an inducible Ucp1-CreERT mouse line in 2013 (Rosenwald et al. 2013). This mouse line allowed to induce genetic recombination by the drug tamoxifen specifically in brown adipocytes. In 2017, Schreiber and colleagues published the first tamoxifen inducible brown fat-specific knockout of a gene. By applying this new mouse model, they could rule out that it was the absence of ATGL in brown adipocytes that disrupted the non-shivering thermogenic capacity of BAT (Schreiber et al. 2017). Until today, there are a few publications utilizing the tamoxifen-inducible Ucp1-CreERT mouse models (Lasar et al. 2018; Sanchez-Gurmaches et al. 2018). In 2014, the first use of a brown fat-specific doxycycline-inducible Ucp1-rtTA mouse line was published by Sun and colleagues to overexpress vascular endothelial growth factor A (VEGF-A) in BAT (Sun et al. 2014). By creating a genetic mouse model that combines the transgenes Ucp1-rtTA and TRE-Cre, the Scherer laboratory was not only able to knockout Connexin 43 but also to first perform a doxycycline-inducible gene knockout only in brown adipocytes (Zhu et al. 2016).

8 Lessons from Fat but Non-brown Fat-Specific Gene Targeting and Tracing

In the following, we will highlight some points, which can be deduced from the genetic manipulation of adipose tissue with the aim to study brown fat function. In this context, some studies have reported the use of conditional ablation of specific genes by employing the adiponectin-Cre mice, which targets the whole adipose tissue, to study brown adipocyte function. As described above, targeting such a large and important organ as the adipose tissue can lead to an altered systemic metabolism, which in turn can impact the function of brown adipocytes. One example is the study of Labbé et al. (2016), who used the adiponectin-Cre to delete Raptor in adipose tissue and studied the gene's role in brown adipose tissue function. Raptor is the regulatory-associated protein of mTOR. The Raptor knockout is a generally accepted means of disrupting mTOR complex 1 (mTORC1) formation and function. Labbé and colleagues showed that mice with a conditional knockout of Raptor have reduced mitochondrial biogenesis in and expansion of BAT. Furthermore, they found reduced capacity of non-shivering thermogenesis in response to cold in mice lacking Raptor in all mature adipocytes. They conclude that mTORC1 would play a key role in the process of brown adipocyte recruitment and that the absence of mTORC1 would be responsible for shifting brown adipocytes' metabolism from oxidizing lipids to metabolizing glucose. Even before this study was published, it had been reported by Lee and colleagues that adipocyte-specific Raptor knockout caused the development of a lipodystrophy, which was associated with hepatomegaly, hepatic steatosis, and systematic insulin intolerance (Lee et al. 2016). One has to point out that the absence of functional white adipose tissue in these mice forces an atypical allocation of fat into the liver and other organs. One of these organs could be the brown adipose tissue. Evidence from the conditional knockout of insulin receptor (IR) with adiponectin-Cre suggested that storage of fat is increased in BAT upon lipodystrophy. In this mouse model, the lipoatrophy affected preferentially white adipose tissue depots, and a contrasting whitening of brown adipose tissue was reported (Qiang et al. 2016). Furthermore, one major part of the phenotype on non-shivering thermogenic capacity of the Raptor knockout in all adipocytes could be explained by the finding that lipodystrophy itself reduces the thermogenic potential of mice, as demonstrated by an inducible lipodystrophic mouse model (Sakaguchi et al. 2017). Based on these data, it seems more plausible that whitening of BAT is in part caused by the lipoatrophic phenotype and it remains to be seen how much the direct effect of mTORC1 disruption in brown adipocytes is involved. Indeed the hypothesis that the whitening and dysfunction of BAT are an indirect effect of systemic metabolism can be confirmed by results from a mouse model of brown fat-specific knockout of the Raptor gene (Lee et al. 2016). BAT-specific knockout of Raptor resulted in reduced BAT mass and lipid content. In accordance with this, the knockout of IR using the Ucp1-Cre animal model was shown to reduce brown adipose tissue mass and lipid content (*58*).

The notion that it is of great importance, to carefully analyze the cell-specific impact of a particular gene in brown adipose tissue function, was very elegantly demonstrated by a recent paper of Schreiber and colleagues. The authors analyzed the role of adipose tissue triglyceride lipase (ATGL), the key lipolytic enzyme in white and brown adipose tissue, in the context of non-shivering thermogenesis. Based on the data from the global ATGL ko mice, which are completely cold-intolerant, it was generally believed that release of fatty acids from intracellular fat stores through ATGL would be a key determinant of brown adipocyte functionality. Using BAT-specific ablation of ATGL in brown adipose tissue, the authors demonstrate that, although ATGL loss impairs BAT lipolysis and alters BAT morphology, it does not compromise the β3-adrenergic thermogenic response or cold-induced non-shivering thermogenesis. Instead they could demonstrate through the use of a complete adipose tissue ablation of ATGL that non-shivering thermogenesis depends on the lipolysis of white adipose tissue to generate energy substrates and that the cold intolerance in ATGL global ko mice is due to severe cardiomyopathy. Taken together, these above-discussed studies illustrate the tight dependence of brown adipose tissue function on white adipose tissue lipolytic capacity and clearly underscore the necessity to ablate genes of interest specifically in brown adipocytes to unequivocally show their cell autonomous function.

9 Indicator Mice of Brown Fat Function

Besides their use in gene ablation, Cre-mice and other models have been used to study brown fat formation and development. In line with this, the term "indicator mouse" has been coined, which describes a mouse that has been genetically modified with the goal of enabling the detection of a biological process in a simplified manner. Concerning brown fat visualization, there have been different attempts to develop a technique that enables in vivo tracing of active brown fat. The most well-known indicator mouse lines for brown fat activation are the Ucp1-GFP mouse line and the ThermoMouse.

Activation of brown adipocyte function as well as the number of brown adipocytes can be monitored by the amount of *Ucp1*. One basic strategy can be summarized by controlling the expression of a fluorescent protein through the *Ucp1* gene promoter elements. In the first study, which applied genetics to transiently indicate *Ucp1* expression in vivo, our lab expressed an enhanced green fluorescent protein (eGFP) (Rosenwald et al. 2013). As preceding studies in non-mitotic mouse cells had shown, the half-life of eGFP was around 6 h (Corish and Tyler-Smith 1999). When brown adipocytes are activated, they increase eGFP expression. After decreasing its eGFP expression, the eGFP is degraded by the cell, and the fluorescence signal weakens. In detail, the Ucp1-GFP mouse was created by bacterial artificial chromosome (BAC) recombineering (Johansson et al. 2010). The transgene encodes a primate heparin-binding epidermal growth factor-like growth factor (HB-EGF)/eGFP fusion protein. The fusion protein is located to the plasma membrane and thereby can be easily distinguished from other fluorescent proteins that are located in the cytosol. Based upon the work of Yang and colleagues, it seems possible to use the Ucp1-GFP mouse to study in vivo brown fat activity (Yang et al. 2000). The imaging of eGFP localization in the subcutaneous fat depots requires transilluminated epifluorescence microscope or a fluorescence light box and a thermoelectrically cooled color charge-coupled camera.

The ThermoMouse applies the transgenic expression of luciferase2 under the brown fat-specific promoters of the *Ucp1* gene to visualize the size and activity of the brown fat tissue (Galmozzi et al. 2014). In detail, Galmozzi and colleagues used BAC recombineering techniques (Warming et al. 2005) to ensure brown fat-specific expression of the luciferase2; they inserted the luciferase2-T2A-tdTomato transgene cassette at the initiation codon of exon 1 of the *Ucp1* gene. The BAC was 98.6 kb long (bMQ353d13, Source BioScience) and contained the entire *Ucp1* gene locus. The transfer of recombineered BAC DNA into the germline of FVB mouse strain was mediated by microinjection into single-cell embryos. The ThermoMouse enabled researchers to quantify brown fat function and the amount of brown fat present in vivo through luciferase2, which serves as a transient biomarker (Liu et al. 2017). After the injection of D-luciferin (10 μL g^{-1}) into a living mouse, bioluminescence can be measured using an in vivo optical imaging system. This approach allows longitudinal studies of individual mice over longer times. Unfortunately, it is only possible to obtain the male ThermoMouse, because the transgene randomly integrated into the Y chromosome. To allow the study of both sexes, Mao and colleagues have developed a similar mouse model in which the transgene is not integrated into an allosome (Mao et al. 2017). In addition, their strategy of transgenic mouse creation was different. In contrast to the approach of Galmozzi and colleagues, Mao and colleagues produced a knock-in mouse line by homologous recombination the endogenous *Ucp1* locus in embryonic stem cells. The targeted insertion method excludes effects of regulatory sequences in the genome environment at the side of random integration. The end of exon 6 of the endogenous *Ucp1* gene was chosen as the side for targeted insertion of luciferase-coding sequence.

10 White/Brown/Brite Adipocyte-Specific Labeling

The first genetically distinction of the different characters of adipocytes has been made by Moulin and colleagues in 2001 (Moulin et al. 2001). They used an Ucp1-Cre line to drive constitutive expression of LacZ in all adipocytes that at a point in time expressed Ucp1. Although their findings seem very simple at first sight, the concept they established with a solid experiment was of great importance for starting the still ongoing debate about the mixed character of adipose tissue. By showing in the adult Ucp1-Crex^{fl}stop^{fl}LacZ mouse at 21°C, which most adipocytes in white adipose tissues did not express LacZ, they were able to prove that there are two distinct lines of mature adipocytes. Thus, they not only established the existence of a distinct brown adipocyte cell type but indirectly also defined the white adipocyte as a distinct cell type. It could have been possible that all adipocytes express *Ucp1* at a certain stage of development. This could have led to the conclusion that what we define as a white adipocyte is a brown adipocyte, which has filled its lipid storage leading to the formation of a unilocular lipid droplet. Since then many concepts of different adipocyte cell types have been developed. Although the nomenclature has not been unified yet, scientists widely accepted the definition of a third adipocyte cell type as mentioned in the introduction. A brite adipocyte can be defined as a cell in predominantly white adipose tissue depots, which expresses *Ucp1* and contains multilocular lipid droplets. While the nomenclature controversy seems to be an argument about which name sounds better, it actually also demonstrates disagreements regarding aspects of underlying theoretical concepts. As of now, "beige" adipocyte is the most widely distributed term for the third type of adipocytes. The term was coined building on the idea that beige adipocytes are a separate cell type that derives from a distinct precursor lineage. Critics of the term "beige" point out that its concept does not account sufficiently for the transient character of that cell type, which is supported by findings that demonstrated the interconvertibility of white adipocytes into beige adipocytes. While the classical adipocyte cell types brown and white cells are easily detectable, the appearance of brite adipocytes is heavily dependent on environmental cues. Brite adipocytes form upon cold exposure, excessive activation of sympathetic nerve system, or stimulation of fat tissue with beta-3 adrenergic receptor agonists (Bartelt and Heeren 2014). The concept of interconvertibility (or transdifferentiation) of white adipocytes into brite adipocytes has been proposed by Cinti and colleagues upon findings from comprehensive anatomical studies of the adipose organ housed at different environmental temperatures (Himms-Hagen et al. 2000; Vitali et al. 2012). Transdifferentiation is synonymous with the term "direct lineage reprogramming," which can be described as a form of cellular reprogramming that converts a mature cell type into another different mature cell type without the transitioning through an undifferentiated progenitor stage (Jopling et al. 2011). To use the term transdifferentiation when describing white to brite adipocyte cell, interconversion has long been abandoned by most scientists, because it seemed premature due to lack of molecular evidence.

Technological advances were required before the question of interconversion could be addressed further (Wang et al. 2014). One such technological advancement was the creation and combination of the previously described Ucp1-CreERT and Ucp1-GFP mouse models (Lee and Cowan 2013). Technically this was mediated by the creation of a mouse model that combined transient green fluorescent protein labeling of acutely *Ucp1*-expressing cells (Ucp1-GFP transgene) with the constitutive red fluorescence labeling of all brite adipocytes (Ucp1-CreERTx^{fl}stop^{fl}tdRFP transgenes) at a first cold exposure of mice (Rosenwald et al. 2013). The first cold exposure was followed by a warm adaptation phase of the mice during which the brite adipocytes interconverted back to the white phenotype. After that warm adaptation of 6 weeks, during which the tamoxifen was washed out, the second cold exposure showed that a big fraction of all brite adipocytes have already been activated during the first cold exposure. Since the main readout was fluorescence microscopy, it is hard to quantify the percentage of adipocytes that interconverted from mature cells. In another study, Wang and colleagues showed the interconversion of mature adipocytes with the help of their elegant AdipoChaser mouse model (Wang et al. 2013). The AdipoChaser mouse builds upon the Tet-On system and contains three transgenes that allow the inducibility of Cre recombinase function by doxycycline in mature adipocytes. Since the washout time is shorter than that of tamoxifen, they could easily trace how preexisting mature adipocytes interconvert into the brite state. Today, the route of the brite cell recruitment via precursor differentiation has received most of the attention (Hepler and Gupta 2017; Sanchez-Gurmaches et al. 2012). This might be due to the development of novel and sophisticated tools to study precursor differentiation, in vivo (Hepler et al. 2017). Similarly, the discovery of transcription factors, which mediate adipocyte cell type identity formation, represents key steps in the understanding of the adipose organ (Cohen et al. 2014; Duteil et al. 2016; Harms et al. 2014; Rajakumari et al. 2013; Sambeat et al. 2016; Shao et al. 2016).

To conclude, the question of brite adipocyte formation remains one of the most interesting questions in the field of brown adipose tissue research, due to the fact that detailed knowledge of the involved processes might lead to strategies that would allow to convert white into brown adipose tissue with the ultimate goal to enhance systemic energy expenditure of an organism. In this context, in vivo studies applying the different genetic tools, which have been illustrated in this chapter, might enable the comprehensive profiling of the three different adipocytes.

References

Bartelt A, Bruns OT, Reimer R, Hohenberg H, Ittrich H, Peldschus K, Kaul MG, Tromsdorf UI, Weller H, Waurisch C (2011) Brown adipose tissue activity controls triglyceride clearance. Nat Med 17:200

Bartelt A, Heeren J (2012) The holy grail of metabolic disease: brown adipose tissue. Curr Opin Lipidol 23:190–195

Bartelt A, Heeren J (2014) Adipose tissue browning and metabolic health. Nat Rev Endocrinol 10:24

Cannon B, Nedergaard J (2004) Brown adipose tissue: function and physiological significance. Physiol Rev 84:277–359

Cassard-Doulcier A-M, Gelly C, Bouillaud F, Ricquier D (1998) A 211-bp enhancer of the rat uncoupling protein-1 (UCP-1) gene controls specific and regulated expression in brown adipose tissue. Biochem J 333:243

Cassard-Doulcier A-M, Gelly C, Fox N, Schrementi J, Raimbault S, Klaus S, Forest C, Bouillaud F, Ricquier D (1993) Tissue-specific and beta-adrenergic regulation of the mitochondrial uncoupling protein gene: control by cis-acting elements in the 5′-flanking region. Mol Endocrinol 7:497–506

Cohen P, Levy JD, Zhang Y, Frontini A, Kolodin DP, Svensson KJ, Lo JC, Zeng X, Ye L, Khandekar MJ (2014) Ablation of PRDM16 and beige adipose causes metabolic dysfunction and a subcutaneous to visceral fat switch. Cell 156:304–316

Copeland NG, Jenkins NA (2001) Recombineering: a powerful new tool for mouse functional genomics. Nat Rev Genet 2:769

Corish P, Tyler-Smith C (1999) Attenuation of green fluorescent protein half-life in mammalian cells. Protein Eng 12:1035–1040

Das K, Lin Y, Widen E, Zhang Y, Scherer PE (2001) Chromosomal localization, expression pattern, and promoter analysis of the mouse gene encoding adipocyte-specific secretory protein Acrp30. Biochem Biophys Res Commun 280:1120–1129

Deroo BJ, Korach KS (2006) Estrogen receptors and human disease. J Clin Invest 116:561–570

Duteil D, Tosic M, Lausecker F, Nenseth HZ, Müller JM, Urban S, Willmann D, Petroll K, Messaddeq N, Arrigoni L (2016) Lsd1 ablation triggers metabolic reprogramming of brown adipose tissue. Cell Rep 17:1008–1021

Eguchi J, Wang X, Yu S, Kershaw EE, Chiu PC, Dushay J, Estall JL, Klein U, Maratos-Flier E, Rosen ED (2011) Transcriptional control of adipose lipid handling by IRF4. Cell Metab 13:249–259

Elmasri H, Karaaslan C, Teper Y, Ghelfi E, Weng M, Ince TA, Kozakewich H, Bischoff J, Cataltepe S (2009) Fatty acid binding protein 4 is a target of VEGF and a regulator of cell proliferation in endothelial cells. FASEB J 23:3865–3873

Feil R, Wagner J, Metzger D, Chambon P (1997) Regulation of Cre recombinase activity by mutated estrogen receptor ligand-binding domains. Biochem Biophys Res Commun 237:752–757

Flemming W (1870) On the histogenesis of fixed cells and fat cells in connective tissue. Centrabl Med Wiss 31:481–483

Fu Y, Luo N, Lopes-Virella MF (2000) Oxidized LDL induces the expression of ALBP/aP2 mRNA and protein in human THP-1 macrophages. J Lipid Res 41:2017–2023

Furuhashi M, Tuncman G, Görgün CZ, Makowski L, Atsumi G, Vaillancourt E, Kono K, Babaev VR, Fazio S, Linton MF (2007) Treatment of diabetes and atherosclerosis by inhibiting fatty-acid-binding protein aP2. Nature 447:959

Galmozzi A, Sonne SB, Altshuler-Keylin S, Hasegawa Y, Shinoda K, Luijten IH, Chang JW, Sharp LZ, Cravatt BF, Saez E (2014) ThermoMouse: an in vivo model to identify modulators of UCP1 expression in brown adipose tissue. Cell Rep 9:1584–1593

Gessner K (1551) Conradi Gesneri medici Tigurine Historiae Animalium: Lib. I De Quadrupedibus viviparis, 842

Gossen M, Freundlieb S, Bender G, Muller G, Hillen W, Bujard H (1995) Transcriptional activation by tetracyclines in mammalian cells. Science 268:1766–1769
Guerra C, Navarro P, Valverde AM, Arribas M, Brüning J, Kozak LP, Kahn CR, Benito M (2001) Brown adipose tissue–specific insulin receptor knockout shows diabetic phenotype without insulin resistance. J Clin Invest 108:1205–1213
Harms MJ, Ishibashi J, Wang W, Lim H-W, Goyama S, Sato T, Kurokawa M, Won K-J, Seale P (2014) Prdm16 is required for the maintenance of brown adipocyte identity and function in adult mice. Cell Metab 19:593–604
Hatai S (1902) On the presence in human embryos of an interscapular gland corresponding to the so-called hibernating gland of lower mammals. Anat Anz 21:369–373
Hepler C, Gupta RK (2017) The expanding problem of adipose depot remodeling and postnatal adipocyte progenitor recruitment. Mol Cell Endocrinol 445:95–108
Hepler C, Vishvanath L, Gupta RK (2017) Sorting out adipocyte precursors and their role in physiology and disease. Genes Dev 31:127–140
Himms-Hagen J, Melnyk A, Zingaretti M, Ceresi E, Barbatelli G, Cinti S (2000) Multilocular fat cells in WAT of CL-316243-treated rats derive directly from white adipocytes. Am J Phys Cell Phys 279:C670–C681
Holland WL, Xia JY, Johnson JA, Sun K, Pearson MJ, Sharma AX, Quittner-Strom E, Tippetts TS, Gordillo R, Scherer PE (2017) Inducible overexpression of adiponectin receptors highlight the roles of adiponectin-induced ceramidase signaling in lipid and glucose homeostasis. Mol Metab 6:267–275
Hotamisligil GS, Shargill NS, Spiegelman BM (1993) Adipose expression of tumor necrosis factor-alpha: direct role in obesity-linked insulin resistance. Science 259:87–91
Iwaki M, Matsuda M, Maeda N, Funahashi T, Matsuzawa Y, Makishima M, Shimomura I (2003) Induction of adiponectin, a fat-derived antidiabetic and antiatherogenic factor, by nuclear receptors. Diabetes 52:1655–1663
Jimenez V, Muñoz S, Casana E, Mallol C, Elias I, Jambrina C, Ribera A, Ferre T, Franckhauser S, Bosch F (2013) In vivo adeno-associated viral vector–mediated genetic engineering of white and brown adipose tissue in adult mice. Diabetes 62:4012–4022
Jo J, Gavrilova O, Pack S, Jou W, Mullen S, Sumner AE, Cushman SW, Periwal V (2009) Hypertrophy and/or hyperplasia: dynamics of adipose tissue growth. PLoS Comput Biol 5:e1000324
Johansson T, Broll I, Frenz T, Hemmers S, Becher B, Zeilhofer HU, Buch T (2010) Building a zoo of mice for genetic analyses: a comprehensive protocol for the rapid generation of BAC transgenic mice. Genesis 48:264–280
Johnson P, Hirsch J (1972) Cellularity of adipose depots in six strains of genetically obese mice. J Lipid Res 13:2–11
Jopling C, Boue S, Belmonte JCI (2011) Dedifferentiation, transdifferentiation and reprogramming: three routes to regeneration. Nat Rev Mol Cell Biol 12:79
Kahn SE, Hull RL, Utzschneider KM (2006) Mechanisms linking obesity to insulin resistance and type 2 diabetes. Nature 444:840
Kim HB, Kong M, Kim TM, Suh YH, Kim W-H, Lim JH, Song JH, Jung MH (2006) NFATc4 and ATF3 negatively regulate adiponectin gene expression in 3T3-L1 adipocytes. Diabetes 55:1342–1352
Kong X, Banks A, Liu T, Kazak L, Rao RR, Cohen P, Wang X, Yu S, Lo JC, Tseng Y-H (2014) IRF4 is a key thermogenic transcriptional partner of PGC-1α. Cell 158:69–83
Krishnan J, Danzer C, Simka T, Ukropec J, Walter KM, Kumpf S, Mirtschink P, Ukropcova B, Gasperikova D, Pedrazzini T (2012) Dietary obesity-associated Hif1α activation in adipocytes restricts fatty acid oxidation and energy expenditure via suppression of the Sirt2-NAD+ system. Genes Dev 26:259–270
Krombach F, Münzing S, Allmeling A-M, Gerlach JT, Behr J, Dörger M (1997) Cell size of alveolar macrophages: an interspecies comparison. Environ Health Perspect 105:1261

Labbé SM, Mouchiroud M, Caron A, Secco B, Freinkman E, Lamoureux G, Gélinas Y, Lecomte R, Bossé Y, Chimin P (2016) mTORC1 is required for brown adipose tissue recruitment and metabolic adaptation to cold. Sci Rep 6:37223

Lasar D, Rosenwald M, Kiehlmann E, Balaz M, Tall B, Opitz L, Lidell ME, Zamboni N, Krznar P, Sun W (2018) Peroxisome proliferator activated receptor gamma controls mature brown adipocyte inducibility through glycerol kinase. Cell Rep 22:760–773

Lee MY, Tse H-F, Siu C-W, Zhu S-G, Man RY, Vanhoutte PM (2007) Genomic changes in regenerated porcine coronary arterial endothelial cells. Arterioscler Thromb Vasc Biol 27:2443–2449

Lee PL, Tang Y, Li H, Guertin DA (2016) Raptor/mTORC1 loss in adipocytes causes progressive lipodystrophy and fatty liver disease. Mol Metab 5:422–432

Lee Y-K, Cowan CA (2013) White to brite adipocyte transition and back again. Nat Cell Biol 15:568

Liu P, Ji Y, Yuen T, Rendina-Ruedy E, DeMambro VE, Dhawan S, Abu-Amer W, Izadmehr S, Zhou B, Shin AC (2017) Blocking FSH induces thermogenic adipose tissue and reduces body fat. Nature 546:107–112

Makowski L, Boord JB, Maeda K, Babaev VR, Uysal KT, Morgan MA, Parker RA, Suttles J, Fazio S, Hotamisligil GS (2001) Lack of macrophage fatty-acid–binding protein aP2 protects mice deficient in apolipoprotein E against atherosclerosis. Nat Med 7:699

Mao L, Nie B, Nie T, Hui X, Gao X, Lin X, Liu X, Xu Y, Tang X, Yuan R (2017) Visualization and quantification of browning using a Ucp1-2A-Luciferase knock-in mouse model. Diabetes 66:407–417

Mitsui Y, Schneider EL (1976) Relationship between cell replication and volume in senescent human diploid fibroblasts. Mech Ageing Dev 5:45–56

Moulin K, Truel N, André M, Arnauld E, Nibbelink M, Cousin B, Dani C, Pénicaud L, Casteilla L (2001) Emergence during development of the white-adipocyte cell phenotype is independent of the brown-adipocyte cell phenotype. Biochem J 356:659

Moullan N, Mouchiroud L, Wang X, Ryu D, Williams EG, Mottis A, Jovaisaite V, Frochaux MV, Quiros PM, Deplancke B (2015) Tetracyclines disturb mitochondrial function across eukaryotic models: a call for caution in biomedical research. Cell Rep 10:1681–1691

Müller S, Kulenkampff E, Wolfrum C (2015) Adipose tissue stem cells. Metabolic control. Springer, Berlin, pp 251–263

O'Neill SM, Hinkle C, Chen S-J, Sandhu A, Hovhannisyan R, Stephan S, Lagor WR, Ahima RS, Johnston JC, Reilly MP (2014) Targeting adipose tissue via systemic gene therapy. Gene Ther 21:653

Orban PC, Chui D, Marth JD (1992) Tissue-and site-specific DNA recombination in transgenic mice. Proc Natl Acad Sci U S A 89:6861–6865

Osoegawa K, Tateno M, Woon PY, Frengen E, Mammoser AG, Catanese JJ, Hayashizaki Y, de Jong PJ (2000) Bacterial artificial chromosome libraries for mouse sequencing and functional analysis. Genome Res 10:116–128

Park S-K, Oh S-Y, Lee M-Y, Yoon S, Kim K-S, Kim J-W (2004) CCAAT/enhancer binding protein and nuclear factor-Y regulate adiponectin gene expression in adipose tissue. Diabetes 53:2757–2766

Qiang G, Kong HW, Xu S, Pham HA, Parlee SD, Burr AA, Gil V, Pang J, Hughes A, Gu X (2016) Lipodystrophy and severe metabolic dysfunction in mice with adipose tissue-specific insulin receptor ablation. Mol Metab 5:480–490

Rajakumari S, Wu J, Ishibashi J, Lim H-W, Giang A-H, Won K-J, Reed RR, Seale P (2013) EBF2 determines and maintains brown adipocyte identity. Cell Metab 17:562–574

Rajewsky K, Gu H, Kühn R, Betz UA, Müller W, Roes J, Schwenk F (1996) Conditional gene targeting. J Clin Invest 98:600–603

Rickert RC, Roes J, Rajewsky K (1997) B lymphocyte-specific, Cre-mediated mutagenesis in mice. Nucleic Acids Res 25:1317–1318

Rosengren S, Henson PM, Worthen GS (1994) Migration-associated volume changes in neutrophils facilitate the migratory process in vitro. Am J Phys Cell Phys 267:C1623–C1632
Rosenwald M, Perdikari A, Rülicke T, Wolfrum C (2013) Bi-directional interconversion of brite and white adipocytes. Nat Cell Biol 15:659
Sakaguchi M, Fujisaka S, Cai W, Winnay JN, Konishi M, O'Neill BT, Li M, García-Martín R, Takahashi H, Hu J (2017) Adipocyte dynamics and reversible metabolic syndrome in mice with an inducible adipocyte-specific deletion of the insulin receptor. Cell Metab 25:448–462
Sambeat A, Gulyaeva O, Dempersmier J, Tharp KM, Stahl A, Paul SM, Sul HS (2016) LSD1 interacts with Zfp516 to promote UCP1 transcription and brown fat program. Cell Rep 15:2536–2549
Sanchez-Gurmaches J, Guertin DA (2014) Adipocytes arise from multiple lineages that are heterogeneously and dynamically distributed. Nat Commun 5:4099
Sanchez-Gurmaches J, Hung C-M, Sparks CA, Tang Y, Li H, Guertin DA (2012) PTEN loss in the Myf5 lineage redistributes body fat and reveals subsets of white adipocytes that arise from Myf5 precursors. Cell Metab 16:348–362
Sanchez-Gurmaches J, Tang Y, Jespersen NZ, Wallace M, Calejman CM, Gujja S, Li H, Edwards YJ, Wolfrum C, Metallo CM (2018) Brown fat AKT2 is a cold-induced kinase that stimulates ChREBP-mediated de Novo lipogenesis to optimize fuel storage and thermogenesis. Cell Metab 27:195–209.e6
Schmid-Schonbein G, Shih YY, Chien S (1980) Morphometry of human leukocytes. Blood 56:866–875
Schreiber R, Diwoky C, Schoiswohl G, Feiler U, Wongsiriroj N, Abdellatif M, Kolb D, Hoeks J, Kershaw EE, Sedej S (2017) Cold-induced thermogenesis depends on ATGL-mediated lipolysis in cardiac muscle, but not brown adipose tissue. Cell Metab 26:753–763.e7
Segawa K, Matsuda M, Fukuhara A, Morita K, Okuno Y, Komuro R, Shimomura I (2009) Identification of a novel distal enhancer in human adiponectin gene. J Endocrinol 200:107–116
Shan T, Liu W, Kuang S (2013) Fatty acid binding protein 4 expression marks a population of adipocyte progenitors in white and brown adipose tissues. FASEB J 27:277–287
Shao M, Ishibashi J, Kusminski CM, Wang QA, Hepler C, Vishvanath L, MacPherson KA, Spurgin SB, Sun K, Holland WL (2016) Zfp423 maintains white adipocyte identity through suppression of the beige cell thermogenic gene program. Cell Metab 23:1167–1184
Shaw HB (1901) A contribution to the study of the morphology of adipose tissue. J Anat Physiol 36:1
Spalding KL, Arner E, Westermark PO, Bernard S, Buchholz BA, Bergmann O, Blomqvist L, Hoffstedt J, Näslund E, Britton T (2008) Dynamics of fat cell turnover in humans. Nature 453:783
Sternberg N, Hamilton D (1981) Bacteriophage P1 site-specific recombination: I. Recombination between loxP sites. J Mol Biol 150:467–486
Sun K, Kusminski CM, Luby-Phelps K, Spurgin SB, An YA, Wang QA, Holland WL, Scherer PE (2014) Brown adipose tissue derived VEGF-A modulates cold tolerance and energy expenditure. Mol Metab 3:474–483
Turpin SM, Nicholls HT, Willmes DM, Mourier A, Brodesser S, Wunderlich CM, Mauer J, Xu E, Hammerschmidt P, Brönneke HS (2014) Obesity-induced CerS6-dependent C 16: 0 ceramide production promotes weight gain and glucose intolerance. Cell Metab 20:678–686
Urs S, Harrington A, Liaw L, Small D (2006) Selective expression of an aP2/fatty acid binding Protein4-Cre transgene in non-adipogenic tissues during embryonic development. Transgenic Res 15:647–653
Vitali A, Murano I, Zingaretti MC, Frontini A, Ricquier D, Cinti S (2012) The adipose organ of obesity-prone C57BL/6J mice is composed of mixed white and brown adipocytes. J Lipid Res 53:619–629
Wang QA, Scherer PE (2014) The AdipoChaser mouse: a model tracking adipogenesis in vivo. Adipocytes 3:146–150

Wang QA, Scherer PE, Gupta RK (2014) Improved methodologies for the study of adipose biology: insights gained and opportunities ahead. J Lipid Res 55:605–624
Wang QA, Tao C, Gupta RK, Scherer PE (2013) Tracking adipogenesis during white adipose tissue development, expansion and regeneration. Nat Med 19:1338
Wang ZV, Deng Y, Wang QA, Sun K, Scherer PE (2010) Identification and characterization of a promoter cassette conferring adipocyte-specific gene expression. Endocrinology 151:2933–2939
Warming S, Costantino N, Court DL, Jenkins NA, Copeland NG (2005) Simple and highly efficient BAC recombineering using galK selection. Nucleic Acids Res 33:e36
Widdowson EM (1976) The response of the sexes to nutritional stress. Proc Nutr Soc 35:175–180
Xia JY, Sun K, Hepler C, Ghaben AL, Gupta RK, An YA, Holland WL, Morley TS, Adams AC, Gordillo R (2017) Acute loss of adipose tissue-derived adiponectin triggers immediate metabolic deterioration in mice. Diabetologia:1–10
Yang M, Baranov E, Jiang P, Sun F-X, Li X-M, Li L, Hasegawa S, Bouvet M, Al-Tuwaijri M, Chishima T (2000) Whole-body optical imaging of green fluorescent protein-expressing tumors and metastases. Proc Natl Acad Sci 97:1206–1211
Ye R, Wang QA, Tao C, Vishvanath L, Shao M, McDonald JG, Gupta RK, Scherer PE (2015) Impact of tamoxifen on adipocyte lineage tracing: inducer of adipogenesis and prolonged nuclear translocation of Cre recombinase. Mol Metab 4:771–778
Zhang F, Hao G, Shao M, Nham K, An Y, Wang Q, Zhu Y, Kusminski CM, Hassan G, Gupta RK (2018) An adipose tissue atlas: an image-guided identification of human-like BAT and beige depots in rodents. Cell Metab 27:252–262.e3
Zhu Y, Gao Y, Tao C, Shao M, Zhao S, Huang W, Yao T, Johnson JA, Liu T, Cypess AM (2016) Connexin 43 mediates white adipose tissue beiging by facilitating the propagation of sympathetic neuronal signals. Cell Metab 24:420–433

Aging of Brown and Beige/Brite Adipose Tissue

Antonia Graja, Sabrina Gohlke, and Tim J. Schulz

Contents

1 Introduction 56
2 General Aspects of White and Brown Adipose Tissue Aging 58
3 Molecular Characteristics of Brown Adipose Tissue Aging 60
4 Cellular Senescence: Stem Cell Dysfunction 60
5 Extracellular Matrix and Adipose Tissue Aging 62
6 Neuronal Regulation of Aging Brown Adipocytes 62
7 Age-Related Changes to the Endocrine Regulation of Brown Adipocytes 63
8 Contribution of Inflammatory Processes to BAT Atrophy 64
9 Conclusion 65
References 66

Abstract

Brown adipose tissue aging and the concomitant loss of thermogenic capacity have been linked to an inability to maintain normal energy homeostasis in late life. Similarly, the ability of white fat to convert into brite/beige adipose tissue declines. This may ultimately exacerbate the progression of age-related metabolic pathologies, such as insulin resistance and obesity. The depletion of all types of brown adipocytes during aging is well-established and has been described in rodent models as well as humans. We here review the available literature on the

A. Graja · S. Gohlke
Department of Adipocyte Development and Nutrition, German Institute of Human Nutrition Potsdam-Rehbrücke, Nuthetal, Germany

T. J. Schulz (✉)
Department of Adipocyte Development and Nutrition, German Institute of Human Nutrition Potsdam-Rehbrücke, Nuthetal, Germany

University of Potsdam, Institute of Nutritional Science, Nuthetal, Germany

German Center for Diabetes Research (DZD), Munich-Neuherberg, Germany
e-mail: tim.schulz@dife.de

A. Pfeifer et al. (eds.), *Brown Adipose Tissue*,
Handbook of Experimental Pharmacology 251, https://doi.org/10.1007/164_2018_151

potential mechanisms leading to cell-autonomous and microenvironment-related aspects of brown adipocyte dysfunction. Among these, cellular senescence, mitochondrial impairment, and deteriorating changes to the local and endocrine microenvironments have been proposed. An important goal of aging research is to develop approaches that may not only extend life expectancy but also prolong health-span. These efforts may also be aimed at maintaining metabolic health throughout life by targeting brown adipocyte function.

Keywords
Aging · Brown adipose tissue · Progenitor cells · Senescence · Stem cell niche

1 Introduction

Aging is accompanied by a functional decline of many organ systems and physiological processes that ultimately result in various age-related health issues. Importantly, the prevalence of over 60-year-old individuals as percentage of the worlds' population is expected to almost double from 12 to 22% between 2015 and 2050, altogether emphasizing the need to better understand the contributing factors that affect life expectancy and health-span (World Health Organization 2015).

The development of obesity is aggravated during aging, which in itself is an independent risk factor for this metabolic condition. Thus, the age-dependent accumulation of body fat is well-documented in the literature (Bazzocchi et al. 2013; Chumlea et al. 2002; Kyle et al. 2001). Specifically, old age associates with an increase in abdominal fat, as well as ectopic fat in the liver, skeletal muscles, and the long bones, among other tissues (Kuk et al. 2009; Kyle et al. 2001; Slawik and Vidal-Puig 2006). The incidence of obesity, which is mainly a consequence of increased energy consumption over reduced expenditure, has nearly tripled since 1975: More than 1.9 billion adults (18 years and older) are overweight with a body mass index (BMI) greater than 25, and over 650 million are obese, defined by a BMI greater than 30 (World Health Organization 2017) (Fig. 1).

White adipose tissue (WAT) stores the majority of excess energy and is by far more abundant than the second type of fat, the brown adipose tissue (BAT) (Cannon and Nedergaard 2004). Although both types of adipocytes are able to store triglycerides, they differ substantially in several molecular and physiological parameters, such as lipid droplet conformation, mitochondrial content and activity, and localization of the nucleus (Schulz et al. 2011). Additionally, brown adipocytes uniquely express uncoupling protein 1 (UCP1) which resides in the inner mitochondrial membrane (Heaton et al. 1978). UCP1 uncouples the mitochondrial proton gradient, which is generated during oxidative phosphorylation (OXPHOS), from ATP production, and thereby enables the generation of heat in brown adipocytes. While this process is thought to be a main contributor to non-shivering thermogenesis, several recent studies have suggested that some portions of non-shivering thermogenesis may be independent of UCP1 activity or even BAT (Bal et al. 2012; Kazak et al. 2015; Keipert et al. 2017).

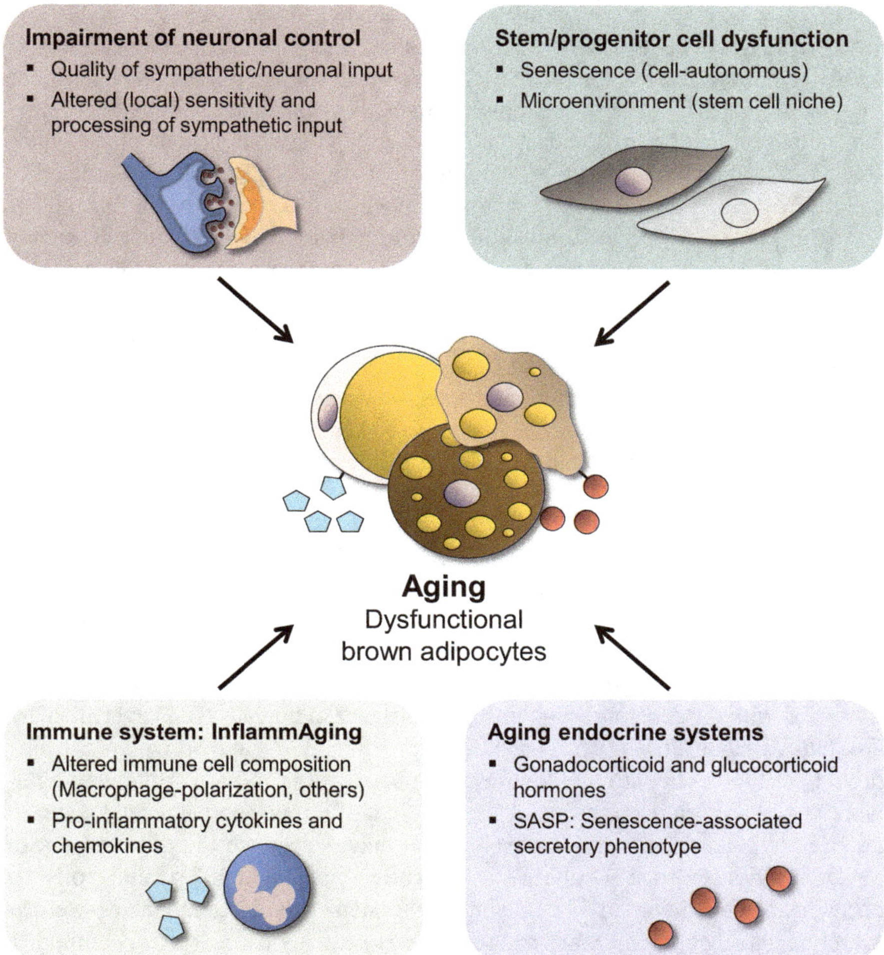

Fig. 1 Depiction of the main molecular processes that regulate brown and brite adipose tissue aging, starting with impaired neuronal control, stem cell dysfunction, impaired endocrine control mechanisms, and increased inflammatory exposure during aging (inflammaging)

In addition to the classical brown and white adipocytes, prolonged cold exposure or adrenergic stimulation results in a browning of WAT into the beige or brite (brown-in-white) adipose tissue, which is recognized as a population of brown-like adipocytes that are interspersed between the white adipocytes of most WAT depots (Ishibashi and Seale 2010; Petrovic et al. 2010). The extent of conversion appears to be depot specific and may be either due to transdifferentiation of mature white adipocytes or de novo differentiation of a population of committed beige/brite preadipocytes and/or a pool of bi-potent adipogenic progenitors in WAT (Barbatelli et al. 2010; Himms-Hagen et al. 2000; Lee et al. 2012; Rosenwald et al. 2013; Shao et al. 2016; Vishvanath et al. 2016; Wu et al. 2012).

The mechanisms that control UCP1-dependent thermogenesis have been reviewed in great detail elsewhere (for instance, Cannon and Nedergaard 2004). Briefly, thermogenesis is initiated by the release of the neurotransmitter hormone norepinephrine (NE) from the terminal endings of sympathetic neurons that innervate adipose tissues. This leads to activation of β3-adrenergic receptors, which triggers a classical G-protein-coupled receptor signaling cascade of adenylate cyclase activation, increases in intracellular levels of cyclic adenosine monophosphate (cAMP), and activation of protein kinase A. This kinase subseqently phosphorylates important substrates such as p38-mitogen-activated protein kinase (p38-MAPK), lipid droplet-associated protein perilipin (PLIN), and hormone-sensitive lipase (HSL). As a result, free fatty acids are mobilized from lipid droplets to activate UCP1 and energy metabolism.

The presence of active BAT in humans has long been thought to be restricted to newborns and infants that are more vulnerable to heat loss than adults. BAT rapidly declines during adolescence and is no longer detectable in significant amounts at a very old age (Heaton 1972). However, it is by now well recognized that metabolically active BAT occurs and contributes to energy homeostasis of healthy adult humans (Cypess et al. 2009; Nedergaard et al. 2007; Saito et al. 2009; van Marken Lichtenbelt et al. 2009; Virtanen et al. 2009; Zingaretti et al. 2009). It has been shown that repeated exposure to mild cold counteracts the development of obesity in a close correlation with activation of BAT and energy expenditure (Cypess et al. 2015; Yoneshiro et al. 2013). Conversely, reduced metabolic activity of BAT is associated with the development of insulin resistance in humans (Bakker et al. 2014; Koksharova et al. 2017). Thus, maintaining active BAT and browning capacity throughout life may contribute to a healthy phenotype in old age. Here we review the body of knowledge on the age-dependent loss of brown and beige adipocytes. It has been hypothesized that this development may exacerbate the development of metabolic diseases such as obesity and diabetes. We highlight several potential mechanisms in relation to brown adipogenic stem cell dysfunction, as well as endocrine and central mechanisms that may result in loss of brown adipocyte formation and function.

2 General Aspects of White and Brown Adipose Tissue Aging

Many aspects of adipose tissue biology display age-related dysfunction. Aside from changes in depot and fat cell sizes, a decrease in insulin responsiveness and the response to lipolytic agents and an increased inflammatory cytokine production have been discussed (Kirkland et al. 2002; Kuk et al. 2009; Tchkonia et al. 2007). Fat accumulation displays characteristic dynamics throughout life with an initial increase of body fat content mainly in the abdominal regions, followed by a frailty-associated decline that occurs in very old ages (Atlantis et al. 2008; Chumlea et al. 2002; Visser et al. 2003). A reduced metabolic rate during aging is well described, although it is still a matter of debate whether this is a cause or effect of adiposity (St-Onge and Gallagher 2010). Since a reduced metabolic rate may to a certain extent be a consequence of reduced brown adipocyte function, it stands to reason that this may also contribute to the development of adiposity.

During aging, lipid mobilization is impaired due to changes in the gene expression of apolipoproteins. This is further accompanied by a reduced ability for adaptation to metabolic stimuli, such as fasting, due to a blunted activation of lipoprotein lipase (Araki et al. 2004). When adipose tissue cannot meet the demand for energy storage, triglycerides accumulate in non-adipose tissues as ectopic fat. The resulting lipotoxicity could thereby lead to local insulin resistance in the liver and skeletal muscles and insufficient insulin secretion in the pancreas (Kusminski et al. 2009; McNelis and Olefsky 2014; Sun et al. 2011). Under normal physiological conditions, adipocytes are relatively resistant to lipotoxicity. However, adipogenic progenitor cells become more susceptible to lipotoxic reactions with age, resulting in elevated apoptosis rates that contribute to adipose tissue dysfunction (Guo et al. 2007).

The hallmark characteristics describing BAT aging are atrophy, a functional decline of mitochondria, and reduced UCP1 activation (Valle et al. 2008a). Accordingly, a decline in mitochondrial energy metabolism and accumulation of mitochondrial DNA mutations are important contributors to human aging, although it has yet to be determined whether excessive formation of reactive oxygen species is an integral part of this process (Lapointe and Hekimi 2010; Schulz et al. 2007; Wallace 1999). In addition to telomere shortening, free oxygen radical reactions cause progressive accumulation of oxidative damage to DNA, lipids, and proteins and thereby may promote several pathological features (Harman 1988). Nevertheless, mitochondrial reactive oxygen species are not entirely detrimental and exert beneficial effects through a process known as mitochondrial hormesis (mitohormesis) (Schulz et al. 2007). Indeed, acutely activated thermogenesis leads to a marked increase of mitochondrial ROS in BAT and furthermore supports UCP1-dependent thermogenesis by altering the redox status of cysteine thiols in UCP1 (Chouchani et al. 2016). Related to this, reactive oxygen species may also be involved in the regulation of mitochondrial biogenesis (St-Pierre et al. 2006).

Macrophage infiltration and excessive triglyceride accumulation promote a white adipocyte-like phenotype that gradually replaces the brown fat cells (Bargut et al. 2016). In line with this, the browning capacity of WAT decreases with age, which may further contribute to the age-related increase of body fat (Rogers et al. 2012). Loss of UCP1 by genetic ablation led to an exacerbation of age-related adiposity in mice (Kontani et al. 2005). It should be noted that this effect was only evident in mice fed a high fat diet and did not shorten overall life expectancy. These data suggest that compensatory mechanisms may be in effect under these housing conditions that are sufficient to rescue detrimental consequences due to lifelong UCP1 deficiency (Kontani et al. 2005), which reiterate the conclusions on the presence of UCP1-independent thermogenesis (Bal et al. 2012; Kazak et al. 2015; Keipert et al. 2017). While many of the molecular analyses of BAT aging have been conducted in rodent models, it has been demonstrated repeatedly that aging is a negative determinant of human BAT mass and activity (Brendle et al. 2018; Cypess et al. 2009; Mattson 2010; McDonald and Horwitz 1999; Pfannenberg et al. 2010; Yoneshiro et al. 2011).

3 Molecular Characteristics of Brown Adipose Tissue Aging

The age-related decline of BAT is associated with the inability of aged rodents to maintain body temperature and to adapt to short-term and chronic cold exposure. On the cellular level, this appears to be mainly due to a blunted increase in BAT mass, reduced cell proliferation, and UCP1 content (Florez-Duquet et al. 1998; McDonald et al. 1989). It is interesting to note that this phenotype is dimorphic as it is much more prevalent in males. Indeed, the mitochondrial ultrastructure in brown adipocytes of female rodents appears to be much more efficiently set up for thermogenesis (Rodriguez-Cuenca et al. 2002). In line with these observations, greater BAT mass has also been described in women compared to men throughout life (Cypess et al. 2009; Pfannenberg et al. 2010).

On the molecular level, the density of β-adrenergic receptors in aged BAT is reduced (Scarpace et al. 1988). The authors themselves argue that this could be due to reduced receptor densities on cell surfaces or a dilution effect due to an age-related change in the overall cellular composition. The study further describes a lower level of adenylate cyclase activity, suggesting that aging brown adipocytes are less sensitive to adrenergic stimulation (Scarpace et al. 1988). Transcription control through forkhead box protein A3 (Foxa3) has recently been proposed as a potential regulatory factor of BAT and brite fat aging. In this study, a mouse model with genetic ablation of *Foxa3* displayed sustained browning capacity and BAT gene expression at higher levels throughout life compared to aging control animals. Knockout animals showed extended life expectancy which was potentially due to protection from age-related obesity (Ma et al. 2014). In addition, total DNA content per depot and the amount of protein per DNA content are decreased in BAT of aged rats (Valle et al. 2008a). Accordingly, reduced mitochondrial protein content was found but an increase in mitochondrial DNA content as a compensatory mechanism in rats. These changes were accompanied by reduced UCP1 protein levels and UCP1-dependent as well as UCP1-independent mitochondrial respiration (Valle et al. 2008a) and a decrease in GDP bound to mitochondria per gram of BAT mass (McDonald et al. 1988). Altogether, these observations suggest that loss of function in aging brown adipocytes may be a multifactorial process that involves cell-autonomous defects on the stem cell and mature adipocyte levels. Secondly, mechanisms related to the local microenvironment, for instance, interactions with the extracellular matrix but also more distal effects to the nervous, endocrine, and immune systems, may further regulate aging of BAT.

4 Cellular Senescence: Stem Cell Dysfunction

Adipocytes are regenerated from a pool of mesenchymal stem or progenitor cells that are capable of self-renewal, expansive proliferation, and subsequent differentiation into mature fat cells (Rodeheffer et al. 2008). Although there are different types of adipogenic progenitor cells that give rise to either white, brown, or brite/beige adipocytes (Schulz et al. 2011), all of them share an equal fate – the process of cellular senescence. This mechanism of cellular aging leads to an irreversible arrest of cell proliferation. Accordingly, cell division rates were reduced in BAT of rats aged 26 months during cold exposure and when compared to 6-month-old animals (Florez-Duquet and McDonald 1998). The authors go on to demonstrate that attenuated proliferation is not due to cold- and aging-induced changes in insulin-like growth factor (IGF)-1 concentration, which was unchanged at either age or changes in plasma norepinephrine, which was upregulated in response to aging. Consistent with these data, limited proliferation and differentiation capacities were observed in preadipocytes derived from subcutaneous WAT of aged humans when compared to cells isolated from young subjects (Caso et al. 2013).

On the molecular level, the interruption of cell cycle progression is due to activation of the $p16^{INK4a}$/pRB and the p53/p21 pathways, both known for their tumor suppressor functions (Terzi et al. 2016). For instance, the cell cycle inhibitor $p16^{INK4a}$ controls the clearance of damaged cells, and the genetic ablation of $p16^{INK4a}$-expressing cells delays age-related tissue dysfunction (Baker et al. 2011). Adipogenic progenitor cells derived from gonadal WAT of rats exhibit a progressive defect in proliferation and lipid accumulation with increasing age, which is most pronounced in rats aged 24 months (Kirkland et al. 1994). Accordingly, cell cycle arrest also impairs brown adipogenesis. In one study, fatty acid binding protein-4 (*Fabp4*)-driven overexpression of the cell cycle inhibitor p27 was sufficient to induce cell cycle arrest and to attenuate metabolic adaptations of BAT, but not the formation of beige adipocytes, to cold (Okamatsu-Ogura et al. 2017). It should be emphasized that expression of *Fabp4* is not limited to mature adipocytes but occurs in adipogenic progenitors (Shan et al. 2013). One of the factors influencing the proliferative capacity of cells is the length of telomeres. Shortening of telomere length in turn activates the p53-mediated growth arrest and apoptosis (Epel et al. 2008; Sahin et al. 2011; Yu et al. 1990). In addition to impaired proliferation, senescent cells display several other characteristics, such as changes in chromatin organization and the secretion of pro-inflammatory cytokines and other proteins, which have been termed senescence-associated secretory phenotype (SASP) (Wiley et al. 2016). Release of SASP molecules from senescent cells, such as interleukin-6 (IL6) and tumor necrosis factor alpha (TNFα), among other cytokines and pro-inflammatory molecules, has been implicated in the functional decline of adipose tissue and impaired adipogenesis (Tchkonia et al. 2013). Another example is activin A, which is secreted by senescent cells and induces the JAK/STAT (Janus kinase/signal transducer and activator of transcription) pathway. Accordingly, inhibition of JAK/STAT led to improved function in aged adipose tissue (Xu et al. 2015). Of note, JAK inhibition also enhances browning of white adipose tissue (Moisan et al. 2015). The negative effect of aging on browning of inguinal WAT in response to cold exposure may also be partially due to a progenitor cell defect. Interestingly, the depletion of senescent progenitor cells improved the age-related failure of browning, and senescence of adipogenic progenitors could be reversed by inhibition of p38 MAPK (Berry et al. 2016). The rejuvenation of inguinal WAT-derived progenitor cells by blocking the p38-MAPK pathway is surprising since other studies could show that p38-MAPK inhibition leads to impaired formation of beige adipocytes during sympathetic stimulation, e.g., β3-adrenergic receptor activation (Bordicchia et al. 2012; Cao et al. 2001). A potential explanation for these contradicting observations could be that some subpopulations of inguinal WAT-derived progenitor cells respond differentially to cold exposure or β3-adrenergic stimulation, but this remains to be investigated in further detail.

5 Extracellular Matrix and Adipose Tissue Aging

Aging is often associated with progressive tissue fibrosis, e.g., an excessive accumulation of extracellular matrix (ECM) components, such as collagens and other fibrillary proteins (Wynn 2007). Furthermore, an aging ECM is characterized by a deregulated occurrence of proteolytic enzymes, the matrix metalloproteinases (MMPs), and is often accompanied by a loss of elasticity due to the degradation of elastic fibers (Sherratt 2009). Matrix elasticity in turn drives mesenchymal stem cell specification (Engler et al. 2006) and could thus also affect adipogenic fate commitment. Senescent cells themselves and proteins released as part of the SASP influence the matrix by altered secretion of structural proteins, MMPs, and other proteins (Wiley et al. 2016), which could in turn alter accessibility and responsiveness to growth factors (Sprenger et al. 2010). Moreover, senescent cells express only specific subsets of collagens and laminins which could hamper appropriate ECM assembly. As changes of the MMP profile have been linked to the development of obesity (Chavey et al. 2003; Liu and Hornsby 2007), increased MMP activities and degradation processes could altogether lead to a dysfunctional matrix. While these processes occur in most tissues, very little is known about BAT-specific roles of the ECM. A deletion of microfibril-associated glycoprotein 1 (MAGP1) resulted in defective thermogenesis, and animals displayed adipocyte hypertrophy and metabolic dysfunction (Craft et al. 2014). The authors observed increased levels of transforming growth factor-β (TGFβ), a growth factor that has been linked to metabolic perturbation and impaired browning (Yadav et al. 2011). In summary, age-related alterations to the brown adipose ECM have not been assessed in great detail and may exert significant effects on brown adipocyte formation and metabolism by regulation of growth factors and cytokines.

6 Neuronal Regulation of Aging Brown Adipocytes

Sympathetic nervous system (SNS) activity increases with age in mice as well as in humans, which would argue against an overall, age-related reduction of sympathetic input to brown adipocytes (Gabaldon et al. 2003; Kawate et al. 1993; Seals and Esler 2000). This is further supported by the observation that sympathetic signals to BAT were increased in lean older humans compared to lean young individuals. At the same time, sympathetic drive, e.g., the SNS-dependent activation of BAT, was lower in lean aged compared to lean young study participants, but not in obese young individuals (Bahler et al. 2016). Age-dependent changes of the SNS seem to be tissue-specific: for instance, similar processes were reported in the skeletal muscle (Jones et al. 1997), while age-related degeneration of SNS innervation in the bone marrow impairs hematopoietic stem cell function (Maryanovich et al. 2018). Altogether these reports suggest that the local sensitivity to the SNS may be impaired within adipose tissues (Seals and Bell 2004). In line with this notion, the density of adrenergic receptors is reduced in aged BAT, and similar levels of sympathetic stimulation result in lower thermogenesis in aged rodents in comparison to young animals (Scarpace et al. 1988, 1992). A potential explanation consolidating these observations could be related to an altered quality, rather than overall quantity, of the neuronal input to brown adipocytes. The hypothalamus plays a central role in the regulation of energy balance. Increases in fat mass are invariably accompanied by higher leptin levels, an adipokine secreted predominantly by white adipocytes. Notably, neurons in the preoptic area of the hypothalamus express the leptin receptor. They control BAT-induced thermogenesis and are sufficient to inhibit energy expenditure (Yu et al. 2016). In addition, stimulation of agouti-related peptide (AgRP) neurons, located in the arcuate nucleus of the hypothalamus, causes a myogenic reprogramming of BAT that is associated with defective thermogenesis and development of insulin resistance in BAT (Steculorum et al. 2016). In addition to this, the deletion of hypothalamic neuropeptide Y (NPY) promotes the development of brown adipocytes (Chao et al. 2011). Notably, NPY expression is also increased in stromal-vascular cells derived from WAT of mice fed a high fat diet and participates in obesity-induced adipose tissue inflammation (Singer et al. 2013). Conversely, orexin, a neuropeptide which is produced in the hypothalamus, has the potential to restore the age-related decline of BAT (Sellayah and Sikder 2014). In summary, the neuronal control of thermogenesis is highly complex, and alterations of several distinct neuronal populations during aging could translate into impaired BAT function.

7 Age-Related Changes to the Endocrine Regulation of Brown Adipocytes

One model of endocrine control of BAT aging has been proposed previously and integrates changes to the gonadocorticoid and glucocorticoid hormone systems which may drive the conversion of brown into white-like adipocytes, e.g., the whitening of BAT (Cannon and Nedergaard 2004; Nedergaard and Cannon 2010). This hypothesis proposes a positive effect of sex hormones on brown adipocyte function which would serve to maintain normal levels of brown adipogenesis in the young. The age-dependent decline of the circulating levels of androgen and estrogen hormones may in turn contribute to a diminished brown adipogenic stimulus and ultimately reduced thermogenesis (Rodriguez-Cuenca et al. 2007). At the same time, glucocorticoids reduce *Ucp1* expression and BAT activity in mice and rats (Soumano et al. 2000; Valle et al. 2008b). Since glucocorticoid levels appear to remain relatively stable throughout life, this would suggest that the loss of sex hormones in aged animals may unmask the negative regulatory effect of glucocorticoids (Nedergaard and Cannon 2010). The situation is somewhat more complex in human BAT: acute administration of glucocorticoids induces BAT-dependent thermogenesis (Ramage et al. 2016), whereas another study has found a negative effect of glucocorticoids during chronic treatments, suggesting that timing and duration of glucocorticoid exposure may be an essential determining factor (Thuzar et al. 2017). Aside from these two endocrine systems, thyroid hormones are well-known controllers of metabolism and brown adipocyte function (Cannon and Nedergaard 2004). In rats, aging correlates inversely with the thyroid hormone T3 (Valle et al. 2008b). Related to this observation, the activity of thyroxine 5-deiodinase (T5D) declines with age, suggesting that aging also leads to perturbations of this endocrine system (Gabaldon et al. 1995).

8 Contribution of Inflammatory Processes to BAT Atrophy

Closely related to hormonal control, aging is accompanied by a reduced immune system functionality which has been termed immuno-senescence (Baylis et al. 2013). Such changes are characterized by increased secretion of pro-inflammatory cytokines from T lymphocytes and other immune cells and a decreased phagocytic capability (Weiskopf et al. 2009). Interestingly, it has been shown that T cells in gonWAT of mice contribute to a pro-inflammatory environment during aging (Lumeng et al. 2011). Increased levels of circulating cytokines, such as interleukin-1β, interleukin-6, as well as interleukin-10, and TNFα, were also found in older humans (Alvarez-Rodriguez et al. 2012). An increase in TNFα levels and receptor signaling has been shown to hamper brown adipogenesis where TNFα induces apoptosis of multilocular brown adipocytes, inhibits expression of *Adrb3* and *Ucp1*, and decreases the thermoadaptive response of brown adipocytes to cold (Nisoli et al. 1997, 2000).

Obesity is generally associated with a systemic low-grade inflammation that is characterized by excessive accumulation of adipose tissue resident pro-inflammatory macrophages and other immune cells. Macrophages can be roughly divided into two main categories, classically activated macrophages (M1-like phenotype) and alternatively activated macrophages (M2-like phenotype). M1-like macrophages are activated by interferon gamma (IFNγ) and promote to a pro-inflammatory profile as well as an increased ROS production. M2-like, or wound-healing, macrophages are instead activated by interleukin-4 and interleukin-13 (Doyle et al. 1994; Stein et al. 1992) and promote tissue repair and suppress inflammation (Gordon and Taylor 2005). During obesity, a shift toward a profile of more classically activated pro-inflammatory macrophages can be observed (Lumeng et al. 2007). In BAT, such an excess infiltration of macrophages negatively affects *Ucp1* expression (Sakamoto et al. 2013). In addition to macrophages, the numbers of mast cells and natural killer (NK) lymphocytes also increase in obese compared to lean adipose tissue and can also negatively affect brown adipocytes (Liu et al. 2009; Ohmura et al. 2010). It has been proposed that alternatively activated macrophages are a source of catecholamines and could thereby promote browning of WAT (Nguyen et al.

2011). However, this finding seems to be controversial since a recent study in mice with a peripheral deletion of tyrosine hydroxylase, an enzyme controlling catecholamine synthesis, demonstrated that alternatively activated macrophages do not express tyrosine hydroxylase and therefore would not contribute to adaptive thermogenesis (Fischer et al. 2017). The extent to which age-related changes to inflammatory processes in BAT inhibit its thermogenic function remains only partially understood and warrants further investigation in the future.

9 Conclusion

Aging is a multifactorial process that relies on a cell-autonomous loss of function, for instance, due to mutations and impairment of mitochondria that affect adipogenic stem cells as well as mature, terminally differentiated brown and white adipocytes. Secondly, external events contribute to a dysfunctional microenvironment. This may lead to the deterioration of the stem cell niche and thereby could hamper brown adipocyte regeneration as a normal part of tissue turnover or in response to stimulatory cues, such as cold. In addition to this, an altered microenvironment may provide inhibitory input to mature brown adipocytes, for instance, in the form of pro-inflammatory cytokines or changes to the extracellular matrix. Caloric restriction (CR), a mechanism that attenuates many aspects of aging in multiple species, also induces browning and BAT function (Fabbiano et al. 2016; Valle et al. 2008a). This effect may involve microRNA processing through the protein DICER, which is downregulated during aging and can be restored by CR in BAT and WAT. Accordingly, genetic ablation of *Dicer1* in murine adipose tissue also impaired brown adipogenesis (Mori et al. 2012, 2014). From other branches of aging research come experimental approaches that may also be employed to improve brown adipogenesis in aged animals and individuals, such as the removal of senescent cells form the tissue (Baker et al. 2011). Interestingly, exercise counteracts diet-induced cellular senescence in adipose tissue (Schafer et al. 2016). This effect may also be achieved by treatment with senolytics, e.g., reagents that can specifically target and remove senescent cells from the tissue (Zhu et al. 2015). These promising new approaches may also be a feasible strategy when developing strategies to counter the age-related impairment of brown adipogenesis and, as a potential consequence of this, the progression of metabolic disorders.

Acknowledgements This work was supported by the European Research Council (ERC-StG 311082), the Emmy Noether Program of the German Research Foundation (DFG; grant SCHU 2445/2-1), and a grant from the German Ministry of Education and Research (BMBF) and the State of Brandenburg (DZD grant 82DZD00302).

References

Alvarez-Rodriguez L, Lopez-Hoyos M, Munoz-Cacho P et al (2012) Aging is associated with circulating cytokine dysregulation. Cell Immunol 273:124–132. https://doi.org/10.1016/j.cellimm.2012.01.001

Araki S, Okazaki M, Goto S (2004) Impaired lipid metabolism in aged mice as revealed by fasting-induced expression of apolipoprotein mRNAs in the liver and changes in serum lipids. Gerontology 50:206–215. https://doi.org/10.1159/000078349

Atlantis E, Martin SA, Haren MT et al (2008) Lifestyle factors associated with age-related differences in body composition: the Florey Adelaide Male Aging Study. Am J Clin Nutr 88:95–104

Bahler L, Verberne HJ, Admiraal WM et al (2016) Differences in sympathetic nervous stimulation of brown adipose tissue between the young and old, and the lean and obese. J Nucl Med 57:372–377. https://doi.org/10.2967/jnumed.115.165829

Baker DJ, Wijshake T, Tchkonia T et al (2011) Clearance of p16Ink4a-positive senescent cells delays ageing-associated disorders. Nature 479:232–236. https://doi.org/10.1038/nature10600

Bakker LE, Boon MR, van der Linden RA et al (2014) Brown adipose tissue volume in healthy lean south Asian adults compared with white Caucasians: a prospective, case-controlled observational study. Lancet Diabetes Endocrinol 2:210–217. https://doi.org/10.1016/S2213-8587(13)70156-6

Bal NC, Maurya SK, Sopariwala DH et al (2012) Sarcolipin is a newly identified regulator of muscle-based thermogenesis in mammals. Nat Med 18:1575–1579. https://doi.org/10.1038/nm.2897

Barbatelli G, Murano I, Madsen L et al (2010) The emergence of cold-induced brown adipocytes in mouse white fat depots is determined predominantly by white to brown adipocyte transdifferentiation. Am J Physiol Endocrinol Metab 298:E1244–E1253. https://doi.org/10.1152/ajpendo.00600.2009

Bargut TC, Silva-e-Silva AC, Souza-Mello V et al (2016) Mice fed fish oil diet and upregulation of brown adipose tissue thermogenic markers. Eur J Nutr 55:159–169. https://doi.org/10.1007/s00394-015-0834-0

Baylis D, Bartlett DB, Patel HP et al (2013) Understanding how we age: insights into inflammaging. Longev Healthspan 2:8. https://doi.org/10.1186/2046-2395-2-8

Bazzocchi A, Diano D, Ponti F et al (2013) Health and ageing: a cross-sectional study of body composition. Clin Nutr 32:569–578. https://doi.org/10.1016/j.clnu.2012.10.004

Berry DC, Jiang Y, Arpke RW et al (2016) Cellular aging contributes to failure of cold-induced beige adipocyte formation in old mice and humans. Cell Metab. https://doi.org/10.1016/j.cmet.2016.10.023

Bordicchia M, Liu D, Amri EZ et al (2012) Cardiac natriuretic peptides act via p38 MAPK to induce the brown fat thermogenic program in mouse and human adipocytes. J Clin Invest 122:1022–1036. https://doi.org/10.1172/JCI59701

Brendle C, Werner MK, Schmadl M et al (2018) Correlation of brown adipose tissue with other body fat compartments and patient characteristics: a retrospective analysis in a large patient cohort using PET/CT. Acad Radiol 25:102–110. https://doi.org/10.1016/j.acra.2017.09.007

Cannon B, Nedergaard J (2004) Brown adipose tissue: function and physiological significance. Physiol Rev 84:277–359. https://doi.org/10.1152/physrev.00015.2003

Cao W, Medvedev AV, Daniel KW et al (2001) Beta-adrenergic activation of p38 MAP kinase in adipocytes: cAMP induction of the uncoupling protein 1 (UCP1) gene requires p38 MAP kinase. J Biol Chem 276:27077–27082. https://doi.org/10.1074/jbc.M101049200

Caso G, McNurlan MA, Mileva I et al (2013) Peripheral fat loss and decline in adipogenesis in older humans. Metabolism 62:337–340. https://doi.org/10.1016/j.metabol.2012.08.007

Chao PT, Yang L, Aja S et al (2011) Knockdown of NPY expression in the dorsomedial hypothalamus promotes development of brown adipocytes and prevents diet-induced obesity. Cell Metab 13:573–583. https://doi.org/10.1016/j.cmet.2011.02.019

Chavey C, Mari B, Monthouel MN et al (2003) Matrix metalloproteinases are differentially expressed in adipose tissue during obesity and modulate adipocyte differentiation. J Biol Chem 278:11888–11896. https://doi.org/10.1074/jbc.M209196200
Chouchani ET, Kazak L, Jedrychowski MP et al (2016) Mitochondrial ROS regulate thermogenic energy expenditure and sulfenylation of UCP1. Nature 532:112–116. https://doi.org/10.1038/nature17399. http://www.nature.com/nature/journal/v532/n7597/abs/nature17399.html#supplementary-information
Chumlea WC, Guo SS, Kuczmarski RJ et al (2002) Body composition estimates from NHANES III bioelectrical impedance data. Int J Obes Relat Metab Disord 26:1596–1609. https://doi.org/10.1038/sj.ijo.0802167
Craft CS, Pietka TA, Schappe T et al (2014) The extracellular matrix protein MAGP1 supports thermogenesis and protects against obesity and diabetes through regulation of TGF-beta. Diabetes 63:1920–1932. https://doi.org/10.2337/db13-1604
Cypess AM, Lehman S, Williams G et al (2009) Identification and importance of brown adipose tissue in adult humans. N Engl J Med 360:1509–1517. https://doi.org/10.1056/NEJMoa0810780
Cypess AM, Weiner LS, Roberts-Toler C et al (2015) Activation of human brown adipose tissue by a beta3-adrenergic receptor agonist. Cell Metab 21:33–38. https://doi.org/10.1016/j.cmet.2014.12.009
Doyle AG, Herbein G, Montaner LJ et al (1994) Interleukin-13 alters the activation state of murine macrophages in vitro: comparison with interleukin-4 and interferon-gamma. Eur J Immunol 24:1441–1445. https://doi.org/10.1002/eji.1830240630
Engler AJ, Sen S, Sweeney HL et al (2006) Matrix elasticity directs stem cell lineage specification. Cell 126:677–689. https://doi.org/10.1016/j.cell.2006.06.044
Epel ES, Merkin SS, Cawthon R et al (2008) The rate of leukocyte telomere shortening predicts mortality from cardiovascular disease in elderly men. Aging 1:81–88. https://doi.org/10.18632/aging.100007
Fabbiano S, Suarez-Zamorano N, Rigo D et al (2016) Caloric restriction leads to browning of white adipose tissue through type 2 immune signaling. Cell Metab 24:434–446. https://doi.org/10.1016/j.cmet.2016.07.023
Fischer K, Ruiz HH, Jhun K et al (2017) Alternatively activated macrophages do not synthesize catecholamines or contribute to adipose tissue adaptive thermogenesis. Nat Med 23:623. https://doi.org/10.1038/nm.4316
Florez-Duquet M, McDonald RB (1998) Cold-induced thermoregulation and biological aging. Physiol Rev 78:339–358
Florez-Duquet M, Horwitz BA, McDonald RB (1998) Cellular proliferation and UCP content in brown adipose tissue of cold-exposed aging Fischer 344 rats. Am J Phys 274:R196–R203
Gabaldon AM, Florez-Duquet ML, Hamilton JS et al (1995) Effects of age and gender on brown fat and skeletal muscle metabolic responses to cold in F344 rats. Am J Phys 268:R931–R941. https://doi.org/10.1152/ajpregu.1995.268.4.R931
Gabaldon AM, Gavel DA, Hamilton JS et al (2003) Norepinephrine release in brown adipose tissue remains robust in cold-exposed senescent Fischer 344 rats. Am J Physiol Regul Integr Comp Physiol 285:R91–R98. https://doi.org/10.1152/ajpregu.00494.2002
Gordon S, Taylor PR (2005) Monocyte and macrophage heterogeneity. Nat Rev Immunol 5:953–964. https://doi.org/10.1038/nri1733
Guo W, Pirtskhalava T, Tchkonia T et al (2007) Aging results in paradoxical susceptibility of fat cell progenitors to lipotoxicity. Am J Physiol Endocrinol Metab 292:E1041–E1051. https://doi.org/10.1152/ajpendo.00557.2006
Harman D (1988) Free radicals in aging. Mol Cell Biochem 84:155–161. https://doi.org/10.1007/BF00421050
Heaton JM (1972) The distribution of brown adipose tissue in the human. J Anat 112:35–39
Heaton GM, Wagenvoord RJ, Kemp A Jr et al (1978) Brown-adipose-tissue mitochondria: photoaffinity labelling of the regulatory site of energy dissipation. Eur J Biochem 82:515–521

Himms-Hagen J, Melnyk A, Zingaretti MC et al (2000) Multilocular fat cells in WAT of CL-316243-treated rats derive directly from white adipocytes. Am J Physiol Cell Physiol 279: C670–C681

Ishibashi J, Seale P (2010) Medicine. Beige can be slimming. Science 328:1113–1114. https://doi.org/10.1126/science.1190816

Jones PP, Davy KP, Alexander S et al (1997) Age-related increase in muscle sympathetic nerve activity is associated with abdominal adiposity. Am J Phys 272:E976–E980. https://doi.org/10.1152/ajpendo.1997.272.6.E976

Kawate R, Talan MI, Engel BT (1993) Aged C57BL/6J mice respond to cold with increased sympathetic nervous activity in interscapular brown adipose tissue. J Gerontol 48:B180–B183. https://doi.org/10.1093/geronj/48.5.B180

Kazak L, Chouchani ET, Jedrychowski MP et al (2015) A creatine-driven substrate cycle enhances energy expenditure and thermogenesis in beige fat. Cell 163:643–655. https://doi.org/10.1016/j.cell.2015.09.035

Keipert S, Kutschke M, Ost M et al (2017) Long-term cold adaptation does not require FGF21 or UCP1. Cell Metab 26:437–446 e435. https://doi.org/10.1016/j.cmet.2017.07.016

Kirkland JL, Hollenberg CH, Kindler S et al (1994) Effects of age and anatomic site on preadipocyte number in rat fat depots. J Gerontol 49:B31–B35

Kirkland JL, Tchkonia T, Pirtskhalava T et al (2002) Adipogenesis and aging: does aging make fat go MAD? Exp Gerontol 37:757–767

Koksharova E, Ustyuzhanin D, Philippov Y et al (2017) The relationship between brown adipose tissue content in supraclavicular fat depots and insulin sensitivity in patients with type 2 diabetes mellitus and prediabetes. Diabetes Technol Ther 19:96–102. https://doi.org/10.1089/dia.2016.0360

Kontani Y, Wang Y, Kimura K et al (2005) UCP1 deficiency increases susceptibility to diet-induced obesity with age. Aging Cell 4:147–155. https://doi.org/10.1111/j.1474-9726.2005.00157.x

Kuk JL, Saunders TJ, Davidson LE et al (2009) Age-related changes in total and regional fat distribution. Ageing Res Rev 8:339–348. https://doi.org/10.1016/j.arr.2009.06.001

Kusminski CM, Shetty S, Orci L et al (2009) Diabetes and apoptosis: lipotoxicity. Apoptosis 14:1484–1495. https://doi.org/10.1007/s10495-009-0352-8

Kyle UG, Genton L, Hans D et al (2001) Age-related differences in fat-free mass, skeletal muscle, body cell mass and fat mass between 18 and 94 years. Eur J Clin Nutr 55:663–672. https://doi.org/10.1038/sj.ejcn.1601198

Lapointe J, Hekimi S (2010) When a theory of aging ages badly. Cell Mol Life Sci 67:1–8. https://doi.org/10.1007/s00018-009-0138-8

Lee YH, Petkova AP, Mottillo EP et al (2012) In vivo identification of bipotential adipocyte progenitors recruited by beta3-adrenoceptor activation and high-fat feeding. Cell Metab 15:480–491. https://doi.org/10.1016/j.cmet.2012.03.009

Liu D, Hornsby PJ (2007) Senescent human fibroblasts increase the early growth of xenograft tumors via matrix metalloproteinase secretion. Cancer Res 67:3117–3126. https://doi.org/10.1158/0008-5472.CAN-06-3452

Liu J, Divoux A, Sun J et al (2009) Genetic deficiency and pharmacological stabilization of mast cells reduce diet-induced obesity and diabetes in mice. Nat Med 15:940–945. https://doi.org/10.1038/nm.1994

Lumeng CN, Bodzin JL, Saltiel AR (2007) Obesity induces a phenotypic switch in adipose tissue macrophage polarization. J Clin Invest 117:175–184. https://doi.org/10.1172/JCI29881

Lumeng CN, Liu J, Geletka L et al (2011) Aging is associated with an increase in T cells and inflammatory macrophages in visceral adipose tissue. J Immunol 187:6208–6216. https://doi.org/10.4049/jimmunol.1102188

Ma X, Xu L, Gavrilova O et al (2014) Role of forkhead box protein A3 in age-associated metabolic decline. Proc Natl Acad Sci U S A 111:14289–14294. https://doi.org/10.1073/pnas.1407640111

Maryanovich M, Zahalka AH, Pierce H et al (2018) Adrenergic nerve degeneration in bone marrow drives aging of the hematopoietic stem cell niche. Nat Med. https://doi.org/10.1038/s41591-018-0030-x

Mattson MP (2010) Perspective: does brown fat protect against diseases of aging? Ageing Res Rev 9:69–76. https://doi.org/10.1016/j.arr.2009.11.004

McDonald RB, Horwitz BA (1999) Brown adipose tissue thermogenesis during aging and senescence. J Bioenerg Biomembr 31:507–516

McDonald RB, Horwitz BA, Hamilton JS et al (1988) Cold- and norepinephrine-induced thermogenesis in younger and older Fischer 344 rats. Am J Phys 254:R457–R462

McDonald RB, Day C, Carlson K et al (1989) Effect of age and gender on thermoregulation. Am J Phys 257:R700–R704

McNelis JC, Olefsky JM (2014) Macrophages, immunity, and metabolic disease. Immunity 41:36–48. https://doi.org/10.1016/j.immuni.2014.05.010

Moisan A, Lee YK, Zhang JD et al (2015) White-to-brown metabolic conversion of human adipocytes by JAK inhibition. Nat Cell Biol 17:57–67. https://doi.org/10.1038/ncb3075

Mori MA, Raghavan P, Thomou T et al (2012) Role of microRNA processing in adipose tissue in stress defense and longevity. Cell Metab 16:336–347. https://doi.org/10.1016/j.cmet.2012.07.017

Mori MA, Thomou T, Boucher J et al (2014) Altered miRNA processing disrupts brown/white adipocyte determination and associates with lipodystrophy. J Clin Invest 124:3339–3351. https://doi.org/10.1172/JCI73468

Nedergaard J, Cannon B (2010) The changed metabolic world with human brown adipose tissue: therapeutic visions. Cell Metab 11:268–272. https://doi.org/10.1016/j.cmet.2010.03.007

Nedergaard J, Bengtsson T, Cannon B (2007) Unexpected evidence for active brown adipose tissue in adult humans. Am J Physiol Endocrinol Metab 293:E444–E452. https://doi.org/10.1152/ajpendo.00691.2006

Nguyen KD, Qiu Y, Cui X et al (2011) Alternatively activated macrophages produce catecholamines to sustain adaptive thermogenesis. Nature 480:104–108. https://doi.org/10.1038/nature10653

Nisoli E, Briscini L, Tonello C et al (1997) Tumor necrosis factor-alpha induces apoptosis in rat brown adipocytes. Cell Death Differ 4:771–778. https://doi.org/10.1038/sj.cdd.4400292

Nisoli E, Briscini L, Giordano A et al (2000) Tumor necrosis factor alpha mediates apoptosis of brown adipocytes and defective brown adipocyte function in obesity. Proc Natl Acad Sci U S A 97:8033–8038

Ohmura K, Ishimori N, Ohmura Y et al (2010) Natural killer T cells are involved in adipose tissues inflammation and glucose intolerance in diet-induced obese mice. Arterioscler Thromb Vasc Biol 30:193–199. https://doi.org/10.1161/ATVBAHA.109.198614

Okamatsu-Ogura Y, Fukano K, Tsubota A et al (2017) Cell-cycle arrest in mature adipocytes impairs BAT development but not WAT browning, and reduces adaptive thermogenesis in mice. Sci Rep 7:6648. https://doi.org/10.1038/s41598-017-07206-8

Petrovic N, Walden TB, Shabalina IG et al (2010) Chronic peroxisome proliferator-activated receptor gamma (PPARgamma) activation of epididymally derived white adipocyte cultures reveals a population of thermogenically competent, UCP1-containing adipocytes molecularly distinct from classic brown adipocytes. J Biol Chem 285:7153–7164. https://doi.org/10.1074/jbc.M109.053942

Pfannenberg C, Werner MK, Ripkens S et al (2010) Impact of age on the relationships of brown adipose tissue with sex and adiposity in humans. Diabetes 59:1789–1793. https://doi.org/10.2337/db10-0004

Ramage LE, Akyol M, Fletcher AM et al (2016) Glucocorticoids acutely increase brown adipose tissue activity in humans, revealing species-specific differences in UCP-1 regulation. Cell Metab 24:130–141. https://doi.org/10.1016/j.cmet.2016.06.011

Rodeheffer MS, Birsoy K, Friedman JM (2008) Identification of white adipocyte progenitor cells in vivo. Cell 135:240–249. https://doi.org/10.1016/j.cell.2008.09.036

Rodriguez-Cuenca S, Pujol E, Justo R et al (2002) Sex-dependent thermogenesis, differences in mitochondrial morphology and function, and adrenergic response in brown adipose tissue. J Biol Chem 277:42958–42963. https://doi.org/10.1074/jbc.M207229200
Rodriguez-Cuenca S, Monjo M, Frontera M et al (2007) Sex steroid receptor expression profile in brown adipose tissue. Effects of hormonal status. Cell Physiol Biochem 20:877–886. https://doi.org/10.1159/000110448
Rogers NH, Landa A, Park S et al (2012) Aging leads to a programmed loss of brown adipocytes in murine subcutaneous white adipose tissue. Aging Cell 11:1074–1083. https://doi.org/10.1111/acel.12010
Rosenwald M, Perdikari A, Rulicke T et al (2013) Bi-directional interconversion of brite and white adipocytes. Nat Cell Biol 15:659–667. https://doi.org/10.1038/ncb2740
Sahin E, Colla S, Liesa M et al (2011) Telomere dysfunction induces metabolic and mitochondrial compromise. Nature 470:359–365. https://doi.org/10.1038/nature09787
Saito M, Okamatsu-Ogura Y, Matsushita M et al (2009) High incidence of metabolically active brown adipose tissue in healthy adult humans: effects of cold exposure and adiposity. Diabetes 58:1526–1531. https://doi.org/10.2337/db09-0530
Sakamoto T, Takahashi N, Sawaragi Y et al (2013) Inflammation induced by RAW macrophages suppresses UCP1 mRNA induction via ERK activation in 10T1/2 adipocytes. Am J Physiol Cell Physiol 304:C729–C738. https://doi.org/10.1152/ajpcell.00312.2012
Scarpace PJ, Mooradian AD, Morley JE (1988) Age-associated decrease in beta-adrenergic receptors and adenylate cyclase activity in rat brown adipose tissue. J Gerontol 43:B65–B70
Scarpace PJ, Matheny M, Bender BS et al (1992) Impaired febrile response with age: role of thermogenesis in brown adipose tissue. Proc Soc Exp Biol Med 200:353–358
Schafer MJ, White TA, Evans G et al (2016) Exercise prevents diet-induced cellular senescence in adipose tissue. Diabetes 65:1606–1615. https://doi.org/10.2337/db15-0291
Schulz TJ, Zarse K, Voigt A et al (2007) Glucose restriction extends Caenorhabditis elegans life span by inducing mitochondrial respiration and increasing oxidative stress. Cell Metab 6:280–293. https://doi.org/10.1016/j.cmet.2007.08.011
Schulz TJ, Huang TL, Tran TT et al (2011) Identification of inducible brown adipocyte progenitors residing in skeletal muscle and white fat. Proc Natl Acad Sci U S A 108:143–148. https://doi.org/10.1073/pnas.1010929108
Seals DR, Bell C (2004) Chronic sympathetic activation: consequence and cause of age-associated obesity? Diabetes 53:276–284
Seals DR, Esler MD (2000) Human ageing and the sympathoadrenal system. J Physiol 528:407–417. https://doi.org/10.1111/j.1469-7793.2000.00407.x
Sellayah D, Sikder D (2014) Orexin restores aging-related brown adipose tissue dysfunction in male mice. Endocrinology 155:485–501. https://doi.org/10.1210/en.2013-1629
Shan T, Liu W, Kuang S (2013) Fatty acid binding protein 4 expression marks a population of adipocyte progenitors in white and brown adipose tissues. FASEB J 27:277–287. https://doi.org/10.1096/fj.12-211516
Shao M, Ishibashi J, Kusminski CM et al (2016) Zfp423 maintains white adipocyte identity through suppression of the beige cell thermogenic gene program. Cell Metab 23:1167–1184. https://doi.org/10.1016/j.cmet.2016.04.023
Sherratt MJ (2009) Tissue elasticity and the ageing elastic fibre. Age (Dordr) 31:305–325. https://doi.org/10.1007/s11357-009-9103-6
Singer K, Morris DL, Oatmen KE et al (2013) Neuropeptide Y is produced by adipose tissue macrophages and regulates obesity-induced inflammation. PLoS One 8:e57929. https://doi.org/10.1371/journal.pone.0057929
Slawik M, Vidal-Puig AJ (2006) Lipotoxicity, overnutrition and energy metabolism in aging. Ageing Res Rev 5:144–164. https://doi.org/10.1016/j.arr.2006.03.004
Soumano K, Desbiens S, Rabelo R et al (2000) Glucocorticoids inhibit the transcriptional response of the uncoupling protein-1 gene to adrenergic stimulation in a brown adipose cell line. Mol Cell Endocrinol 165:7–15

Sprenger CC, Plymate SR, Reed MJ (2010) Aging-related alterations in the extracellular matrix modulate the microenvironment and influence tumor progression. Int J Cancer 127:2739–2748. https://doi.org/10.1002/ijc.25615

Steculorum SM, Ruud J, Karakasilioti I et al (2016) AgRP neurons control systemic insulin sensitivity via myostatin expression in brown adipose tissue. Cell 165:125–138. https://doi.org/10.1016/j.cell.2016.02.044

Stein M, Keshav S, Harris N et al (1992) Interleukin 4 potently enhances murine macrophage mannose receptor activity: a marker of alternative immunologic macrophage activation. J Exp Med 176:287–292

St-Onge MP, Gallagher D (2010) Body composition changes with aging: the cause or the result of alterations in metabolic rate and macronutrient oxidation? Nutrition 26:152–155. https://doi.org/10.1016/j.nut.2009.07.004

St-Pierre J, Drori S, Uldry M et al (2006) Suppression of reactive oxygen species and neurodegeneration by the PGC-1 transcriptional coactivators. Cell 127:397–408. https://doi.org/10.1016/j.cell.2006.09.024

Sun K, Kusminski CM, Scherer PE (2011) Adipose tissue remodeling and obesity. J Clin Invest 121:2094–2101. https://doi.org/10.1172/JCI45887

Tchkonia T, Pirtskhalava T, Thomou T et al (2007) Increased TNFalpha and CCAAT/enhancer-binding protein homologous protein with aging predispose preadipocytes to resist adipogenesis. Am J Physiol Endocrinol Metab 293:E1810–E1819. https://doi.org/10.1152/ajpendo.00295.2007

Tchkonia T, Thomou T, Zhu Y et al (2013) Mechanisms and metabolic implications of regional differences among fat depots. Cell Metab 17:644–656. https://doi.org/10.1016/j.cmet.2013.03.008

Terzi MY, Izmirli M, Gogebakan B (2016) The cell fate: senescence or quiescence. Mol Biol Rep 43:1213–1220. https://doi.org/10.1007/s11033-016-4065-0

Thuzar M, Law WP, Ratnasingam J et al (2017) Glucocorticoids suppress brown adipose tissue function in humans: a double-blind placebo-controlled study. Diabetes Obes Metab 20:840. https://doi.org/10.1111/dom.13157

Valle A, Guevara R, Garcia-Palmer FJ et al (2008a) Caloric restriction retards the age-related decline in mitochondrial function of brown adipose tissue. Rejuvenation Res 11:597–604. https://doi.org/10.1089/rej.2007.0626

Valle A, Santandreu FM, Garcia-Palmer FJ et al (2008b) The serum levels of 17beta-estradiol, progesterone and triiodothyronine correlate with brown adipose tissue thermogenic parameters during aging. Cell Physiol Biochem 22:337–346. https://doi.org/10.1159/000149812

van Marken Lichtenbelt WD, Vanhommerig JW, Smulders NM et al (2009) Cold-activated brown adipose tissue in healthy men. N Engl J Med 360:1500–1508. https://doi.org/10.1056/NEJMoa0808718

Virtanen KA, Lidell ME, Orava J et al (2009) Functional brown adipose tissue in healthy adults. N Engl J Med 360:1518–1525. https://doi.org/10.1056/NEJMoa0808949

Vishvanath L, MacPherson KA, Hepler C et al (2016) Pdgfrbeta(+) mural preadipocytes contribute to adipocyte hyperplasia induced by high-fat-diet feeding and prolonged cold exposure in adult mice. Cell Metab 23:350–359. https://doi.org/10.1016/j.cmet.2015.10.018

Visser M, Pahor M, Tylavsky F et al (2003) One- and two-year change in body composition as measured by DXA in a population-based cohort of older men and women. J Appl Physiol (1985) 94:2368–2374. https://doi.org/10.1152/japplphysiol.00124.2002

Wallace DC (1999) Mitochondrial diseases in man and mouse. Science 283:1482–1488

Weiskopf D, Weinberger B, Grubeck-Loebenstein B (2009) The aging of the immune system. Transpl Int 22:1041–1050. https://doi.org/10.1111/j.1432-2277.2009.00927.x

Wiley CD, Velarde MC, Lecot P et al (2016) Mitochondrial dysfunction induces senescence with a distinct secretory phenotype. Cell Metab 23:303–314. https://doi.org/10.1016/j.cmet.2015.11.011

World Health Organization (2015) Ageing and health. http://www.who.int/mediacentre/factsheets/fs404/en/

World Health Organization (2017) Obesity and overweight. http://www.who.int/mediacentre/factsheets/fs311/en/

Wu J, Bostrom P, Sparks LM et al (2012) Beige adipocytes are a distinct type of thermogenic fat cell in mouse and human. Cell 150:366–376. https://doi.org/10.1016/j.cell.2012.05.016

Wynn TA (2007) Common and unique mechanisms regulate fibrosis in various fibroproliferative diseases. J Clin Invest 117:524–529. https://doi.org/10.1172/JCI31487

Xu M, Palmer AK, Ding H et al (2015) Targeting senescent cells enhances adipogenesis and metabolic function in old age. elife 4:e12997. https://doi.org/10.7554/eLife.12997

Yadav H, Quijano C, Kamaraju AK et al (2011) Protection from obesity and diabetes by blockade of TGF-beta/Smad3 signaling. Cell Metab 14:67–79. https://doi.org/10.1016/j.cmet.2011.04.013

Yoneshiro T, Aita S, Matsushita M et al (2011) Age-related decrease in cold-activated brown adipose tissue and accumulation of body fat in healthy humans. Obesity (Silver Spring) 19:1755–1760. https://doi.org/10.1038/oby.2011.125

Yoneshiro T, Aita S, Matsushita M et al (2013) Recruited brown adipose tissue as an antiobesity agent in humans. J Clin Invest 123:3404–3408. https://doi.org/10.1172/JCI67803

Yu GL, Bradley JD, Attardi LD et al (1990) In vivo alteration of telomere sequences and senescence caused by mutated tetrahymena telomerase RNAs. Nature 344:126–132. https://doi.org/10.1038/344126a0

Yu S, Qualls-Creekmore E, Rezai-Zadeh K et al (2016) Glutamatergic preoptic area neurons that express leptin receptors drive temperature-dependent body weight homeostasis. J Neurosci 36:5034–5046. https://doi.org/10.1523/jneurosci.0213-16.2016

Zhu Y, Tchkonia T, Pirtskhalava T et al (2015) The Achilles' heel of senescent cells: from transcriptome to senolytic drugs. Aging Cell 14:644–658. https://doi.org/10.1111/acel.12344

Zingaretti MC, Crosta F, Vitali A et al (2009) The presence of UCP1 demonstrates that metabolically active adipose tissue in the neck of adult humans truly represents brown adipose tissue. FASEB J 23:3113–3120. https://doi.org/10.1096/fj.09-133546

Adipogenesis in Primary Cell Culture

Therese Juhlin Larsen, Naja Zenius Jespersen, and Camilla Scheele

Contents

1 Introduction 74
2 Biopsy 75
3 Proliferation 76
  3.1 Priming/Adipocyte Commitment 76
  3.2 Flow Cytometry for Evaluation of Cell Culture Purity 77
4 Confluence 78
5 Initiation of Differentiation 78
6 Differentiation 79
7 Validation of Cell Cultures 80
8 Considerations 81
  8.1 Cell Passaging 81
  8.2 Differentiation With or Without Serum 81
  8.3 Inducing Brown Adipogenesis with External Factors 81
References 81

T. J. Larsen · N. Z. Jespersen
The Centre of Inflammation and Metabolism and Centre for Physical Activity Research
Rigshospitalet, University Hospital of Copenhagen, Copenhagen, Denmark

Danish Diabetes Academy, Odense University Hospital, Odense, Denmark

Faculty of Health and Medical Sciences, University of Copenhagen, Copenhagen, Denmark

C. Scheele (✉)
Novo Nordisk Foundation Center for Basic Metabolic Research, University of Copenhagen, Copenhagen, Denmark

The Centre of Inflammation and Metabolism and Centre for Physical Activity Research
Rigshospitalet, University Hospital of Copenhagen, Copenhagen, Denmark
e-mail: cs@sund.ku.dk

A. Pfeifer et al. (eds.), *Brown Adipose Tissue*,
Handbook of Experimental Pharmacology 251, https://doi.org/10.1007/164_2018_142

Abstract

Obesity involves a contrasting expansion of the energy-storing white fat and loss of functionally competent brown fat, an energy-consuming thermogenic adipose. Leveraging our understanding of white and brown adipocyte recruitment and investigating factors that regulate these processes might reveal novel targets for counteracting obesity. In vitro differentiation of primary preadipocytes mimics many of the morphological and transcriptional events occurring during adipogenesis in vivo. Moreover, preadipocytes isolated from a specific depot maintain features of their originating niche. This makes in vitro adipogenesis a valuable model for identifying differential regulation patterns between brown and white adipogenesis. In this chapter, we describe step-by-step how to isolate brown and white preadipocytes from human tissue biopsies and how to culture and differentiate them in vitro. We discuss this process, what to consider, and how this in vitro system can be used to model in vivo adipogenesis.

Keywords

Adipogenesis · Brown fat · Human brown adipocytes · In vitro differentiation · Obesity · White fat

1 Introduction

Primary adipogenesis in vitro mimics the in vivo phenomenon, in terms of cell growth arrest and remodeling, with a subsequent lipid accumulation (Cornelius et al. 1994; Stacey et al. 2009). In vitro differentiation of preadipocytes isolated from human tissue biopsies or mouse fat depots is a well-established model for adipogenesis and adipose development and is heavily studied with the anticipation of identifying mechanisms controlling adipose expansion and thereby identifying key elements that could counteract obesity. During the recent decade, two key findings have unraveled an exciting novel avenue in adipose and obesity research. Firstly, it was discovered that white fat can acquire a more brown-like phenotype known as beige (Seale et al. 2011) or brite fat (Petrovic et al. 2010) in response to cold or to stimulation with a peroxisome proliferator-activated receptor γ (PPARγ) agonist. Secondly, it was found that functional brown fat exist in adult humans (Cypess et al. 2009; Van Marken Lichtenbelt et al. 2009; Virtanen et al. 2009; Zingaretti et al. 2009; Saito et al. 2009) and that preadipocytes can be isolated from adult human brown fat and cultured and differentiated in vitro while maintaining key features of brown fat characteristic (Jespersen et al. 2013). As brown fat is energy-consuming, identifying mechanisms that enhance brown fat biogenesis and perhaps even induce a brite/beige phenotype in white fat is an attractive research field. Potential mechanisms and regulators of adipose-type-specific differentiation can be studied in vitro, by isolating preadipocytes and following them throughout differentiation into mature lipid droplet-containing cells. Importantly, isolated preadipocytes retain an intrinsic memory of the adipose type of their origin, i.e., white or brown adipose, despite being exposed to identical culturing conditions

(Néchad et al. 1983; Jespersen et al. 2013), and thus represent a model for studies of adipose-type-specific adipogenesis (Hiraike et al. 2017) as well as an in vitro model for investigating adipose-type-specific mechanisms (Gnad et al. 2014; Mottillo et al. 2016; Sanchez-Gurmaches et al. 2018). When investigating basic cellular mechanisms, it can be useful to select one cell strain and just repeat the experiment for technical replicates as would be the procedure with a cell line. However, in many cases it will be preferable to include some biological variation in the experiment and instead study cells derived from several individuals. When using this approach, we would include cell cultures from at least five different donors in our experiments. Notably, we have observed great variability in cell differentiation capacity between cultures (unpublished data). Therefore it is usually necessary to prepare a larger amount of cell cultures than needed for the actual experiment. For example, we would recommend to perform ten cell isolations to ensure obtaining at least five brown adipocyte cultures that differentiate well, as defined by formation of lipid droplets covering at least 40% of the surface in the culture dish.

2 Biopsy

In this chapter, we will focus on human primary cell cultures prepared through enzymatic digestion of fat biopsies. This approach is based on the standardized protocols for isolating precursor cells from fat depots in rat and mice (Néchad et al. 1983). Subcutaneous fat biopsies from, e.g., the abdominal region of adult humans can be obtained using the Bergström approach with a biopsy needle connected to suction (Bergstrom 1975), while deep neck fat biopsies from the supraclavicular region, representing human brown fat, are easiest accessible during a surgical procedure (Jespersen et al. 2013). A drawing based on microscopy pictures taken at different stages during differentiation demonstrates the progressive morphological changes from proliferating preadipocyte to mature, lipid droplet-containing adipocyte (Fig. 1). Biopsies are immediately stored in ice cold DMEM/F-12 (Gibco) + 1% penicillin-streptomycin (P/S) (10,000 U/mL Gibco) until initiation of the isolation. A few hours of incubation is acceptable but should be kept consistent between isolations. Biopsies are digested in a buffer containing 1 mg/ml collagenase II (Sigma-Aldrich) and fatty acid-free bovine serum albumin (15 mg/ml) (Sigma-Aldrich) during gentle shaking at 37°C for 20 min. This suspension is filtered through a cell strainer (size 70 μm), for removal of cell debris. Following 5 min of incubation to allow for the fat to be settled on the surface, the layer below the fat is collected and filtered (size 30 μm), this time to separate the cells before the cells are spun down for 7 min at 800 g. The supernatant is discarded, and cells are resuspended in DMEM/F-12 + 1% P/S and spun again at 7 min at 800 g. The supernatant is again discarded, and the pellet is resuspended in growth medium [DMEM/F-12, 1% P/S, and 10% fetal bovine serum (FBS) (Gibco)] and plated in a T25 culture flask. Growth medium is changed the following day and then every other day until the fat precursor cells reach 80% confluence. Cells are then split into a 100 mm dish, using TrypLE™ Express Enzyme (Life Technologies). When

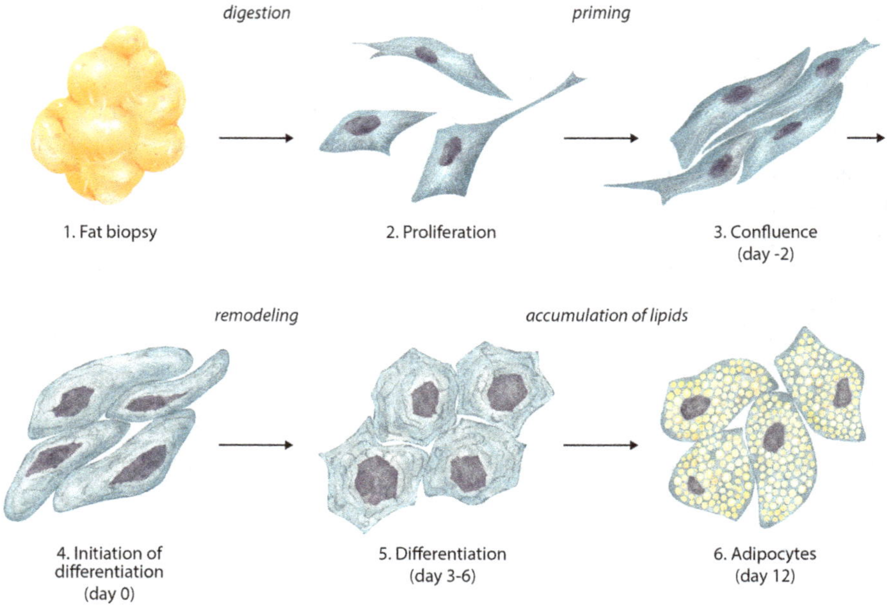

Fig. 1 In vitro adipogenesis. Preadipocytes are isolated from fat biopsies through collagenase digestion. Liberated preadipocytes are expanded in a culture dish and are primed for adipogenic commitment during proliferation by adding FGF-1 to the culture media. Cells are grown to confluence which makes them line up in a more organized manner. Following two days of confluence, a remodeling of the cytoskeleton and extracellular matrix can be observed, and here we add the differentiation media. Further remodeling occurs, and the cytoskeleton appears to form a net-like structure before lipid droplets finally are accumulated by the end of the differentiation

cells again reach 80% confluence, they are frozen in three aliquots in 10% dimethyl sulfoxide + 90% FBS and preserved in cryotubes in liquid nitrogen until thawed and expanded for cell experiments.

3 Proliferation

3.1 Priming/Adipocyte Commitment

Cryopreserved preadipocytes are thawed and cultured in DMEM/F-12, 1% P/S, and 10% FBS. The timeline for the in vitro differentiation, including specifications of the compounds added to the media, is provided in Fig. 2. During the proliferative stage, fibroblast growth factor 1 (FGF-1), provided by the microvasculature in vivo, is included for enhancing cell proliferation (Mori et al. 2008) and for promoting adipogenic commitment (Hutley et al. 2004). Although still a matter of investigation, transcription factors important at this early stage include the brown fat directing transcription factor early b cell factor 2 (EBF2) (Rajakumari et al. 2013). Extensive

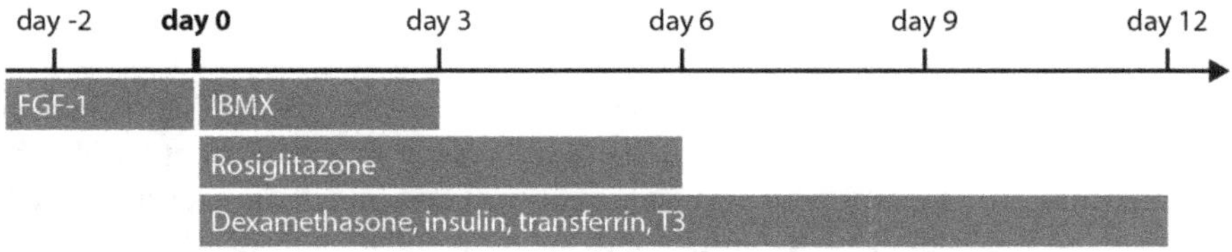

Fig. 2 Differentiation time axis. Proliferating preadipocytes are refreshed with new media every second day and are cultured with FGF-1 added in the media until two days following confluence. Here, the differentiation media is added which is changed every third day. IBMX is kept in the media the first 3 days only, while Rosiglitazone is removed after the first 6 days. Dexamethasone, insulin, transferrin and T3 are included in the differentiation media throughout the 12-day differentiation protocol

descriptions of the transcriptional regulation during brown and white adipogenesis has recently been provided (Loft et al. 2017; Mota de Sá et al. 2017).

3.2 Flow Cytometry for Evaluation of Cell Culture Purity

A consideration in relation to working with primary cell cultures is the composition and heterogeneity of the cell cultures. Primary precursor cells are isolated from an adipose tissue biopsy, and all the cell types constituting the stromal vascular fraction besides fat precursor cells (also called adipocyte progenitor cells or preadipocytes), e.g., endothelial cells, fibroblasts, neurons, and various types of immune cell such as macrophages, might potentially contaminate the cell cultures and propagate along with the subcultured fat precursor cells. When working with primary adipocyte cultures, we have experienced that the individual cell cultures from distinct donors differentiated and accumulated fat droplets with different efficiencies, which could relate to different levels of priming and cell determination but could alternatively indicate that non-adipogenic cell types were present in the cultures. Flow cytometry analysis is an efficient way to evaluate the purity of isolated cell populations. Cells with a high adipogenic potential are characterized as positive for the cell surface markers CD90 and CD166, both mesenchymal stem cell markers, and lacking the cell surface markers CD31 (epithelial cell marker) and CD45 (hematopoietic cell marker) (Schultz et al. 2014). Both fibroblasts and adipocytes arise from mesenchymal stem cells, and as there is no good marker for fibroblasts which is not also present on preadipocytes, it is difficult to exclude a fibroblast contamination of the cell cultures with flow cytometry. However, differentiation capacity and subsequent functional evaluation of the cell cultures will also illustrate the adipogenic status of the cell cultures.

4 Confluence

Growing the preadipocytes into confluence initiate a cellular remodeling including a reorganization of the nuclear lamina and cytoskeleton (Verstraeten et al. 2011). Upon reaching confluence, cells are incubated for additionally 2 days, ensuring cell growth arrest and optimizing the response to the differentiation media. During these 2 days, visible changes in the cell morphology can be observed (Fig. 1).

5 Initiation of Differentiation

Adipogenic differentiation takes place via a sequential transcription cascade, during which various transcription factors activate and repress each other (Lefterova and Lazar 2009; Siersbæk et al. 2012). In this phase, the master regulator of adipogenesis, PPARγ, is induced. Other key adipogenic transcription factors include CCAAT-enhancer binding protein α (C/EBPα) and C/EBPβ. These factors are important for both brown and white adipogenesis.

Two days following 100% confluency, differentiation is induced with the addition of a differentiation cocktail (Table 1). In the literature, most protocols, including ours, are relatively similar in the composition of adipogenic differentiation promoting factors and consist of insulin, dexamethasone, and the cyclic adenosine monophosphate (cAMP)-increasing agent 3-isobutyl-1-methylxanthine (IBMX) (Scott et al. 2011; Park et al. 2017). Doses however vary substantially, especially between mouse and human protocols (Scott et al. 2011). Our differentiation media also contain the thyroid hormone triiodothyronine (T3), transferrin, and the PPARγ agonist, rosiglitazone. Two key events are kick-starting the early differentiation:

- Increased intracellular cAMP levels mediated by the phosphodiesterase inhibitor, IBMX
- Stimulation of the glucocorticoid receptor by dexamethasone

The rise in cAMP activates the protein kinase A (PKA) signaling pathway and leads to cAMP response element-binding protein (CREB) phosphorylation and activation, followed by increased transcription of adipogenic transcription factors such as C/EBPβ (Zhang et al. 2004; Cristancho and Lazar 2011; Siersbæk et al. 2012) and decreased expression of anti-adipogenic factors such as Wnt10b, the major Wnt signaling ligand in adipocytes (Fox et al. 2008; Cristancho and Lazar 2011; Cawthorn et al. 2012).

Activation of the glucocorticoid receptor increases both C/EBPβ and C/EBPδ, and their transient expression is central for the subsequent expression of the central transcription factors of adipogenesis: C/EBPα and PPARγ (Rosen and MacDougald 2006). The key role of PPARγ in adipogenesis is illustrated by the fact that almost all other pro-adipogenic factors, including C/EBPβ and C/EBPδ, at least partly exert their function by interacting with PPARγ and/or C/EBPα. Conversely,

Table 1 Compounds added to the differentiation media

Compound	Concentration	Role in differentiation
Dexamethasone	100 nM	A corticosteroid that acts on the glucocorticoid receptor. Early potent inductor of adipogenesis
3-Isobutyl-1-methylxanthine (IBMX)	540 μM	Phosphodiesterase inhibitor which increase intracellular cAMP levels. Stimulates the PKA signaling pathway which is required for transcriptional activation of PPARγ
Insulin	100 nM	Activates PPARγ. Induces lipogenesis. Essential for lipid accumulation and terminal differentiation for both white and brown adipocytes. Increases the number of cells which develop into mature brown and white adipocytes in a given culture (Klaus et al. 1995)
Rosiglitazone	200 nM	Activates PPARγ directly. Insulin sensitizing effect of adipocyte (Scott et al. 2011). Also induces browning, partly through PRDM16 stabilization (Ohno et al. 2012)
Transferrin	10 μg/ml	Transports iron, which is essential in many cellular processes, into cells. Also possesses iron-independent growth factor properties and modulate growth and differentiation across many cell types (Gomme and McCann 2005). Iron deficiency as well as excess impairs lipid accumulation and decreases expression of adipogenic genes during adipocyte differentiation both in 3T3-L1 cells and primary human preadipocytes (Moreno-Navarrete et al. 2014)
3,3′,5-Triiodo-L-thyronine sodium salt (T3)	2 nM	Stimulates adipogenesis by regulating transcription factors and genes involved in lipogenesis, thermogenesis, and mitochondrial function. The active form of the thyroid hormone. Binds to the nuclear thyroid hormone receptor, which have been shown to modulate adipogenesis via cross talk with PPARγ in WAT (Lu and Cheng 2010)

anti-adipogenic factors act largely by repressing PPARγ expression or activity (Rosen and MacDougald 2006).

6 Differentiation

During the last part of the differentiation, lipid droplets are accumulating and can be visually observed (Fig. 1). While in vivo, brown and white adipocytes can be distinguished by their unilocular appearance in white and multilocular phenotype in brown adipose tissue, this is not visible in most cell culture models where both white and brown adipocytes display a multilocular phenotype (Jespersen et al. 2013). We have observed that early events as priming with FGF-1 and incubating with the PPARγ agonist, rosiglitazone, the first 6 days of the protocol (Fig. 2) is crucial for maximal lipid accumulation (unpublished data). The expression of PPARγ during later stages of adipogenesis is autoregulated by its own robust induction and through interaction with C/EBPα, C/EBPβ, and retinoid x receptors (RXR) to maintain the terminally differentiated adipocyte stage (Cristancho and Lazar 2011).

7 Validation of Cell Cultures

To validate the brown adipocyte phenotype, we stimulate the cells with norepinephrine. For gene expression analysis, cells are "starved" prior to the stimulation, in fresh cell culture media (DMEM/F12) with P/S, but without any other supplements. Following 2 h of starvation, 1 μM NE is added into the wells (without changing media). Following 4 h of incubation, RNA is harvested using Trizol, and cDNA synthesis is performed. Genes that we observe to be higher expressed in brown adipocytes and induced upon NE stimulation include UCP1, PPARGC1A, and DIO2. Genes that are not necessarily induced but are higher expressed in brown adipocytes compared to white (isolated from subcutaneous fat in the abdominal region) include EBF2, PRDM16, and CIDEA. In the white adipocytes, HOXC8 and HOXC9 are higher expressed compared to brown adipocytes (Jespersen et al. 2013). The murine orthologue, Hoxc8, has been shown to inhibit brown adipogenesis of white fat progenitor cells by repressing C/EBPβ (Mori et al. 2012). To evaluate the functional brown fat phenotype, we utilize the Seahorse Bioscience XF96 platform, to access NE-induced oxygen consumption. We specifically use the Seahorse XF Cell Mito Stress Test Kit (Agilent) and inject 1 μM NE through the injection port of the instrument, resulting in ~twofold increase in mitochondrial oxygen consumption rate (unpublished data).

Some of the brown adipocyte cultures will not respond to NE in terms of gene expression or oxygen consumption despite differentiating into lipid droplet-containing adipocytes. Whether these cells represent white adipocytes is something we currently investigate using global gene expression analysis at single-cell level across adipose depots. Among the primary white adipocytes from the subcutaneous fat depot, the expression of UCP1 is low, and we have not observed a consistent beige/brite phenotype. Although markers for beige/brite adipocytes have been identified in murine models (Walden et al. 2012; Wu et al. 2012), the picture is not clear-cut in human primary cell cultures as marker genes for "classical brown" and "beige/brite" are overlapping in human supraclavicular brown fat (Jespersen et al. 2013; Scheele et al. 2014). Importantly, in humans, the visceral white fat might be a more relevant depot to investigate for identifying beige/brite adipocytes as the abdominal visceral adipose can acquire a brown fat phenotype in vivo in humans during conditions of highly elevated levels of local and circulating levels of NE (Frontini et al. 2013; Søndergaard et al. 2015; Scheele and Nielsen 2017).

8 Considerations

8.1 Cell Passaging

Non-immortalized cells have limited propagation and passage capacity (Zhao et al. 2012). Repetitive passaging may alter the cell phenotype and inflict selection on a population of cells. Thus, it is desirable to use fat precursor cell strains in as low a passage as possible and only compare cultures in roughly the same passage number. It is furthermore important to understand that the passage number is an unprecise estimate of cell doublings as biopsy sizes will determine the starting number of cells.

8.2 Differentiation With or Without Serum

Adding serum to the cell culture media can provide multiple factors affecting the adipogenesis. However, for the approach described in the current chapter, we utilize serum-free culture media during the differentiation and instead add compounds with known enhancing effects on adipogenesis. With this approach, the method can be kept consistent across serum batches and should be more easily reproducible between different laboratories.

8.3 Inducing Brown Adipogenesis with External Factors

As we have discussed in this chapter, brown and white adipocytes retain a memory of their originating niche and will display representative phenotypes upon differentiation. However, adipocytes are flexible and as also mentioned, white preadipocytes can acquire a brown-like phenotype upon high levels of an externally added PPARγ agonist, e.g., rosiglitazone (Petrovic et al. 2010). Since this discovery a large number of factors have been identified to be involved in adipocyte "browning." Interestingly, many browning factors are produced by the brown adipocytes themselves and are known as "batokines" (Villarroya et al. 2016). An extensive mapping of these batokines and their mechanism of action will provide increased insight in how to switch between brown and white adipose during adipogenesis.

Acknowledgments The Center for Physical Activity Research (CFAS), Rigshospitalet, is supported by a grant from TrygFonden. The Novo Nordisk Foundation Center for Basic Metabolic Research (http://www.metabol.ku.dk) is supported by an unconditional grant from the Novo Nordisk Foundation to University of Copenhagen.

References

Bergstrom J (1975) Percutaneous needle biopsy of skeletal muscle in physiological and clinical research. Scand J Clin Lab Invest 35:609–616

Cawthorn WP, Bree AJ, Yao Y et al (2012) Wnt6, Wnt10a and Wnt10b inhibit adipogenesis and stimulate osteoblastogenesis through a β-catenin-dependent mechanism. Bone 50:477–489. https://doi.org/10.1016/j.bone.2011.08.010

Cornelius P, MacDougald OA, Lane MD (1994) Regulation of adipocyte development. Annu Rev Nutr 14:99–129. https://doi.org/10.1146/annurev.nu.14.070194.000531

Cristancho AG, Lazar MA (2011) Forming functional fat: a growing understanding of adipocyte differentiation. Nat Rev Mol Cell Biol 12:722–734. https://doi.org/10.1038/nrm3198

Cypess AM, Lehman S, Williams G et al (2009) Identification and importance of brown adipose tissue in adult humans. N Engl J Med 360:1509–1517. https://doi.org/10.1056/NEJMoa0810780

Fox KE, Colton LA, Erickson PF et al (2008) Regulation of cyclin D1 and Wnt10b gene expression by cAMP-responsive element-binding protein during early adipogenesis involves differential promoter methylation. J Biol Chem 283:35096–35105. https://doi.org/10.1074/jbc.M806423200

Frontini A, Vitali A, Perugini J et al (2013) White-to-brown transdifferentiation of omental adipocytes in patients affected by pheochromocytoma. Biochim Biophys Acta 1831:950–959. https://doi.org/10.1016/j.bbalip.2013.02.005

Gnad T, Scheibler S, von Kügelgen I et al (2014) Adenosine activates brown adipose tissue and recruits beige adipocytes via A2A receptors. Nature 516:395–399. https://doi.org/10.1038/nature13816

Gomme PT, McCann KB (2005) Transferrin: structure, function and potential therapeutic actions. Drug Discov Today 10:267–273. https://doi.org/10.1016/S1359-6446(04)03333-1

Hiraike Y, Waki H, Yu J et al (2017) NFIA co-localizes with PPARγ and transcriptionally controls the brown fat gene program. Nat Cell Biol 19:1081–1092. https://doi.org/10.1038/ncb3590

Hutley L, Shurety W, Newell F et al (2004) Fibroblast growth factor 1: a key regulator of human adipogenesis. Diabetes 53:3097–3106. https://doi.org/10.2337/diabetes.53.12.3097

Jespersen NZ, Larsen TJ, Peijs L et al (2013) A classical brown adipose tissue mrna signature partly overlaps with brite in the supraclavicular region of adult humans. Cell Metab 17:798–805. https://doi.org/10.1016/j.cmet.2013.04.011

Klaus S, Ely M, Encke D, Heldmaier G (1995) Functional assessment of white and brown adipocyte development and energy metabolism in cell culture. Dissociation of terminal differentiation and thermogenesis in brown adipocytes. J Cell Sci 108(Pt 1):3171–3180

Lefterova MI, Lazar MA (2009) New developments in adipogenesis. Trends Endocrinol Metab 20:107–114. https://doi.org/10.1016/j.tem.2008.11.005

Loft A, Forss I, Mandrup S (2017) Genome-wide insights into the development and function of thermogenic adipocytes. Trends Endocrinol Metab 28:104–120. https://doi.org/10.1016/j.tem.2016.11.005

Lu C, Cheng S-Y (2010) Thyroid hormone receptors regulate adipogenesis and carcinogenesis via crosstalk signaling with peroxisome proliferator-activated receptors. J Mol Endocrinol 44:143–154. https://doi.org/10.1677/JME-09-0107

Moreno-Navarrete JM, Ortega F, Moreno M et al (2014) Fine-tuned iron availability is essential to achieve optimal adipocyte differentiation and mitochondrial biogenesis. Diabetologia 57:1957–1967. https://doi.org/10.1007/s00125-014-3298-5

Mori M, Nakagami H, Rodriguez-Araujo G et al (2012) Essential role for miR-196a in brown adipogenesis of white fat progenitor cells. PLoS Biol 10:e1001314. https://doi.org/10.1371/journal.pbio.1001314

Mori S, Wu CY, Yamaji S et al (2008) Direct binding of integrin αvβ3 to FGF1 plays a role in FGF1 signaling. J Biol Chem 283:18066–18075. https://doi.org/10.1074/jbc.M801213200

Mota de Sá P, Richard AJ, Hardy H, Stephens JM (2017) Transcriptional regulation of adipogenesis. Compr Physiol 7:635–674

Mottillo EP, Desjardins EM, Crane JD et al (2016) Lack of adipocyte AMPK exacerbates insulin resistance and hepatic steatosis through brown and beige adipose tissue function. Cell Metab 24:118–129. https://doi.org/10.1016/j.cmet.2016.06.006

Néchad M, Kuusela P, Carneheim C et al (1983) Development of brown fat cells in monolayer culture. I. Morphological and biochemical distinction from white fat cells in culture. Exp Cell Res 149:105–118. https://doi.org/10.1016/0014-4827(83)90384-1
Ohno H, Shinoda K, Spiegelman BM, Kajimura S (2012) PPARγ agonists induce a white-to-brown fat conversion through stabilization of PRDM16 protein. Cell Metab 15:395–404. https://doi.org/10.1016/j.cmet.2012.01.019
Park Y, Wang L, Giampietro A et al (2017) Distinct roles of transcription factors KLF4, Krox20, and peroxisome proliferator-activated receptor γ in adipogenesis. Mol Cell Biol 37:1–11. https://doi.org/10.1128/MCB.00554-16
Petrovic N, Walden TB, Shabalina IG et al (2010) Chronic peroxisome proliferator-activated receptor gamma (PPARG) activation of epididymally derived white adipocyte cultures reveals a population of thermogenically competent, UCP1-containing adipocytes molecularly distinct from classic brown adipocytes. J Biol Chem 285:7153–7164. https://doi.org/10.1074/jbc.M109.053942
Rajakumari S, Wu J, Ishibashi J et al (2013) EBF2 determines and maintains brown adipocyte identity. Cell Metab 17:562–574. https://doi.org/10.1016/j.cmet.2013.01.015
Rosen ED, MacDougald OA (2006) Adipocyte differentiation from the inside out. Nat Rev Mol Cell Biol 7:885–896. https://doi.org/10.1038/nrm2066
Saito M, Okamatsu-ogura Y, Matsushita M et al (2009) High incidence of metabolically active brown adipose tissue in healthy adult humans: effects of cold exposure and adiposity. Diabetes 58:1526–1531. https://doi.org/10.2337/db09-0530
Sanchez-Gurmaches J, Tang Y, Jespersen NZN et al (2018) Brown fat AKT2 is a cold-induced kinase that stimulates ChREBP-mediated de novo lipogenesis to optimize fuel storage and thermogenesis. Cell Metab 27:195–209. https://doi.org/10.1016/j.cmet.2017.10.008
Scheele C, Larsen TJ, Nielsen S (2014) Novel nuances of human brown fat. Adipocytes 3:54–57. https://doi.org/10.4161/adip.26520
Scheele C, Nielsen S (2017) Metabolic regulation and the anti-obesity perspectives of human brown fat. Redox Biol 12:770–775. https://doi.org/10.1016/j.redox.2017.04.011
Schultz NS, Broholm C, Gillberg L et al (2014) Impaired leptin gene expression and release in cultured preadipocytes isolated from individuals born with low birth weight. Diabetes 63:111–121. https://doi.org/10.2337/db13-0621
Scott MA, Nguyen VT, Levi B, James AW (2011) Current methods of adipogenic differentiation of mesenchymal stem cells. Stem Cells Dev 20:1793–1804. https://doi.org/10.1089/scd.2011.0040
Seale P, Conroe HM, Estall J et al (2011) Prdm16 determines the thermogenic program of subcutaneous white adipose tissue in mice. J Clin Invest 121:96–105. https://doi.org/10.1172/JCI44271
Siersbæk R, Nielsen R, Mandrup S (2012) Transcriptional networks and chromatin remodeling controlling adipogenesis. Trends Endocrinol Metab 23:56–64. https://doi.org/10.1016/j.tem.2011.10.001
Søndergaard E, Gormsen LC, Christensen MH et al (2015) Chronic adrenergic stimulation induces brown adipose tissue differentiation in visceral adipose tissue. Diabet Med 32:e4–e8. https://doi.org/10.1111/dme.12595
Stacey DH, Hanson SE, Lahvis G et al (2009) In vitro adipogenic differentiation of preadipocytes varies with differentiation stimulus, culture dimensionality, and scaffold composition. Tissue Eng Part A 15:3389–3399. https://doi.org/10.1089/ten.tea.2008.0293
Van Marken Lichtenbelt WD, Vanhommerig JW, Smulders NM et al (2009) Cold-activated brown adipose tissue in healthy men. N Engl J Med 360:1500–1508. https://doi.org/10.1056/NEJMoa0808718
Verstraeten VLRM, Renes J, Ramaekers FCS et al (2011) Reorganization of the nuclear lamina and cytoskeleton in adipogenesis. Histochem Cell Biol 135:251–261. https://doi.org/10.1007/s00418-011-0792-4
Villarroya F, Cereijo R, Villarroya J, Giralt M (2016) Brown adipose tissue as a secretory organ. Nat Rev Endocrinol 13:26–35. https://doi.org/10.1038/nrendo.2016.136

Virtanen KA, Lidell ME, Orava J et al (2009) Functional brown adipose tissue in healthy adults. N Engl J Med 360:1518–1525. https://doi.org/10.1056/NEJMoa0808949

Walden TB, Hansen IR, Timmons JA et al (2012) Recruited vs. nonrecruited molecular signatures of brown, "brite," and white adipose tissues. AJP Endocrinol Metab 302:E19–E31. https://doi.org/10.1152/ajpendo.00249.2011

Wu J, Boström P, Sparks LMM et al (2012) Beige adipocytes are a distinct type of thermogenic fat cell in mouse and human. Cell 150:366–376. https://doi.org/10.1016/j.cell.2012.05.016

Zhang JW, Klemm DJ, Vinson C, Lane MD (2004) Role of CREB in transcriptional regulation of CCAAT/enhancer-binding protein β gene during adipogenesis. J Biol Chem 279:4471–4478. https://doi.org/10.1074/jbc.M311327200

Zhao Y, Waldman SD, Flynn LE (2012) The effect of serial passaging on the proliferation and differentiation of bovine adipose-derived stem cells. Cells Tissues Organs 195:414–427. https://doi.org/10.1159/000329254

Zingaretti MC, Crosta F, Vitali A et al (2009) The presence of UCP1 demonstrates that metabolically active adipose tissue in the neck of adult humans truly represents brown adipose tissue. FASEB J 23:3113–3120. https://doi.org/10.1096/fj.09-133546

In Vitro Models for Study of Brown Adipocyte Biology

Mark Christian

Contents

1 Introduction 86
2 In Vitro Adipocyte Research Tools 86
3 Commonly Used In Vitro Models for Studying Brown Adipocyte Biology 87
  3.1 Characteristics of Brown Adipocytes 87
  3.2 Brown Adipose Tissue Explant Culture 87
  3.3 Primary Culture 88
  3.4 Brown Adipocyte Cell Lines 88
4 Conditionally Immortalised Brown Adipocyte Cell Lines 90
5 Self-Immortalised Adipocyte Cell Lines 91
6 Mouse Embryo Fibroblasts 91
7 Human Multipotent Adipose-Derived Stem Cells 92
8 Pluripotent Stem Cells 93
9 Conclusion 93
References 93

Abstract

Brown adipocytes are the key cell type in brown adipose tissue (BAT) that express the genes required for heat production through the process of thermogenesis. Brown adipocyte cell culture models are important for researching the molecular pathways that control cell autonomous processes. In vitro tools for the study of brown adipocytes include BAT explant cultures and BAT primary cultures that are first proliferated and then differentiated. A number of stable brown preadipocyte cell lines have been generated by the expression transforming factors such as SV40 T antigen. The application of these cell lines reduces the requirement for animal tissue which is needed for primary culture and explants. Furthermore, brown adipocyte cell lines that effectively recapitulate the

M. Christian (✉)
Division of Biomedical Sciences, Warwick Medical School, University of Warwick, Coventry, UK
e-mail: M.Christian@warwick.ac.uk

A. Pfeifer et al. (eds.), *Brown Adipose Tissue*,
Handbook of Experimental Pharmacology 251, https://doi.org/10.1007/164_2018_122

properties of brown adipocytes permit large-scale experimental procedures that are generally unfeasible with primary cultures that undergo a restricted number of cell divisions. Cell lines are valuable for applications such as large-scale endogenous protein expression, ChIP assay, and procedures requiring antibiotic selection over several cell divisions including stable exogenous gene expression and CRISR/Cas9 gene editing.

Keywords
Brown adipocytes · Brown fat · Cell lines

1 Introduction

Brown adipocytes function to maintain body temperature by the expression of a set key of factors that facilitate cellular metabolic events including uncoupling protein 1 (UCP1)-dependent thermogenesis. The dissipation of energy by brown adipocytes is in stark contrast to white adipocytes which serve as the body's primary site of energy storage. In addition to the presence of brown adipocytes, brown adipose tissue (BAT) is a heterogeneous tissue containing fibroblasts, preadipocytes, macrophages (Wolf et al. 2017), and endothelial cells (Tran et al. 2012). The adipose organ is composed of discrete visceral and subcutaneous depots that contain different amounts of brown adipocytes that serve to generate heat by thermogenesis and white adipocytes, specialised cells for energy storage. Adipose tissue is a highly dynamic organ that responds to stimuli from the external and internal environment. For example, upon cold exposure profound changes occur in white adipose tissue (WAT) resulting in the appearance of brown-in-white (BRITE) or beige adipocytes (Wu et al. 2012). These BRITE adipocytes present brown fat features, such as multilocular lipid droplets and inducible UCP1 expression.

As obesity is a major global health issue due to its associated risk for type 2 diabetes, hypertension, cardiovascular disease, and some cancers, it is imperative that researchers identify new approaches to improve metabolic health. Following the identification of functional brown fat deposits in adult humans (Cypess et al. 2009), there has been a recent resurgence in the study of brown adipocyte biology. As BAT contributes to energy expenditure, discovering treatments that promote the recruitment and activation brown fat have great therapeutic potential to combat weight gain. Hence, there is a requirement for experimental tools to study the molecular events that control brown adipocyte function.

2 In Vitro Adipocyte Research Tools

In vitro models are essential for researchers to study the cell autonomous actions of brown adipocytes. Adipose tissue is composed of fully differentiated adipocytes and numerous other cell types. Currently, most investigations of adipocyte function utilise cell lines such as 3T3-L1 cells. Although this research tool has advanced the understanding of fundamental mechanisms underpinning adipocyte differentiation, it cannot fully model the diverse adipose depots found in vivo. The 3T3-L1 cell line is considered a model representative of white adipocytes and is not suitable for the study of brown adipocyte biology. A significant limitation of 3T3-L1 cells, derived from the mouse embryo fibroblast Swiss 3T3-M line, is their aneuploidy which could lead to genes expressed at different levels to that of adipocytes in vivo (Smas and Sul 1995). Primary cultures prepared from anatomically distinct adipose tissue depots are more physiologically relevant adipocyte models but require large numbers of animals to achieve reproducible, statistically significant data.

3 Commonly Used In Vitro Models for Studying Brown Adipocyte Biology

3.1 Characteristics of Brown Adipocytes

Brown adipocytes in vivo are characterised histologically by the presence of multilocular adipocytes and by the expression of a selection of genes that include UCP1 and cell death-inducing DFFA-like effector A (CIDEA) (Christian 2015). White adipocytes in contrast present with a single large (unilocular) lipid droplet that almost fills the cell and low or absent expression of UCP1. In contrast, the morphology of adipocyte lipid droplets in vitro cannot, in general, be used to discriminate brown and white adipocytes. This is evident using the "ceiling culture" procedure whereby dissociated mature adipocytes are placed in culture flasks completely filled with medium. This allows the floating adipocytes to adhere to the top surface after which the flask is inverted and the cells are subject to standard culture procedures. Using this method, it was found that initially the mature white adipocytes retained their unilocular appearance. However, after several days both rodent and human cells became multilocular or assumed a fibroblast-like appearance, containing a large number of small lipid droplets (Sugihara et al. 1986). Further approaches are required to facilitate the in vivo lipid droplet morphologies in an in vitro context. 3D culture in a matrix such as collagen hydrogen (Emont et al. 2015) may assist in providing the correct culture environment as opposed to standard 2D cell culture.

3.2 Brown Adipose Tissue Explant Culture

The culture of whole adipose tissue (explant cultures) has certain distinct advantages over the use of cell culture models. As BAT is heterogeneous, explant cultures provide a valuable method for understanding the physiology and pathophysiology of this tissue. By maintaining the existing cross-talk between different cell types, organ cultures preserve the paracrine signals, allowing the investigation of secreted factors and their regulation in a physiological microenvironment. The procedure to establish explant cultures requires fresh BAT to be cut into small pieces (1 mm) after careful dissection to remove visible vessels and connective tissue. The explants can then be maintained under standard culture conditions with DMEM and 10% fetal bovine serum (FBS). This type of experimental approach has been used to investigate a wide range of aspects of brown adipocyte biology (Gnad et al. 2014; Lee et al. 2016; Pulinilkunnil et al. 2011) and has even been used in conjunction with siRNA transfection to silence gene expression (Puri et al. 2007).

3.3 Primary Culture

Preadipocytes are a major component of the brown fat stromal vascular fraction (SVF). They are easily obtained from BAT and, under the appropriate conditions, can be induced to differentiate into mature adipocytes. Primary brown adipocytes have proved to be an excellent model for studying brown adipocyte biology. The procedure to prepare primary brown preadipocytes requires freshly dissected BAT. In the mouse, this is taken from the interscapular depot alone or pooled with the axillary and cervical BAT depots (de Jong et al. 2015). The tissue is then finely minced before being digested with collagenase to degrade the extracellular matrix. The SVF consists of the cell populations that can be pelleted by centrifugation as opposed to the mature adipocytes that float on the dissociation medium.

Although many labs use subtly different methodologies to generate and differentiate preadipocyte cultures, a comprehensive description of a standard protocol is detailed in Cannon and Nedergaard (2001).

3.4 Brown Adipocyte Cell Lines

Several immortalised human and mouse brown adipocyte cell lines have been generated and used to study brown adipocyte function. These offer certain advantages compared to primary cultures including a reduction in the requirement for freshly isolated BAT. Whereas primary cultures will undergo a restricted number of cell divisions, immortalised cell lines will undergo continuous culture. However, validation is required to ensure that the cells retain the properties that effectively recapitulate the properties of brown adipocytes. Commonly, the expression of the brown fat marker gene Ucp1 is used to validate the cell phenotype. However, caution should be taken when interpreting an in vitro model of brown adipocytes as even cell lines that are considered white in nature, e.g. 3T3-L1 cells have the capacity to express Ucp1 mRNA and respond to adrenergic activators (Miller et al. 2015). Therefore, researchers should consider additional validation of brown adipocyte models such as determining the expression of UCP1 protein.

Brown adipocyte cell lines facilitate assessment of reproducibility and permit experimental procedures requiring continuous culture and expansion of cell numbers that are generally unfeasible with primary cultures. These include applications such as large-scale endogenous protein expression, chromatin immunoprecipitation assay, and applications that require antibiotic selection over several cell divisions including stable exogenous gene/reporter gene expression and CRISPR/Cas9-mediated genetic manipulations. A selection of brown adipocyte cell lines that have been reported and characterised are described below including specific differentiation treatments.

3.4.1 T37i

The T37i immortalised cell line was derived from the hibernoma of a transgenic mouse expressing SV40 T antigen under the control of the human mineralocorticoid receptor promoter (Zennaro et al. 1998). This cell line differentiates when treated with only insulin (20 nM) and triiodothyronine (T3, 2 nM) after reaching confluency.

3.4.2 WT-1

The mouse brown adipocyte cell line WT-1 was generated from the interscapular BAT of newborn FVB mice and immortalised by retrovirus-mediated expression of SV40 T antigen (Klein et al. 1999). For differentiation, the preadipocytes are grown to confluence in culture medium supplemented with 20 nM insulin and 1 nM T3. The confluent cells are then incubated in medium supplemented with insulin (20 nM), T3 (1 nM), isobutylmethylxanthine (IBMX, 0.5 mM), dexamethasone (0.5 μM), and indomethacin (0.125 mM) for 24 h. The cells are then maintained in medium containing only insulin and T3 until fully differentiated (7 days after induction of differentiation).

3.4.3 HIB 1B

HIB 1B is an immortalised cell line established from the hibernoma of a transgenic mouse expressing SV40 transforming genes linked to the adipocyte-specific regulatory region of the adipocyte P2 (aP2) gene (Klaus et al. 1994; Ross et al. 1992). HIB 1B cells are cultured in a mixture of Dulbecco's modified Eagle's medium (DMEM)/Ham's F-12 nutritive medium (1:1, v/v) supplemented with 10% FBS and biotin (0.016 mM). Differentiation is induced by plating the cells in the same medium supplemented with 20 nM insulin and 1 nM T3 and allowing the cells to become confluent. When confluent, cells are treated with medium supplemented with IBMX (0.5 mM), hydrocortisone (0.5 μM), and indomethacin (0.125 mM) (Ross et al. 1992). Three days following the induction of differentiation, the cells are treated with medium supplemented with 20 nM insulin and 1 nM T3 for a further 7 days until cells are fully differentiated.

3.4.4 PAZ6

The PAZ6 cell line is reported to be a human adipocyte cell line that has properties of brown adipocytes, notably UCP1 expression (Guennoun et al. 2015; Kazantzis et al. 2012). It was obtained by immortalising cells of the vascular stromal fraction from human infant BAT through introduction of genes coding for the SV40 T and t antigens under the control of the human vimentin promoter (Zilberfarb et al. 1997). For differentiation, confluent PAZ6 cells (Zilberfarb et al. 1997) are treated with DMEM/Ham's F12 medium (1:1, v/v) supplemented with 7.5% donor calf serum and 2.5% newborn calf serum, IBMX (0.25 mM), dexamethasone (0.1 μM), insulin (850 nM), T3 (1 nM), pioglitazone (1 μM), biotin (33 μM), pantothenate (17 μM), and Hepes (15 mM). After 4 days, this medium is replaced with medium without insulin, dexamethasone, and serum but supplemented with 10 μg/ml transferrin and differentiation continued for 15–21 days. Whether this cell line is appropriate for the study of human brown adipocyte biology requires further validation as the increase in UCP1 following differentiation is low (Zilberfarb et al. 1997). Similarly, the induction in UCP1 after norepinephrine treatment is below that normally reported in murine brown adipocyte cells.

4 Conditionally Immortalised Brown Adipocyte Cell Lines

The immortalisation of cells with at temperature sensitive transforming factor can offer improvements over standard protocols. For example, the thermolabile SV40 T antigen tsA58 provides a means to generate cultures that are immortalised at the permissive temperature of 33°C, whereas at higher temperatures (37°C) the T antigen is ineffective. This has a distinct advantage, in relation to adipocyte study, over the wild-type SV40 T antigen which functions by deactivating proteins such as Retinoblastoma (Rb). Preadipocytes lacking Rb differentiate preferentially into brown adipocytes (Scime et al. 2005) and wild-type SV40 T antigen can even block differentiation of white adipocytes (Cherington et al. 1988). Hence, at the non-permissive temperature (37°C), the tsA58 is unstable and the differentiation process is not affected by the transforming gene. Such conditionally immortalised cells have been generated by the culture of cells prepared from interscapular BAT deposits of the "ImmortoMouse" line (*H-2Kb*-tsA58) (Jat et al. 1991) expressing a gamma interferon-inducible tsA58 (Debevec et al. 2007; Hallberg et al. 2008). Cells were allowed to proliferate in the presence of 2 ng/ml gamma interferon at 33°C and differentiated in the absence of gamma interferon at 37°C. Although this protocol is suitable to generate brown adipocyte cell lines, it is not applicable for most labs as it requires access to the ImmortoMouse line. Furthermore, several crosses may be required to obtain homozygosity for the study of the effects of knockout or transgenic manipulations. An alternative approach used to successfully prepare adipocyte cell lines from BAT as well as subcutaneous, gonadal, and mesenteric WAT is to transduce the primary cultures generated from the fat depots with a retrovirus to express the tsA58 followed by selection with antibiotic (Rosell et al. 2014). The tsA58 expressing cells are cultured under the same condition as those prepared from the ImmortoMouse except that gamma interferon is not included. The advantage of this method is that it can be used to transform primary brown adipocyte cultures prepared from any wild-type or genetically modified mouse strain. For example, it has been successfully used to prepare cell lines from the BAT of the RIP140 knockout mouse (Kiskinis et al. 2014). The differentiation protocol is similar to other brown adipocyte cell lines with confluent cell lines at 37°C treated for 48 h with DMEM/F12 medium supplemented with 10% FBS, insulin (170 nM), T3 (1 nM), dexamethasone (250 nM), IBMX (500 μM), indomethacin (125 μM), biotin (33 μM), and calcium pantothenate (17 μM). After induction, the cells are maintained in the above medium without the inclusion of dexamethasone, IBMX, and indomethacin. Cells are fully differentiated after 8 days.

5 Self-Immortalised Adipocyte Cell Lines

Brown, white, and BRITE adipocyte cell lines have been established by the classical 3T3 immortalisation protocol by serial dilution of cultures from the SVF of BAT and subcutaneous WAT (Wu et al. 2012). This protocol enables selection of clones based upon their capacity to proliferate and, combined with subsequent selection for capacity to differentiate into adipocytes, isolate cell lines for studying adipocyte biology. It was found that, when differentiated, distinct cell lines derived from subcutaneous WAT displayed properties of BRITE or white adipocytes. The clonal lines were differentiated with a standard protocol in which confluent cells were induced to differentiate in DMEM/F12 supplemented with 10% FBS, dexamethasone (5 μM), insulin (0.5 μg/ml), IBMX (0.5, mM), rosiglitazone (1 μM), and T3 (1 nM). After 4 days of induction treatment, cells were maintained in media containing insulin (0.5 μg/ml), T3 (1 nM), and 10% FBS until they were fully differentiated. As preadipocytes are likely to undergo a limited number of cell divisions, the cell lines generated could potentially be of fibroblast origin akin to 3T3-L1 adipocytes.

6 Mouse Embryo Fibroblasts

Mouse embryonic fibroblasts (MEFs) are a useful tool to investigate the genetic control of differences between brown and white adipocytes. They are generated from multipotent cells of early mouse embryos, generally at embryonic day 13. Primary MEFs are easy to establish and culture, and proliferate rapidly to produce large numbers of cells from a single embryo. The majority of MEF cultures will be restricted in their ability to proliferate and tend to reach senescence after around 12 passages. However, MEFs can be immortalised by serial passaging or transformation with SV40 large T antigen. Due to the heterogeneity of embryonic tissue, there is variation in the ability of MEF cultures to differentiate into adipocytes of between 10 and 70% (Rosen and MacDougald 2006). Adipogenesis of primary MEFs can be induced by a standard protocol of treating 2-day post-confluent cells with insulin (170 nM), IBMX (0.5 mM), dexamethasone (250 nM), and rosiglitazone (2.5 nM) in DMEM/F12 supplemented with 10% FBS (Leonardsson et al. 2004; Van De Pette et al. 2016). After 48 h, the cells are maintained in medium containing only insulin and rosiglitazone for an additional 6 days.

Studies with MEFs prepared from genetically modified mice have revealed important information on the factors that regulate transition to the BRITE phenotype. MEFs prepared from embryos lacking the nuclear receptor corepressor RIP140 present a brown/BRITE pattern of gene expression with elevated levels of Ucp1 and Cidea compared to wild-type cells (Christian et al. 2005). MEFs prepared from embryos lacking the pocket protein Rb differentiate into brown/BRITE adipocytes (Hansen et al. 2004). The imprinted gene Cdkn1c has been found to have a key role in the brown adipocyte lineage (Van De Pette et al. 2016). In MEFs generated from elegant mouse genetic models that introduce additional Cdkn1c gene copies, it was found that loss of imprinting greatly enhanced brown fat gene expression. Hence, MEF cell lines are a very useful tool to study the molecular control of brown, BRITE, and white adipogenesis.

7 Human Multipotent Adipose-Derived Stem Cells

Human multipotent adipose-derived stem (hMADS) cells are mesenchymal stem cells derived from human adipose tissue of young donors which exhibit self-renewal ability, the absence of tumorigenicity, and the ability to differentiate into a range of different cell types (Elabd et al. 2007; Rodriguez et al. 2005; Zaragosi et al. 2006). In the literature, there is some confusion around the nomenclature of adipose tissue-derived multipotent stem cells. Historically, the stromal cells isolated from adipose tissue were termed preadipocytes. More recently, it has become appreciated that the cellular population that adhere to plastic include multipotent mesenchymal stem cells, preadipocytes, vascular cells (pericytes and endothelial progenitors), fibroblasts, smooth muscle cells, immune cells, and blood cells. The mesenchymal stem cell population have the capacity of differentiating towards the adipocyte, chondrocyte, neuronal, and osteocyte lines. Furthermore, hMADS cells can be converted into functional brown-like adipocytes (Elabd et al. 2009). These cells have been prepared from adipose tissue, of donors 1 month to 7 years old, by digestion with collagenase. After centrifugation, the mature adipocytes and fat are removed and the pelleted SVF is resuspended. The hMADS cells are isolated following their adherence to uncoated tissue culture plastic in low glucose DMEM supplemented with 10% FBS and culture for 160–200 population doublings (Rodriguez et al. 2005). The cells are maintained in 2.5 ng/ml hFGF-2 (Pisani et al. 2011). For differentiation to adipocytes, confluent hMADS cells are cultured in DMEM/Ham's F12 media supplemented with transferrin (10 μg/ml), insulin (0.85 μM), T3 (0.2 nM), DEX (1 μM), and IBMX (500 μM) (Pisani et al. 2011). After 3 days, the medium is changed (with DEX and IBMX omitted). Cells are fully differentiated after 15–16 days. By modulating the exposure of the differentiating cells to rosiglitazone (100 nM) during days 3–16, the cells can preferentially become white or UCP1-expressing brown/BRITE adipocytes. White and BRITE adipocytes can be generated by treatment between days 3–9 and 3–16, respectively (Pisani et al. 2011).

8 Pluripotent Stem Cells

As embryonic stem cells are pluripotent, they can be differentiated into adipocytes and therefore are a valuable research tool to investigate the process of adipogenesis and adipocyte biology. Adipocytes have been induced from mouse and human embryonic stem cells as well as induced pluripotent stem cells (iPSCs). To obtain adipocytes from mouse embryonic stem cells (mESCs), embryoid bodies are prepared in hanging drops containing 10^3 cells cultured for 2 days (Dani et al. 1997). The embryoid bodies are then transferred to a culture plate and maintained for 3 days in suspension, in the presence of all-trans retinoic acid (10^{-8} M). Subsequently, the embryoid bodies are cultured for a further 2 days in suspension in the absence of retinoic acid before being allowed to attach to gelatin-coated plates and provided with differentiation treatment (85 nM insulin, 2 nM T3, and 10% FBS). Differentiation is assessed at day 27.

By expressing the molecular regulators of adipogenesis and "browning" the cells can be directed to a brown phenotype. PPARγ in combination with PRDM16 and C/EBPβ human pluripotent stem cells have been programmed to differentiate into brown adipocytes (Ahfeldt et al. 2012).

9 Conclusion

In vitro experimental models of brown adipocytes are a key resource for researching the biological features of BAT. Although the use of primary cultures is an important and useful approach, generation of immortalised cell lines more easily permits the assessment of experiment reproducibility and reduces the requirement for fresh tissue. Furthermore, cell lines can be used for procedures such as CRISPR/Cas9 genetic engineering which require antibiotic selection and is generally unsuited to primary preadipocyte cultures that will undergo a restricted number of cell divisions. The application of brown adipocyte cell lines provides a means to dissect the molecular pathways that control the biological processes, such as thermogenesis, that are central to BAT function.

References

Ahfeldt T, Schinzel RT, Lee YK, Hendrickson D, Kaplan A, Lum DH, Camahort R, Xia F, Shay J, Rhee EP, Clish CB, Deo RC, Shen T, Lau FH, Cowley A, Mowrer G, Al-Siddiqi H, Nahrendorf M, Musunuru K, Gerszten RE, Rinn JL, Cowan CA (2012) Programming human pluripotent stem cells into white and brown adipocytes. Nat Cell Biol 14:209–219. https://doi.org/10.1038/ncb2411

Cannon B, Nedergaard J (2001) Cultures of adipose precursor cells from brown adipose tissue and of clonal brown-adipocyte-like cell lines. Methods Mol Biol 155:213–224. https://doi.org/10.1385/1-59259-231-7:213

Cherington V, Brown M, Paucha E, St Louis J, Spiegelman BM, Roberts TM (1988) Separation of simian virus 40 large-T-antigen-transforming and origin-binding functions from the ability to block differentiation. Mol Cell Biol 8:1380–1384

Christian M (2015) Transcriptional fingerprinting of "browning" white fat identifies NRG4 as a novel adipokine. Adipocytes 4:50–54. https://doi.org/10.4161/adip.29853

Christian M, Kiskinis E, Debevec D, Leonardsson G, White R, Parker MG (2005) RIP140-targeted repression of gene expression in adipocytes. Mol Cell Biol 25:9383–9391. https://doi.org/10.1128/MCB.25.21.9383-9391.2005

Cypess AM, Lehman S, Williams G, Tal I, Rodman D, Goldfine AB, Kuo FC, Palmer EL, Tseng YH, Doria A, Kolodny GM, Kahn CR (2009) Identification and importance of brown adipose tissue in adult humans. N Engl J Med 360:1509–1517. https://doi.org/10.1056/NEJMoa0810780

Dani C, Smith AG, Dessolin S, Leroy P, Staccini L, Villageois P, Darimont C, Ailhaud G (1997) Differentiation of embryonic stem cells into adipocytes in vitro. J Cell Sci 110 (Pt 11):1279–1285

de Jong JM, Larsson O, Cannon B, Nedergaard J (2015) A stringent validation of mouse adipose tissue identity markers. Am J Physiol Endocrinol Metab 308:E1085–E1105. https://doi.org/10.1152/ajpendo.00023.2015

Debevec D, Christian M, Morganstein D, Seth A, Herzog B, Parker M, White R (2007) Receptor interacting protein 140 regulates expression of uncoupling protein 1 in adipocytes through specific peroxisome proliferator activated receptor isoforms and estrogen-related receptor alpha. Mol Endocrinol 21:1581–1592. https://doi.org/10.1210/me.2007-0103

Elabd C, Chiellini C, Massoudi A, Cochet O, Zaragosi LE, Trojani C, Michiels JF, Weiss P, Carle G, Rochet N, Dechesne CA, Ailhaud G, Dani C, Amri EZ (2007) Human adipose tissue-derived multipotent stem cells differentiate in vitro and in vivo into osteocyte-like cells. Biochem Biophys Res Commun 361:342–348. https://doi.org/10.1016/j.bbrc.2007.06.180

Elabd C, Chiellini C, Carmona M, Galitzky J, Cochet O, Petersen R, Penicaud L, Kristiansen K, Bouloumie A, Casteilla L, Dani C, Ailhaud G, Amri EZ (2009) Human multipotent adipose-derived stem cells differentiate into functional brown adipocytes. Stem Cells 27:2753–2760. https://doi.org/10.1002/stem.200

Emont MP, Yu H, Jun H, Hong X, Maganti N, Stegemann JP, Wu J (2015) Using a 3D culture system to differentiate visceral adipocytes in vitro. Endocrinology 156:4761–4768. https://doi.org/10.1210/en.2015-1567

Gnad T, Scheibler S, von Kugelgen I, Scheele C, Kilic A, Glode A, Hoffmann LS, Reverte-Salisa L, Horn P, Mutlu S, El-Tayeb A, Kranz M, Deuther-Conrad W, Brust P, Lidell ME, Betz MJ, Enerback S, Schrader J, Yegutkin GG, Muller CE, Pfeifer A (2014) Adenosine activates brown adipose tissue and recruits beige adipocytes via A2A receptors. Nature 516:395–399. https://doi.org/10.1038/nature13816

Guennoun A, Kazantzis M, Thomas R, Wabitsch M, Tews D, Seetharama Sastry K, Abdelkarim M, Zilberfarb V, Strosberg AD, Chouchane L (2015) Comprehensive molecular characterization of human adipocytes reveals a transient brown phenotype. J Transl Med 13:135. https://doi.org/10.1186/s12967-015-0480-6

Hallberg M, Morganstein DL, Kiskinis E, Shah K, Kralli A, Dilworth SM, White R, Parker MG, Christian M (2008) A functional interaction between RIP140 and PGC-1alpha regulates the expression of the lipid droplet protein CIDEA. Mol Cell Biol 28:6785–6795. https://doi.org/10.1128/MCB.00504-08

Hansen JB, Jorgensen C, Petersen RK, Hallenborg P, De Matteis R, Boye HA, Petrovic N, Enerback S, Nedergaard J, Cinti S, te Riele H, Kristiansen K (2004) Retinoblastoma protein functions as a molecular switch determining white versus brown adipocyte differentiation. Proc Natl Acad Sci U S A 101:4112–4117. https://doi.org/10.1073/pnas.0301964101

Jat PS, Noble MD, Ataliotis P, Tanaka Y, Yannoutsos N, Larsen L, Kioussis D (1991) Direct derivation of conditionally immortal cell lines from an H-2Kb-tsA58 transgenic mouse. Proc Natl Acad Sci U S A 88:5096–5100

Kazantzis M, Takahashi V, Hinkle J, Kota S, Zilberfarb V, Issad T, Abdelkarim M, Chouchane L, Strosberg AD (2012) PAZ6 cells constitute a representative model for human brown pre-adipocytes. Front Endocrinol 3:13. https://doi.org/10.3389/fendo.2012.00013

Kiskinis E, Chatzeli L, Curry E, Kaforou M, Frontini A, Cinti S, Montana G, Parker MG, Christian M (2014) RIP140 represses the "brown-in-white" adipocyte program including a futile cycle of triacylglycerol breakdown and synthesis. Mol Endocrinol 28:344–356. https://doi.org/10.1210/me.2013-1254

Klaus S, Choy L, Champigny O, Cassard-Doulcier AM, Ross S, Spiegelman B, Ricquier D (1994) Characterization of the novel brown adipocyte cell line HIB 1B. Adrenergic pathways involved in regulation of uncoupling protein gene expression. J Cell Sci 107(Pt 1):313–319

Klein J, Fasshauer M, Ito M, Lowell BB, Benito M, Kahn CR (1999) Beta(3)-adrenergic stimulation differentially inhibits insulin signaling and decreases insulin-induced glucose uptake in brown adipocytes. J Biol Chem 274:34795–34802

Lee P, Bova R, Schofield L, Bryant W, Dieckmann W, Slattery A, Govendir MA, Emmett L, Greenfield JR (2016) Brown adipose tissue exhibits a glucose-responsive thermogenic biorhythm in humans. Cell Metab 23:602–609. https://doi.org/10.1016/j.cmet.2016.02.007

Leonardsson G, Steel JH, Christian M, Pocock V, Milligan S, Bell J, So PW, Medina-Gomez G, Vidal-Puig A, White R, Parker MG (2004) Nuclear receptor corepressor RIP140 regulates fat accumulation. Proc Natl Acad Sci U S A 101:8437–8442. https://doi.org/10.1073/pnas.0401013101

Miller CN, Yang JY, England E, Yin A, Baile CA, Rayalam S (2015) Isoproterenol increases uncoupling, glycolysis, and markers of Beiging in mature 3T3-L1 adipocytes. PLoS One 10:e0138344. https://doi.org/10.1371/journal.pone.0138344

Pisani DF, Djedaini M, Beranger GE, Elabd C, Scheideler M, Ailhaud G, Amri EZ (2011) Differentiation of human adipose-derived stem cells into "Brite" (Brown-in-White) adipocytes. Front Endocrinol 2:87. https://doi.org/10.3389/fendo.2011.00087

Pulinilkunnil T, He H, Kong D, Asakura K, Peroni OD, Lee A, Kahn BB (2011) Adrenergic regulation of AMP-activated protein kinase in brown adipose tissue in vivo. J Biol Chem 286:8798–8809. https://doi.org/10.1074/jbc.M111.218719

Puri V, Chakladar A, Virbasius JV, Konda S, Powelka AM, Chouinard M, Hagan GN, Perugini R, Czech MP (2007) RNAi-based gene silencing in primary mouse and human adipose tissues. J Lipid Res 48:465–471. https://doi.org/10.1194/jlr.D600033-JLR200

Rodriguez AM, Pisani D, Dechesne CA, Turc-Carel C, Kurzenne JY, Wdziekonski B, Villageois A, Bagnis C, Breittmayer JP, Groux H, Ailhaud G, Dani C (2005) Transplantation of a multipotent cell population from human adipose tissue induces dystrophin expression in the immunocompetent mdx mouse. J Exp Med 201:1397–1405. https://doi.org/10.1084/jem.20042224

Rosell M, Kaforou M, Frontini A, Okolo A, Chan YW, Nikolopoulou E, Millership S, Fenech ME, Macintyre D, Turner JO, Moore JD, Blackburn E, Gullick WJ, Cinti S, Montana G, Parker MG, Christian M (2014) Brown and white adipose tissues: intrinsic differences in gene expression and response to cold exposure in mice. Am J Physiol Endocrinol Metab 306:E945–E964. https://doi.org/10.1152/ajpendo.00473.2013

Rosen ED, MacDougald OA (2006) Adipocyte differentiation from the inside out. Nat Rev Mol Cell Biol 7:885–896. https://doi.org/10.1038/nrm2066

Ross SR, Choy L, Graves RA, Fox N, Solevjeva V, Klaus S, Ricquier D, Spiegelman BM (1992) Hibernoma formation in transgenic mice and isolation of a brown adipocyte cell line expressing the uncoupling protein gene. Proc Natl Acad Sci U S A 89:7561–7565

Scime A, Grenier G, Huh MS, Gillespie MA, Bevilacqua L, Harper ME, Rudnicki MA (2005) Rb and p107 regulate preadipocyte differentiation into white versus brown fat through repression of PGC-1alpha. Cell Metab 2:283–295. https://doi.org/10.1016/j.cmet.2005.10.002

Smas CM, Sul HS (1995) Control of adipocyte differentiation. Biochem J 309(Pt 3):697–710

Sugihara H, Yonemitsu N, Miyabara S, Yun K (1986) Primary cultures of unilocular fat cells: characteristics of growth in vitro and changes in differentiation properties. Differentiation 31:42–49

Tran KV, Gealekman O, Frontini A, Zingaretti MC, Morroni M, Giordano A, Smorlesi A, Perugini J, De Matteis R, Sbarbati A, Corvera S, Cinti S (2012) The vascular endothelium of the adipose tissue gives rise to both white and brown fat cells. Cell Metab 15:222–229. https://doi.org/10.1016/j.cmet.2012.01.008

Van De Pette M, Tunster SJ, McNamara GI, Shelkovnikova T, Millership S, Benson L, Peirson S, Christian M, Vidal-Puig A, John RM (2016) Cdkn1c boosts the development of brown adipose tissue in a murine model of Silver Russell Syndrome. PLoS Genet 12:e1005916. https://doi.org/10.1371/journal.pgen.1005916

Wolf Y, Boura-Halfon S, Cortese N, Haimon Z, Sar Shalom H, Kuperman Y, Kalchenko V, Brandis A, David E, Segal-Hayoun Y, Chappell-Maor L, Yaron A, Jung S (2017) Brown-adipose-tissue macrophages control tissue innervation and homeostatic energy expenditure. Nat Immunol 18:665–674. https://doi.org/10.1038/ni.3746

Wu J, Bostrom P, Sparks LM, Ye L, Choi JH, Giang AH, Khandekar M, Virtanen KA, Nuutila P, Schaart G, Huang K, Tu H, van Marken Lichtenbelt WD, Hoeks J, Enerback S, Schrauwen P, Spiegelman BM (2012) Beige adipocytes are a distinct type of thermogenic fat cell in mouse and human. Cell 150:366–376. https://doi.org/10.1016/j.cell.2012.05.016

Zaragosi LE, Ailhaud G, Dani C (2006) Autocrine fibroblast growth factor 2 signaling is critical for self-renewal of human multipotent adipose-derived stem cells. Stem Cells 24:2412–2419. https://doi.org/10.1634/stemcells.2006-0006

Zennaro MC, Le Menuet D, Viengchareun S, Walker F, Ricquier D, Lombes M (1998) Hibernoma development in transgenic mice identifies brown adipose tissue as a novel target of aldosterone action. J Clin Invest 101:1254–1260. https://doi.org/10.1172/JCI1915

Zilberfarb V, Pietri-Rouxel F, Jockers R, Krief S, Delouis C, Issad T, Strosberg AD (1997) Human immortalized brown adipocytes express functional beta3-adrenoceptor coupled to lipolysis. J Cell Sci 110(Pt 7):801–807

Brown-Like Adipocyte Progenitors Derived from Human iPS Cells: A New Tool for Anti-obesity Drug Discovery and Cell-Based Therapy?

Xi Yao, Barbara Salingova, and Christian Dani

Contents

1 Introduction 98
2 Capacity of Human-Induced Pluripotent Stem Cells to Generate BAs 99
3 Pathways Governing hiPSC-BAP Adipogenic Capacity 99
4 Perspectives of hiPSCs in the Obesity Field 100
References 103

Abstract

Alternative strategies are urgently required to fight obesity and associated metabolic disorders including diabetes and cardiovascular diseases. Brown and brown-like adipocytes (BAs) store fat, but in contrast to white adipocytes, activated BAs are equipped to dissipate energy stored. Therefore, BAs represent promising cell targets to counteract obesity. However, the scarcity of BAs in adults is a major limitation for a BA-based therapy of obesity, and the notion to increase the BA mass by transplanting BA progenitors (BAPs) in obese patients recently emerged. The next challenge is to identify an abundant and reliable source of BAPs. In this chapter, we describe the capacity of human-induced pluripotent stem cells (hiPSCs) to generate BAPs able to differentiate at a high efficiency with no gene transfer. This cell model represents an unlimited source of human BAPs that in a near future may be a suitable tool for both therapeutic transplantation and for the discovery of novel efficient and safe anti-obesity drugs. The generation of a relevant cell model, such as hiPSC-BAs in 3D adipospheres enriched with macrophages and endothelial cells to better mimic the microenvironment within the adipose tissue, will be the next critical step.

X. Yao · B. Salingova · C. Dani (✉)
Faculté de Médecine, Université Nice Sophia Antipolis, iBV, UMR CNRS/INSERM, Nice, Cedex 2, France
e-mail: dani@unice.fr

A. Pfeifer et al. (eds.), *Brown Adipose Tissue*,
Handbook of Experimental Pharmacology 251, https://doi.org/10.1007/164_2018_115

Keywords
Adipocyte progenitors · Brown adipocytes · Cell-based therapy · Drug discovery · Human-induced pluripotent stem cells · Obesity

1 Introduction

Obesity and associated metabolic disorders such as diabetes and cardiovascular diseases are major health problems. Obesity results from an imbalance between calorie intake and energy expenditure, and three types of adipocytes are the main regulators of this balance. White adipocytes are involved in energy storage and their accumulation marks obesity. Therapies based to reduce energy intake are difficult to follow in our modern life, and current anti-obesity drugs cause important side effects for the patients that limits their use. Bariatric surgery has proven efficiency for obesity, although long-term complications and obesity relapse may appear. Therefore, alternative strategies to increase energy expenditure with the identification of new anti-obesity targets are urgently required. In contrast to white adipocytes, classical brown adipocytes and brown-like adipocytes (BAs) dispersed in white adipose tissues are specialized in energy expenditure. Upon activation, BAs consume metabolic substrates and burn fat via the mitochondrial uncoupling protein-1 (UCP-1). Moreover, the ability of BAs to actively drain circulating glucose and triglycerides to oxidize them can prevent hyperglycemia and hypertriglyceridemia. Therefore, BAs represent promising cell targets to counteract obesity in human. However, there are major limitations for a BA-based treatment of obesity. BAs hold a minor fraction of adipose tissue in human and disappear from most areas with age, persisting only around deeper organs. In addition, BA activity is lower in overweight and obese individuals than in leans (Cypess et al. 2013; van Marken Lichtenbelt et al. 2009). Hence, the notion to increase the BA mass by transplanting BA progenitors (BAPs) in obese patients as a therapeutic alternative to counteract obesity and its associated metabolic complications recently emerged. The proof-of-concept has been validated in murine models as it has been reported that implants of mouse BAT or of human BAs developed from capillary networks were able to restore normoglycemia in diabetic mice and to reduce obesity in Ob/Ob mice (Gunawardana and Piston 2010; Liu et al. 2015; Min et al. 2016; Stanford et al. 2013). It is interesting to note that small amount of transplanted materials was sufficient to display a beneficial effect, reflecting that in addition to acting as a glucose and energy sink, BAs secrete adipokines that could also contribute to metabolic effects (Villarroya et al. 2017).

Therefore, a reliable source of human BAs is urgently needed. Induced pluripotent stem cells appear as a suitable source of BAs for both cell transplantation to increase the BA mass in patients and for anti-obesity drug discovery.

2 Capacity of Human-Induced Pluripotent Stem Cells to Generate BAs

As embryonic stem cells (ESCs) are derived from frozen extra embryos after in vitro fertilization, induced pluripotent stem cells (iPSCs) generated from adult cells represent an abundant source of multiple cell types of therapeutic interest for drug screening as well as for transplantation (Shi et al. 2016; Takahashi et al. 2007). However, in contrast to human ESCs human iPSCs are generated from somatic cells. Their derivation does not require destruction of embryos, thus avoiding ethical problems. Nakao's group was the first to demonstrate the capacity of hiPSCs to generate white adipocytes (Taura et al. 2009). Then, Nishio et al. (2012) developed a procedure to generate functional brown adipocytes at a high efficiency using a hematopoietic cocktail to induce hiPSC differentiation. Interestingly, hiPSC-brown adipocytes were able to improve glucose tolerance after transplantation in mice underlying the therapeutic potential of hiPSCs in the obesity field. However, numerous issues have to be investigated before a therapeutic use of hiPSC-BAs, notably the purification of BA progenitors (BAPs) and their differentiation into functional adipocytes at a high level. In fact, in both works mentioned above, total differentiated hiPSC populations, but not purified progenitors, were transplanted into mice. Indeed, differentiated hiPSC cultures can be enriched with adipocytes but still contain other cell types that are unsuitable for transplantation, including undifferentiated hiPSCs that can form teratomas. An alternative to eliminate hiPSC capacity to form teratomas consists in purifying progenitors of interest during hiPSC differentiation. This has been done by Ahfeldt et al. who purified adipose progenitors (APs) from hiPSCs able to undergo differentiation into BAs but only following forced expression of three adipogenic master genes (Ahfeldt et al. 2012). Mohsen-Kanson et al. reported a procedure to selectively derive white and brown-like APs from hiPSCs. Adipogenic cocktails usually used to induce the differentiation of APs derived from adult adipose tissues failed to differentiate hiPSC-BAPs and forced expression of a master adipogenic gene was also required for their differentiation (Mohsen-Kanson et al. 2014). The need to genetically modify hiPSC-APs clearly illustrates their weak adipogenic potential that represents a bottleneck hampering their clinical use. This feature has been observed by us and others using different approaches to derive progenitors from hiPSCs (Hafner and Dani 2014). These data suggested that the commitment of hiPSC-BAPs toward adipocytes requires activation and/or inhibition of pathways different from those regulating differentiation of BAPs derived from adult adipose tissues.

3 Pathways Governing hiPSC-BAP Adipogenic Capacity

Critical pathways regulating hiPSC-BAPs differentiation were recently identified (Hafner et al. 2016a, b; Mohsen-Kanson et al. 2014). It has been shown that Smad2, a potent anti-adipogenic pathway (Dani 2013), is activated during the differentiation of hiPSC-BAPs, likely thanks to the autocrine expression of TGFb1 and of activin

A. Interestingly, treatment of hiPSC-BAPs with the activin/TGFb inhibitor SB431542 switched on the differentiation process. Other factors such as ascorbic acid, EGF, and hydrocortisone have also been shown to be required for hiPSC-BAPs differentiation. A detailed protocol for the differentiation of hiPSC-BAPs is described in Hafner et al. (2016b). The molecular pathways mediating the effects of these compounds in hiPSC-BAPs remain to be elucidated. Ascorbic acid was previously shown to enhance the differentiation of 3T3-L1 mouse cell line, and recent studies attribute this effect to the regulation of type VI collagen (Kim et al. 2013; Liu et al. 2017). However, the requirement of ascorbic acid for UCP-1 expression would deserve further investigations. EGF is known to stimulate adult-BAPs proliferation (Hebert et al. 2009; Holmstrom et al. 2008); however, its role in the differentiation of human BAPs is not known. The stimulation of EGR1 (early growth response-1) observed upon treatment of hiPSC-BAPs with EGF is of interest. EGR1 is a zinc finger transcription factor expressed in adult adipose tissue, and it has been reported that Egr1 deficiency induces WAT browning by releasing EGR1-mediated UCP-1 transcription repression in mice (Milet et al. 2017; Zhang et al. 2013). The role of EGR1 in EGF-induced hiPSC-BAPs differentiation remains to be determined, but these data suggest a dual role of EGF, i.e., required for the commitment of BAPs toward adipocytes and a break for fully expression of UCP-1. Finally, it appears that the choice of the basal medium enriched with the different factors indicated above is not trivial. Indeed, no differentiation was observed when hiPSC-BAPs were maintained in DMEM. In contrast, hiPSC-BAPs could undergo differentiation at a high level when maintained in a medium developed to support proliferation and differentiation of endothelial cells (Hafner et al. 2016a, b). Interestingly, a common basal medium is used for differentiation of functional endothelial cells, hematopoietic cells, and brown adipocytes from human iPSCs (Hafner et al. 2016a, b; Nishio et al. 2012; Orlova et al. 2014). The reasons why a medium supporting endothelial cells is critical for hiPSC-BAPs differentiation are unknown. Studies showed that adipocytes and endothelial cells surrounding might originate from a common progenitor that could go through adipogenesis or angiogenesis depending on the microenvironment. There are other evidences showing that human adipose-derived stem cells have feature of endothelial progenitor cells (Planat-Benard et al. 2004). Striking similarity between the differentiation conditions suggests a close developmental origin, and the fact that adipocytes may originate from subsets of endothelial cells could be an explanation (Sanchez-Gurmaches and Guertin 2013). Altogether, these data show a functional link between adipogenesis and angiogenesis, and the differentiation of hiPSC-BAPs can constitute a platform for the identification of new pathways regulating the generation of human BAs.

4 Perspectives of hiPSCs in the Obesity Field

The lack of a relevant physiologic cell model for preclinical testing of drugs activating human BAs is a major limitation for a BA-based treatment of obesity. The cellular models mainly used for drug discovery are murine cell lines. They may not provide relevant models as recent findings support species differences for key adipogenic regulatory pathways (Lindroos et al. 2013). Despite their gold standard state, primary human cells derived from adult adipose tissues can be studied only for short periods before they reach senescence and differentiation failure precluding reproducible assays. Because iPSCs display a high self-renewal capacity, they can be cultured for long term and expanded into large numbers under completely defined conditions. In addition, human iPSCs are able to undergo differentiation into hepatocytes and cardiomyocytes, and therefore they are outstanding tools in early preclinical phase of drug discovery to perform toxicity assessment. A myriad of molecules has been tested as anti-obesity drugs. To date, the marketed drugs for BAs activation showed limited efficacy and/or displayed substantial adverse effects (Giordano et al. 2016). Among the reasons involved in this weak efficacy, one can also mention the limitation of the drug testing models due to the culture conditions that do not mimic the phenotype of the cells and their physiological microenvironment within the adipose tissue. Cells are classically grown as monolayer, which poorly reflects the in vivo situation (Horvath et al. 2016). In contrast, the cell–cell and cell–extracellular matrix interactions are promoted in 3D configurations. Therefore, 3D cultures represent a bridge between traditional cell culture and live tissue to overcome the limitations that impaired discovery of efficient and safe drugs. hiPSC-BAPs can form 3D aggregates, also known as spheroids, able to differentiate into UCP-1-expressing adipocytes, therefore named as adipospheres (Yao and Dani, personal observation). Three-dimensional adipospheres formed from hiPSC-BAs could serve as a platform to identify pathways governing the recruitment and the activation of human BAs and for anti-obesity drug discovery (see Fig. 1). On the other hand, the use of iPSC-derived cells in clinic has been approved by the FDA for at least two diseases, for iPSC-myogenic progenitors to be delivered to muscular dystrophic patients, and phase I/II results are showing promising safety data as well as possible efficacy when used in ocular diseases. Therefore, BA adipospheres from obese-iPSCs can be proposed as therapeutic cells (see Fig. 1). Interestingly, it has been recently demonstrated that the therapeutic potential of cells is dramatically improved when transplanted as 3D structures (Petrenko et al. 2017). Indeed, autologous transplantations are not financially feasible at present, but a recent study reported the possibility of allogenic iPSCs transplants, which has a tremendous potential for clinic use (Shiba et al. 2016). Human iPSCs can be cultured for long term and expanded into large numbers under complete defined conditions. Thanks to their high self-renewal capacity, it is not a challenge to produce a sufficient number of cells for an effective treatment. However, crucial points for future applications in humans remain to be solved. Recent studies have clearly established that hiPSC clones can display variable capacities to differentiate into specific lineages (Nishizawa et al. 2016) which complicate the development of universal protocols of differentiation. The identification of the origins of this functional variability is still a file of investigations (Ortmann and Vallier 2017). An alternative to bypass the hiPSC variability consists in the purification of progenitors from differentiating hiPSCs. Several approaches have been used for the purification of hiPSC-BAPs (Hafner and Dani 2014). The

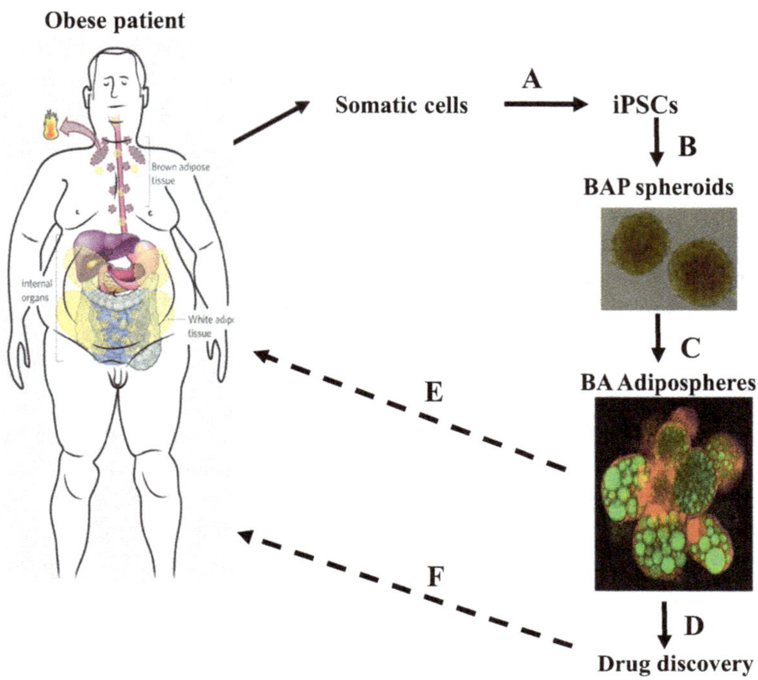

Fig. 1 The main steps of the generation of hiPSC-BA adipospheres for anti-obesity drug discovery and cell-based therapy. (**a**) Generation of iPSCs from obese patients; (**b**) derivation of brown-like adipose progenitors and cell culture in spheroids; (**c**) spheroids induction of differentiation into functional BAs and to mimic the physio- or pathological microenvironments; (**d**) transplantation of BA adipospheres to increase the BA mass in obese; (**e**) use the physiological model as a basis for drug discovery or drug repositioning and drug safety; (**f**) discovery of drugs which could attend patient drug treatment to activate BAs. The cartoon of the man comes from Farmer (2009). Photos of BAP spheroids and of BA adipospheres come from experiments performed by X. Yao in C. Dani's lab

purification of hiPSC-BAPs from different hiPSC clones displays various efficiencies, but we showed that progenitors purified from different hiPSC sources have a similar capacity to undergo differentiation when using a unique adipogenic protocol (Hafner et al. 2016a, b). This result strongly supports that the progenitor purification step is a requirement. An advantage to purify progenitors is also to eliminate undifferentiated hiPSCs prone to form teratoma after transplantation. Another issue is the generation of mutations during hiPSC expansion. Fine genetic analysis and preclinical animal studies should be able to address the safety of hiPSC-BAs to be transplanted, including potential teratoma formation and other cancer development. Finally, the stability of the phenotype of transplanted BAs has to be monitored as dedifferentiation of BAs or transdifferentiation of BAs into white adipocytes after transplantation cannot be ruled out. Tissue engineering aiming to develop tissue-like structures that mimic the in vivo situation as closely as possible should prevent this risk. As hiPSCs can differentiate into cells present in the adipose tissue, BA adipospheres could be enriched with macrophages and endothelial cells (ECs) to mimic an adipose-like structure and to improve their therapeutic potential. Endothelial growth media supports hiPSC-BAPs differentiation (see above), suggesting the feasibility to coculture BA adipospheres with ECs. Pre-vascularized organoids derived from hiPSCs have been generated including the liver, skeletal, and cardiac muscle. All constructs show incorporation into vascular network and a better viability of the graft after transplantation in animal models (Caspi et al. 2007; Takebe et al. 2013).

Numerous other issues have also to be solved before a therapeutic use of iPSCs in the obesity field, but the derivation of BAPs having a high level of differentiation without gene transfer and their capacity to form 3D adipospheres open the opportunity of using hiPSCs advantages for anti-obesity therapy and for a better understanding of the interactions between different cell types of adipose tissue for BAs recruitment and activation.

References

Ahfeldt T, Schinzel RT, Lee YK, Hendrickson D, Kaplan A, Lum DH, Camahort R, Xia F, Shay J, Rhee EP, Clish CB, Deo RC, Shen T, Lau FH, Cowley A, Mowrer G, Al-Siddiqi H, Nahrendorf M, Musunuru K, Gerszten RE, Rinn JL, Cowan CA (2012) Programming human pluripotent stem cells into white and brown adipocytes. Nat Cell Biol 14:209–219

Caspi O, Lesman A, Basevitch Y, Gepstein A, Arbel G, Habib IH, Gepstein L, Levenberg S (2007) Tissue engineering of vascularized cardiac muscle from human embryonic stem cells. Circ Res 100:263–272

Cypess AM, White AP, Vernochet C, Schulz TJ, Xue R, Sass CA, Huang TL, Roberts-Toler C, Weiner LS, Sze C, Chacko AT, Deschamps LN, Herder LM, Truchan N, Glasgow AL, Holman AR, Gavrila A, Hasselgren PO, Mori MA, Molla M, Tseng YH (2013) Anatomical localization, gene expression profiling and functional characterization of adult human neck brown fat. Nat Med 19:635–639

Dani C (2013) Activins in adipogenesis and obesity. Int J Obes (Lond) 37:163–166

Farmer SR (2009) Obesity: be cool, lose weight. Nature 458:839–840

Giordano A, Frontini A, Cinti S (2016) Convertible visceral fat as a therapeutic target to curb obesity. Nat Rev Drug Discov 15:405–424

Gunawardana SC, Piston DW (2010) Reversal of type 1 diabetes in mice by brown adipose tissue transplant. Diabetes 61:674–682

Hafner AL, Dani C (2014) Human induced pluripotent stem cells: a new source for brown and white adipocytes. World J Stem Cells 6:467–472

Hafner AL, Contet J, Ravaud C, Yao X, Villageois P, Suknuntha K, Annab K, Peraldi P, Binetruy B, Slukvin II, Ladoux A, Dani C (2016a) Brown-like adipose progenitors derived from human induced pluripotent stem cells: identification of critical pathways governing their adipogenic capacity. Sci Rep 6:32490

Hafner A-L, Mohsen-Kanson T, Dani C (2016b) A protocol for the differentiation of brown adipose progenitors derived from human induced pluripotent stem cells at a high efficiency with no gene transfer. Nat Protocol Exchange. Doi: https://doi.org/10.1038/protex.2016.067

Hebert TL, Wu X, Yu G, Goh BC, Halvorsen YD, Wang Z, Moro C, Gimble JM (2009) Culture effects of epidermal growth factor (EGF) and basic fibroblast growth factor (bFGF) on cryopreserved human adipose-derived stromal/stem cell proliferation and adipogenesis. J Tissue Eng Regen Med 3:553–561

Holmstrom TE, Mattsson CL, Falting JM, Nedergaard J (2008) Differential signalling pathways for EGF versus PDGF activation of Erk1/2 MAP kinase and cell proliferation in brown pre-adipocytes. Exp Cell Res 314:3581–3592

Horvath P, Aulner N, Bickle M, Davies AM, Nery ED, Ebner D, Montoya MC, Ostling P, Pietiainen V, Price LS, Shorte SL, Turcatti G, von Schantz C, Carragher NO (2016) Screening out irrelevant cell-based models of disease. Nat Rev Drug Discov 15:751–769

Kim B, Choi KM, Yim HS, Lee MG (2013) Ascorbic acid enhances adipogenesis of 3T3-L1 murine preadipocyte through differential expression of collagens. Lipids Health Dis 12:182

Lindroos J, Husa J, Mitterer G, Haschemi A, Rauscher S, Haas R, Groger M, Loewe R, Kohrgruber N, Schrogendorfer KF, Prager G, Beck H, Pospisilik JA, Zeyda M, Stulnig TM, Patsch W, Wagner O, Esterbauer H, Bilban M (2013) Human but not mouse adipogenesis is critically dependent on LMO3. Cell Metab 18:62–74

Liu X, Wang S, You Y, Meng M, Zheng Z, Dong M, Lin J, Zhao Q, Zhang C, Yuan X, Hu T, Liu L, Huang Y, Zhang L, Wang D, Zhan J, Jong Lee H, Speakman JR, Jin W (2015) Brown adipose tissue transplantation reverses obesity in Ob/Ob mice. Endocrinology 156:2461–2469

Liu C, Huang K, Li G, Wang P, Liu C, Guo C, Sun Z, Pan J (2017) Ascorbic acid promotes 3T3-L1 cells adipogenesis by attenuating ERK signaling to upregulate the collagen VI. Nutr Metab (Lond) 14:79

van Marken Lichtenbelt WD, Vanhommerig JW, Smulders NM, Drossaerts JM, Kemerink GJ, Bouvy ND, Schrauwen P, Teule GJ (2009) Cold-activated brown adipose tissue in healthy men. N Engl J Med 360:1500–1508

Milet C, Bleher M, Allbright K, Orgeur M, Coulpier F, Duprez D, Havis E (2017) Egr1 deficiency induces browning of inguinal subcutaneous white adipose tissue in mice. Sci Rep 7:16153

Min SY, Kady J, Nam M, Rojas-Rodriguez R, Berkenwald A, Kim JH, Noh HL, Kim JK, Cooper MP, Fitzgibbons T, Brehm MA, Corvera S (2016) Human "brite/beige" adipocytes develop from capillary networks, and their implantation improves metabolic homeostasis in mice. Nat Med 22:312–318

Mohsen-Kanson T, Hafner AL, Wdziekonski B, Takashima Y, Villageois P, Carriere A, Svensson M, Bagnis C, Chignon-Sicard B, Svensson PA, Casteilla L, Smith A, Dani C (2014) Differentiation of human induced pluripotent stem cells into brown and white adipocytes: role of Pax3. Stem Cells 32:1459–1467

Nishio M, Yoneshiro T, Nakahara M, Suzuki S, Saeki K, Hasegawa M, Kawai Y, Akutsu H, Umezawa A, Yasuda K, Tobe K, Yuo A, Kubota K, Saito M, Saeki K (2012) Production of functional classical brown adipocytes from human pluripotent stem cells using specific hemopoietin cocktail without gene transfer. Cell Metab 16:394–406

Nishizawa M, Chonabayashi K, Nomura M, Tanaka A, Nakamura M, Inagaki A, Nishikawa M, Takei I, Oishi A, Tanabe K, Ohnuki M, Yokota H, Koyanagi-Aoi M, Okita K, Watanabe A, Takaori-Kondo A, Yamanaka S, Yoshida Y (2016) Epigenetic variation between human induced pluripotent stem cell lines is an Indicator of differentiation capacity. Cell Stem Cell 19:341–354

Orlova VV, van den Hil FE, Petrus-Reurer S, Drabsch Y, Ten Dijke P, Mummery CL (2014) Generation, expansion and functional analysis of endothelial cells and pericytes derived from human pluripotent stem cells. Nat Protoc 9:1514–1531

Ortmann D, Vallier L (2017) Variability of human pluripotent stem cell lines. Curr Opin Genet Dev 46:179–185

Petrenko Y, Sykova E, Kubinova S (2017) The therapeutic potential of three-dimensional multipotent mesenchymal stromal cell spheroids. Stem Cell Res Ther 8:94

Planat-Benard V, Silvestre JS, Cousin B, Andre M, Nibbelink M, Tamarat R, Clergue M, Manneville C, Saillan-Barreau C, Duriez M, Tedgui A, Levy B, Penicaud L, Casteilla L (2004) Plasticity of human adipose lineage cells toward endothelial cells: physiological and therapeutic perspectives. Circulation 109:656–663

Sanchez-Gurmaches J, Guertin DA (2013) Adipocyte lineages: tracing back the origins of fat. Biochim Biophys Acta 1842:340–351

Shi Y, Inoue H, Wu JC, Yamanaka S (2016) Induced pluripotent stem cell technology: a decade of progress. Nat Rev Drug Discov 16:115–130

Shiba Y, Gomibuchi T, Seto T, Wada Y, Ichimura H, Tanaka Y, Ogasawara T, Okada K, Shiba N, Sakamoto K, Ido D, Shiina T, Ohkura M, Nakai J, Uno N, Kazuki Y, Oshimura M, Minami I, Ikeda U (2016) Allogeneic transplantation of iPS cell-derived cardiomyocytes regenerates primate hearts. Nature 538:388–391

Stanford KI, Middelbeek RJ, Townsend KL, An D, Nygaard EB, Hitchcox KM, Markan KR, Nakano K, Hirshman MF, Tseng YH, Goodyear LJ (2013) Brown adipose tissue regulates glucose homeostasis and insulin sensitivity. J Clin Invest 123:215–223

Takahashi K, Tanabe K, Ohnuki M, Narita M, Ichisaka T, Tomoda K, Yamanaka S (2007) Induction of pluripotent stem cells from adult human fibroblasts by defined factors. Cell 131:861–872

Takebe T, Sekine K, Enomura M, Koike H, Kimura M, Ogaeri T, Zhang RR, Ueno Y, Zheng YW, Koike N, Aoyama S, Adachi Y, Taniguchi H (2013) Vascularized and functional human liver from an iPSC-derived organ bud transplant. Nature 499:481–484

Taura D, Noguchi M, Sone M, Hosoda K, Mori E, Okada Y, Takahashi K, Homma K, Oyamada N, Inuzuka M, Sonoyama T, Ebihara K, Tamura N, Itoh H, Suemori H, Nakatsuji N, Okano H, Yamanaka S, Nakao K (2009) Adipogenic differentiation of human induced pluripotent stem cells: comparison with that of human embryonic stem cells. FEBS Lett 583:1029–1033

Villarroya F, Gavalda-Navarro A, Peyrou M, Villarroya J, Giralt M (2017) The lives and times of brown adipokines. Trends Endocrinol Metab 28:855–867

Zhang J, Zhang Y, Sun T, Guo F, Huang S, Chandalia M, Abate N, Fan D, Xin HB, Chen YE, Fu M (2013) Dietary obesity-induced Egr-1 in adipocytes facilitates energy storage via suppression of FOXC2. Sci Rep 3:1476

Brown Adipose Tissue in Human Infants

Martin E. Lidell

Contents

1 Introduction 108
2 Anatomical Location and Morphology of Fetal/Infant Brown Adipose Tissue 109
2.1 The Early Studies of Fetal/Infant Brown Adipose Tissue 109
2.2 The Global Distribution and Appearance of Fetal/Infant Brown Adipose Tissue ... 111
2.3 Studies of Infant Brown Adipose Tissue Using Modern Noninvasive Imaging Techniques 113
3 The Physiology of Infant Brown Adipose Tissue 115
4 The Molecular Characteristics of Infant Brown Adipose Tissue 118
5 Conclusions 119
References 120

Abstract

Adapting to the cold extrauterine environment after birth is a great challenge for the newborn. Due to their high surface area-to-volume ratio, infants tend to lose more heat to the environment as compared to adults. In addition, human newborns lack sufficiently developed skeletal muscle mass to maintain body temperature through shivering thermogenesis, an important source of heat in cold-exposed adults. Evolution has provided humans and other placental mammals with brown adipose tissue (BAT), a tissue that converts chemically stored energy, in the form of fatty acids and glucose, into heat through non-shivering thermogenesis. The thermogenic activity of this tissue is significant for the human infant's ability to maintain a sufficiently high core body temperature. Although BAT has been studied in human infants for more than a century, the literature covering different aspects of the tissue is rather limited. The aim of

M. E. Lidell (✉)
Department of Medical Biochemistry and Cell Biology, Institute of Biomedicine, The Sahlgrenska Academy, University of Gothenburg, Gothenburg, Sweden
e-mail: martin.lidell@medgen.gu.se

A. Pfeifer et al. (eds.), *Brown Adipose Tissue*,
Handbook of Experimental Pharmacology 251, https://doi.org/10.1007/164_2018_118

this review is to summarize the literature and describe what is actually known about the tissue and its importance for early human life.

Keywords
Brown adipose tissue · Fetal · Human · Infant

1 Introduction

One of the key challenges a newborn mammal faces after birth is to cope with the temperature in the extrauterine environment that is often well below their thermoneutral zone. Evolution has provided different mammalian species with various attributes that help their newborns to maintain a constant body temperature above ambient temperature. One important feature that modern eutherian (placental) mammals have acquired is brown adipose tissue (BAT) (Oelkrug et al. 2015). This tissue contains thermogenic adipocytes that, in response to signals triggered by cold exposure, are able to generate heat by a process referred to as non-shivering thermogenesis. There are two types of thermogenic adipocytes; brown adipocytes that are typically found in the classical BAT organ that is located in the interscapular region, and beige adipocytes (also referred to as inducible brown or brite adipocytes) that appear in otherwise white adipose tissue (WAT) depots in response to prolonged cold exposure or stimulation with β3-adrenergic agonists (Harms and Seale 2013). The two cell types share many features; they both contain multilocular lipid droplets and are very rich in mitochondria that contain uncoupling protein 1 (UCP1) in their inner membrane. Despite their similarities, the brown and beige adipocytes appear to be derived from different cell lineages. Brown adipocytes have been shown to originate from precursor cells that at some time point have expressed the myogenic marker gene *Myf5*, suggesting an ontogenetic relationship between brown adipocytes and skeletal muscle cells (Seale et al. 2008). Beige adipocytes on the other hand are derived from a *Myf5*-negative cell population related to that of mural and vascular smooth muscle cells (Long et al. 2014; Vishvanath et al. 2016). The different origins of brown and beige as well as white adipocytes are also reflected by the fact that they express distinct sets of cell selective marker genes (Wu et al. 2012). In addition, in contrast to brown adipocytes, beige adipocytes have a low basal expression level of *Ucp1*. However, in response to β-adrenergic signaling the *Ucp1* expression in these cells can reach a level similar to that in brown adipocytes (Wu et al. 2012).

Classical BAT can be viewed as a specialized heat-generating organ that is activated when the organism senses cold and needs to induce heat production to maintain core body temperature. Thermogenesis in BAT is primarily activated by the sympathetic nervous system; reflected by the tissue being densely innervated by adrenergic nerve fibers (Smith and Horwitz 1969). Peripheral temperature-sensitive receptors relay sensation of cold to hypothalamus that integrates the signals and activates BAT through increased adrenergic nerve activity (Kooijman et al. 2015). Norepinephrine released from the sympathetic nerve endings activates β-adrenergic receptors found on the surface of brown adipocytes, thereby inducing lipolysis of stored triglycerides (Cannon and Nedergaard 2004). The released fatty acids serve both as fuel for the thermogenesis and as activators of the main thermogenic executer in BAT, UCP1 (Fedorenko et al. 2012), a protein uniquely found in the inner mitochondrial membrane of thermogenic adipocytes (Cannon et al. 1982; Lin and Klingenberg 1980). When activated, UCP1 uncouples cellular respiration from ATP production by short-circuiting the proton gradient over the inner mitochondrial membrane, thereby generating heat. Hence, BAT has the ability to transform chemically stored energy into heat, a feature that has caught a lot of attention as it makes it an interesting drug target of potential use in the fight against obesity and obesity-related diseases, such as type 2 diabetes (Lidell et al. 2014).

BAT is also highly vascularized, and cold exposure significantly increases the perfusion of the tissue (Orava et al. 2011). The increased blood flow both helps supplying the tissue with sufficient amounts of oxygen for uncoupled respiration, and to distribute the generated heat to the rest of the body.

Thermogenesis in BAT is particularly important for small animals and infants, who due to their large surface area-to-volume ratio are prone to lose more heat and therefore have a greater demand for thermogenesis as compared to larger animals (Silva 2006). In addition, newborns often lack a skeletal muscle mass sufficient for maintaining body temperature through shivering thermogenesis, an important source of heat in cold-exposed adults. In the literature covering BAT and its properties, it is often stated that the tissue is important also for the human newborn's ability to maintain normal body temperature. In most cases, the statement is not supported by any references, and it therefore appears to be considered common knowledge that the tissue is an important thermoregulator during human infancy. This chapter reviews the literature related to human fetal/infant BAT and pinpoints what is actually known about the tissue, and its importance for early human life.

2 Anatomical Location and Morphology of Fetal/Infant Brown Adipose Tissue

2.1 The Early Studies of Fetal/Infant Brown Adipose Tissue

Although BAT was described in hibernating animals as early as the mid-sixteenth century, it was not until the early twentieth century that the tissue was generally recognized in humans. Due to its presence in most hibernating mammals, the tissue was initially referred to as the hibernating gland. However, the name soon became a bit ambiguous as the tissue was also found in many nonhibernating mammals such as rats and cats (Sheldon 1924). One of the first indications of the tissue being present also in humans came from the work of Hatai (1902). During his studies on the structure and histogenesis of the hibernating gland, he got access to five human fetuses (varying in length from 75 to 260 mm). When examining these, he "much to his surprise" found a tissue that closely resembled the hibernating gland of lower mammals, both in terms of location and gross morphology. Hatai found the tissue in

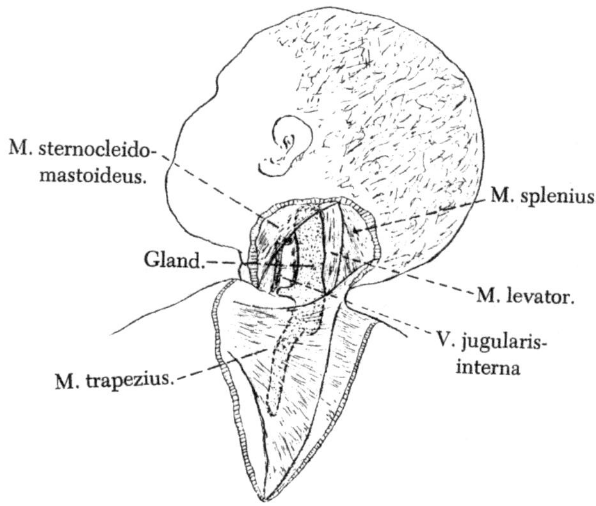

Fig. 1 The first illustration of the gross morphology of human fetal brown adipose tissue (BAT). The illustration shows the BAT depot described by Hatai in 1902. The author provisionally named the tissue the interscapular gland, and described it as "a long, narrow paired organ, lying partly along the neck and partly occupying the scapular region." Illustration from Hatai (1902)

all fetal specimens, but not in an adult subject, and described the tissue as "a long, narrow paired organ, lying partly along the neck and partly occupying the scapular region." He provisionally named the human tissue the interscapular gland (*glandula interscapularis*), a name that was commonly used the years that followed. A reproduction of Hatai's original illustration showing the anatomical site and gross morphology of the tissue is seen in Fig. 1. In 1902, Bonnot published the results from a more extensive study on human samples (Bonnot 1908). In addition to a set of samples from adults, 12 human embryos/fetuses (32–280 mm in length; uterine position) were also examined. Like Hatai, Bonnot focused his study on the shoulder and neck areas, and his results concerning the gross morphology of the fetal tissue were well in agreement with those presented by Hatai. Bonnot concluded that the whole tissue mass was lobulated, and presented with a size almost proportional to the length of the embryo (being about 25% of the head and body length). A histologic examination of the tissue showed that its lobulated appearance was due to connective tissue septa that divided the tissue into asymmetrical units, a feature adding to the overall gland-like appearance of the tissue. Furthermore, Bonnot reported that a large, irregularly shaped, and reticular cell type, partially filled with fat, was present within the lobules, a cell morphology matching that of brown adipocytes. The arrangement of multilocular parenchymal cells in lobular structures separated by septa of connective tissue was also noticed by Shattock a few years later (Shattock 1909). In a human fetus of about 5 months, he found these lobular structures adjacent to the great vessels of the neck. He also noticed that the multilocular cells within the lobules clearly differed from the unilocular cells found in the neighboring subcutaneous fat and concluded that two forms of adipose tissue may exist.

Although, the majority of the early studies on human fetal/infant BAT were concentrated on the shoulder and neck areas; anatomical regions harboring the hibernating gland in lower mammals, some studies suggested the existence of polygonal multilocular cells in gland-like structures at other sites, such as the subpleural, axillary, and perirenal regions (Rasmussen 1923; Shaw 1901). However, it was not until about half a century later that studies dedicated to map the overall distribution of BAT were performed.

2.2 The Global Distribution and Appearance of Fetal/Infant Brown Adipose Tissue

During the 1960s and 1970s, several ambitious investigations focusing on the localization of BAT depots were conducted. Although first used by Hammar already in 1895 (Hammar 1895), it was also now that the term BAT (or brown fat) started to be more widely used instead of the previous gland-related names.

Aherne and Hull collected adipose tissue samples from a wide range of sites during 25 successive necropsies of infants whose gestational age varied from 29 to 40 weeks, and postnatal age 3 min to 4 weeks (Aherne and Hull 1966). Based on the morphology of the cells in the sampled tissue, they assigned them as WAT or BAT. In all infants, they found BAT in the following locations: (1) the interscapular region (here, a thin diamond-shaped sheet of BAT was found under the subcutaneous WAT), (2) as many small masses around the muscles and blood vessels of the neck (the main depot followed the internal jugular vein and the common carotid artery), (3) the axillae (these depots were connected to the ones in the neck through extensions passing under the clavicles), (4) around the trachea, esophagus, and large vessels in mediastinum, spreading laterally with the intercostal arteries, and (5) intra-abdominally around structures such as the aorta, paravertebral autonomic ganglia, kidneys, and adrenals. Hence, with the exception of the interscapular depot, most BAT appeared to be situated deep within the body, around important structures of the neck, chest, and abdomen.

A few years later, Merklin performed a study aiming at identifying all fetal BAT depots (Merklin 1974). Although he found some small additional BAT depots in locations such as the anterior abdominal, and retropubic areas, and behind the sternum, the outcome of his study was well in agreement with that earlier presented for human infants by Aherne and Hull (1966). Merklin concluded that the fetal BAT depots collectively resembled a high collared vest that covered the cervical, thoracic, and abdominal viscera. As had been noticed in earlier studies, he also observed that, in addition to the multilocular brown adipocytes, the different BAT depots also contained a varying proportion of white-like unilocular adipocytes. BAT depots such as the perirenal and posterior cervical depots presented with a very low proportion of these white-like adipocytes, while others such as the interscapular and axillary depots contained a more mixed cell population with multilocular and unilocular adipocytes in almost equal proportions. Notably, Merklin did not exclude the possibility that also the unilocular adipocytes were of the brown type, as they generally were smaller and contained a wider and more granular cytoplasm as compared to true white adipocytes.

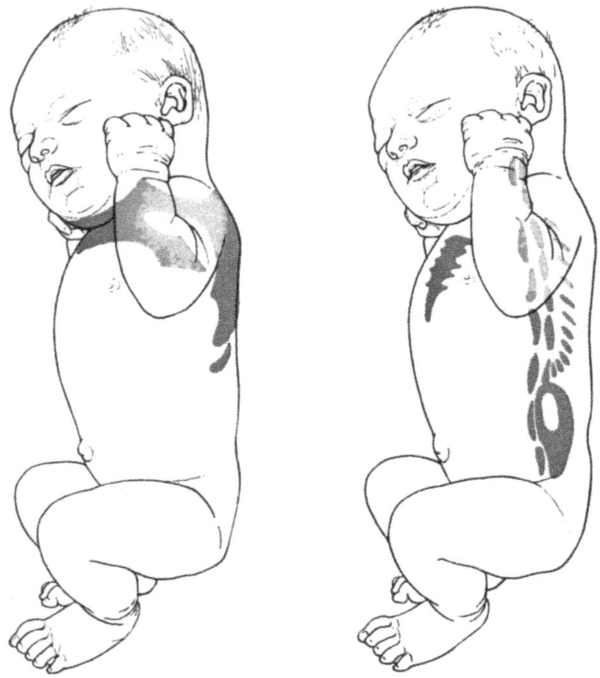

Fig. 2 The main BAT depots in the human newborn. The illustrations specifically depict the BAT depots present in the interscapular, neck, and axillary regions (left), and the smaller deposits behind the sternum, and along the spine (right). Illustration from Dawkins and Hull (1965). Reproduced with permission. Copyright © (1965) Scientific American, a division of Nature America, Inc. All rights reserved

In 1972, Heaton performed a study on the presence of BAT at 18 different locations, from infancy to late adult life (Heaton 1972). She showed that BAT present with a wide distribution during the first decade of life and thereafter starts to disappear from most areas. The depots most resistant to regression appeared to be the ones around the kidneys, adrenals, and aorta as well as those in the neck region and mediastinum, where BAT could be found even in the eighth decade of life.

All the studies mentioned above had one limitation in common: as no cell-specific marker proteins had been deciphered yet, the authors identified BAT solely on the morphological features of the cells within the tissue. Although, the combined cell features, polygonal shape, reticular cytoplasm, and multilocular lipid droplets, are a very good indicator of a brown adipocyte phenotype, there is a risk of misinterpreting a less active, and therefore lipid filled, brown adipocyte as being white. Despite this limitation, the results from the studies mentioned above constitute the foundation for our knowledge of the global distribution of BAT in human fetuses and infants. A schematic overview of the main BAT depots identified in human infants is seen in Fig. 2.

With the discovery of UCP1 in the late 1970s (Heaton et al. 1978; Ricquier and Kader 1976), and the subsequent generation of the first antisera against the protein (Cannon et al. 1982; Lean and James 1983), an important tool that allowed a more certain identification of BAT was attained. However, this tool has so far not been extensively used in histological studies of human fetal or infant BAT. Kortelainen et al. used an antiserum raised against rat UCP1 to analyze human BAT obtained at medicolegal autopsies (Kortelainen et al. 1993). They confirmed that the multilocular adipocytes present in BAT collected from areas around the carotid arteries, subscapular region, and around the thoracic aorta of infants stained positive for UCP1. Interestingly, some small unilocular adipocytes present in the tissues also stained positive, suggesting that these cells might represent less active brown (or beige) adipocytes. UCP1-positive unilocular cells have recently also been found among the multilocular adipocytes in axillary BAT of human fetuses (Velickovic et al. 2014). Hence, as there is an apparent risk of misinterpreting a less active, and therefore unilocular, brown, or beige adipocyte as being white when determining its cell identity on the basis of morphological appearance only, the morphological evaluation should preferentially be performed in conjunction with UCP1 immunostaining when possible.

2.3 Studies of Infant Brown Adipose Tissue Using Modern Noninvasive Imaging Techniques

In recent years, several noninvasive imaging strategies have been used in studies of BAT in human adults (Sampath et al. 2016). Positron emission tomography (PET) with ^{18}F-fluorodeoxyglucose (^{18}F-FDG) as a tracer in combination with computed tomography (CT) has been the most commonly used strategy so far. This modality is attractive as it gives information about the distribution and the metabolic activity of BAT, reflected by the level of tracer uptake in the tissue. Clearly, this is a useful feature when looking for regimens with the potential of affecting BAT activity and recruitment. However, the fact that only metabolically active, and not inactive BAT can be observed is a limitation of the modality. Another limitation of the technique is that it includes the use of radioactive tracers. Hence, its use is primarily restricted to the field of oncology, where it is a key imaging modality for cancer imaging, assisting in diagnosis, staging, and follow-up of treatment success. As in adults, children often show an increased ^{18}F-FDG uptake in the neck, supraclavicular and axillary regions, mediastinum, paravertebral and perinephric areas, and anterior abdominal wall, suggesting the presence of metabolically active BAT in these locations (Gelfand et al. 2005; Hong et al. 2011). No studies presenting data from ^{18}F-FDG-PET/CT imaging performed on human infants appear to have been published so far.

Scintigraphy and single-photon emission computed tomography (SPECT) are other imaging technologies that can be utilized to identify BAT. ^{123}I-metaiodobenzylguanidine (^{123}I-MIBG) is a radioisotope that accumulates in adrenergic tissues and is used when studying the sympathetic innervation of the heart, and to detect neuroendocrine tumors such as pheochromocytomas and neuroblastomas. On the indication that ^{123}I-MIBG scintigraphy of neuroendocrine tumors sometimes displays symmetrical accumulation of the radioisotope in the neck and shoulder regions of children, Okuyama et al. performed a study in which they showed that ^{123}I-MIBG accumulates in the adrenergic nervous system of interscapular BAT of rats (Okuyama et al. 2002). In a retrospective study, examining a multitude of ^{123}I-MIBG scintigraphic studies performed on infants and children who had been treated for neuroendocrine tumors, it was shown that the radioisotope accumulated in the nape of the neck region of 12% of the patients (Okuyama et al. 2003). The fact that the ^{123}I-MIBG uptake is blocked by surgical interruption of thoracic and cervical sympathetic innervation supports the view that the tissue with elevated radioisotope uptake is BAT (Gelfand 2004). ^{99m}Tc-tetrofosmin, a radioisotope absorbed by functional mitochondria, is another example of a tracer accumulating in the shoulder and neck areas of human adults and children, potentially detecting mitochondria-rich BAT (Fukuchi et al. 2003). Like PET, the latter technologies involve the use of radioactive substances, making them less suitable for studies on healthy subjects, particularly infants and children, and studies performed on such groups would need to use safer imaging techniques.

During the last years, magnetic resonance imaging (MRI), a safe modality not involving the use of radioactive tracers or ionizing radiation, has been applied in studies of human BAT in both adults and infants. This is a modality that generally does not depend on the tissue being active and can consequently be used also in postmortem studies. Several MRI based studies have identified BAT by the fat fraction method. In a postmortem study, Hu et al. used this method to identify BAT in a 3-month-old infant (Hu et al. 2012). The authors focused their study on the neck region where they could detect a bilateral tissue in the supraclavicular area that presented with an intermediate-fat fraction. The BAT identity was confirmed by the histological appearance of tissue sampled from the areas assumed to contain BAT, thereby establishing the feasibility of the imaging method for identifying the tissue. Using the fat fraction method, BAT has also been identified in the interscapular region (Hu et al. 2013, 2014; Lidell et al. 2013). Lidell et al. performed postmortem MRI of eight human infants (aged between 0 and 74 weeks) and found an adipose depot in the interscapular region that presented with an intermediate-fat fraction, as expected for BAT (Fig. 3 shows a representative reconstruction of the depot) (Lidell et al. 2013). From a subset of these subjects, tissue samples were collected from the locations indicative of BAT. Histological analyses of the sampled tissue showed that it contained densely packed multilocular UCP1-positive adipocytes. In addition, gene expression analyses revealed that the tissue showed a high *UCP1* expression. Hence, the overall analyses clearly confirmed the BAT identity of the tissue identified by MRI.

Collectively, the modern noninvasive techniques appear to have the ability to visualize most of the BAT depots discovered in the old postmortem studies that identified the tissue by its gross morphology and histological appearance. However, small BAT depots and clusters of thermogenic adipocytes present in otherwise WAT are more difficult to detect by these modern techniques, and histological analyses of sampled tissue will continue to be an integral component of studies of BAT.

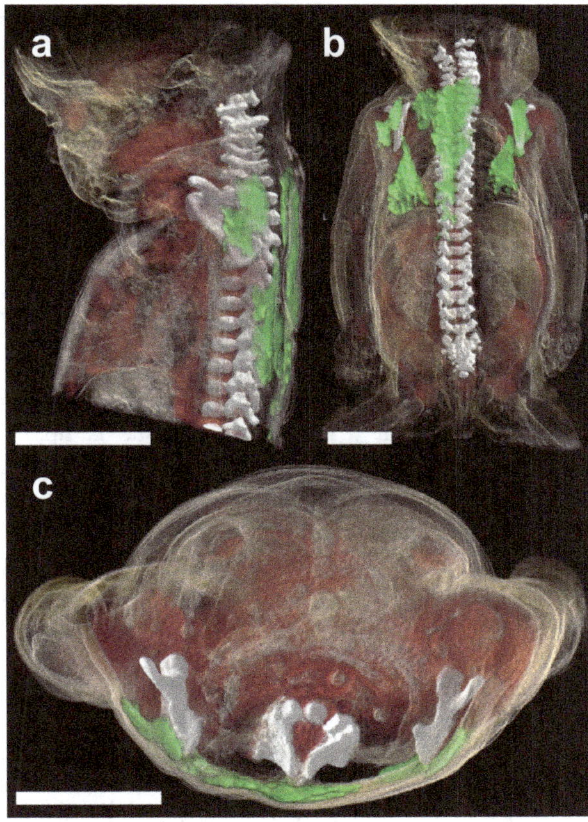

Fig. 3 Postmortem magnetic resonance imaging of infant interscapular BAT. (**a–c**) Three-dimensional reconstruction of interscapular BAT in a human infant. The volume-rendered data set is shown in the sagittal (**a**), coronal (**b**), and axial (**c**) planes. The interscapular BAT is highlighted in green. Scale bars, 5 cm. Reprinted with permission from Macmillan Publishers Ltd: Nature Medicine (Lidell et al. 2013), Copyright © 2013

3 The Physiology of Infant Brown Adipose Tissue

From the discussion above, it is clear that human infants have morphologically identifiable BAT depots. So, what human data exists to support a thermogenic capacity of the tissue during infancy? Silverman et al. were probably first to provide data suggesting the presence of a thermogenic tissue, presumably BAT, in the interscapular region of infants (Silverman et al. 1964). They showed that when exposing infants to a moderately cold environment, the skin temperature over the nape of the neck remained relatively constant while it fell at other exposed body areas. In a few cases, the skin temperature in the neck region even increased in response to the cold exposure. Similar results were later attained by Grausz, who showed that the interscapular skin temperature increased in 13 out of 16 cold exposures, and rose as much as 0.5°C above the colonic temperature during the treatment (Grausz 1970). Furthermore, when assessing cold-exposed infants by thermography, Rylander et al. noted a tendency of heat gain in the nape region after about 10–15 min of cold exposure, and after 30 min the region was the warmest area on the back (Rylander et al. 1972). Collectively, the studies above suggest that newborn human infants respond to a decreased environmental temperature by increasing their heat production in an anatomical location coinciding with that of the vital thermogenic interscapular BAT organ in lower mammals.

Human infants increase their oxygen consumption in response to cold or norepinephrine infusion without shivering (Brück 1961; Hey 1969; Karlberg et al. 1962). Dawkins and Scopes showed that the increased oxygen consumption seen when exposing human newborns to mild cold associates with increased levels of circulating glycerol, without an accompanying rise in plasma free fatty acids, suggesting that cold exposure induces lipolysis in BAT that uses the liberated fatty acids for local thermogenesis (Dawkins and Scopes 1965). The suggestion that cold exposure induces local lipolysis in BAT is supported by histological analyses of BAT from necropsies of infants nursed at different ambient temperatures. Such studies have shown that, while the tissue is lipid filled in infants nursed at thermoneutrality (34–35°C), it is devoid of fat in infants conventionally swaddled and nursed at room temperature (23–27°C) (reviewed in Smith and Horwitz 1969). These results are in line with those recently presented by Hu et al. who used MRI to study a group of newborns who had been subjected to hypothermia therapy (lowering of body core temperature by 3–4°C below normal for 48 h) as a treatment for hypoxic–ischemic encephalopathy (Hu et al. 2014). The authors found that the fat-signal fraction was significantly lower in the supraclavicular BAT depots of the hypothermia-treated infants as compared to those in the untreated control group, suggesting a lower triglyceride content in the BAT of the cold-exposed group. A similar trend, although not statistically significant, was also seen in the interscapular BAT depot, but not in subcutaneous WAT. The authors concluded that it is likely that the hypothermia therapy triggers BAT-mediated non-shivering thermogenesis, which subsequently depletes the tissue of its intracellular triglyceride stores.

Interestingly, an infant's nutritional status might affect its ability to generate heat in BAT. Brooke et al. examined a group of malnourished Jamaican infants and young children, before and after recovery, to see if malnutrition impaired their response to cold (Brooke et al. 1973). They found that, while malnourished, the patients failed to increase their heat production above resting levels when subjected to mild cold, and consequently their body temperature fell. However, following recovery from malnourishment, the patients increased their heat production by 20% during cold exposure and were able to maintain stable body temperature. Since none of the patients shivered during the cold exposure, the thermogenesis was apparently of the non-shivering type. The authors also performed a postmortem study in which the histology of interscapular BAT of malnourished and well-nourished infants and young children was examined. While the tissue sampled from malnourished subjects was fat-depleted, the tissue sampled from well-nourished ones was not. Based on their collective data, the authors concluded that malnourishment, and the associated lipid depletion in BAT, makes the tissue unable to produce all the heat needed for maintaining body temperature when the infant is subjected to cold.

If BAT is of high importance for an infant's ability to maintain core body temperature, one would expect a hampered function of the tissue to have devastating consequences. That functional BAT, and intact sympathetic signaling via β-adrenergic receptors is vital for thermoregulation in small mammals such as rodents is undisputed. Both *Ucp1*-deficient mice (Enerbäck et al. 1997), and mice with reduced BAT content due to transgenic overexpression of the cell toxic diphtheria toxin A-chain specifically in *Ucp1*-expressing adipocytes (Lowell et al. 1993), are sensitive to cold, indicating that their thermoregulation is defective. The same is true for both mice lacking all the β-adrenergic receptors (Bachman et al. 2002), and mice not capable of synthesizing norepinephrine and epinephrine due to inactivation of the gene encoding the enzyme dopamine β-hydroxylase (Thomas and Palmiter 1997). For obvious reasons, similar studies cannot be performed on humans. However, there are some indications that malfunctioning BAT might have serious consequences also for an infant's health. Sudden infant death syndrome (SIDS) is characterized by the sudden death of a seemingly healthy infant during sleep. The etiology of SIDS is complex and includes interactions between multiple factors (Kinney and Thach 2009). One of the risk factors for SIDS has been suggested to be thermal stress (Sawczenko and Fleming 1996), and dysregulated activity in BAT has been hypothesized to play a role in some cases of the syndrome. Based on the indication that the specific UCP1 content in axillary BAT seemed to be lower in subjects with SIDS as cause of death, as compared to infants who had died from other causes, Douglas proposed that a lowered capacity for thermogenesis in BAT could play a role in the etiology of SIDS (Douglas 1992). This view was in line with the results attained in a large postmortem study performed a few years earlier (Emery and Dinsdale 1978). Here, it was shown that periadrenal BAT of young infants, aged 2–20 weeks, who had died from SIDS, contained a higher proportion of unilocular adipocytes, indicating a less active tissue, as compared to the tissue collected from infants in the control group. However, no evidence for a link between SIDS and polymorphisms in the genes encoding UCP1 or the β3-adrenergic receptor, both polymorphisms suggested to associate with a lower ability for BAT thermogenesis, have been found (Fatemi et al. 2002). Interestingly, an increased thermogenic activity in BAT has also been suggested to be involved in the etiology of SIDS. In a report describing two cases of SIDS in which the infants' core temperatures were above 40°C on arrival to the morgue, Lean and Jennings collected tissue from the axillary, cervical, interscapular, and perirenal BAT depots (Lean and Jennings 1989). Measurements on mitochondria isolated from the tissues indicated active thermogenesis with uncoupled respiration, as measured by guanosine diphosphate binding and oxygen uptake. The authors proposed that inappropriate thermogenesis in BAT in the pyrexial state might have contributed to the deaths, and that thermogenesis in BAT, occurring inappropriately in a warm, well-insulated infant could be a cause of some cases of SIDS. Taken together, the notion of a connection between dysregulated thermogenesis in BAT and SIDS is interesting. However, the data supporting such connection are rather weak, and further studies evaluating the potential link are therefore warranted.

4 The Molecular Characteristics of Infant Brown Adipose Tissue

Few studies have investigated the molecular features of infant BAT in detail. However, some molecular data have been attained. The development of antisera against UCP1 enabled analyses of the specific UCP1 content in different BAT depots. In a postmortem study of BAT from almost 50 subjects, about half being infants, it was shown that the UCP1 content, as measured by a radioimmunoassay, was higher in the mitochondria of axillary as compared to perirenal BAT (Lean et al. 1986). In addition, variations in the UCP1 content were seen between the different age groups. Adults as well as preterm and stillborn infants had lower UCP1 levels than older infants and children, an observation consistent with altered requirements and capacities for thermogenesis at different stages of human development. In another postmortem study of newborn infants, gestational age 25–40 weeks, interscapular BAT was analyzed for its specific UCP1 content (Houstek et al. 1993). All samples contained significant amounts of UCP1 and the specific content of the protein almost doubled in the tissue between weeks 25 and 32, and remained nearly constant thereafter. It was also shown that the activity of type II iodothyronine deiodinase (DIO2), an enzyme that converts thyroxine (T4) to triiodothyronine (T3), the more active ligand for the thyroid hormone receptors, was active in interscapular BAT already at gestational week 25, and that the activity of the enzyme increased over the gestational period examined in a similar way as that of the UCP1 content in the tissue. From animal studies, it is clear that DIO2-mediated local production of T3 in BAT, and the subsequent activation of signaling through thyroid hormone receptors, is of great importance for adaptive thermogenesis in the tissue (de Jesus et al. 2001). Taken together, based on the UCP1 content and DIO2 activity in interscapular BAT, it appears that the thermogenic capacity of the tissue develops rather early, and by the beginning of the last trimester it is likely that it has acquired a significant thermogenic potential.

As mentioned above, it is now clear that there are two types of UCP1 expressing adipocytes, the classical brown and the beige adipocytes, that originate from distinct cell lineages. The studies reaching this conclusion included lineage tracing experiments (Seale et al. 2008), and molecular characterization of clonal cell lines established from precursor cells isolated from the stromal vascular fractions of interscapular BAT and inguinal adipose tissue of mice (Wu et al. 2012). The energy expending feature of thermogenic adipocytes is of potential therapeutic use against obesity and obesity-related diseases such as type 2 diabetes. Therefore, the existence of two different thermogenic cell types, which potentially can be recruited and/or activated by different stimuli, is of great interest, as they might represent two separate targets for therapeutic intervention. Hence, it is important to establish if humans, as rodents, possess the two cell types. In rodents, classical brown adipocytes are typically found in the interscapular BAT depot. The observations from the discussion above indicate that human infants have BAT in the interscapular region, and it is probably here it is most likely to find classical brown adipocytes in humans. Recently, a postmortem study of human infants showed that BAT sampled from this region guided by its intermediate-fat fraction, as determined by MRI, presented with features similar to that of the interscapular BAT organ of rodents (Lidell et al. 2013). Like the classical brown adipocytes present in the rodent tissue, the multilocular and UCP1-positive cells in the infant tissue were densely packed and delineated from the subcutaneous WAT by a layer of connective tissue. In addition, the cells presented with a gene expression signature more similar to that of classical brown than beige adipocytes. Hence, it appears like at least human infants possess classical brown adipocytes. This view is also supported by a late study in which mesenchymal stem cells, attained from interscapular BAT of human fetuses, were shown to differentiate into thermogenically competent, UCP1-containing adipocytes with a classical brown-like gene expression profile, when subjected to an adipogenic differentiation cocktail (Di Franco et al. 2016). Importantly, as in mice, the identity of the thermogenic cells in different adipose depots appears to differ in human infants. Sharp et al. studied the molecular features of BAT sampled from supraclavicular areas, posterior mediastinum, and retroperitoneal, intra-abdominal, and mesenteric depots (Sharp et al. 2012). They found that nearly all sampled BAT abundantly expressed beige-selective genes, while the expression of classical brown-selective genes was barely detectable, suggesting that the BAT was of the beige type. The beige nature of human supraclavicular BAT had previously been suggested in adults (Wu et al. 2012), and at the time it was even questioned if humans possess classical brown adipocytes at all (Cannon and Nedergaard 2012). However, apart from the studies mentioned above, showing that infant interscapular BAT is likely to contain this cell type, there are also indications of it being present in the deep neck region of adults (Cypess et al. 2013). Hence, it is likely that humans, as rodents, have both classical brown and beige adipocytes.

5 Conclusions

The collective data from studies of BAT in human fetuses and infants indicate that the tissue is widely distributed during these developmental stages, and that the tissue is particularly developed in the axillae, neck, and upper back regions as well as deep internally, around important structures such as the kidneys. The thermogenic capacity of the tissue, as measured by its UCP1 content, develops with gestational age and reaches its maximum in infancy and early childhood when the demands for thermogenesis can be expected to be especially high. When subjected to cold, human infants lack a sufficiently developed skeletal muscle mass for maintaining body temperature through shivering thermogenesis. Instead, they respond to cold exposure by inducing non-shivering thermogenesis. As the response is associated with increased lipolysis in BAT, and induction of heat production in regions harboring the tissue, it is reasonable to assume that the cold-induced thermogenesis occurs in BAT. This notion is also supported by numerous animal studies that additionally have shown that thermogenesis in BAT is essential for maintaining a stable core body temperature in a cold environment.

Over the last years, our knowledge of rodent BAT, and factors that affect its recruitment and activation, have increased substantially. However, despite these advances, little is still known about the human tissue on a molecular level, and dedicated studies trying to establish if the knowledge attained in the animal studies can be translated to humans are warranted. Importantly, the infant data that are available imply that humans, as rodents, harbor two types of thermogenic adipocytes, the classical brown and the beige adipocytes. Further evaluations of the molecular features of the human cells would benefit from establishing clonal cell lines from the different fetal and infant BAT depots. Such clonal cell lines would also be useful tools in studies aiming to identify factors with the capacity of activating and/or expanding the thermogenic potential of human BAT; that is, factors that would pose potential targets for therapy against obesity and obesity-related diseases such as type 2 diabetes.

Acknowledgements The author is supported by the Swedish Research Council, Wilhelm and Martina Lundgren's Foundation, and The Royal Society of Arts and Sciences in Gothenburg.

References

Aherne W, Hull D (1966) Brown adipose tissue and heat production in the newborn infant. J Pathol Bacteriol 91:223–234

Bachman ES, Dhillon H, Zhang CY, Cinti S, Bianco AC, Kobilka BK, Lowell BB (2002) betaAR signaling required for diet-induced thermogenesis and obesity resistance. Science 297:843–845

Bonnot E (1908) The interscapular gland. J Anat Physiol 43:43–58

Brooke OG, Harris M, Salvosa CB (1973) The response of malnourished babies to cold. J Physiol 233:75–91

Brück K (1961) Temperature regulation in newborn infant. Biol Neonat 3:65–81

Cannon B, Nedergaard J (2004) Brown adipose tissue: function and physiological significance. Physiol Rev 84:277–359

Cannon B, Nedergaard J (2012) Cell biology: neither brown nor white. Nature 488:286–287

Cannon B, Hedin A, Nedergaard J (1982) Exclusive occurrence of thermogenin antigen in brown adipose tissue. FEBS Lett 150:129–132

Cypess AM, White AP, Vernochet C, Schulz TJ, Xue R, Sass CA, Huang TL, Roberts-Toler C, Weiner LS, Sze C, Chacko AT, Deschamps LN, Herder LM, Truchan N, Glasgow AL, Holman AR, Gavrila A, Hasselgren PO, Mori MA, Molla M, Tseng YH (2013) Anatomical localization, gene expression profiling and functional characterization of adult human neck brown fat. Nat Med 19:635–639

Dawkins MJ, Hull D (1965) The production of heat by fat. Sci Am 213:62–67

Dawkins MJ, Scopes JW (1965) Non-shivering thermogenesis and brown adipose tissue in the human new-born infant. Nature 206:201–202

de Jesus LA, Carvalho SD, Ribeiro MO, Schneider M, Kim SW, Harney JW, Larsen PR, Bianco AC (2001) The type 2 iodothyronine deiodinase is essential for adaptive thermogenesis in brown adipose tissue. J Clin Invest 108:1379–1385

Di Franco A, Guasti D, Squecco R, Mazzanti B, Rossi F, Idrizaj E, Gallego-Escuredo JM, Villarroya F, Bani D, Forti G, Vannelli GB, Luconi M (2016) Searching for classical brown fat in humans: development of a novel human fetal brown stem cell model. Stem Cells 34:1679–1691

Douglas RJ (1992) Could a lowered level of uncoupling protein in brown adipose tissue mitochondria play a role in SIDS aetiology? Med Hypotheses 37:100–102

Emery JL, Dinsdale F (1978) Structure of periadrenal brown fat in childhood in both expected and cot deaths. Arch Dis Child 53:154–158

Enerbäck S, Jacobsson A, Simpson EM, Guerra C, Yamashita H, Harper ME, Kozak LP (1997) Mice lacking mitochondrial uncoupling protein are cold-sensitive but not obese. Nature 387:90–94

Fatemi A, Item C, Stockler-Ipsiroglu S, Ipsiroglu O, Sperl W, Patsch W, Strobl W (2002) Sudden infant death: no evidence for linkage to common polymorphisms in the uncoupling protein-1 and the beta3-adrenergic receptor genes. Eur J Pediatr 161:337–339

Fedorenko A, Lishko PV, Kirichok Y (2012) Mechanism of fatty-acid-dependent UCP1 uncoupling in brown fat mitochondria. Cell 151:400–413

Fukuchi K, Ono Y, Nakahata Y, Okada Y, Hayashida K, Ishida Y (2003) Visualization of interscapular brown adipose tissue using (99 m)Tc-tetrofosmin in pediatric patients. J Nucl Med 44:1582–1585

Gelfand MJ (2004) 123I-MIBG uptake in the neck and shoulders of a neuroblastoma patient: damage to sympathetic innervation blocks uptake in brown adipose tissue. Pediatr Radiol 34:577–579

Gelfand MJ, O'Hara SM, Curtwright LA, Maclean JR (2005) Pre-medication to block [(18)F]FDG uptake in the brown adipose tissue of pediatric and adolescent patients. Pediatr Radiol 35:984–990

Grausz JP (1970) Interscapular skin temperatures in the newborn infant. J Pediatr 76:752–756

Hammar JA (1895) Zur Kenntniss des fettgewebes. Arch Mikrosk Anat 45:512–574

Harms M, Seale P (2013) Brown and beige fat: development, function and therapeutic potential. Nat Med 19:1252–1263

Hatai S (1902) On the presence in human embryos of an interscapular gland corresponding to the so-called hibernating gland of lower mammals. Anat Anz 21:369–373

Heaton JM (1972) The distribution of brown adipose tissue in the human. J Anat 112:35–39

Heaton GM, Wagenvoord RJ, Kemp A Jr, Nicholls DG (1978) Brown-adipose-tissue mitochondria: photoaffinity labelling of the regulatory site of energy dissipation. Eur J Biochem 82:515–521

Hey EN (1969) The relation between environmental temperature and oxygen consumption in the new-born baby. J Physiol 200:589–603

Hong TS, Shammas A, Charron M, Zukotynski KA, Drubach LA, Lim R (2011) Brown adipose tissue 18F-FDG uptake in pediatric PET/CT imaging. Pediatr Radiol 41:759–768

Houstek J, Vizek K, Pavelka S, Kopecky J, Krejcova E, Hermanska J, Cermakova M (1993) Type II iodothyronine 5′-deiodinase and uncoupling protein in brown adipose tissue of human newborns. J Clin Endocrinol Metab 77:382–387

Hu HH, Tovar JP, Pavlova Z, Smith ML, Gilsanz V (2012) Unequivocal identification of brown adipose tissue in a human infant. J Magn Reson Imaging 35:938–942

Hu HH, Yin L, Aggabao PC, Perkins TG, Chia JM, Gilsanz V (2013) Comparison of brown and white adipose tissues in infants and children with chemical-shift-encoded water-fat MRI. J Magn Reson Imaging 38:885–896

Hu HH, Wu TW, Yin L, Kim MS, Chia JM, Perkins TG, Gilsanz V (2014) MRI detection of brown adipose tissue with low fat content in newborns with hypothermia. Magn Reson Imaging 32:107–117

Karlberg P, Moore RE, Oliver TK Jr (1962) The thermogenic response of the newborn infant to noradrenaline. Acta Paediatr 51:284–292

Kinney HC, Thach BT (2009) The sudden infant death syndrome. N Engl J Med 361:795–805

Kooijman S, van den Heuvel JK, Rensen PC (2015) Neuronal control of brown fat activity. Trends Endocrinol Metab 26:657–668

Kortelainen ML, Pelletier G, Ricquier D, Bukowiecki LJ (1993) Immunohistochemical detection of human brown adipose tissue uncoupling protein in an autopsy series. J Histochem Cytochem 41:759–764

Lean ME, James WP (1983) Uncoupling protein in human brown adipose tissue mitochondria. Isolation and detection by specific antiserum. FEBS Lett 163:235–240

Lean ME, Jennings G (1989) Brown adipose tissue activity in pyrexial cases of cot death. J Clin Pathol 42:1153–1156
Lean ME, James WP, Jennings G, Trayhurn P (1986) Brown adipose tissue uncoupling protein content in human infants, children and adults. Clin Sci (Lond) 71:291–297
Lidell ME, Betz MJ, Dahlqvist Leinhard O, Heglind M, Elander L, Slawik M, Mussack T, Nilsson D, Romu T, Nuutila P, Virtanen KA, Beuschlein F, Persson A, Borga M, Enerbäck S (2013) Evidence for two types of brown adipose tissue in humans. Nat Med 19:631–634
Lidell ME, Betz MJ, Enerbäck S (2014) Brown adipose tissue and its therapeutic potential. J Intern Med 276:364–377
Lin CS, Klingenberg M (1980) Isolation of the uncoupling protein from brown adipose tissue mitochondria. FEBS Lett 113:299–303
Long JZ, Svensson KJ, Tsai L, Zeng X, Roh HC, Kong X, Rao RR, Lou J, Lokurkar I, Baur W, Castellot JJ Jr, Rosen ED, Spiegelman BM (2014) A smooth muscle-like origin for beige adipocytes. Cell Metab 19:810–820
Lowell BB, SS V, Hamann A, Lawitts JA, Himms-Hagen J, Boyer BB, Kozak LP, Flier JS (1993) Development of obesity in transgenic mice after genetic ablation of brown adipose tissue. Nature 366:740–742
Merklin RJ (1974) Growth and distribution of human fetal brown fat. Anat Rec 178:637–645
Oelkrug R, Polymeropoulos ET, Jastroch M (2015) Brown adipose tissue: physiological function and evolutionary significance. J Comp Physiol B 185:587–606
Okuyama C, Sakane N, Yoshida T, Shima K, Kurosawa H, Kumamoto K, Ushijima Y, Nishimura T (2002) (123)I- or (125)I-metaiodobenzylguanidine visualization of brown adipose tissue. J Nucl Med 43:1234–1240
Okuyama C, Ushijima Y, Kubota T, Yoshida T, Nakai T, Kobayashi K, Nishimura T (2003) 123I-Metaiodobenzylguanidine uptake in the nape of the neck of children: likely visualization of brown adipose tissue. J Nucl Med 44:1421–1425
Orava J, Nuutila P, Lidell ME, Oikonen V, Noponen T, Viljanen T, Scheinin M, Taittonen M, Niemi T, Enerbäck S, Virtanen KA (2011) Different metabolic responses of human brown adipose tissue to activation by cold and insulin. Cell Metab 14:272–279
Rasmussen AT (1923) The so-called hibernating gland. J Morphol 38:147–205
Ricquier D, Kader JC (1976) Mitochondrial protein alteration in active brown fat: a soidum dodecyl sulfate-polyacrylamide gel electrophoretic study. Biochem Biophys Res Commun 73:577–583
Rylander E, Pribylova H, Lind J (1972) A thermographic study of infants exposed to cold. Acta Paediatr Scand 61:42–48
Sampath SC, Sampath SC, Bredella MA, Cypess AM, Torriani M (2016) Imaging of brown adipose tissue: state of the art. Radiology 280:4–19
Sawczenko A, Fleming PJ (1996) Thermal stress, sleeping position, and the sudden infant death syndrome. Sleep 19:S267–S270
Seale P, Bjork B, Yang W, Kajimura S, Chin S, Kuang S, Scime A, Devarakonda S, Conroe HM, Erdjument-Bromage H, Tempst P, Rudnicki MA, Beier DR, Spiegelman BM (2008) PRDM16 controls a brown fat/skeletal muscle switch. Nature 454:961–967
Sharp LZ, Shinoda K, Ohno H, Scheel DW, Tomoda E, Ruiz L, Hu H, Wang L, Pavlova Z, Gilsanz V, Kajimura S (2012) Human BAT possesses molecular signatures that resemble beige/brite cells. PLoS One 7:e49452
Shattock SG (1909) On normal tumour-like formations of fat in man and the lower animals. Proc R Soc Med 2:207–270
Shaw HB (1901) A contribution to the study of the morphology of adipose tissue. J Anat Physiol 36:1–13
Sheldon EF (1924) The so-called hibernating gland in mammals: a form of adipose tissue. Anat Rec 28:331–347
Silva JE (2006) Thermogenic mechanisms and their hormonal regulation. Physiol Rev 86:435–464
Silverman WA, Zamelis A, Sinclair JC, Agate FJ (1964) Warm nap of the newborn. Pediatrics 33:984–987

Smith RE, Horwitz BA (1969) Brown fat and thermogenesis. Physiol Rev 49:330–425
Thomas SA, Palmiter RD (1997) Thermoregulatory and metabolic phenotypes of mice lacking noradrenaline and adrenaline. Nature 387:94–97
Velickovic K, Cvoro A, Srdic B, Stokic E, Markelic M, Golic I, Otasevic V, Stancic A, Jankovic A, Vucetic M, Buzadzic B, Korac B, Korac A (2014) Expression and subcellular localization of estrogen receptors alpha and beta in human fetal brown adipose tissue. J Clin Endocrinol Metab 99:151–159
Vishvanath L, MacPherson KA, Hepler C, Wang QA, Shao M, Spurgin SB, Wang MY, Kusminski CM, Morley TS, Gupta RK (2016) Pdgfrbeta+ mural preadipocytes contribute to adipocyte hyperplasia induced by high-fat-diet feeding and prolonged cold exposure in adult mice. Cell Metab 23:350–359
Wu J, Boström P, Sparks LM, Ye L, Choi JH, Giang AH, Khandekar M, Virtanen KA, Nuutila P, Schaart G, Huang K, Tu H, van Marken Lichtenbelt WD, Hoeks J, Enerbäck S, Schrauwen P, Spiegelman BM (2012) Beige adipocytes are a distinct type of thermogenic fat cell in mouse and human. Cell 150:366–376

Part II

Molecular Mechanisms of BAT Function and Signaling

Evolution of UCP1

Michael J. Gaudry, Kevin L. Campbell, and Martin Jastroch

Contents

1 Brown Adipose Tissue Thermogenesis 128
2 The Rise of UCP1 129
3 The Fall of UCP1 134
4 Translational Value 136
References 137

Abstract

Brown adipose tissue (BAT), the specialized heat-producing organ found in many placental mammals including humans, may be accessible for clinical drug intervention to help combat metabolic diseases. Understanding the biology of BAT and its thermogenic uncoupling protein 1 (UCP1) will benefit from an assessment of its evolution, answering where UCP1 originated and how it has been modified and integrated into cellular energy metabolism. Here, we review topical insights regarding the molecular evolution of UCP1—also reconstructing the proximate and ultimate factors selecting for brown fat thermogenesis in placental mammals. This new thinking on "old" events will assist our understanding of how thermogenic mitochondrial uncoupling was integrated into the physiology of the brown adipocyte. Recent comparative studies examining the occurrence of UCP1 in vertebrates not only identified the ancient (pre-mammal) rise of UCP1 but also its repeated downfall during mammalian evolution as

M. J. Gaudry · K. L. Campbell
Department of Biological Sciences, University of Manitoba, Winnipeg, MB, Canada

M. Jastroch (✉)
Institute for Diabetes and Obesity, Helmholtz Diabetes Center at Helmholtz Zentrum München, Neuherberg, Germany

German Center for Diabetes Research (DZD), München-Neuherberg, Germany
e-mail: martin.jastroch@helmholtz-muenchen.de

A. Pfeifer et al. (eds.), *Brown Adipose Tissue*,
Handbook of Experimental Pharmacology 251, https://doi.org/10.1007/164_2018_116

evidenced by multiple independent gene loss and/or inactivation events. Together with the comparative physiology of various species, we may be able to find conditions that favor UCP1 thermogenesis and, learning from these insights, identify molecular networks that will be useful to pharmacologically stimulate the tissue.

Keywords
Brown adipose tissue · Evolution · Metabolic disease · Thermogenesis · Uncoupling protein

1 Brown Adipose Tissue Thermogenesis

Brown adipose tissue (BAT) is a specialized organ in placental mammals that enables non-shivering thermogenesis (NST) via molecular mechanisms centered in the mitochondria that lower metabolic efficiency. In addition to the dense mitochondrial content of this tissue imparting its brown coloration, BAT is both advantageously situated near vital organs of the body (i.e., interscapular, subscapular, dorso-cervical and axillary regions, as well as near the kidneys) and highly vascularized allowing for effective transfer of heat to the circulatory system (Oelkrug et al. 2015). The rapid energy turnover in brown adipocytes is enabled by high mitochondrial concentrations of uncoupling protein 1 (UCP1). UCP1 short-circuits the proton-motive force that typically drives ATP synthesis, increasing substrate oxidation and, consequently, enhancing cellular heat output (Cannon and Nedergaard 2004). BAT is widely accepted to enable both small-bodied mammals and the neonates of many larger-bodied species to survive acute and chronic cold challenges, as well as to facilitate rewarming in lineages that utilize torpor (Oelkrug et al. 2010, 2013; Cannon and Nedergaard 2004; Nicol et al. 2009). While multilocular lipid droplets of brown adipocytes provide enhanced surface area to facilitate rapid lipolysis and catabolism (Keipert and Jastroch 2014), heat output of BAT can also be maintained via glucose oxidation without the contribution of lipolysis (Shin et al. 2017). This indifference regarding substrate preference further highlights the predominant role of UCP1 for mitochondrial energy turnover. However, a pivotal role for UCP1 in thermogenesis is not universal among mammals (Gaudry et al. 2017; Keipert et al. 2017; Meyer et al. 2012; Golozoubova et al. 2001; Ukropec et al. 2006), as alternative molecular mechanisms to produce heat via ATP-consuming processes or other endogenous uncouplers have been proposed (Ikeda et al. 2017; Kazak et al. 2017b; Long et al. 2016). These thermogenic pathways offer new pharmacological potential, but notably, their measured physiological impacts have so far been exclusively restricted to UCP1 knockout mice (Kazak et al. 2017b; Long et al. 2016). For pigs, which naturally lack UCP1 (Berg et al. 2006), and mice, an ATP-consuming calcium futile cycle through SERCA/ryanodine receptor activity has been proposed as an elegant alternative heat-generating mechanism in beige adipose tissue (Ikeda et al. 2017). While pharmacological as well as gain- and loss-of-function experiments targeting SERCA and ryanodine receptors show the impact on mitochondrial respiration rates in beige adipocytes, the bioenergetic data of the study surprisingly reveal no impact on ATP-linked respiration, which is in stark contrast to the suggested ATP-dependent model of thermogenesis. Notably, increased ATP consumption would stimulate both glycolytic and oxidative ATP production, at least based upon our current understanding of cellular bioenergetics. Further experiments are thus required to delineate whether calcium futile cycling contributes to thermogenesis in pigs or whether this mechanism remains fishy (in the sense that it was originally proposed for and remains limited to the cranial heater organ of billfishes, swordfish, and the butterfly mackerel; Morrissette et al. 2003). It should be taken into consideration that the disturbance of calcium homeostasis in the Ikeda study may have affected calcium-sensitive dehydrogenases (e.g., pyruvate dehydrogenase), providing a simple explanation for the observed cellular phenomena. Additional research is required to determine whether these and other alternative mechanisms of heat production have any physiological significance in nature, or not. With regard to mitochondrial uncoupling, other UCPs have been suggested to thermogenically compensate if BAT-mediated NST becomes impaired, but this idea has been refuted several times (Golozoubova et al. 2001; Nedergaard and Cannon 2003), mainly due to inconclusive evidence regarding uncoupling activity (Cadenas et al. 2002; Shabalina et al. 2010: Nabben et al. 2011). A potential role for UCP3 in thermogenesis by uncoupled respiration has been recently revived in (some) pig breeds lacking UCP1 (Lin et al. 2017), but the conclusions have to be taken with similar caution as the observed uncoupling cannot be reconciled with our current knowledge on mitochondrial respiratory control (Jastroch et al. 2018). Notably, UCP1-ablated mice further render BAT dysfunctional by inflammation and electron transport deficiencies in the cold (Oelkrug et al. 2010; Kazak et al. 2017a; Keipert et al. 2017), most likely also compromising ATP output. Thus, it remains questionable whether BAT of UCP1 knockout mice represents a good model to investigate some proposed ATP-dependent thermogenic pathways. ATP-dependent thermogenic pathways in beige adipose tissue may be significant if glycolytic and oxidative ATP production can provide sufficient capacities. Furthermore, it should not be forgotten that heat generated by muscle shivering is able to compensate for the lack of brown fat NST (Golozoubova et al. 2001).

2 The Rise of UCP1

Utilizing a comparative approach to investigate the evolution of UCP1 over millions of years could provide valuable insights of both its function and capacity for medical intervention. UCP1, first discovered in the late 1970s (Ricquier and Kader 1976; Heaton et al. 1978), was long believed to have originated with the evolution of eutherian BAT as a unique strategy to defend high body temperatures in the cold (Cannon and Nedergaard 2004). However, tracing orthologous *UCP1* loci by comparative genomics unambiguously revealed the presence of this six-exon gene in teleost fishes, discernible from *UCP2* and *UCP3* paralogs by its conserved syntenic arrangement (i.e., 5′-*TBC1D9-UCP1-ELMOD2*-3′; Fig. 1; Jastroch et al. 2005).

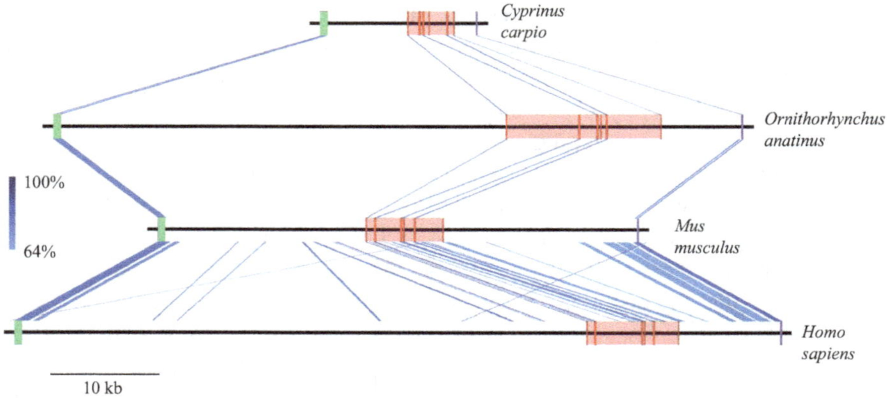

Fig. 1 Sequence identity comparison made in Easyfig 2.2.2 displaying the conserved synteny of the *UCP1* gene among representative vertebrate species (common carp [*Cyprinus carpio*], platypus [*Ornithorhynchus anatinus*], house mouse [*Mus musculus*], and human [*Homo sapiens*]) with respective accession numbers of LHQP01013372.1, NW_001794248.1, AC122890.4, and LOQN01003215.1. *UCP1* exons are symbolized by solid red bars with intervening intron sequences shaded in red. Terminal exons of flanking genes *TBC1D9* and *ELMOD2* are symbolized with green and purple boxes, respectively

This unequivocally revealed that *UCP1* predated the evolution of endothermy and was present in a common ancestor of ray- and lobe-finned fishes (~420 million years ago [MYA]), thus spurring a re-examination of its evolutionary origin.

Common carp (*Cyprinus carpio*) *UCP1* mRNA is detected in a wide range of tissues including the liver, brain, intestine, and kidney, but interestingly not adipose tissue (Jastroch et al. 2005). The high expression of the fish *UCP1* orthologue in the liver of the common carp has been independently confirmed by other studies (Bermejo-Nogales et al. 2014; Murakami et al. 2015; Wen et al. 2015). Interestingly, the regulation of fish UCP1 by environmental cues differs vastly between organs as cold-exposed carp downregulate *UCP1* mRNA levels in the liver (Jastroch et al. 2005), but increase expression in brain tissue (Jastroch et al. 2007). In particular, this latter upregulation in certain areas of the brain fostered speculation of local thermogenesis in the neuronal tissues of carp (Jastroch et al. 2007). A similar role for UCP1 has also been suggested in the brain of hibernating thirteen-lined ground squirrels (*Spermophilus tridecemlineatus*; Laursen et al. 2015), but these results await confirmation by other studies. Additionally, one has to take into account that local thermogenesis would not only require UCP1, but also high oxidative capacity and rapid metabolic fluxes. Although liver mitochondrial uncoupling in carp coincides with the presence of *UCP1* mRNA (Jastroch et al. 2007), the involvement of other mitochondrial anion carriers, such as GDP-sensitive uncoupling activity of the adenine nucleotide transporter (Khailova et al. 2006), has not yet been ruled out. In short, future studies are required to fully elucidate the myriad functions of UCP1 in ectothermic vertebrates and nonplacental mammals (monotremes and marsupials), though its physiological roles are likely important as it is highly conserved and evolving under strong purifying selection within these lineages (Gaudry et al. 2017). Similarly, *UCP2* and especially *UCP3* are remarkably well conserved among vertebrates (Gaudry et al. 2017), and while these paralogous members of the UCP gene family have been proposed to fulfill several functions including catalyzing mild uncoupling to mitigate the production of harmful reactive oxygen species, to date, a consensus has not been reached on their functions, and any suggestions of their thermogenicity remain unsubstantiated (Brand and Esteves 2005; Echtay 2007; Mailloux and Harper 2011; Lin et al. 2017).

When we focus on the most basal branch of the mammalian family tree, monotremes, RNA-seq BLAST searches reveal the presence of *UCP1* translation in a surprisingly wide range of tissue types (Gaudry and Campbell 2017). Indeed, *UCP1* transcripts are found in platypus (*Ornithorhynchus anatinus*) testis, ovary, liver, kidney, heart, and brain tissue, which is reminiscent of the expression pattern seen in ectothermic vertebrates. By contrast, available published evidence suggests marsupial UCP1 may be restricted to adipose depots, which often appear brownish in coloration and are located in the pectoral regions of juvenile gray short-tailed opossums (*Monodelphis domestica*) and the interscapular regions of the fat-tailed dunnart (*Sminthopsis crassicaudata*; Jastroch et al. 2008). Furthermore, this latter species utilizes daily torpor and upregulates UCP1 expression following cold exposure, raising the possibility of a heat-producing role. Yet, so far, evidence for thermogenesis by marsupial UCP1 is lacking, as the fat-tailed dunnart was later shown to be incapable of adaptive NST through classical noradrenaline injection (Polymeropoulos et al. 2012). It remains unknown whether monotreme or marsupial UCP1 permits mitochondrial proton leak as it does in eutherians. Notably, a 5′ enhancer box located ~3–5 kb upstream of the *UCP1* gene, believed to contribute to high UCP1 expression in eutherian BAT, first arose in a stem eutherian ancestor and is absent in non-eutherian mammals (Jastroch et al. 2008; Gaudry and Campbell 2017). This difference may partly underlie comparatively lower *UCP1* transcription in marsupial adipose depots relative to that of comparably sized eutherians (Rousset et al. 2004). While this "brownish" adipose tissue may still contribute to heat production in this lineage, further studies are required to clarify this and other potential physiological roles.

In sharp contrast to ectothermic vertebrates and the platypus, eutherian UCP1 is predominantly expressed in brown adipocytes and, under certain physiological conditions (e.g., cold stress), within white adipose tissue depots. These latter UCP1-positive cells are referred to as "beige" or "brite" (brown-in-white) adipocytes, as they take on further brown adipocyte-like characteristics such as increased mitochondrial density and multilocularity (Harms and Seale 2013). Apart from some reports on UCP1 expression in neurons and thymocytes (Carroll et al. 2005; Adams et al. 2008; Laursen et al. 2015), eutherian UCP1 expression appears highly tissue-specific. Moreover, UCP1 in BAT mitochondria is exceedingly concentrated and can represent as much as 8% of mitochondrial membrane proteins in brown adipocytes (Rousset et al. 2004), some 100- to 1000×-fold higher than expression levels of UCP2 and UCP3 in other tissues (Brand and Esteves 2005).

Among a wide range of eutherian mammals, the role of UCP1 has been well-documented to be thermogenic and, for some species, contribute a crucial survival advantage in cold environments, including mouse knockout models that confirm an essential role of UCP1 to acute cold challenges. For instance, UCP1 is implicated in BAT-mediated NST of the rock elephant shrew (*Elephantulus myurus*; Mzilikazi et al. 2007) and lesser hedgehog tenrec (*Echinops telfairi*; Oelkrug et al. 2013), two affiliates of the Afrotherian superorder. Some features of BAT in Afrotherian species may be considered archetypal, such as no regulation of BAT mass and UCP1 content in the elephant shrews, or the unusual abdominal location of BAT in the tenrec. This finding is in line with recent work that reveals marked differences in putative promoter and enhancer elements in non-model versus murid rodent model systems (Gaudry and Campbell 2017). Indeed, some promoter elements (e.g., CRE-4 and CCAAT box) postulated to contribute to *UCP1* transcription in the mouse and rat are absent in non-murid species. Also, while the enhancer is generally conserved, relatively few motifs (CRE-3, PPRE, and RARE-3) thought to be key for *UCP1* expression in rodents display high identity to homologous sequences in other lineages, highlighting the likely evolution of differential transcriptional control mechanisms within the eutherian radiation. In members of both the Laurasiatheria and Euarchontoglires superorders, BAT-mediated NST has been detailed in a wide range of studies as reviewed by Oelkrug et al. (2015). Notably, the Euarchontoglires superorder also includes humans, who are known to express BAT as newborns and even into adulthood (Nedergaard et al. 2007; Cypress et al. 2009; van Marken Lichtenbelt et al. 2009; Virtanen et al. 2009). Thus, due to its widespread occurrence across the eutherian radiation, the rise of adaptive UCP1 thermogenesis presumably originated following the marsupial/placental divergence. Recent fossil-calibrated timetrees (Springer et al. 2017; Liu et al. 2017) accordingly place this event within a ~100 million year window between the Middle Jurassic and Late Cretaceous (~180 and 80 MYA). The global climate over much of this period is generally characterized as a warm-to-hot "greenhouse" with no polar glaciations and only small latitudinal temperature gradients, though the early Middle Jurassic may have been cooler with latitudinal gradients more in line with the present (O'Brien et al. 2017; Alberti et al. 2017). Ancestral reconstructions also infer that the Late Cretaceous crown eutherian ancestor was a small-bodied (<245 g) scansorial insectivore with an altricial reproductive strategy (O'Leary et al. 2013) and was probably heterothermic (Lovegrove 2012). Therefore, despite likely having evolved in a much warmer world, BAT may have arisen as a mechanism to maintain elevated body temperatures in newborns (Rowlatt et al. 1971), for offspring thermoincubation (Oelkrug et al. 2013), and/or as a more efficient method of rewarming from bouts of torpor (Oelkrug et al. 2011). Although UCP1-mediated NST has even been proposed to have facilitated the expansion of modern eutherian groups to cold ecological niches (and temperature seasonality) that emerged in the early Oligocene (~34 MYA), it should be noted that the advent of thermogenic BAT predates these events by 50–150 million years.

Molecular phylogenetic studies reveal long stem eutherian branch lengths in both UCP1 nucleotide and amino acid trees relative to those of UCP2 and UCP3 (Fig. 2), demonstrating an elevated substitution rate that may have led to the hypothesized

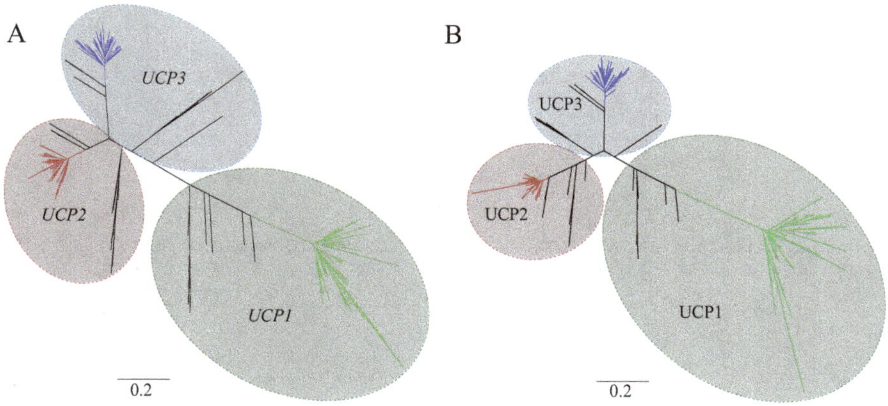

Fig. 2 UCP gene trees of nucleotide coding sequences (**a**) and virtually translated amino acid sequences (**b**) from the data set of Gaudry and Campbell (2017; $n = 448$). Approximately-maximum-likelihood phylogenetic trees were built using FastTree 2.1.5 with default settings. Branch lengths denote number of substitutions per site. Branches of non-eutherians are displayed in black, while eutherian branches are colored in green (UCP1), red (UCP2), and blue (UCP3). Note the lengths of the stem eutherian branches are substantially longer for UCP1 than both UCP2 and UCP3 indicating an increased rate of nucleotide and amino acid substitution

novel function of UCP1 that permitted physiologically significant proton translocation across the mitochondrial inner membrane (Saito et al. 2008; Hughes et al. 2009; Gaudry and Campbell 2017). Saito et al. (2008) attributed this apparent gain of thermogenic function to directional/positive selection, although follow-up studies with improved (larger and more diverse) data sets indicate non-synonymous (dn) to synonymous (ds) nucleotide substitution ratios along the stem eutherian branch of 0.5–0.6, which is more in line with relaxed evolutionary constraints rather than positive selection (defined as dn/ds > 1; Hughes et al. 2009; Gaudry et al. 2017). Additionally, while eutherian UCP1s exhibit numerous unique amino acid residues relative to non-eutherian mammals, to date, no evidence demonstrates that these substitutions can be accredited to positive selection (Fig. 3; Hughes et al. 2009; Gaudry et al. 2017). However, future studies may be able to target these residues to determine if they do in fact alter proton translocation across the mitochondrial inner membrane and can serve as potential sites of medical intervention. Comparative cryo-EM or crystal structures will also be key in identifying important amino acid interactions and structure-function relationships underlying the variability in uncoupling activity among vertebrate UCP1s. Unfortunately, attempts to resolve the physical structure of UCP1 have failed so far, and uncertainty remains with many mutational loss-of-function studies, as protein integrity cannot be confirmed. However, it may be possible to scale UCP1 functional features by naturally diversified sequences, which circumvent mutational integrity problems as nature's blueprint has purified functional sequences for the in vivo condition.

```
                    1        10        20        30        40        50
                    |........|.........|.........|.........|.........|
                                                                     S
Marsupial_consensus MVGLKPSDVPPTPGVKFLGAGAAACIADLVTFPLDTAKVRLQIQGEAQTT
Eutherian_consensus MVGPTASDVHPTMGVKIFSAGVAACVADVITFPLDTAKVRLQIQGECQTS
Homo_sapiens        MGGLTASDVHPTLGVQLFSAGIAACLADVITFPLDTAKVRLQVQGECPTS
Mus_musculus        MVNPTTSEVQPTMGVKIFSAGVSACLADIITFPLDTAKVRLQIQGEGQAS

                             60        70        80        90       100
                    .........|.........|.........|.........|.........|
                    E           I
Marsupial_consensus GAVRYKGVLGTIVTLVKTEGPRSLYSGLHAGLQRQMSFASIRIGLYDTAK
Eutherian_consensus SAIRYKGVLGTITTLAKTEGPMKLYSGLPAGLQRQISFASLRIGLYDTVQ
Homo_sapiens        SVIRYKGVLGTITAVVKTEGRMKLYSGLPAGLQRQISSASLRIGLYDTVQ
Mus_musculus        STIRYKGVLGTITTLAKTEGLPKLYSGLPAGIQRQISFASLRIGLYDSVQ

                            110       120       130       140       150
                    .........|.........|.........|.........|.........|
                                                             N
Marsupial_consensus QFYNNGRET-AGIGSRILAGCTTGGLAVIVAQPTDVVKVRLQAQSSLSGA
Eutherian_consensus EFFTAGKETTPSLGSKISAGLTTGGVAVFIGQPTEVVKVRLQAQSHLHGL
Homo_sapiens        EFLTAGKETAPSLGSKILAGLTTGGVAVFIGQPTEVVKVRLQAQSHLHGI
Mus_musculus        EYFSSGRETPASLGNKISAGLMTGGVAVFIGQPTEVVKVRMQAQSHLHGI

                            160       170       180       190       200
                    .........|.........|.........|.........|.........|
                                   S   A
Marsupial_consensus KPRYTGTFHAYKTIATEEGTRGLWKGTTPNVTRNAIVNSAELVTYDLIKE
Eutherian_consensus KPRYTGTYNAYRIIATTEGLTGLWKGTTPNLMRNVIINCTELVTYDLMKE
Homo_sapiens        KPRYTGTYNAYRIIATTEGLTGLWKGTTPNLMRSVIINCTELVTYDLMKE
Mus_musculus        KPRYTGTYNAYRVIATTESLSTLWKGTTPNLMRNVIINCTELVTYDLMKG

                            210       220       230       240       250
                    .........|.........|.........|.........|.........|

Marsupial_consensus NLLKYNLLTDNLPCHFVSAFGAGFCTTVVASPVDVVKTRYMNSPPGQYTS
Eutherian_consensus ALVKNKILADDVPCHLVSALIAGFCTTVLSSPVDVVKTRFINSPPGQYTS
Homo_sapiens        AFVKNNILADDVPCHLVSALIAGFCATAMSSPVDVVKTRFINSPPGQYKS
Mus_musculus        ALVNNKILADDVPCHLLSALVAGFCTTLLASPVDVVKTRFINSLPGQYPS

                            260       270       280       290       300    307
                    .........|.........|.........|.........|.........|......|
                                 L                                   G
Marsupial_consensus APKCAWTMLTREGPTAFYKGFVPSFLRLGSWNVVMFVSYEQLKRAMMRSRPTIDCAT
Eutherian_consensus VPNCAMTMLTKEGPTAFFKGFVPSFLRLGSWNVIMFVCFEQLKRELMKSRQTVDCAT
Homo_sapiens        VPNCAMKVFTNEGPTAFFKGLVPSFLRLGSWNVIMFVCFEQLKRELSKSRQTMDCAT
Mus_musculus        VPSCAMSMYTKEGPTAFFKGFVASFLRLGSWNVIMFVCFEQLKKELMKSRQTVDCTT
```

Fig. 3 Deduced UCP1 amino acid alignment of the marsupial consensus sequence, eutherian consensus sequence, human (*Homo sapiens*), and mouse (*Mus musculus*). Consensus sequences were generated from the Gaudry and Campbell (2017) data set by a simple majority and excluding any eutherian *UCP1* pseudogenes. Equally represented amino acids between the four marsupial species (*Monodelphis domestica*, *Macropus eugenii*, *Sminthopsis crassicaudata*, *Sarcophilus harrisii*) are shown in red. Amino acids highlighted in blue have been tested as candidates for positive selection, but these hypotheses were statistically rejected using likelihood ratio tests (Hughes et al. 2009; Gaudry et al. 2017)

3 The Fall of UCP1

Despite the documented benefits of UCP1-mediated NST in several (small) eutherian species, members of a variety of lineages are accomplished endotherms despite lacking a functional UCP1. For instance, birds possess even higher body temperatures than mammals and defend their body temperature by shivering and possibly other poorly characterized mechanisms. The detection of a *UCP* mRNA, termed avian UCP, fostered early speculation that it may catalyze thermogenesis similar to eutherian UCP1. However, the molecular and physiological evidence remains correlative (Raimbault et al. 2001; Vianna et al. 2001), and functional data in bird mitochondria do not support the typical mode of UCP1 action and may instead be related to reactive oxygen species (Talbot et al. 2004; Criscuolo et al. 2005). Conserved synteny revealed that avian *UCP* is an orthologue of *UCP3* and suggested that *UCP1* and *UCP2* are absent in birds (Emre et al. 2007). In fact, *UCP1* appears to have been excised from the genome in a common ancestor of the Sauropsida lineage (reptiles and birds; Mezentseva et al. 2008; McGaugh and Schwartz 2017).

Among mammals, suids (pigs and kin) were the first described lineage to lack a functional copy of the *UCP1* gene (Berg et al. 2006). Exons 3–5 of this pseudogene have been deleted, and the remaining exons (1, 2, and 6) are plagued by nonsense and frameshift mutations (Berg et al. 2006). A recent attempt to override genomic evidence with immunological detection (Mostyn et al. 2014) has been criticized for methodological shortcomings (Jastroch and Andersson 2015). Delineating experiments using specific pig UCP1 antibodies further confirmed that this inactivated gene does not result in a translated protein (Hou et al. 2017a). Taken together, this work provides a molecular explanation of why piglets lack BAT (Herpin et al. 2002, Hou et al. 2017b), are vulnerable to cold temperatures, and rely upon shivering thermogenesis and behavioral adaptations, such as maternal nest building, to defend against hypothermia. While Berg et al. (2006) estimated this pseudogene to have arisen ~20 MYA, newborn peccaries also reportedly lack BAT (Rowlatt et al. 1971); thus it seems plausible that a shared inactivation event may have occurred prior to the radiation of Suoidea ~37 MYA. More recent work by Lin et al. (2017) suggested that certain cold-adapted pig breeds compensate for the lack of UCP1 and BAT by upregulating UCP3 expression in thermogenic white/beige adipocytes in response to cold. However, the interpretation that UCP3 contributes to an increased rate of proton leak may be a misconception stemming from a lack of proton leak kinetics data that require simultaneous measurement of mitochondrial membrane potential and oxygen consumption (Jastroch et al. 2018). Thus, UCP3 has not actually been shown to contribute to proton conductance in pigs.

The progressive downfall of *UCP1* throughout the course of eutherian evolution was further traced in two studies by McGaugh and Schwartz (2017) and Gaudry et al. (2017). The latter study employed a comparative phylogenetic approach to reveal that members within nearly half of all traditional eutherian orders (8 of 18) lack functional *UCP1*. These findings corroborate reports that failed to detect discernable BAT depots in neonates of each of these groups (Rowlatt et al. 1971). Notably, as these ancient inactivations were accompanied by dn/ds ratios indicative of neutral evolution, Gaudry et al. (2017) were further able to estimate pseudogenization dates using nucleotide substitution models and phylogenetic bracketing techniques, thereby correlating these independent inactivations to sharp reductions in metabolic intensity or rapid evolutionary increases in body size (and hence species diversity). For instance, ancient (likely Cretaceous to early Paleocene) independent *UCP1* inactivations in both Xenarthrans (anteaters, sloths, and armadillos) and pangolins are presumably linked to the adoption of energetically diffuse diets that favored energy-conserving reductions in body temperature and metabolic turnover. By contrast, independent *UCP1* inactivations in the ancestors of proboscideans (elephants and mammoths), hyraxes, sirenians (sea cows), equids (horses and zebras), and cetaceans (whales and dolphins) all temporally coincide with magnitude-scale increases in body size of each lineage. For the terrestrial clades, these shifts in body size were likely driven in part by progressive global cooling in the ~30 million years following the Paleocene-Eocene thermal maximum some 55 MYA (Gaudry et al. 2017); i.e., *UCP1* was primarily inactivated in response to planetary *cooling* as opposed to a relaxation of thermogenic needs in tropic environments as was previously proposed for pigs (Berg et al. 2006). The evolution of larger body size evident from the fossil record of these lineages reduced surface area to volume ratios, allowing for more efficient retention of body heat and, theoretically, reduced their need for BAT-mediated NST. This interpretation is consistent with a strong negative correlation between body size and NST capacity that has been observed for eutherians by Oelkrug et al. (2015). Heldmaier (1971) similarly predicted no thermal benefit of BAT in species above 10 kg. It is thus perhaps unsurprising that a number of large-bodied species that lack UCP1 (mammoths, Steller's sea cows, horses, cetaceans, and even extinct ground sloths) were able to expand their ranges into Arctic and sub-Arctic environments. It should be stressed, however, that *UCP1* pseudogenization did not come without evolutionary costs as it presumably hindered the re-evolution of small body size, thereby limiting current species diversity in UCP1 lacking clades (Gaudry et al. 2017). It is also of note that some large-bodied eutherian species (e.g., rhinoceroses, hippos, giraffes, camels) possess *UCP1* genes with putatively translatable open reading frames, though whether or not the protein is expressed in BAT or beige adipocytes remains unknown. Future functional assays of UCP1 from these species may reveal amino acid substitutions accumulated through neutral evolution or relaxed selection pressures that suppress maximal proton leak or other uncoupling attributes relative to species that are known to rely heavily upon BAT-mediated thermogenesis. For example, UCP1 of extant camels contains a four-residue deletion bordering the putative GDP-binding domain that may compromise functional control of thermogenesis (Gaudry et al. 2017). Overall, the fall of *UCP1* highlights the need to further investigate alternate heat-producing mechanisms or greater heat retention capacities (e.g., arteriovenous rete) in these lineages that may compensate for the lack of BAT-mediated NST, as well as the importance of BAT in eutherians that have retained it throughout evolution, such as humans.

4 Translational Value

The most relevant mammals to the medical community are humans, *Homo sapiens*. Over the last decades, and presumably as a consequence of the modern lifestyle in our society, there has been an increasing prevalence of metabolic diseases such as obesity and type 2 diabetes. The metabolic syndrome of positive energy balance that leads to corpulence, inflammation, and insulin resistance is usually mimicked in mice to gain further insights into this condition. To overcome disease-promoting aspects, the scientific community concurs that increasing energy expenditure would benefit humans by adjusting energy balance and improving systemic lipid and glucose metabolism. The discovery of BAT in adult humans (Nedergaard et al. 2007; Cypress et al. 2009; van Marken Lichtenbelt et al. 2009; Virtanen et al. 2009) fostered major efforts to harness adipose tissue thermogenesis to this end, and current efforts are geared toward the recruitment and activation of brown and beige cells. In addition to cold exposure, many promising "browning" agents have been discovered (e.g., butyrate metabolite, lactate, BMPs, adenosine) that may have potential to be targeted in humans (Roberts et al. 2014; Carrière et al. 2014; Xue et al. 2014; Okla et al. 2015; Gnad et al. 2014). These avenues await further consolidation, and future research of signaling pathways controlling the differentiation of beige adipocytes may yield additional sites of possible therapeutic/pharmacological intervention.

However, the dominant experimental model organism is the house mouse (*Mus musculus*), and our knowledge from this species is translated to the human condition. Restricting our research to laboratory mice, however, may bare some caveats that can only be rectified by expanding the range of experimental species. The mouse reflects very different thermoregulatory demands compared to humans, which can be up to three orders heavier than mice. Furthermore, the relative amount of brown fat in adult humans is minor, decreasing in mass after neonatal life (Cannon and Nedergaard 2004; Nedergaard et al. 2007). Thus, the amount of UCP1 in adult humans is rather small, with further decreases in obese subjects (Wang et al. 2015). Furthermore, from the viewpoint of a comparative physiologist, medical research performs a 1-vs-1 species comparison, and we cannot be sure which signaling pathways in mice are specialized for this diminutive life-form and which signaling pathways apply to the energy-wasting function that can be applied to humans. Indeed, *UCP1* transcriptional control mechanisms may vary substantially between humans and mice (Gaudry and Campbell 2017), highlighting the need for broad comparative studies that provide translatable insights to human medicine. For all these caveats, nature's diversity provides blueprints for the understanding of BAT, in particular in humans. Embracing Krogh's principle "for such a large number of problems there will be some animal of choice, or a few such animals, on which it can be most conveniently studied" could be very true to solve the metabolic pandemic with mechanisms of brown adipose tissue.

References

Adams AE, Hanrahan O, Nolan DN, Voorheis HP, Fallon P, Porter RK (2008) Images of mitochondrial UCP1 in mouse thymocytes using confocal microscopy. Biochim Biophys Acta 1777:115–117

Alberti M, Fürsich FT, Abdelhady AA, Anderson N (2017) Middle to late Jurassic equatorial seawater temperatures and latitudinal temperature gradients based on stable isotopes of brachiopods and oysters from Gebel Maghara, Egypt. Palaeogeogr Palaeoclimatol Palaeoecol 468:301–313

Berg F, Gustafson U, Andersson L (2006) The uncoupling protein 1 gene (UCP1) is disrupted in the pig lineage: a genetic explanation for poor thermoregulation in piglets. PLoS Genet 2:e129

Bermejo-Nogales A, Claduch-Giner JA, Pérez-Sánchez J (2014) Tissue specific gene expression of uncoupling protein 2 (UCP2) by hypoxia and nutrient availability in gilthead sea bream (*Sparus aurata*): implications on the physiological significance of UCP1-3 variants. Fish Physiol Biochem 40:751–762

Brand MD, Esteves TC (2005) Physiological functions of the mitochondrial uncoupling proteins UCP2 and UCP3. Cell Metab 2:85–93

Cadenas S, Echtay KS, Harper JA, Jekabsons MB, Buckingham JA, Grau E, Abuin A, Chapman H, Clapham JC, Brand MD (2002) The basal proton conductance of skeletal muscle mitochondria from transgenic mice overexpressing or lacking uncoupling protein-3. J Biol Chem 25:2773–2778

Cannon B, Nedergaard J (2004) Brown adipose tissue: function and physiological significance. Physiol Rev 84:277–359

Carrière A, Jeanson Y, Berger-Müller S, André M, Chenouard V, Arnaud E, Barreau C, Walther R, Galinier A, Wdziekonski B, Villageois P, Louche K, Collas P, Moro C, Dani C, Villarroya F, Casteilla L (2014) Browning of white adipose cells by intermediate metabolites: an adaptive mechanism to alleviate redox pressure. Diabetes 63:3253–3265

Carroll AM, Haines LR, Pearson TW, Fallon PG, Walsh CM, Brennan CM, Breen EP, Porter RK (2005) Identification of a functioning mitochondrial uncoupling protein 1 in thymus. J Biol Chem 280:15534–15543

Criscuolo F, del Mar Gonzalez-Barroso M, Le Maho Y, Ricquier D, Bouillaud F (2005) Avian uncoupling protein expressed in yeast mitochondria prevents endogenous free radical damage. Proc Biol Sci 272:803–810

Cypress AM, Lehman S, Williams G, Tal I, Rodman D, Goldfine AB, Kuo FC, Palmer EL, Tseng YH, Doria A, Kolodny GM, Kahn CR (2009) Identification and importance of brown adipose tissue in adult humans. N Engl J Med 360:1509–1517

Echtay K (2007) Mitochondrial uncoupling proteins-what is their physiological role? Free Radic Biol Med 43:1351–1371

Emre Y, Hurtaud C, Ricquier D, Bouillaud F, Hughes J, Criscuolo F (2007) Avian UCP: the Killjoy in the evolution of the mitochondrial uncoupling proteins. J Mol Evol 65:392–402

Gaudry MJ, Campbell KL (2017) Evolution of UCP1 transcriptional regulatory elements across the mammalian phylogeny. Front Physiol. https://doi.org/10.3389/fphys.2017.00670

Gaudry MJ, Jastroch M, Treberg JR, Hofreiter M, Paijmans JLA, Starrett J, Wales N, Signore AV, Springer MS, Campbell KL (2017) Inactivation of thermogenic UCP1 as a historical contingency in multiple placental mammal clades. Sci Adv 3:e1602878

Gnad T, Scheibler S, von Kügelgen I, Scheele C, Kilić A, Glöde A, Hoffmann LS, Reverte-Salisa L, Horn P, Mutlu S, El-Tayeb A, Kranz M, Deuther-Conrad W, Brust P, Lidell ME, Betz MJ, Enerbäck S, Schrader J, Yegutkin GG, Müller CE, Pfeifer A (2014) Adenosine activates brown adipose tissue and recruits beige adipocytes via A2A receptors. Nature 516(7531):395–399

Golozoubova V, Hohtola E, Matthias A, Jacobsson A, Cannon B, Nedergaard J (2001) Only UCP1 can mediate adaptive nonshivering thermogenesis in the cold. FASEB J 15:2048–2050

Harms M, Seale P (2013) Brown and beige fat: development, function and therapeutic potential. Nat Med 19:1252–1263

Heaton GM, Wagenvoord RJ, Kemp A, Nicholls DG (1978) Brown-adipose-tissue mitochondria: photoaffinity labelling of the regulatory site of energy dissipation. Eur J Biochem 82:515–521

Heldmaier G (1971) Zitterfreie Wärmebildung und Körpergröße bei Säugetieren. Z Vergl Physiol 73:222–247

Herpin P, Damon M, Le Dividich J (2002) Development of thermoregulation and neonatal survival in pigs. Livest Prod Sci 78:25–45

Hou L, Shi J, Cao L, Xu G, Hu C, Wang C (2017a) Pig has no uncoupling protein 1. Biochem Biophys Res Commun 487:795–800

Hou L, Hu CY, Wang C (2017b) Pig has no brown adipose tissue. FASEB J 31:lb695

Hughes DA, Jastroch M, Stoneking M, Klingenspor M (2009) Molecular evolution of UCP1 and the evolutionary history of mammalian non-shivering thermogenesis. BMC Evol Biol 9:4

Ikeda K, Kang Q, Yoneshiro T, Camporez JP, Maki H, Homma M, Shinoda K, Chen Y, Lu X, Maretich P, Tajima K, Ajuwon KW, Soga T, Kajimura S (2017) UCP1-independent signaling involving SERCA2b-mediated calcium cycling regulates beige fat thermogenesis and systemic glucose homeostasis. Nat Med 23:1454

Jastroch M, Andersson L (2015) When pigs fly, UCP1 makes heat. Mol Metab 4:359–362

Jastroch M, Wuertz S, Kloas W, Klingenspor M (2005) Uncoupling protein in fish uncovers an ancient evolutionary history of mammalian nonshivering thermogenesis. Physiol Genomics 22:150–156

Jastroch M, Buckingham JA, Helwig M, Klingenspor M, Brand MD (2007) Functional characterization of UCP1 in the common carp: uncoupling activity in liver mitochondria and cold-induced expression in the brain. J Comp Physiol B 177:743–752

Jastroch M, Withers KW, Taudien S, Frappell PB, Helwig M, Fromme T, Hirschberg V, Heldmaier G, McAllan BM, Firth BT, Burmester T, Platzer M, Klingenspor M (2008) Marsupial uncoupling protein 1 sheds light on the evolution of mammalian nonshivering thermogenesis. Physiol Genomics 32:161–169

Jastroch M, Oelkrug R, Keipert S (2018) Insights into brown adipose tissue evolution and fuction from non-model organisms. J Exp Biol 221:jeb169425

Kazak L, Chouchani ET, Stavrovskaya IG, Lu GZ, Jedrychowski MP, Egan DF, Kumari M, Kong X, Erickson BK, Szpyt J, Rosen ED, Murphy MP, Kristal BS, Gygi SP, Spiegelman BM (2017a) UCP1 deficiency causes brown fat respiratory chain depletion and sensitizes mitochondria to calcium overload-induced dysfunction. Proc Natl Acad Sci U S A 114(30):7981–7986

Kazak L, Chouchani ET, Lu GZ, Jedrychowski MP, Bare CJ, Mina AI, Kumari M, Zhang S, Vuckovic I, Laznik-Bogoslavski D, Dzeja P, Banks AS, Rosen ED, Spiegelman BM (2017b) Genetic depletion of adipocyte creatine metabolism inhibits diet-induced thermogenesis and drives obesity. Cell Metab 3:660–671

Keipert S, Jastroch M (2014) Brite/beige fat and UCP1 – is it thermogenesis? Biochim Biophys Acta Bioenerg 1837:1075–1082

Keipert S, Kutschke M, Ost M, Schwarzmayr T, van Schothorst EM, Lamp D, Brachthauser L, Hamp I, Mazibuko SE, Hartwig S, Lehr S, Graf E, Plettenburg O, Neff F, Tschöp MH, Jastroch M (2017) Long-term cold adaptation does not require FGF21 or UCP1. Cell Metab 26:437–446

Khailova LS, Prikhodko EA, Dedukhova VI, Mokhova EN, Popov VN, Skulachev VP (2006) Participation of ATP/ADP antiporter in oleate- and oleate hyperoxide-induced uncoupling suppressed by GDP and carboxyatractylate. Biochim Biophys Acta 1757:1324–1329

Laursen WJ, Mastrotto M, Pesta D, Funk OH, Goodman JB, Merriman DK, Ingolia N, Shulman GI, Bagriantsev SN, Gracheva EO (2015) Neuronal UCP1 expression suggests a mechanism for local thermogenesis during hibernation. Proc Natl Acad Sci U S A 112:1607–1612

Lin J, Cao C, Tao C, Ye R, Dong M, Zheng Q, Wang C, Jiang X, Yan C, Li K, Speakman JR, Wang Y, Jin W, Zhao J (2017) Cold adaptation in pigs depends on UCP3 in beige adipocytes. J Mol Cell Biol 9:364–375

Liu L, Zhang J, Rheindt FE, Lei F, Qu Y, Wang Y, Zhang Y, Sullivan C, Nie W, Wang J, Yang F, Chen J, Edwards SV, Meng J, Wu S (2017) Genomic evidence reveals a radiation of placental mammals uninterrupted by the KpG boundary. Proc Natl Acad Sci U S A 114:E7282–E7290

Long JZ, Svensson KJ, Bateman LA, Lin H, Kamenecka T, Lokurkar IA, Lou J, Rao RR, Chang MR, Jedrychowski MP, Paulo JA (2016) The secreted enzyme PM20D1 regulates lipidated amino acid uncouplers of mitochondria. Cell 14:424–435

Lovegrove BG (2012) The evolution of endothermy in Cenozoic mammals: a plesiomorphic-apomorphic continuum. Biol Rev Camb Philos Soc 87:218–162

Mailloux RJ, Harper ME (2011) Uncoupling proteins and the control of mitochondrial reactive oxygen species production. Free Radic Biol Med 51:1106–1115

van Marken Lichtenbelt WD, Vanhommerig JW, Smulders NM, Drossaerts JM, Kemerink GJ, Bouvy ND, Schrauwen P, Teule GJ (2009) Cold-activated brown adipose tissue in healthy men. N Engl J Med 360:1500–1508

McGaugh S, Schwartz TS (2017) Here and there, but not everywhere: repeated loss of uncoupling protein 1 in amniotes. Biol Lett 13:20160749

Meredith RW, Janečka JE, Gatesy J, Ryder OA, Fisher CA, Teeling EC, Goodbla A, Eizirik E, Simão TLL, Stadler T, Rabosky DL (2011) Impacts of the cretaceous terrestrial revolution and KPg extinction on mammal diversification. Science 334:521–524

Meyer CW, Willershauser M, Jastroch M, Rourke BC, Fromme T, Oelkrug R, Heldmaier G, Klingenspor M (2012) Adaptive thermogeneisis and thermal conductance in wilde-type and UCP1-KO mice. Am J Physiol Regul Integr Comp Physiol 299:R1396–R1406

Mezentseva NV, Kumaratilake JS, Newman SA (2008) The brown adipocyte differentiation pathway in birds: an evolutionary road not taken. BMC Biol 6:17

Morrissette JM, Franck JP, Block BA (2003) Characterization of ryanodine receptor and Ca2+-ATPase isoforms in the thermogenic heater organ of blue marlin (*Makaira nigricans*). J Exp Biol 206:805–812

Mostyn A, Attig L, Larcher T, Dou S, Chavatte-Palmer P, Boukthir M, Gertler A, Djiane J, Symonds ME, Abdennebi-Najar L (2014) UCP1 is present in porcine adipose tissue and is responsive to postnatal leptin. J Endocrinol 223:M31–M38

Murakami M, Ohi M, Ishikawa S, Shirai M, Horiguchi H, Nishino Y, Funaba M (2015) Adaptive expression of uncoupling protein 1 in carp liver and kidney in response to changes in ambient temperature. Comp Biochem Physiol A Mol Integr Physiol 185:142–149

Mzilikazi N, Jastroch M, Meyer CW, Klingenspor M (2007) The molecular and biochemical basis of nonshivering thermogenesis in an African endemic mammal, *Elephantulus myurus*. Am J Physiol Regul Integr Comp Physiol 293:R2120–R2127

Nabben M, Shabalina IG, Moonen-Kornips E, van Beurden D, Schwrauwen P, Nedergaard J, Hoeks J (2011) Biochim Biophys Acta 1807:1095–1105

Nedergaard J, Cannon B (2003) The “novel” “uncoupling” proteins UCP2 and UCP3: what do they really do? Pros and cons for suggested functions. Exp Physiol 88:65–84

Nedergaard J, Bengtsson T, Cannon B (2007) Unexpected evidence for active brown adipose tissue in adult humans. Am J Physiol Endocrinol Metab 293:E444–E452

Nicol SC, Andersen NA, Arruda AP, Ruf T (2009) Rewarming rates of two large hibernators: comparison of a monotreme and a eutherian. J Therm Biol 34:155–159

O’leary MA, Bloch JI, Flynn JJ, Gaudin TJ, Giallombardo A, Giannini NP, Goldberg SL, Kraatz BP, Luo ZX, Meng J, Ni X, Novacek MJ, Perini FA, Randall ZS, Rougier GW, Sargis EJ, Silcox MT, Simmons NB, Spaulding M, Velazco PM, Weksler M, Wible JR, Cirranello AL (2013) The placental mammal ancestor and the post-K-Pg radiation of placentals. Science 339:662–667

O'Brien CL, Robinson SA, Pancost RD, Damsté JSS, Schouten S, Lunt DJ, Alsenz H, Bornemann A, Bottini C, Brassell SC, Farnsworth A, Forster A, Huber BT, Inglis GN, Jenkyns HC, Linner C, Littler K, Markwick P, McAnena A, Mutterlos J, Naafs DA, Püttman W, Sluijs A, van Helmond NAGM, Vellekoop J, Wagner T, Wrobel NE (2017) Cretaceous sea-surface temperature evolution: constraints from TEX86 and planktonic foraminiferal oxygen isotopes. Earth Sci Rev 172:224–247

Oelkrug R, Kutschke M, Meyer CW, Heldmaier G, Jastroch M (2010) Uncoupling protein 1 decreases superoxide production in brown adipose tissue mitochondria. J Biol Chem 285:21961–21968

Oelkrug R, Heldmaier G, Meyer CW (2011) Torpor patterns, arousal rates, and temporal organization of torpor entry in wildtype and UCP1-ablated mice. J Comp Physiol B 181:137–145

Oelkrug R, Exner NGC, Exner C, Lee Y, Ganjam GK, Kutschke M, Müller S, Stöhr S, Tschöp MH, Crichton PG, Heldmaier G, Jastroch M, Meyer CW (2013) Brown fat in a protoendothermic mammal fuels eutherian evolution. Nat Commun 4:2140

Oelkrug R, Polymeropoulos ET, Jastroch M (2015) Brown adipose tissue: physiological function and evolutionary significance. J Comp Physiol B 185:587–606

Okla M, Ha JH, Temel RE, Chung S (2015) BMP7 drives human adipogenic stem cells into metabolically active beige adipocytes. Lipids 50:111–120

Polymeropoulos ET, Jastroch M, Frappell PB (2012) Absence of adaptive nonshivering thermogenesis in a marsupial, the fat-tailed dunnart (*Sminthopsis crassicaudata*). J Comp Physiol B 182:393–401

Raimbault S, Dridi S, Denjean F, Lachuer J, Couplan E, Bouillaud F, Bordas A, Duchamp C, Taouis M, Ricquier D (2001) An uncoupling protein homologue putatively involved in facultative muscle thermogenesis in birds. Biochem J 353:441–444

Ricquier D, Kader JC (1976) Mitochondrial protein alteration in active brown fat: a sodium dodecyl sulfate-polyacrylamide gel electrophoretic study. Biochem Biophys Res Commun 73:577–583

Roberts LD, Boström P, O'Sullivan JF, Schinzel RT, Lewis GD, Dejam A, Lee YK, Palma MJ, Calhoun S, Georgiadi A, Chen MH, Ramachandran VS, Larson MG, Bouchard C, Rankinen T, Souza AL, Clish CB, Wang TJ, Estall JL, Soukas AA, Cowan CA, Spiegelman BM, Gerszten RE (2014) β-Aminoisobutyric acid induces browning of white fat and hepatic β-oxidation and is inversely correlated with cardiometabolic risk factors. Cell Metab 19:96–108

Rousset S, Alves-Guerra MC, Mozo J, Miroux B, Cassard-Doulcier AM, Bouillaud F, Ricquier D (2004) The biology of mitochondrial uncoupling proteins. Diabetes 53:S130–S135

Rowlatt U, Mrosovsky N, English A (1971) A comparative survey of brown fat in the neck and axilla of mammals at birth. Biol Neonate 17:53–83

Saito S, Saito CT, Shingai R (2008) Adaptive evolution of the uncoupling protein 1 gene contributed to the acquisition of novel nonshivering thermogenesis in ancestral eutherian mammals. Gene 408:37–44

Shabalina IG, Hoeks J, Kramarova TV, Schrauwen P, Cannon B, Nedergaard J (2010) Cold tolerance of UCP1-ablated mice: a skeletal muscle mitochondria switch toward lipid oxidation with marked UCP3 up-regulation not associated with increased basal, fatty acid- or ROS-induced uncoupling or enhanced GDP effects. Biochim Biophys Acta 1797:968–980

Shin H, Ma Y, Chanturiya T, Cao Q, Wang Y, Kadegowda AK, Jackson R, Rumore D, Xue B, Shi H, Gavrilova O (2017) Lipolysis in brown adipocytes is not essential for cold-induced thermogenesis in mice. Cell Metab 26:764–777

Springer MS, Emerling CA, Meredith RW, Janecka JE, Eizirik E, Murphy WJ (2017) Waking the undead: implications of a soft explosive model for the timing of placental mammal diversification. Mol Phylogenet Evol 106:86–102

Talbot DA, Duchamp C, Rey B, Hanuise N, Rouanet JL, Sibille B, Brand MD (2004) Uncoupling protein and ATP/ADP carrier increase mitochondrial proton conductance after cold adaptation of king penguins. J Physiol 558:123–135

Ukropec J, Anunciado RP, Ravussin Y, Hulver MW, Kozak LP (2006) UCP1-independent thermogenesis in white adipose tissue of cold-acclimated Ucp1-/- mice. J Biol Chem 281 (42):31894–31908. Epub 2006 Aug 16

Vianna CR, Hagen T, Zhang CY, Bachman E, Boss O, Gereben B, Moriscot AS, Lowell BB, Bicudo JE, Bianco AC (2001) Cloning and functional characterization of an uncoupling protein homolog in hummingbirds. Physiol Genomics 5:137–145

Virtanen KA, Lidell ME, Orava J, Heglind M, Westergren R, Niemi T, Taittonen M, Laine J, Savisto NJ, Enerbäck S, Nuutila P (2009) Functional brown adipose tissue in healthy adults. N Engl J Med 360:1518–1525

Wang Q, Zhang M, Xu M, Gu W, Xi Y, Qi L, Li B, Wang W (2015) Brown adipose tissue activation is inversely related to central obesity and metabolic parameters in adult human. PLoS One 10:e0123795

Wen ZY, Liang XF, He S, Li L, Shen D, Tao YX (2015) Molecular cloning and tissue expression of uncoupling protein 1, 2 and 3 genes in Chinese perch (*Siniperca chuatsi*). Comp Biochem Physiol B Biochem Mol Biol 185:24–33

Xue R, Wan Y, Zhang S, Zhang Q, Ye H, Li Y (2014) Role of bone morphogenetic protein 4 in the differentiation of brown fat-like adipocytes. Am J Physiol Endocrinol Metab 306:E363–E372

The Mechanism FA-Dependent H^+ Transport by UCP1

Ambre M. Bertholet and Yuriy Kirichok

Contents

1 Introduction 144
2 Molecular Mechanisms of UCP1: Different Models 145
3 Direct Patch-Clamp Analysis of UCP1 Currents Across the IMM of Brown and Beige Fat 146
3.1 FA Are Required for H^+ Leak via UCP1 148
3.2 FA as Transport Substrates of UCP1 149
4 A New Model of the FA-Dependent H^+ Leak via UCP1 151
5 FA Remove Purine Nucleotide Inhibition of UCP1 152
6 Converging UCP1 Function and SLC25 Structural Data 153
6.1 A Consensus Structure for All SLC25 Members 153
6.2 Preliminary Considerations for UCP1 Structure–Function Relations 154
6.3 Changes in UCP1 Conformation During Transport 155
7 Concluding Remarks 156
References 157

Abstract

Uncoupling protein 1 (UCP1) is an integral protein of the inner mitochondrial membrane (IMM) that is expressed specifically in brown and beige fat depots. UCP1 is responsible for the production of heat to control core body temperature, the regulation of fat metabolism, and the energy balance. As an uncoupling protein, UCP1 transports H^+ across the IMM in presence of long-chain fatty acids (FA), which makes brown fat mitochondria produce heat at the expense of ATP. However, the exact mechanism of UCP1 action has remained difficult to elucidate, because direct methods for studying currents generated by UCP1 were unavailable. Recently, the patch-clamp technique was successfully applied to brown and beige fat mitochondria to directly study H^+ currents across the IMM

A. M. Bertholet (✉) · Y. Kirichok (✉)
Department of Physiology, University of California San Francisco, San Francisco, CA, USA
e-mail: Ambre.Bertholet@ucsf.edu; Yuriy.Kirichok@ucsf.edu

A. Pfeifer et al. (eds.), *Brown Adipose Tissue*,
Handbook of Experimental Pharmacology 251, https://doi.org/10.1007/164_2018_138

and characterize UCP1 function. A new model of the UCP1 mechanism was proposed based on the patch-clamp analysis. In this model, both FA anions (FA^-) and H^+ are transport substrates of UCP1, and UCP1 operates as a non-canonical FA^-/H^+ symporter. Here, we summarize recent findings obtained with the patch-clamp technique that describe how UCP1 can transport not only H^+ but also FA^-.

Keywords
Beige fat · Brown fat · Fatty acid · Mitochondrial uncoupling · Mitochondrial uncoupling protein 1 · Patch-clamp technique · Purine nucleotides · Thermogenesis · UCP1

1 Introduction

Brown fat and the more recently identified beige fat are tissues specialized in adaptive thermogenesis and fat burning in mammals (Cannon and Nedergaard 2004; Cohen and Spiegelman 2015; Wu et al. 2012). The H^+ leak across the inner mitochondrial membrane (IMM) of brown and beige fat is responsible for the thermogenesis (Bertholet et al. 2017; Cannon and Nedergaard 2004; Enerback et al. 1997; Feldmann et al. 2009; Nicholls and Locke 1984; Shabalina et al. 2008). This leak dissipates the mitochondrial H^+ gradient and converts the energy of substrate oxidation into heat. The mitochondrial H^+ leak in brown and beige fat is mediated by uncoupling protein 1 (UCP1) (Aquila et al. 1985; Bertholet et al. 2017; Bouillaud et al. 1986; Fedorenko et al. 2012; Klingenberg and Winkler 1985; Krauss et al. 2005; Lin and Klingenberg 1982; Rial et al. 1983; Ricquier and Kader 1976; Shabalina et al. 2013), which belongs to the SLC25 superfamily of mitochondrial solute carriers (Palmieri 2014). UCP1 is activated by long-chain fatty acids (FA) and inhibited by Mg^{2+}-free purines nucleotides (Nicholls and Lindberg 1973; Nicholls and Locke 1984). Despite the importance of UCP1 for the maintenance of core body temperature and the control of fat metabolism (Cannon and Nedergaard 2004; Feldmann et al. 2009; Kozak et al. 2010), the mechanism by which UCP1 increases the permeability of the IMM for H^+ in presence of long-chain FA remains controversial (Cannon and Nedergaard 2004; Klingenberg 2017). The main barrier to understanding the mechanism of the UCP1-mediated H^+ leak has been the lack of direct methods to measure this leak and the inability to strictly control experimental conditions. Application of the patch-clamp technique to mitochondria for the first time allowed high-resolution functional analysis of UCP1 in its native membrane environment in both brown and beige fat (Bertholet et al. 2017; Bertholet and Kirichok 2017; Fedorenko et al. 2012). This method led to refinement of the mechanism by which FA and purine nucleotides regulate the H^+ leak through UCP1. Here we describe a new model of UCP1 operation based on the patch-clamp analysis and recently obtained structural data for UCP1 and other SLC25 members.

2 Molecular Mechanisms of UCP1: Different Models

Despite the fact that free long-chain FA and purine nucleotides were established as the principal UCP1 regulators as early as the 1970s (Nicholls 2001; Nicholls and Lindberg 1973; Nicholls and Locke 1984), the exact mechanism by which FA activate and cytosolic nucleotides inhibit the UCP1-dependent H$^+$ leak is still actively debated. Over the years, several mechanisms of the FA-dependent H$^+$ leak through UCP1 were proposed. The following models have attracted the most attention (Fig. 1).

1. The H$^+$ uniporter (channel) model

   In this model, UCP1 has a basal, FA-independent H$^+$ transport activity, and the allosteric binding of FA to UCP1 is primarily required to competitively remove purine nucleotide inhibition (Cannon and Nedergaard 2004; Gonzalez-Barroso et al. 1998; Jiménez-Jiménez et al. 2006; Rial and Gonzalez-Barroso 2001; Shabalina et al. 2004; Winkler and Klingenberg 1994).

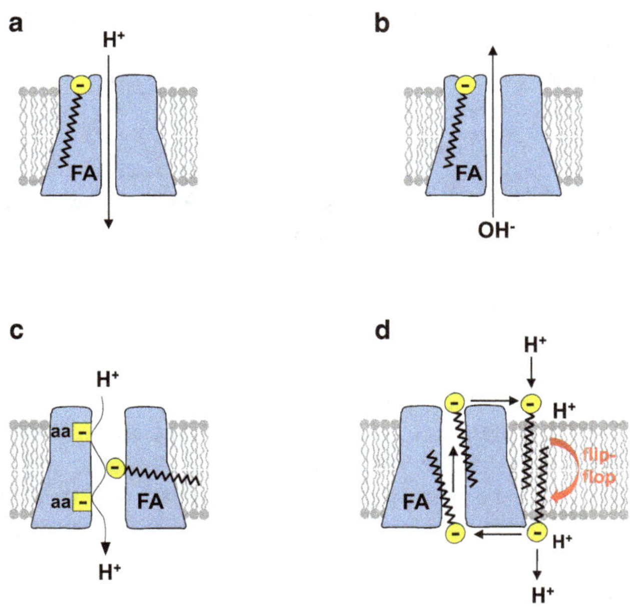

Fig. 1 Proposed models of UCP1 function. (**a**) The H$^+$ channel model. Upon allosteric binding of FA, UCP1 acts as a H$^+$ channel. (**b**) The OH$^-$ channel model. This model is similar to the H$^+$ channel model, but UCP1 transports OH$^-$ in the opposite direction. (**c**) The "H$^+$ buffering" model. This model postulates that UCP1 is a H$^+$ channel, but the FA binds in the translocation pathway. The carboxylic group of the FA enables H$^+$ binding and translocation, possibly along with titratable amino acid residues (aa) of UCP1. (**d**) The "FA-cycling" model. In this model, UCP1 operates as a FA$^-$ carrier, and the H$^+$ transport occurs outside the UCP1 translocation pathway. UCP1 transports FA$^-$ outside mitochondria, where they bind H$^+$ and, in protonated form, cross the IMM to release H$^+$ into the mitochondrial matrix

2. The OH^- uniporter (channel) model
   UCP1 carries not H^+ but OH^- ions and is activated by allosteric binding of FA (Nicholls 2006; Nicholls and Rial 1999).
3. The "H^+ buffering" model
   UCP1 acts as a H^+ channel in which FA bind in the pore and provide their carboxylic groups to complete the H^+ translocation pathway along with the titratable amino acid residues of UCP1 (Klingenberg and Huang 1999).
4. The "FA-cycling" model
   In this model, UCP1 is a long-chain FA^- carrier that indirectly transports H^+. UCP1 carries long-chain FA^- outside the mitochondria where they bind H^+ and, in protonated form, "flip-flop" back across the IMM to release the proton into the mitochondrial matrix (Garlid et al. 1998; Garlid et al. 1996; Skulachev 1991).

However, the mechanism of thermogenic H^+ leak mediated by UCP1 was controversial, and numerous questions remained unanswered. First, it was unclear what species were transported by UCP1. Second, it was uncertain whether UCP1 has a FA-independent "basal" H^+ transport activity. Third, the mechanism by which FA increase H^+ (or OH^-) leak through UCP1 was unresolved. All these questions have been addressed directly by studying the H^+ leak across the IMM with the mitochondrial patch-clamp technique.

3 Direct Patch-Clamp Analysis of UCP1 Currents Across the IMM of Brown and Beige Fat

The patch-clamp technique is applied to mitochondria in essentially the same way as it is to cells, except that the mitochondrion is significantly smaller and has two membranes. Therefore, the outer mitochondrial membrane (OMM) is first mechanically disrupted to release the whole intact IMM (so called mitoplasts, Fig. 2a). UCP1-dependent H^+ currents are then recorded from the whole IMM (mitoplasts) isolated from brown and beige fat of mice. Specifically, after formation of the gigaohm seal between the patch pipette and a mitoplast, the small patch of the IMM under the pipette is disrupted (break-in into the mitoplast) by application of high-amplitude voltage steps to gain access into the mitochondrial matrix from the pipette (Fig. 2b). This whole-mitoplast configuration (also called whole-IMM configuration) allows measurement of currents across the whole IMM with full control of the transmembrane voltage (Fig. 2b, c). Solution compositions on both sides of the IMM, matrix (pipette) and cytosolic (bath), are also fully controlled.

Application of long-chain FA on the cytosolic face of the IMM induces a large H^+ current across the IMM of brown and beige fat mitochondria (Bertholet et al. 2017; Fedorenko et al. 2012). This current is inhibited by purine nucleotides such as GDP and ATP and cannot be detected in UCP1-deficient mice. Thus, it is mediated by UCP1. The reversal potential analysis of the UCP1 current induced by long-chain FA revealed a H^+ selectivity (Fedorenko et al. 2012). These experiments

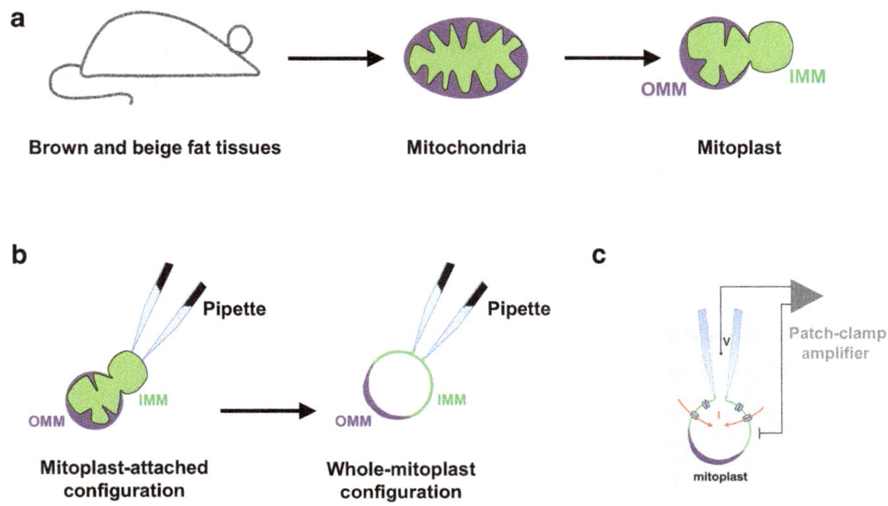

Fig. 2 The patch-clamp technique applied to mitochondria. (**a**) Isolation of mitoplasts, native IMM with attached remnants of the OMM. Mitochondria are isolated from brown and beige fat tissue lysates and exposed to low-pressure French press to rupture the OMM and release the IMM, generating mitoplasts. (**b**) Formation of the whole-mitoplast configuration of the mitochondrial patch-clamp. First, a gigaohm seal is formed between the glass patch pipette and the IMM to obtain the so-called mitoplast-attached configuration. Next, high-amplitude voltage pulses are applied into the pipette to rupture the IMM patch under the pipette (this is referred to as "break-in"). After the break-in, the whole-mitoplast configuration is formed, and the interior of the mitoplast is perfused with the pipette solution. Because the tonicity of the pipette solution is higher than that of the bath solution, the IMM is completely released from the OMM, and the mitoplast assumes a round shape. (**c**) Whole-mitoplast configuration of the mitochondrial patch clamp. Two electrodes, one in the pipette and one in the bath, control the voltage across the IMM and enable recording of a current (I) across the whole IMM. The convention used for the directions of currents flowing across the IMM: inward currents (positive charge flowing into the mitoplast) are negative, while outward currents are positive. Voltages are in the matrix as compared to the bath solution (bath solution is defined as 0 mV)

demonstrated the possibility of recording UCP1 currents directly using the patch-clamp technique in both thermogenic tissues, brown and beige fat. In contrast to indirect analysis of the mitochondrial H^+ leak by respiration in suspensions of isolated mitochondria, this method directly measures the H^+ leak to assess the thermogenic capacity of an individual single mitochondrion.

The application of the patch-clamp technique to brown and beige fat mitochondria demonstrated that UCP1 generates one of the largest H^+ currents ever recorded across any cellular membrane (Bertholet et al. 2017; Fedorenko et al. 2012). These large currents can be explained by the extremely high density of UCP1 expression (~10% of the total mitochondrial protein) (Lin and Klingenberg 1980; Ricquier and Kader 1976).

Interestingly, although all mitochondria isolated from subcutaneous (inguinal) beige fat have robust UCP1-dependent H^+ leak, only about 15% of beige fat mitochondria of abdominal (epididymal) fat have this current (Bertholet et al. 2017). In this study, beige fat mitochondria were isolated upon chronic injection of mice with β3-adrenergic agonist CL316.243 (in control mice injected with saline, the inguinal and epididymal depots contained only white adipocytes, and the same mitochondrial isolation protocol produced no mitochondria). Thus, the mitochondrial patch-clamp analysis revealed two types of beige adipocytes: UCP1 positive and UCP1 negative. UCP1-negative beige adipocytes do not use UCP1 for thermogenesis and appear to employ only mitochondrial creatine-driven futile cycling as an alternative thermogenic mechanism (Bertholet et al. 2017; Kazak et al. 2015).

The UCP1-negative beige adipocytes are similar to UCP1-positive beige adipocytes in several important ways. Both types of beige adipocytes are multilocular, have robust mitochondrial biogenesis and mitochondrial biomass, possess active thermogenic gene program, and have similar OXPHOS profile distinct from that of the classical interscapular brown fat (Bertholet et al. 2017). Finally, the UCP1-negative and UCP1-positive beige adipocytes are both capable of creatine-driven futile cycling (Bertholet et al. 2017). Interestingly, although UCP1 is not detectable on the protein level in UCP1-negative beige adipocytes, Ucp1 mRNA is upregulated in these cells as compared to white fat (Bertholet et al. 2017). When multilocular UCP1-negative cells are compared to white adipocytes, striking differences in cytoplasmic lipid droplet morphology, mitochondrial abundance, and thermogenic gene expression are observed (Bertholet et al. 2017). From all these data, it follows that the multilocular UCP1-negative cells are a new type of a thermogenic beige adipocyte.

To summarize, the mitochondrial patch clamp combined with genetic, histological, and biochemical techniques enables a new level of insight into the cellular and molecular mechanisms of thermogenesis.

3.1 FA Are Required for H^+ Leak via UCP1

It was long recognized that FA stimulate the UCP1-dependent mitochondrial uncoupling and thermogenesis. Moreover, UCP1 reconstituted in artificial lipid membranes requires FA for its activity (Klingenberg 2010). However, isolated brown fat mitochondria remained uncoupled even in the presence of large concentrations of albumin, a FA acceptor (Nicholls 2006). The complete inhibition of UCP1 and recoupling of brown fat mitochondria could only be achieved in the presence of purine nucleotides (Nicholls 2006). Therefore, it was unclear whether FA are required for UCP1 activation or UCP1 has basal, FA-independent H^+ transport activity.

The patch-clamp technique not only allows the study of UCP1 in its native membrane environment but also provides full control over the concentration of membrane FA. Such patch-clamp experiments have demonstrated that the IMM of brown and beige fat possesses a robust phospholipase A2 (PLA2) activity, and it is very effective in activating the H^+ leak through UCP1 (Bertholet et al. 2017; Fedorenko et al. 2012). Furthermore, this PLA2 activity was so strong that application of FA-free albumin only on the cytosolic face of the IMM could not completely inhibit the H^+ current via UCP1. Only high concentrations of albumin applied on both sides of the IMM fully deactivated the H^+ current, and no additional inhibition could be achieved with purine nucleotides. These data demonstrate that UCP1 has no FA-independent basal H^+ transport activity. Previous observations of "FA-independent" or "basal" uncoupling in suspensions of brown fat-isolated mitochondria were very likely associated with the inability to fully extract endogenous FA due to mitochondrial PLA2 activity. These results highlight a potential role of the signaling mechanisms associated with IMM phospholipases in regulating UCP1 in vivo.

3.2 FA as Transport Substrates of UCP1

It was essential to identify the ion species transported by UCP1 to determine the mechanism by which FA activate the UCP1-dependent H^+ leak across the IMM of brown fat. As mentioned previously, the patch-clamp experiments demonstrated that the UCP1-dependent current induced by long-chain FA is highly H^+ selective (Fedorenko et al. 2012). However, to facilitate H^+ current across the IMM, UCP1 may not need to conduct H^+ directly. Indeed, the previously proposed FA-cycling model postulated that UCP1 only transports FA^- while H^+ are carried across the IMM via flip-flop of protonated FA (Fig. 1d).

To reveal a possible FA^- current via UCP1 in isolation from the H^+ current, UCP1 activity was recorded in the presence of various low-pKa FA analogs (such as long-chain alkyl sulfonates) that are unable to bind H^+ at physiological pH and cannot cause UCP1-dependent H^+ translocation. These experiments demonstrated that UCP1 indeed transports low-pKa FA^- (Fedorenko et al. 2012). Interestingly, differences exist in the way UCP1 interacts with short-chain and long-chain FA. Specifically, short-chain FA^- are simply carried by UCP1 across the membrane as they produce steady transmembrane currents in response to a voltage step protocol (Fig. 3a). In contrast, long-chain FA^- are translocated but cannot dissociate from UCP1 due to strong hydrophobic interactions, resulting in limited motion within the membrane and transient currents in response to the same voltage step protocol (Fig. 3b). In addition, long-chain FA^- activate UCP1 currents at significantly lower concentrations than short-chain FA (low micromolar vs. millimolar, respectively), likely due to stronger hydrophobic interactions with UCP1.

Interestingly, long-chain FA^- (the physiological activators of the H^+ leak via UCP1) bind to UCP1 only on the cytosolic side, while the short-chain FA^- can bind on both cytosolic and matrix sides. Even when added at unphysiologically high concentrations, matrix long-chain FA^- failed to induce UCP1 currents in the electrophysiological experiments (Fedorenko et al. 2012). In contrast, short-chain FA could activate UCP1 currents when added on any side of the IMM. This suggests that access to the substrate-binding site on the matrix side of UCP1 is limited and long-chain FA cannot reach it probably due to their larger size.

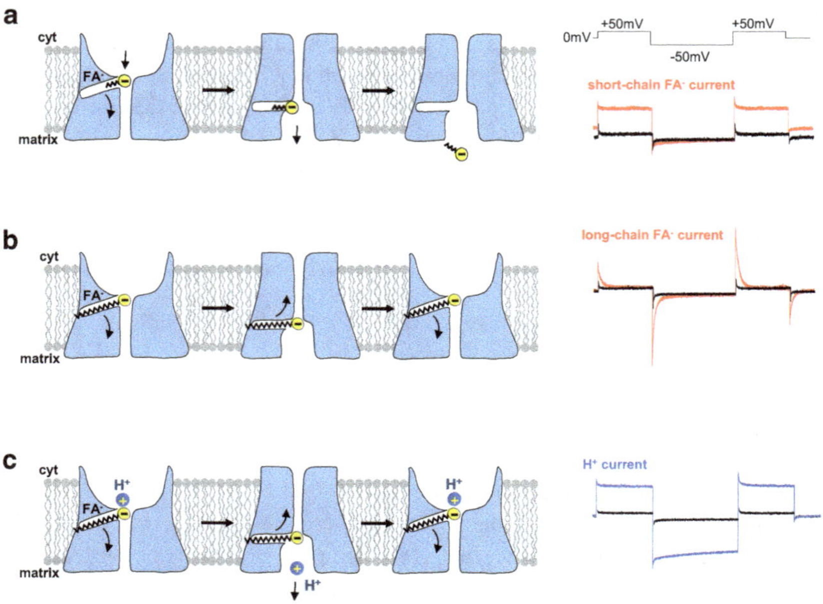

Fig. 3 A new model of the UCP1 mechanism based on the electrophysiological data. (**a**) UCP1 transports short-chain FA$^-$. The polar head of a FA$^-$ binds to the UCP1 SBS, while the carbon tail is stabilized by a hydrophobic pocket within UCP1 or simply protrudes into the lipid bilayer. Short-chain FA$^-$ added on the cytosolic face of the IMM are transported by UCP1 across the IMM (left panel), which induces a steady transmembrane FA$^-$ current. A typical whole-mitoplast recording of a short-chain FA$^-$ current via UCP1 (right panel). The voltage protocol is shown above. This model assumes that UCP1 is a transporter and has two conformation states, with the SBS exposed either to the cytosolic (*c*-state) or matrix (*m*-state) side of the IMM. The access to the SBS is shown to be narrower in the *m*-state than in the *c*-state, because long-chain FA$^-$ cannot bind to UCP1 on the matrix side. (**b**) UCP1 traps long-chain FA$^-$. UCP1 binds and transports long-chain FA$^-$ in the same fashion as short-chain FA$^-$. However, long-chain FA$^-$ establish a much stronger hydrophobic interaction with UCP1 and cannot leave the UCP1 translocation pathway as easily as short-chain FA$^-$ (left panel). Thus, long-chain FA$^-$ shuttle within the UCP1 translocation pathway to produce transient currents in response to changes in the transmembrane voltage (left panel). A typical whole-mitoplast recording of a long-chain FA$^-$ current via UCP1 in response to voltage steps (right panel). To explain why the long-chain FA$^-$ generates transient currents while associated with the SBS, we need to assume that the SBS changes its position within the membrane during the c to m conformational change. Alternatively, if SBS and bound long-chain FA remain in the same position within the membrane, transient currents could be explained by the movement of charged UCP1 amino acid residues during the c-m transition. (**c**) UCP1 operates as a FA-dependent H^+ carrier in the presence of protonatable long-chain FA. When the long-chain FA$^-$ can be protonated at physiological pH, UCP1 acts as a FA$^-$/H^+ symporter (left panel). FA$^-$ and H^+ bind on the cytosolic side of the IMM. Upon the c to m conformational change, H^+ is released on the opposite side of the IMM, while the long-chain FA$^-$ stays associated with UCP1 due to the hydrophobic interactions. The FA$^-$ then returns back to face the cytosolic side and starts a new H^+ translocation cycle. A typical whole-mitoplast recording of a H^+ current activated by protonatable long-chain FA in response to voltage steps (right panel)

Thus, FA^- are UCP1 transport substrates. Although certain aspects of FA^- transport via UCP1 resemble an ion channel mechanism (Fedorenko et al. 2012; Jezek et al. 2010), it is assumed that UCP1, similar to other SLC25 members, is a carrier, and to this extent, UCP1 likely transports FA^- by an alternating access mechanism of transport (Kunji and Robinson 2010; Robinson and Kunji 2006; Robinson et al. 2008). Indeed, FA^- have two radically different moieties, a hydrophobic tail and a polar head, and it would be difficult to imagine a channel that would simultaneously translocate both. FA^- transport requires a transporter.

Finally, the long-chain low-pKa FA^- inhibit the H^+ currents activated by regular long-chain FA, which suggests that (1) FA^- and H^+ transport use the same translocation pathway and (2) H^+ binding to the polar head of the FA^- is essential for the H^+ transport through UCP1.

How do FA, the UCP1 transport substrates, activate the H^+ leak through UCP1?

4 A New Model of the FA-Dependent H^+ Leak via UCP1

The patch-clamp analysis of FA-dependent H^+ leak helped to dissect the UCP1 transport function into H^+ and FA^- currents and redefined the model of the mechanism of UCP1 action. The previously proposed models of UCP1 operation (Fig. 1) could not entirely explain the new data on FA–UCP1 interaction obtained in the patch-clamp electrophysiology experiments. Indeed, because FA are essential for the activation of the H^+ leak and serve as UCP1 transport substrates, the models in which FA activate the H^+ leak via UCP1 allosterically and especially those in which UCP1 has a basal FA-independent H^+ transport activity (the "H^+ channel" or "OH^-" channel in Fig. 1) cannot adequately explain the functional properties of UCP1. Moreover, the "H^+ buffering" model does not incorporate the fact that FA^- are UCP1 transport substrates. Finally, since long-chain FA^- cannot bind UCP1 on the matrix side (Fedorenko et al. 2012), the "FA-cycling" model, which requires binding of long-chain FA^- on the matrix side of the IMM for H^+ translocation into the mitochondrion, does not align with the electrophysiological data.

Therefore, a new model of UCP1 operation had to be proposed, which took into consideration both the new patch-clamp data and the previous findings (Fedorenko et al. 2012). This model, presented in Fig. 3c, assumes that UCP1 operates as a transporter rather than a channel and FA^- serve as a co-substrate for H^+ transport by UCP1. However, as mentioned above, long-chain FA^- are retained within the UCP1 translocation pathway by hydrophobic interactions and essentially serve as cofactors that shuttle within the UCP1 to enable H^+ transport. In accordance with this model, long-chain FA^- and H^+ interact with the UCP1 substrate-binding site (SBS) on the cytosolic face of the IMM (Fig. 3c), which triggers a conformational change. Upon the conformational change, the SBS is exposed to the opposite side of the IMM. H^+ is then released in the mitochondrial matrix, while the FA^- remains anchored to UCP1 due to the hydrophobic interactions established by its long carbon tail. Next, the model postulates that while the FA^- is still associated with UCP1, the reverse conformational change is possible, which will return the FA^- back to the cytosolic face of the IMM to initiate another H^+ transport cycle. Because in this model the FA^- shuttles within UCP1, it was called the "FA shuttling" model of UCP1 operation (Fedorenko et al. 2012). In the "FA shuttling" model, the cotransport of FA^-/H^+ through UCP1 is electroneutral, and the charge translocation is produced by the long-chain FA^- when it returns, after the release of H^+, to the opposite side of the membrane (Fig. 3c).

5 FA Remove Purine Nucleotide Inhibition of UCP1

In intact brown adipocytes, cytosolic purine nucleotides (primarily ATP) bind on the cytosolic side of UCP1 and occlude the translocation pathway to tonically inhibit UCP1-dependent thermogenesis (Berardi et al. 2011; Klingenberg 2010). Upon adrenergic stimulation of brown adipocytes, which triggers UCP1-dependent thermogenesis, the purine nucleotide inhibition is relieved. Two primary candidates have been suggested to overcome the purine nucleotide inhibition: FA and long-chain acyl-CoA. However, results pertaining to the ability of either of these two molecules to overcome purine nucleotide inhibition have been controversial (Huang 2003; Rial et al. 1983; Shabalina et al. 2004; Winkler and Klingenberg 1994).

Patch-clamp experiments demonstrated that long-chain FA overcome the inhibition of UCP1 by ATP (Fedorenko et al. 2012). FA^- are a permeable species and could compete with ATP^{4-} that binds near or within the translocation pathway (Klingenberg 2010). Structurally, FA^- and ATP^{4-} are too different to bind to the same site within UCP1. However, their binding sites may partially overlap or be located in close proximity, so that the electrostatic repulsion between the two negatively charged species results in competition. In the absence of purine nucleotides, even sub-micromolar concentrations of long-chain FA can activate very large UCP1 currents (Fedorenko et al. 2012). Purine nucleotides reduce the sensitivity of UCP1 for long-chain FA, likely to ensure that the UCP1-dependent H^+ leak is not activated by basal cytosolic FA concentrations (before adrenergic stimulation).

Although it has been proposed that acyl-CoA can overcome UCP1 inhibition by purine nucleotides (Cannon et al. 1977; Katiyar and Shrago 1991), patch-clamp data demonstrate that long-chain acyl-CoA inhibits the FA-dependent H^+ leak via UCP1 (Fedorenko et al. 2012). Thus, acyl-CoA is not an activator but rather an inhibitor of UCP1 similar to purine nucleotides. Indeed, the acyl-CoA molecule contains a nucleotide and acyl moieties. Thus, the nucleotide moiety could associate with the purine nucleotide-binding site of UCP1, while the acyl moiety could bind to the hydrophobic pocket intended for the carbon tail of the long-chain FA, resulting in UCP1 inhibition. The physiological significance of the acyl-CoA inhibition remains unclear.

While electrophysiological analysis demonstrated that long-chain FA can overcome ATP inhibition, this may not be the only physiological mechanism for the removal of UCP1 inhibition by purine nucleotides. Other mechanisms such as the elevation of pH in the cytosol (or more locally, in the intermembrane space)

(Nicholls and Locke 1984), by the interaction of UCP1 with cardiolipin (Klingenberg 2010), or by a recently reported redox regulation of UCP1 could also overcome the ATP inhibition in vivo (Chouchani et al. 2016).

6 Converging UCP1 Function and SLC25 Structural Data

6.1 A Consensus Structure for All SLC25 Members

UCP1 belongs to the SLC25 family, a large family of mitochondrial carriers responsible for the transport of various ions and metabolites across the IMM (Palmieri 2014). All SLC25 members have six transmembrane α-helices – three homologous two-transmembrane helix repeats that form a quasi-symmetrical tripartite structure. Although the nature and size of the transported substrates differ among the SLC25 members, the general molecular mechanism of substrate translocation is believed to be the same (Robinson et al. 2008). Hence, structural insight into the general mechanism of operation of SCL25 carriers can help to understand the UCP1 structure–function mechanisms.

Based on sequence and symmetry analyses, all SLC25 family members are predicted to have one SBS located near the center of the membrane (Robinson and Kunji 2006; Robinson et al. 2008). This single SBS is believed to be intermittently exposed to different sides of the IMM as the carrier changes its conformation from a cytosolic to a matrix state during substrate translocation (Kunji and Robinson 2010; Robinson et al. 2008). The structures for the cytosolic state of the transporter are currently available for two SLC25 members: the ADP/ATP carrier (Pebay-Peyroula et al. 2003; Ruprecht et al. 2014) and UCP2 (Berardi et al. 2011). These two structures suggest the existence of a deep aqueous cavity on the cytosolic side with a putative SBS located at the bottom (Fig. 4). The cytosolic state of all other

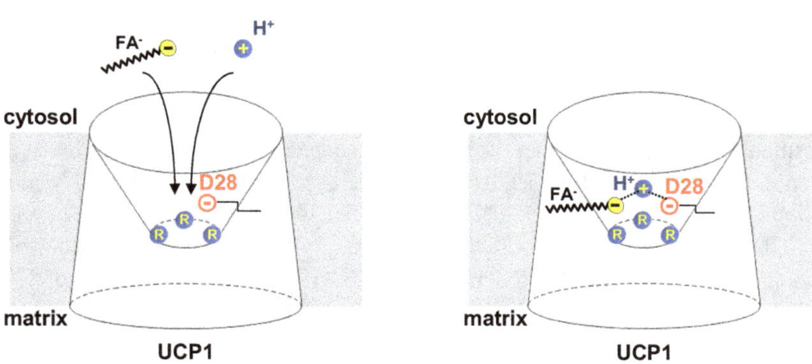

Fig. 4 Predicted SBS of UCP1. Schematic representation of UCP1 in the c-conformation state (SBS is exposed to cytosol). The proposed SBS of UCP1 with key arginines R84, R183, and R277 shown in blue and the titratable residue D28 in red. When FA^- and H^+ bind to the SBS, the FA carboxylic group interacts with the arginines, while H^+ is stabilized between the carboxylic headgroup of FA and D28

carriers is predicted to have a similar architecture (Kunji and Robinson 2010; Robinson and Kunji 2006).

6.2 Preliminary Considerations for UCP1 Structure–Function Relations

UCP1 was proposed to have a SBS consisting of three positively charged arginines (R84, R183, and R277) and a titratable residue (aspartate, D28) (Robinson et al. 2008). Based on the symmetry analysis of this SBS, it was proposed that the intended substrates of UCP1 and other uncoupling proteins are small carboxylic or keto acids, transported in symport with protons (Robinson et al. 2008). The reported transport of short-chain FA by UCP1 (Fig. 3a) correlates well with this prediction. As was suggested (Robinson et al. 2008), the positively charged arginines of the SBS could attract the negatively charged head of FA. However, substitution of any one of the arginines with a neutral amino acid did not seem to affect H^+ transport when mutant UCP1 was expressed in yeast (Echtay et al. 2001). Arguably, neutralization of more than one arginine must be required to completely prevent attraction of FA^- into the SBS and H^+ translocation.

Interestingly, long-chain FA do not fit well into the predicted category of UCP1 substrates. Indeed, because of their long carbon tail, long-chain FA are neither small carboxylic nor keto acids. UCP1 residues that bind the hydrophobic tail of FA should be located near the SBS. However, identification of such residues has been problematic (Robinson et al. 2008). Alternatively, the long hydrophobic tail could simply protrude into and be stabilized within the lipid bilayer. Regardless of the nature of the hydrophobic interactions that long-chain FA establish either with UCP1 or the lipid bilayer, such interactions do exist because the concentration at which FA activate UCP1 currents sharply decreases as the length of the hydrophobic tail increases (Fedorenko et al. 2012).

The importance of a long hydrophobic tail in FA activation of UCP1 uncoupling is consistent between the patch-clamp experiments and previous studies (Fedorenko et al. 2012; Rial et al. 1983; Winkler and Klingenberg 1994). Interestingly, indirect assessment of UCP1 transport activities in suspensions of isolated brown fat mitochondria suggested that UCP1 could mediate a GDP-sensitive Cl^- current (Nicholls and Lindberg 1973; Nicholls and Rial 1999). Therefore, it was proposed that UCP1 could be an anion channel, and its uncoupling activity is due to the transport OH^- rather than H^+ (Nicholls and Rial 1999). The new model of UCP1 function based on electrophysiological studies does not completely exclude transport of Cl^- or OH^-. However, because Cl^- and OH^- cannot establish hydrophobic interactions with UCP1, their binding and transport should be very limited. UCP1-mediated Cl^- currents could not be detected by the patch-clamp technique, but Cl^- could slightly affect the reversal potential of H^+ currents (Fedorenko et al. 2012). The negatively charged D28 may serve as a UCP1 selectivity filter that helps to prevent binding of small anions to the arginines of the SBS. FA^- would also be repelled by D28, but FA binding to UCP1 is additionally stabilized by the hydrophobic interactions, which make translocation of FA^- possible.

The hydrophobic interactions are especially strong for long-chain FA, which easily bind to UCP1 and activate H^+ currents at low micromolar or even sub-micromolar concentrations. How could long-chain FA carry H^+ through UCP1, if their solution pKa is around 5, and the vast majority of them are unprotonated at physiological pH 7? The hydrophobic interactions that anchor long-chain FA within UCP1 could also help to bring the anionic head of long-chain FA in close proximity to D28 (Fig. 4). The close proximity of the carboxylic headgroup of FA (solution pKa ~5) and anionic aspartate D28 (solution pKa ~4) could dramatically increase the pKa of both FA and D28, so that they bind H^+ at physiological pH (Fig. 4). In this preliminary model, both the FA headgroup and D28 are required to stabilize a single H^+ within the SBS, similar to what has been proposed for the mitochondrial phosphate carrier (Kunji and Robinson 2010). Thus, D28 would not only contribute to UCP1 hydrophobic anion selectivity but also H^+ permeation. D28 is known to be essential for H^+ translocation by UCP1 (Echtay et al. 2000).

Low-pKa long-chain FA analogs would bind to the SBS in the same fashion as regular long-chain FA, but the close proximity to D28 would not be able to elevate their pKa enough to enable H^+ binding and translocation. Therefore, low-pKa long-chain FA analogs would be unprotonated within the SBS, resulting in transient FA^- currents in response to the steps in membrane voltage (Fig. 3b).

6.3 Changes in UCP1 Conformation During Transport

It is likely that after the initial binding of the FA^- to the SBS, UCP1 "engulfs" FA and brings it within the membrane electric field (Fig. 5). Indeed, because low-pKa FA^- associated with UCP1 generate transient currents in response to membrane voltage steps (Fig. 3b), they must be located within the membrane electric field, not inside the water-filled cavity (Fedorenko et al. 2012). Furthermore, because FA^- bound to UCP1 translocate in response to changes in transmembrane voltage (Fig. 3b) (Fedorenko et al. 2012), UCP1 must be able to undergo conformational changes when associated with a FA substrate.

The engulfment of the substrate upon binding and the ability to undergo conformational changes thereafter are reminiscent of the "induced transition fit" principle originally proposed for the ATP/ADP carrier (Klingenberg 2008). In accordance with this principle, immediately after the initial binding, the substrate induces a conformational change in the SBS (carrier optimally engulfs the substrate). Such a SBS conformational change enables transitions between the cytosolic and matrix states. Thus, we propose that when a long-chain FA is bound to UCP1, UCP1 can change its conformation state so that the SBS with the attached FA headgroup and H^+ is intermittently exposed to the cytosolic and matrix side to accomplish H^+ transport (Fig. 5).

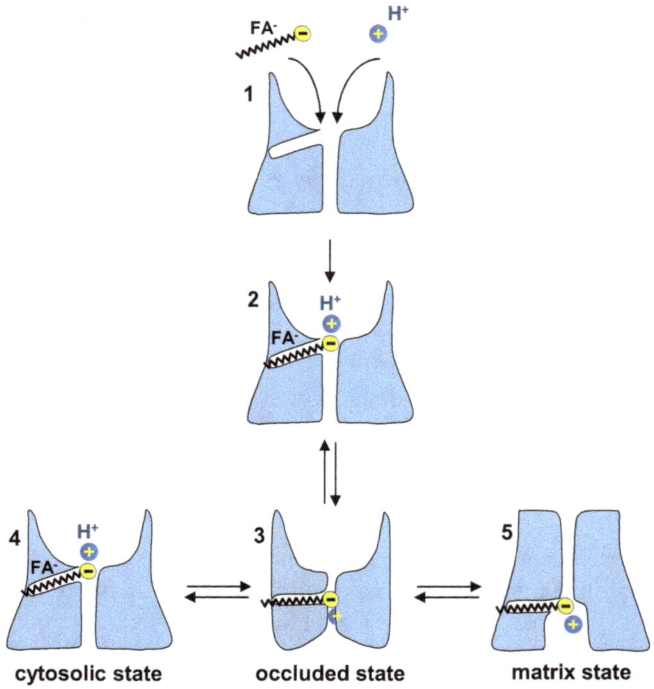

Fig. 5 Proposed UCP1 conformational changes during FA^{-}/H^{+} transport. When long-chain FA^{-} and H^{+} bind to the SBS (1 and 2), UCP1 undergoes a conformational change to "engulf" the substrates. In this state (occluded state, 3), FA^{-} and H^{+} are located in the area that separates the cytosolic cavity and the matrix, within the membrane electric field. The substrate-bound UCP1 (3) spontaneously transitions between the cytosolic state (4) and the matrix state (5), which results in exposure of the SBS to different sides of the IMM to accomplish H^{+} transport

However, UCP1 may be able to transition between the cytosolic and matrix states even in the absence of FA. Electrophysiological experiments demonstrate that low-pKa short-chain FA analogs can be transported by UCP1 when present only on the cytosolic (or matrix) side of the IMM (Fig. 3a) (Fedorenko et al. 2012). Although other interpretations are possible, this could mean that empty (FA-free) UCP1 can transition between matrix and cytosolic conformations without FA^{-}.

7 Concluding Remarks

Direct patch-clamp analysis of currents carried by UCP1 for the first time provided the true functional identity of UCP1 as a carrier for FA^{-} and H^{+}. The wealth of new functional information generated with patch-clamp electrophysiology has resulted in a new model of UCP1 function. However, this model only represents the simplest explanation of the electrophysiological data obtained, and the validity of this model must be further confirmed with structural data. The exact UCP1-binding sites for FA^- and H^+ need to be identified as well. The ability to directly record FA^- and H^+ currents carried by UCP1 using patch-clamp electrophysiology, combined with improved UCP1 structural data/protein crystallography, will answer these important questions and provide further insight into UCP1 structure–function relations.

Acknowledgments This work was supported by NIH grant 5R01GM107710 to Y.K.

References

Aquila H, Link TA, Klingenberg M (1985) The uncoupling protein from brown fat mitochondria is related to the mitochondrial ADP/ATP carrier. Analysis of sequence homologies and of folding of the protein in the membrane. EMBO J 4:2369–2376

Berardi MJ, Shih WM, Harrison SC, Chou JJ (2011) Mitochondrial uncoupling protein 2 structure determined by NMR molecular fragment searching. Nature 476:109–113

Bertholet AM, Kirichok Y (2017) UCP1: a transporter for H+ and fatty acid anions. Biochimie 134:28–34

Bertholet AM, Kazak L, Chouchani ET, Bogaczynska MG, Paranjpe I, Wainwright GL, Betourne A, Kajimura S, Spiegelman BM, Kirichok Y (2017) Mitochondrial patch clamp of beige adipocytes reveals UCP1-positive and UCP1-negative cells both exhibiting futile creatine cycling. Cell Metab 25:811–822.e814

Bouillaud F, Weissenbach J, Ricquier D (1986) Complete cDNA-derived amino acid sequence of rat brown fat uncoupling protein. J Biol Chem 261:1487–1490

Cannon B, Nedergaard J (2004) Brown adipose tissue: function and physiological significance. Physiol Rev 84:277–359

Cannon B, Sundin U, Romert L (1977) Palmitoyl coenzyme A: a possible physiological regulator of nucleotide binding to brown adipose tissue mitochondria. FEBS Lett 74:43–46

Chouchani ET, Kazak L, Jedrychowski MP, Lu GZ, Erickson BK, Szpyt J, Pierce KA, Laznik-Bogoslavski D, Vetrivelan R, Clish CB et al (2016) Mitochondrial ROS regulate thermogenic energy expenditure and sulfenylation of UCP1. Nature 532:112–116

Cohen P, Spiegelman BM (2015) Brown and beige fat: molecular parts of a thermogenic machine. Diabetes 64:2346–2351

Echtay KS, Winkler E, Bienengraeber M, Klingenberg M (2000) Site-directed mutagenesis identifies residues in uncoupling protein (UCP1) involved in three different functions. Biochemistry 39:3311–3317

Echtay KS, Bienengraeber M, Klingenberg M (2001) Role of intrahelical arginine residues in functional properties of uncoupling protein (UCP1). Biochemistry 40:5243–5248

Enerback S, Jacobsson A, Simpson EM, Guerra C, Yamashita H, Harper ME, Kozak LP (1997) Mice lacking mitochondrial uncoupling protein are cold-sensitive but not obese. Nature 387:90–94

Fedorenko A, Lishko PV, Kirichok Y (2012) Mechanism of fatty-acid-dependent UCP1 uncoupling in brown fat mitochondria. Cell 151:400–413

Feldmann HM, Golozoubova V, Cannon B, Nedergaard J (2009) UCP1 ablation induces obesity and abolishes diet-induced thermogenesis in mice exempt from thermal stress by living at thermoneutrality. Cell Metab 9:203–209

Garlid KD, Orosz DE, Modriansky M, Vassanelli S, Jezek P (1996) On the mechanism of fatty acid-induced proton transport by mitochondrial uncoupling protein. J Biol Chem 271:2615–2620

Garlid KD, Jaburek M, Jezek P (1998) The mechanism of proton transport mediated by mitochondrial uncoupling proteins. FEBS Lett 438:10–14

Gonzalez-Barroso MM, Fleury C, Bouillaud F, Nicholls DG, Rial E (1998) The uncoupling protein UCP1 does not increase the proton conductance of the inner mitochondrial membrane by functioning as a fatty acid anion transporter. J Biol Chem 273:15528–15532
Huang SG (2003) Binding of fatty acids to the uncoupling protein from brown adipose tissue mitochondria. Arch Biochem Biophys 412:142–146
Jezek P, Jaburek M, Garlid KD (2010) Channel character of uncoupling protein-mediated transport. FEBS Lett 584:2135–2141
Jiménez-Jiménez J, Zardoya R, Ledesma A, García de Lacoba M, Zaragoza P, Mar González-Barroso M, Rial E (2006) Evolutionarily distinct residues in the uncoupling protein UCP1 are essential for its characteristic basal proton conductance. J Mol Biol 359:1010–1022
Katiyar SS, Shrago E (1991) Differential interaction of fatty acids and fatty acyl CoA esters with the purified/reconstituted brown adipose tissue mitochondrial uncoupling protein. Biochem Biophys Res Commun 175:1104–1111
Kazak L, Chouchani ET, Jedrychowski MP, Erickson BK, Shinoda K, Cohen P, Vetrivelan R, Lu GZ, Laznik-Bogoslavski D, Hasenfuss SC et al (2015) A creatine-driven substrate cycle enhances energy expenditure and thermogenesis in beige fat. Cell 163:643–655
Klingenberg M (2008) The ADP and ATP transport in mitochondria and its carrier. Biochim Biophys Acta 1778:1978–2021
Klingenberg M (2010) Wanderings in bioenergetics and biomembranes. Biochim Biophys Acta 1797:579–594
Klingenberg M (2017) UCP1 – a sophisticated energy valve. Biochimie 134:19–27
Klingenberg M, Huang SG (1999) Structure and function of the uncoupling protein from brown adipose tissue. Biochim Biophys Acta 1415:271–296
Klingenberg M, Winkler E (1985) The reconstituted isolated uncoupling protein is a membrane potential driven H+ translocator. EMBO J 4:3087–3092
Kozak LP, Koza RA, Anunciado-Koza R (2010) Brown fat thermogenesis and body weight regulation in mice: relevance to humans. Int J Obes 34(Suppl 1):S23–S27
Krauss S, Zhang CY, Lowell BB (2005) The mitochondrial uncoupling-protein homologues. Nat Rev 6:248–261
Kunji ER, Robinson AJ (2010) Coupling of proton and substrate translocation in the transport cycle of mitochondrial carriers. Curr Opin Struct Biol 20:440–447
Lin CS, Klingenberg M (1980) Isolation of the uncoupling protein from brown adipose tissue mitochondria. FEBS Lett 113:299–303
Lin CS, Klingenberg M (1982) Characteristics of the isolated purine nucleotide binding protein from brown fat mitochondria. Biochemistry 21:2950–2956
Nicholls DG (2001) A history of UCP1. Biochem Soc Trans 29:751–755
Nicholls DG (2006) The physiological regulation of uncoupling proteins. Biochim Biophys Acta 1757:459–466
Nicholls DG, Lindberg O (1973) Brown-adipose-tissue mitochondria. The influence of albumin and nucleotides on passive ion permeabilities. Eur J Biochem 37:523–530
Nicholls DG, Locke RM (1984) Thermogenic mechanisms in brown fat. Physiol Rev 64:1–64
Nicholls DG, Rial E (1999) A history of the first uncoupling protein, UCP1. J Bioenerg Biomembr 31:399–406
Palmieri F (2014) Mitochondrial transporters of the SLC25 family and associated diseases: a review. J Inherit Metab Dis 37:565–575
Pebay-Peyroula E, Dahout-Gonzalez C, Kahn R, Trezeguet V, Lauquin GJ, Brandolin G (2003) Structure of mitochondrial ADP/ATP carrier in complex with carboxyatractyloside. Nature 426:39–44
Rial E, Gonzalez-Barroso MM (2001) Physiological regulation of the transport activity in the uncoupling proteins UCP1 and UCP2. Biochim Biophys Acta 1504:70–81

Rial E, Poustie A, Nicholls DG (1983) Brown-adipose-tissue mitochondria: the regulation of the 32000-Mr uncoupling protein by fatty acids and purine nucleotides. Eur J Biochem 137:197–203

Ricquier D, Kader JC (1976) Mitochondrial protein alteration in active brown fat: a soidum dodecyl sulfate-polyacrylamide gel electrophoretic study. Biochem Biophys Res Commun 73:577–583

Robinson AJ, Kunji ER (2006) Mitochondrial carriers in the cytoplasmic state have a common substrate binding site. Proc Natl Acad Sci U S A 103:2617–2622

Robinson AJ, Overy C, Kunji ER (2008) The mechanism of transport by mitochondrial carriers based on analysis of symmetry. Proc Natl Acad Sci U S A 105:17766–17771

Ruprecht JJ, Hellawell AM, Harding M, Crichton PG, McCoy AJ, Kunji ERS (2014) Structures of yeast mitochondrial ADP/ATP carriers support a domain-based alternating-access transport mechanism. Proc Natl Acad Sci U S A 111:E426–E434

Shabalina IG, Jacobsson A, Cannon B, Nedergaard J (2004) Native UCP1 displays simple competitive kinetics between the regulators purine nucleotides and fatty acids. J Biol Chem 279:38236–38248

Shabalina IG, Backlund EC, Bar-Tana J, Cannon B, Nedergaard J (2008) Within brown-fat cells, UCP1-mediated fatty acid-induced uncoupling is independent of fatty acid metabolism. Biochim Biophys Acta 1777:642–650

Shabalina IG, Petrovic N, de Jong JM, Kalinovich AV, Cannon B, Nedergaard J (2013) UCP1 in brite/beige adipose tissue mitochondria is functionally thermogenic. Cell Rep 5:1196–1203

Skulachev VP (1991) Fatty acid circuit as a physiological mechanism of uncoupling of oxidative phosphorylation. FEBS Lett 294:158–162

Winkler E, Klingenberg M (1994) Effect of fatty acids on H+ transport activity of the reconstituted uncoupling protein. J Biol Chem 269:2508–2515

Wu J, Bostrom P, Sparks LM, Ye L, Choi JH, Giang AH, Khandekar M, Virtanen KA, Nuutila P, Schaart G et al (2012) Beige adipocytes are a distinct type of thermogenic fat cell in mouse and human. Cell 150:366–376

Role of cAMP and cGMP Signaling in Brown Fat

Laia Reverte-Salisa, Abhishek Sanyal, and Alexander Pfeifer

Contents

1 Importance of the 3′,5′-Cyclic Adenosine Monophosphate Signaling Pathway in Brown Adipocytes ... 162
1.1 Downstream Targets of 3′,5′-Cyclic Adenosine Monophosphate ... 162
1.2 Effects of 3′,5′-Cyclic Adenosine Monophosphate Signaling in Brown Adipocytes ... 164
1.3 Role of 3′,5′-Cyclic Adenosine Monophosphate in Browning of White Adipose Tissue ... 167
2 The 3′,5′-Cyclic Guanosine Monophosphate Signaling and its Role in Brown Adipose Tissue ... 168
2.1 3′,5′-Cyclic Guanosine Monophosphate Production ... 168
2.2 Downstream Targets of 3′,5′-Cyclic Guanosine Monophosphate ... 170
3 Effects of Cyclic Nucleotide Signaling In Vivo ... 171
3.1 Activation of 3′,5′-Cyclic Guanosine Monophosphate and 3′,5′-Cyclic Adenosine Monophosphate Signaling in Mice ... 171
3.2 3′,5′-Cyclic Guanosine Monophosphate Signaling and Obesity ... 172
4 Conclusion ... 173
References ... 173

Abstract

Cold-induced activation of brown adipose tissue (BAT) is mediated by norepinephrine and adenosine that are released during sympathetic nerve activation. Both signaling molecules induce an increase in intracellular levels of 3′,5′-cyclic adenosine monophosphate (cAMP) in murine and human BAT. In brown adipocytes, cAMP plays a central role, because it activates lipolysis, glucose uptake, and thermogenesis. Another well-studied intracellular second messenger is

LR-S and AS contributed equally to this review.

L. Reverte-Salisa · A. Sanyal · A. Pfeifer (✉)
Institute of Pharmacology and Toxicology, University Hospital Bonn, University of Bonn, Bonn, Germany
e-mail: alexander.pfeifer@uni-bonn.de

A. Pfeifer et al. (eds.), *Brown Adipose Tissue*,
Handbook of Experimental Pharmacology 251, https://doi.org/10.1007/164_2018_117

3′,5′-cyclic guanosine monophosphate (cGMP), which closely resembles cAMP. Several studies have shown that intact cGMP signaling is essential for normal adipogenic differentiation and BAT-mediated thermogenesis in mice. This chapter highlights recent observations, demonstrating the physiological significance of cyclic nucleotide signaling in BAT as well as their potential to induce browning of white adipose tissue (WAT) in mice and humans.

Keywords
Brown adipose tissue · Browning · cAMP · cGMP · UCP1

1 Importance of the 3′,5′-Cyclic Adenosine Monophosphate Signaling Pathway in Brown Adipocytes

In 1957, Earl Wilbur Sutherland (Berthet et al. 1957a, b) discovered 3′,5′-cyclic adenosine monophosphate (cAMP). cAMP is a ubiquitous second messenger that plays an important role in a broad spectrum of cells and numerous signaling pathways. Stimulation of G_s protein-coupled receptors, such as β-adrenergic receptors (β-AR) and adenosine receptors, activates adenylate cyclase (AC), triggering conversion of adenosine triphosphate (ATP) to cAMP. In contrast, G_i protein-coupled receptors inhibit AC and thus decrease cAMP levels (Hanoune and Defer 2001). The AC family comprises ten isoforms, nine of them are membrane bound (AC1-9) and one (AC10) is soluble (Hanoune and Defer 2001). Increased intracellular cAMP levels subsequently lead to activation of different cAMP downstream effectors depending on the cell type, receptor, and spatiotemporal context. To date, three major classes of cAMP-activated effectors have been described: protein kinase A (PKA), exchange proteins directly activated by cAMP (EPACs), and cAMP-gated ion channels (CNG).

1.1 Downstream Targets of 3′,5′-Cyclic Adenosine Monophosphate

1.1.1 Protein Kinase A

PKA is a tetramer consisting of two catalytic and two regulatory subunits (Scott et al. 1986; Shabb 2001). The binding of two molecules of cAMP to each regulatory subunit induces the activation of PKA by dissociating the regulatory and the catalytic subunits. Once released, the catalytic subunits are free to phosphorylate Serine (Ser) and Threonine (Thr) residues on target proteins. There are four different catalytic subunits (Cα, Cβ, Cγ, and the related gene Cχ) and four regulatory subunits (RIα, RIβ, RIIα, and RIIβ) (Jahnsen et al. 1988; Uhler et al. 1986a, b, Ringheim and Taylor 1990; Corbin et al. 1977; Zoller et al. 1979; Bechtel et al. 1977; Lee et al. 1983; Scott et al. 1987; Eide et al. 2003). The different regulatory subunits RI and RII rise term to the distinct PKA subtypes, PKA I and PKA II, respectively, which differ in their cAMP sensitivity, expression levels, and localization (Sim and Scott 1999; Tasken and Aandahl 2004; Langeberg and Scott 2015). The subcellular localization and signaling of PKA is determined by A-kinase anchoring proteins (AKAPs) (Gold et al. 2006; Kinderman et al. 2006; Newlon et al. 2001; Perino et al. 2011). The PKA–AKAP complexes allow the cAMP-derived events to be localized within cellular compartments. Namely, AKAPs are responsible to target PKA to different subcellular localizations, optimizing the cAMP signaling transduction (Pidoux and Tasken 2010). Adipose tissue expresses mostly the RIIβ subtype; however, the regulation of the PKA signaling is highly complex. Studies using whole body knockout of RIIβ showed that the loss of RIIβ was compensated by an increased expression of RIα, conferring the mice a lean phenotype and protection against diet-induced obesity (DIO) and insulin resistance (Cummings et al. 1996; Brandon et al. 1998). Studies analyzing the physiological consequences of an activation of the PKA catalytic subunit α in the adipose tissue in vivo observed an improved metabolic status and resistance to DIO, most likely through increased energy expenditure and browning of white adipose tissue (WAT) (Dickson et al. 2016).

1.1.2 Exchange Protein Activated by 3′,5′-Cyclic Adenosine Monophosphate

EPACs have a similar affinity to cAMP as PKA (Dao et al. 2006) and regulate small GTPases promoting the transition to the active GTP-bound form. Two isoforms have been described so far, EPAC1 and EPAC2. Structurally, these isoforms differ only in their catalytic subunit (Rehmann et al. 2006). EPAC1 is expressed ubiquitously, while the expression of EPAC2 appears to be restricted to the brain, liver, pancreas, and adrenal glands (Kawasaki et al. 1998). A recent study showed that EPAC1 regulates leptin levels in 3T3-L1 cells and in WAT (Hu et al. 2016). Although EPAC has been described to contribute to leptin resistance in murine hypothalamus (Hwang et al. 2017; Fukuda et al. 2011) and to regulate insulin secretion in the pancreas (Kasai et al. 1999; Kelley et al. 2009), the role of EPAC in BAT is not known.

1.1.3 3′,5′-Cyclic Adenosine Monophosphate-Gated Ion Channels

cAMP can also signal through CNG such as HCN channels, which act as pacemakers in cardiomyocytes (Kaupp and Seifert 2001). All CNGs respond to both cAMP and 3′,5′-cyclic guanosine monophosphate (cGMP), although depending on the channel and the composition of the channel complex, their specificity and selectivity varies. CNG activity is modulated by diverse stimuli, e.g., phosphorylation, and interestingly, in the continuous presence of cAMP or cGMP they do not desensitize (Bradley et al. 2005; Pifferi et al. 2006). Although not expressed in the adipose tissue, CNGs play a key role in visual and olfactory signal transduction (Kaupp and Seifert 2002).

1.1.4 Cyclic Nucleotide Phosphodiesterases

The activities of all three cAMP effectors are mainly dependent on the availability of intracellular levels of cAMP in different pools. Phosphodiesterases (PDEs) are the main modulators of cyclic nucleotide levels (Bender and Beavo 2006). PDEs break down the phosphodiester bond of cAMP and cGMP, regulating their concentrations and therefore the different cellular processes. There are 11 different PDE families that differ in their structure, activity, and localization, out of which only PDE4, PDE7, and PDE8 are cAMP specific. In addition, PDE1, PDE2, PDE3, PDE10, and PDE11 have dual specificity, modulating levels of both cAMP and cGMP (Bender and Beavo 2006).

PDEs play an important role in brown adipocyte signaling. When cultured in vitro, preadipocytes are induced by adding a nonselective PDE inhibitor such as 3-isobutyl-1-methylxanthine (IBMX) (Jennissen et al. 2013). Insulin activates PDE3B in 3T3-L1 adipocytes in a phosphoinositide 3-kinase-dependent manner, thereby lowering the levels of cellular cAMP (Kitamura et al. 1999). Other PDEs have been described to be regulators of brown adipogenesis. Kraynik and colleagues observed that inhibition of both PDE3 and PDE4 is necessary for the maximal induction of lipolysis and PKA-mediated accumulation of uncoupling protein 1 (UCP1) mRNA (Kraynik et al. 2013).

Inhibition of PDE10A has been shown to increase glucose uptake and thermogenesis, and its expression is increased in BAT of different models of obesity (Hankir et al. 2016). PDE10A modulates the levels of cAMP and cGMP; thus, it was hypothesized that the effects of PDE10A inhibition might be due to a synergistic action of cAMP and cGMP (Hankir et al. 2016).

1.2 Effects of 3′,5′-Cyclic Adenosine Monophosphate Signaling in Brown Adipocytes

BAT is under control of the sympathetic nervous system (SNS). SNS activation regulates lipolysis, metabolic rate, and glucose availability (Bartness et al. 2010) (Louis et al. 2000; Hom et al. 2001). Sympathetic release of norepinephrine (NE) activates β-AR and the subsequent increase in cAMP levels activates brown adipocyte energy expenditure. There are three different subtypes of β-AR: β1-AR, β2-AR, and β3-AR (Deng et al. 1996; Bengtsson et al. 2000). Their expression in BAT varies depending on the differentiation status of brown adipocytes (Granneman and Lahners 1992; Bengtsson et al. 2000). While β1-AR are highly expressed in preadipocytes, β3-AR expression levels are increased in mature adipocytes (Bronnikov et al. 1999). In addition to NE, the SNS releases co-transmitters and Gnad et al. (2014) showed that adenosine is released during SNS activation of BAT. Four receptors for adenosine have been described that either couple to Gs or Gi proteins, respectively: A1/Gi, A2A/Gs, A2B/Gs, and A3/Gi (Sheth et al. 2014). Interestingly, murine and human BAT expresses high levels of A2A and adenosine acts synergistically to NE by increasing cAMP in the BAT of these species (Gnad et al. 2014) (Fig. 1).

1.2.1 Pivotal Role of 3′,5′-Cyclic Adenosine Monophosphate in the Activation of Brown Adipocytes

Lipolysis is the process by which triglycerides are broken down to fatty acids. It is mainly controlled by the enzyme hormone-sensitive lipase (HSL) and it is negatively

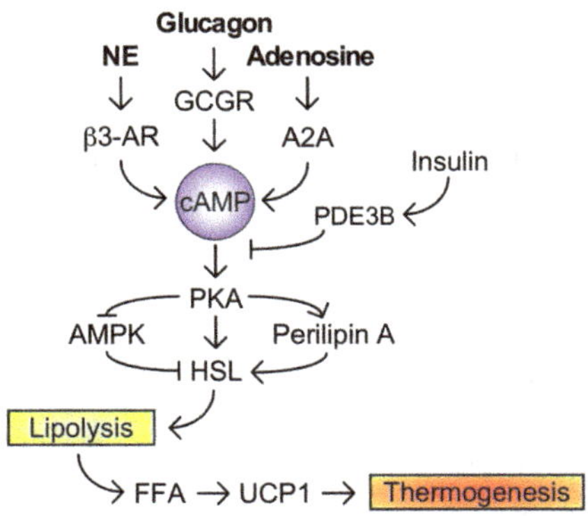

Fig. 1 Pivotal role of cAMP in brown adipocyte activation. *A2A* adenosine receptor 2A, *GCGR* glucagon receptor, *β3-AR* beta-3 adrenergic receptor, *NE* Norepinephrine, *cAMP* 3′,5′-cyclic adenosine monophosphate, *PDE3B* phosphodiesterase 3B, *PKA* protein kinase A, *AMPK* AMP-activated protein kinase, *HSL* hormone-sensitive lipase, *FFA* free fatty acids, *UCP1* uncoupling protein 1

regulated by insulin. The three major lipases driving lipolysis are adipose triglyceride lipase (ATGL), HSL, and monoacylglycerol lipase (MAGL) (Duncan et al. 2007).

The canonical pathway for the activation of lipolysis is the NE-derived stimulation of the β3-AR (Louis et al. 2000; Hom et al. 2001). β3-AR agonists have been shown to activate brown adipose tissue (BAT) in vivo as well as to induce WAT lipolysis (Cypess et al. 2015). Adenosine, which was firstly described to negatively regulate thermogenesis and lipolysis in brown adipocytes from hamsters and rats (Schimmel and McCarthy 1984; Szillat and Bukowiecki 1983; Unelius et al. 1990), was later discovered by Gnad et al. to activate and induce lipolysis of murine and human brown adipocytes via the A2A receptors (Gnad et al. 2014). Glucagon, which induces thermogenesis and increases energy expenditure (Billington et al. 1991; Kinoshita et al. 2014), is released during cold exposure (Seitz et al. 1981) and promotes lipolysis by increasing fatty acid utilization (Howland and Benning 1986). NE, adenosine, and glucagon signal through Gs-coupled receptors and their effects are a consequence of increased cAMP levels and subsequent activation of HSL, leading to triglyceride lipolysis and nonesterified fatty acid (NEFA) release (Anthonsen et al. 1998; Chernick et al. 1986; Perea et al. 1995; Heckemeyer et al. 1983). cAMP signaling promotes lipolysis both in WAT and BAT. Recently, it has been proposed that the main source for BAT activation are the free fatty acids (FFA) released from lipolysis in WAT (Shin et al. 2017; Schreiber et al. 2017; Fedorenko et al. 2012).

The cAMP/PKA pathway has also been proposed to promote lipolysis by inactivating AMP-activated protein kinase α (AMPKα) through phosphorylation at Ser-173 in vitro and in vivo (Djouder et al. 2010). Thus, the stimuli that induce the activation of lipolysis via the PKA/HSL pathway suppress the phosphorylation of HSL by AMPK at Ser-565 (Djouder et al. 2010). The AMPK-derived phosphorylation of HSL is proposed to prevent PKA-mediated phosphorylation and to reduce lipolysis (Daval et al. 2005; Gauthier et al. 2008; Omar et al. 2009). Although the effect of AMPK on HSL activity has been debated (Yin et al. 2003; Koh et al. 2007), the inhibitory role of AMPK on lipolysis seems to be the most accepted (Fig. 1).

The inhibitory effect of insulin on lipolysis is mediated through an inhibition of the cAMP-mediated signaling to HSL via the PI3K/Akt pathway. The PI3K/Akt pathway is responsible for the phosphorylation and activation of the phosphodiesterase PDE3B, inducing the degradation of cAMP (Kitamura et al. 1999). Although PI3K has been proven to mediate the insulin-derived inhibition of lipolysis, Akt might be dispensable (DiPilato et al. 2015). Additionally, Choi et al. proposed that insulin inhibits lipolysis by interfering with the phosphorylation of perilipin by PKA (Choi et al. 2010). The cAMP/PKA-dependent phosphorylation of perilipin A is key for HSL to be translocated to the lipid droplet and be functional (Miyoshi et al. 2006, 2007). Moreover, phosphorylation of perilipin A by PKA facilitates the interaction with HSL, increasing its activity (Sztalryd et al. 2003; Su et al. 2003) (Fig. 1).

Additionally to its role in lipolysis, cAMP promotes glucose transport into brown adipocytes (Chernogubova et al. 2004). The main glucose transporters (GLUT) expressed in BAT are GLUT-1 and GLUT-4, which mediate glucose uptake upon cold exposure (Shimizu et al. 1998; Greco-Perotto et al. 1987; Bryant et al. 2002). Adrenergic stimulation of β3-AR triggers glucose uptake by recruiting and stimulating the PI3K signaling pathway in a cAMP/PKA-dependent manner, similarly to the effects that insulin exerts on glucose uptake (Chernogubova et al. 2004). Moreover, cAMP also promotes expression and translocation of GLUT1 to the plasma membrane through the mTOR pathway in vitro and in vivo (Olsen et al. 2014, 2017) and it has also been suggested that cAMP promotes glucose uptake through activation of AMPK (Hutchinson et al. 2005).

1.2.2 Role of 3′,5′-Cyclic Adenosine Monophosphate in Differentiation of Brown Adipocytes

cAMP signaling is involved in differentiation of murine brown adipocytes by promoting proliferation and the expression of transcription factors involved in early stages of differentiation as well as the expression of key regulators of thermogenesis.

Proliferation of brown adipocytes has been attributed to be driven by β1-AR/cAMP signaling in vitro (Bronnikov et al. 1992; Kozak and Kozak 1994). Using NE, cAMP analogs and forskolin (an AC stimulator), Bronnikov et al. could mimic the response to the in vivo cold-induced increase of cell proliferation (Bronnikov et al. 1992). Indeed, mice lacking β1-AR were not able to induce de novo brown adipogenesis after stimulation with NE (Lee et al. 2016). Fredriksson et al. proposed that cAMP promotes proliferation of murine brown preadipocytes in a PKA-dependent manner, upregulating the Ribonucleotide Reductase Subunit R2 and ultimately signaling through Src-ERK1/2 (Fredriksson and Nedergaard 2002). However, NE can also induce proliferation of rat fetal brown adipocytes through ERK1/2 in a cAMP/PKA-independent manner, since inhibition of PKA did not reduce the NE-derived stimulation of ERK1/2 phosphorylation (Valladares et al. 2000).

Two of the most well-known mediators of early murine brown adipocyte differentiation are the family of transcription factors CCAAT/enhancer-binding proteins (C/EBPs) and peroxisome proliferator-activated receptors (PPARs) (MacDougald and Lane 1995).

C/EBPβ is a key regulator of the transition from preadipocytes to mature brown adipocytes. Moreover, mice lacking C/EBPβ show reduced lipid accumulation and UCP1 expression (Tanaka et al. 1997). cAMP enhances C/EBPβ expression via the cAMP-response element binding protein (CREB) (Zhang et al. 2004). The addition of cAMP elevating agents and glucocorticoids, which increase intracellular levels of cAMP and Ca^{2+}, during the in vitro induction of brown adipogenesis leads to increased C/EBPβ expression for a proper brown adipocyte differentiation (Cao et al. 1991; Darlington et al. 1998; Rosen et al. 2002; Carmona et al. 2005). Peroxisome proliferator-activated receptor gamma (PPARγ) expression is secondary to C/EBPβ as part of a positive loop between C/EBPα and PPARγ itself (Sheyn et al. 2013). PPARγ expression has been reported to be induced by cAMP/PKA and Akt (Kim et al. 2010).

Importantly, cAMP signaling triggers the expression of the thermogenic markers UCP1 and PGC1α. PGC1α regulates the transcription of genes involved in mitochondrial biogenesis (Tiraby and Langin 2003; Tiraby et al. 2003) and its expression is highly induced in response to cold exposure through cAMP signaling and promotes *Ucp1* expression by interacting with other transcription factors. For instance, interferon regulatory factor 4 (IRF4), which is also induced by cAMP, interacts with PGC1α and binds the regulatory regions of *Ucp1* (Kong et al. 2014). PRDM16 also interacts with PGC1α to induce its transcriptional activity (Seale et al. 2007). In addition, cAMP activates the p38 mitogen-activated protein kinase (p38), which induces PGC1α expression (Cao et al. 2001, 2005; Robidoux et al. 2005). The cAMP/PKA/p38 pathway controls the factors that bind to the *Ucp1* gene by two separate ways: (1) phosphorylating the cAMP-dependent transcription factor ATF2 (ATF2) and (2) promoting the transcription of PGC1α.

ATF2 is a transcriptional activator, whose activity depends on its binding to a cAMP-response element. It coordinates the transcription of PGC1α and UCP1, and it has also been proposed to regulate PPARγ in vitro (Lee et al. 2001). Additionally, p38 can activate PGC1α by direct phosphorylation (Puigserver and Spiegelman 2003).

Stimulating the cAMP pathway enhances transcription of the fibroblast growth factor 21 (FGF21) via the PKA/p38/ATF2 pathway. FGF21 is released by brown adipocytes after cold exposure and it has been reported to promote thermogenic activity in an autocrine manner (Hondares et al. 2010, 2011).

1.3 Role of 3′,5′-Cyclic Adenosine Monophosphate in Browning of White Adipose Tissue

Browning of WAT – i.e., the appearance of brown-like "beige" adipocytes – can be stimulated after prolonged cold exposure (Loncar 1991). Beige adipocytes have UCP1-expressing mitochondria, which upon activation contribute to an increase in thermogenesis and energy expenditure (Kim and Plutzky 2016). The physiological response to cold exposure involves an increase in cAMP levels and the subsequent upregulation of UCP1, which can be pharmacologically mimicked by β3-AR agonists. Rats treated with CL 316243, a β3-AR agonist, showed an increase in the number of multilocular cells among WAT depots (Himms-Hagen et al. 2000) and restored levels of genes involved in fatty acid oxidation downregulated in type 2 diabetic mice (Kumar et al. 2015). Another β3-AR agonist, Mirabegron (used for the treatment of hyperactive bladder), stimulated the appearance of beige adipocytes in WAT depots of human subjects and increased glucose uptake in BAT as well as lipolysis of WAT (Cypess et al. 2015). However, Mirabegron had unwanted cardiovascular side effects (Cypess et al. 2015). To avoid undesirable adverse effects, other pathways involving cAMP signaling have been investigated to promote browning.

Adenosine stimulates browning of WAT upon stimulation of the Gs-coupled adenosine receptor A2A, which induces an increase in the intracellular levels of cAMP (Gnad et al. 2014). Slit2 is an extracellular matrix protein that has been mostly studied in the brain. However, Svensson and colleagues showed that Slit2 is not only secreted by beige adipocytes but induces browning of WAT involving increased cAMP levels and the activation of PKA (Svensson et al. 2016). Additionally, the activation of the transient receptor potential cation channel subfamily M member 8 (TRPM8) by menthol induces browning of WAT in a PKA-dependent manner, although it is not well known if it directly involves an increase in cAMP levels (Jiang et al. 2017).

2 The 3′,5′-Cyclic Guanosine Monophosphate Signaling and its Role in Brown Adipose Tissue

cGMP is another important intracellular second messenger (Tsai and Kass 2009). Cyclic GMP was first synthesized in the early 1960s. Soon after, cGMP-degrading enzymes were identified, known as PDEs, which hydrolyze the 3′-5′ phosphodiester bond (Kots et al. 2009). In nature, cGMP was first observed in rat urine (Ashman et al. 1963). Following this discovery, a variety of hormones and neurotransmitters were shown to alter physiological cGMP levels (Kots et al. 2009). This section focuses on the various components making up the cGMP signaling cascade and their effects on brown/beige adipocyte-mediated thermogenesis.

2.1 3′,5′-Cyclic Guanosine Monophosphate Production

The conversion of guanosine triphosphate (GTP) to cGMP is catalyzed by a family of enzymes known as guanylate cyclases (GCs) (Koesling et al. 1991). GCs are broadly classified into two categories: (1) cytoplasmic or soluble GCs (sGCs) (Koesling et al. 1991; McDonald and Murad 1996; Poulos 2006), and (2) membrane-bound particulate or receptor GCs (pGCs), which are receptors for natriuretic peptides (NPRs) (Miyashita et al. 2009; Waldman et al. 1984).

Soluble guanylate cyclases are heterodimeric enzymes made up of an alpha ($sGC\alpha_1$ or $sGC\alpha_2$) and a beta ($sGC\beta_1$) subunit. These subunits have a 32% amino acid sequence homology. Importantly, $sGC\beta_1$ contains a heme moiety at its N-terminal domain (Derbyshire and Marletta 2012). sGCs catalyze cGMP production (from GTP) upon activation by the gaseous ligand, nitric oxide (NO) (Koesling et al. 1991). Synthesized physiologically in several types of cells by enzymes known as NO-synthases (NOS) (Förstermann and Sessa 2012), nitric oxide binds to the heme-containing sGCβ_1 at its reduced state (Fe^{2+}). Binding of NO to the sGCβ_1 subunit leads to conformational changes in sGC that ultimately lead to conversion of GTP to cGMP. At an oxidized state, sGCβ_1 is unresponsive to NO (Denninger and Marletta 1999).

NPRs are membrane-bound GCs that are receptors for the extracellular ligands known as natriuretic peptides (NPs) (Waldman et al. 1984). A-type NP (ANP; secreted by atrial myocytes), B-type NP (BNP; secreted in the ventricles), and C-type NP (CNP; secreted by central nervous system, cardiac fibroblasts, endothelium, and macrophages) (Nishikimi et al. 2006) play an essential role in regulation of blood pressure and natriuresis (Pandit et al. 2011). These NPs bind to the extracellular domain of the homodimeric NPRs and activate the production of cGMP (Miyashita et al. 2009).

In recent years, several reports have demonstrated the importance of guanylate cyclases in adipocyte function. Acting through the NPRs, natriuretic peptides have been shown to enhance insulin sensitivity as well as mitochondrial biogenesis in adipocytes (Bordicchia et al. 2012). In humans, natriuretic peptides induce lipolysis in white adipocytes, mobilizing FFA availability that undergoes β-oxidation in the liver, skeletal muscle, and adipose tissue (Gruden et al. 2014). Interestingly, NPR-induced lipolysis is primate specific and is not observed in rodent or canine adipocytes (Moro and Lafontan 2012). Several in vitro studies on adipocytes have shown that an increase in cGMP production by NPR-A receptor activates PKG, which can phosphorylate perilipin (Moro and Lafontan 2012). Phosphorylation of perilipin leads to physical alterations at the lipid droplet surface, activating lipases, such as ATGL and HSL (Moro and Lafontan 2012). Moreover, NPR activation has been shown to enhance p38-MAPK signaling and phosphorylation of ATF2. This enhances the expression of PPARγ cofactor 1-alpha (PGC-1α) and UCP1, leading to mitochondrial biogenesis and thermogenesis (Bordicchia et al. 2012; Gruden et al. 2014).

Similarly, the importance of sGC in brown adipocytes has also been well studied. It has been demonstrated that exogenous NO stimulation of various cell types leads to enhanced expression of PGC-1α, UCP1, and mitochondrial markers such as nuclear respiratory factor and mitochondrial transcription factor (Nisoli et al. 1998, 2003). Blocking sGC function (in NO-stimulated cells) abrogates these effects, indicating the importance of sGC/cGMP signaling in mediating NO-induced mitochondrial biogenesis (Nisoli et al. 1998, 2003). Likewise, genetic ablation of endothelial NOS (eNOS) in mice leads to reduced mitochondrial content in BAT accompanied with increased lipid accumulation (Nisoli et al. 2003). In accordance, ablation of sGCβ_1 expression diminishes in vitro adipogenic and thermogenic differentiation of brown adipocytes (Hoffmann et al. 2015). In addition, mice lacking sGCβ_1 expression have impaired thermogenesis compared to wild-type littermates (Hoffmann et al. 2015). On the other hand, pharmacological stimulation of sGC using BAY 41-8543 (a precursor to the drug Riociguat) enhances lipid accumulation, mitochondrial biogenesis, and UCP1 expression in brown adipocytes (Hoffmann et al. 2015). Accordingly, enhanced expression of sGCβ_1 has also been demonstrated to increase adipogenesis (lipid accumulation) accompanied by increased UCP1 expression (Jennissen et al. 2012).

2.2 Downstream Targets of 3′,5′-Cyclic Guanosine Monophosphate

As an intracellular second messenger, cGMP acts on three major targets: (a) cyclic nucleotide-gated ion channels (CNGs), (b) cyclic nucleotide PDEs, and (c) cGMP-dependent protein kinases (PKGs).

2.2.1 Cyclic Nucleotide-Gated Ion Channels

CNGs are nonselective cation channels that conduct a mixed inward current, carried by Na^+ and Ca^{2+} ions. These channels are responsive to both cAMP and cGMP, with varying degrees of affinity, depending on the cell type and function (Kaupp and Seifert 2002). CNGs have been shown to play an important role in the functioning of photoreceptors, chemosensory cells (like olfactory sensory neurons), and spermatozoa but have not been described in adipocytes (Kaupp and Seifert 2002).

2.2.2 Cyclic Nucleotide Phosphodiesterases

Inhibiting PDEs has been known to enhance adipocyte lipolysis (Snyder et al. 2005). Studies assessing the effects of inhibiting cGMP-specific PDEs have focused on PDE5, using its inhibitor sildenafil, thereby enhancing cGMP signaling (Moreland et al. 1998). It has been shown that sildenafil promotes adipogenesis in 3T3-L1 cells, through the cGMP–PKG pathway (Mitschke et al. 2013). Sildenafil treatment increases glucose uptake and lipid content as well as enhances expression of adipogenic markers, such as PPARγ and FABP4 in adipocytes (Mitschke et al. 2013; Zhang et al. 2010). Furthermore, treatment of primary murine white adipocytes (in vitro) with sildenafil also induces browning and increases expression of thermogenic markers, such as UCP1 and PGC-1α (Mitschke et al. 2013).

2.2.3 3′,5′-Cyclic Guanosine Monophosphate-Dependent Protein Kinases

The major downstream targets of cGMP in the cardiovascular system and adipocytes are PKGs (Haas et al. 2009; Mitschke et al. 2013; Tsai and Kass 2009). Mammals have two types of PKGs, namely PKG1 and PKG2, encoded by two distinct genes (Wernet et al. 1989). PKG1 is further classified into two subtypes: PKG1α and PKG1β (Jaumann et al. 2012). While PKG1 isoforms are cytosolic, PKG2 is a transmembrane protein (Jaumann et al. 2012; Nádvorník et al. 1999). N-terminal myristoylation is a major determinant of PKG2 membrane association, whereas acetylation of PKG1 N-terminal enhances its solubility and cytoplasmic localization (Vaandrager and de Jonge 1996). Both PKG1 and PKG2 are homodimeric proteins with similar structural features, consisting of an N-terminal regulatory domain (consisting of a homodimerization, an autoinhibitory–autophosphorylation, and a cGMP-binding subdomain) and a C-terminal kinase domain (comprising of Mg^{2+}/ATP-binding and substrate-binding subdomains) (Francis et al. 2010). In an inactive state, the regulatory domain binds (with high affinity) to the catalytic domain, inhibiting any kinase activity. Upon binding of cGMP at the allosteric sites on the regulatory domain, the holoenzyme undergoes conformational changes, freeing the catalytic domain and leading to activation of PKG (Alverdi et al. 2008).

PKG2 is expressed in the kidneys, intestinal mucosa, lungs, and chondrocytes (Hofmann et al. 2006; Tsai and Kass 2009). Genetic ablation of PKG2 in mice imparts resistance to *Escherichia coli* STa, an enterotoxin that stimulates cGMP accumulation and intestinal fluid secretion (Pfeifer et al. 1996). Furthermore, PKG2 knockout mice have retarded bone growth due to reduced endochondral but not membranous ossification resulting in dwarfism (Pfeifer et al. 1996).

Several downstream targets of PKG1 have been reported influencing diverse cellular processes, such as bone development, regulation of cytoskeleton, and vesicular trafficking, as well as regulation of smooth muscle tone (Hofmann et al. 2006). PKG1 has also been identified as the major downstream target of cGMP in adipocytes (Haas et al. 2009; Mitschke et al. 2013) and PKG1 is indispensable for differentiation of adipocytes in vitro (Haas et al. 2009; Mitschke et al. 2013). Similar to the genetic ablation of sGCβ_1, mice lacking PKG1 (PKG1$^{-/-}$) demonstrate severely reduced thermogenic capacity, accompanied by severely reduced UCP1 expression and mitochondrial content in BAT (Haas et al. 2009). Moreover, in vitro differentiation of brown adipocytes from PKG1$^{-/-}$ mice is drastically reduced, compared to wild-type cells. On the contrary, enhancing PKG1 activity, by exogenous stimulation with a cell membrane permeable cGMP analog, has been shown to improve adipogenic and thermogenic differentiation (Haas et al. 2009). In brown adipocytes, PKG1 inhibits the small GTPase RhoA and its associated kinases (ROCKs). This, in turn, reduces ROCK-mediated inhibitory phosphorylation of insulin receptor substrate 1 (IRS-1), thereby, maintaining insulin sensitivity in these cells (Haas et al. 2009). Mice overexpressing PKG1 (PKG1^{tg}) have been shown to be leaner than wild-type mice and are resistant to DIO (Miyashita et al. 2009). PKG1^{tg} mice have increased insulin sensitivity, enhanced energy expenditure (EE), increased BAT mitochondrial content, and increased expression of thermogenic markers, such as UCP1 and PGC-1α. Importantly, activating cGMP signaling is beneficial for the adipogenic and thermogenic programs in cultured human adipocytes (Bordicchia et al. 2012; Haas et al. 2009; Hoffmann et al. 2015; Mitschke et al. 2013; Miyashita et al. 2009).

3 Effects of Cyclic Nucleotide Signaling In Vivo

3.1 Activation of 3′,5′-Cyclic Guanosine Monophosphate and 3′,5′-Cyclic Adenosine Monophosphate Signaling in Mice

The presence of a functional cGMP signaling is pivotal for BAT-mediated thermogenesis in mice. Mice lacking expression of either sGCβ_1 or PKG1 have significantly lower BAT activity than wild-type mice (Haas et al. 2009; Hoffmann et al. 2015). NP-induced lipolysis has long been established in human WAT (Schlueter et al. 2014).

Deletion of the C-type NPR, responsible for natriuretic peptide clearance, enhances the pro-lipolytic effects of exogenously administered NPs in mice (Bordicchia et al. 2012). In addition, BNP overexpression reduces AT weight, accompanied with enhanced mitochondrial biogenesis in white adipocytes resulting in an increased thermogenic energy expenditure in mice (Miyashita et al. 2009). Moreover, mice administered with exogenous BNP have enhanced steady-state energy expenditure as well as increased mitochondrial biogenesis and UCP1 expression in BAT and inguinal WAT (Bordicchia et al. 2012). Overexpression of PKG in mice leads to enhanced glucose tolerance, increases energy expenditure, reduces total body fat content, and enhances BAT mitochondrial biogenesis and thermogenesis (Miyashita et al. 2009). Similarly, administration of Sildenafil enhances browning of subcutaneous WAT in mice, indicated by enhanced UCP1 and PGC-1a expression (Mitschke et al. 2013).

Unlike the effects of ablation of cGMP signaling components, mice lacking the regulatory RIIβ subunit of PKA are resistant to obesity and are more sensitive to β-adrenergic activation. In these animals, the compensatory upregulation of the RIα subunit is responsible for the resistance to obesity, mainly due to its effects in the CNS, since double knockouts for RIIβ and UCP1 display also a lean phenotype and therefore excludes the loss of weight as a consequence of BAT-related oxygen consumption (Cummings et al. 1996; Amieux et al. 1997; Nolan et al. 2004). RIIβ mutant mice are insulin sensitive and resistant to age-induced hyperinsulinemia (Enns et al. 2009). Moreover, double knockout for RIIβ and leptin results in an increase in thermogenesis and energy expenditure as well as decreased food intake (Newhall et al. 2005). On the other hand, mice overexpressing the PKA catalytic subunit showed an improved metabolic profile and resistance to DIO mainly due to an increase in browning of WAT (Dickson et al. 2016). However, studies of an adipose tissue-specific loss of PKA are lacking and whether this loss would affect energy consumption and body weight independently from the CNS-derived effects is yet to be addressed.

3.2 3′,5′-Cyclic Guanosine Monophosphate Signaling and Obesity

A lack of viable therapies against the obesity pandemic has intensified research to identify potential anti-obesity therapeutic avenues. One of the promising targets is enhancing brown/beige adipocyte-mediated thermogenesis (Kusminski et al. 2016). Since pharmacological stimulation of brown adipocyte cAMP signaling by treatment with β3-specific agonists resulted in significant cardiovascular side effects, activating cGMP signaling in the adipose tissue might be an alternative. Therapeutically, two different strategies can be envisioned to increase the bioavailability of cGMP in obese subjects: (1) increase cGMP production by enhancing GC activity and (2) reduce cGMP degradation by inhibiting PDEs (especially PDE5).

Enhancing cGMP production in a diet-induced model of murine obesity has been achieved by increasing cGMP production through NPR activation. Treatment of mice on a high fat diet (HFD) with CD-NP, a synthetic chimeric natriuretic peptide (Lee et al. 2009), has varying effects. Osmotic pump-mediated administration of CD-NP in mice being fed an HFD increases serum adiponectin levels, promotes energy expenditure and WAT inflammation (Glöde et al. 2017). Interestingly, CD-NP treatment also promotes weight gain (body and adipose tissue) and exacerbates insulin resistance and liver steatosis (Glöde et al. 2017). Contrary to CD-NP treatment, stimulating sGC in mice undergoing an HFD regime, using an sGC stimulator, reduces weight gain and body fat content, while increasing insulin sensitivity and energy expenditure, also by promoting UCP1 expression in subcutaneous WAT (Hoffmann et al. 2015). Importantly, sGC stimulation in mice with established (diet-induced) obesity also has a metabolically protective anti-obesity effect (Hoffmann et al. 2015). Therefore, sGC stimulation demonstrates a strong anti-obesity pro-browning potential.

Reducing cGMP degradation through sildenafil treatment of mice on an HFD also demonstrates anti-obesity and pro-thermogenic effects on mouse subcutaneous fat (Mitschke et al. 2013). During an HFD regime, mice treated with sildenafil gained less weight and exhibited a higher insulin sensitivity, than vehicle treated mice (Ayala et al. 2007; Handa et al. 2011). Moreover, inflammation of visceral fat (due to DIO) is reduced in sildenafil treated mice (Handa et al. 2011). This anti-inflammatory effect was also observed in *db*/*db* mice (a genetic model of murine obesity and type 2 diabetes), when administered with sildenafil (Handa et al. 2011). Importantly, sildenafil treatment of overweight humans for a duration of 7 days leads to reduced adipocyte size in subcutaneous WAT and increased expression of PGC-1α and UCP1 (Li et al. 2018).

4 Conclusion

The involvement/recruitment of similar downstream effectors indicates a cross-talk between the cAMP and cGMP cascades, resulting in overlapping functions. Both these pathways enhance lipolysis in human adipocytes through activation of ATGL and HSL. Moreover, the induction of UCP1 by the two cyclic nucleotides appears to use the same pathway: p38-MAPK and the subsequent upregulation of PGC-1α. Moreover, the cAMP–cGMP dual specific PDEs form a junction between the two cyclic nucleotide cascades, facilitating a cross-talk between them. Although, these observations suggest a supplementary role of cAMP and cGMP signaling cascades to promote brown adipocyte thermogenesis and white adipocyte browning, more studies are required to identify the specific roles of cGMP versus cAMP in these cells/processes.

References

Alverdi V, Mazon H, Versluis C, Hemrika W, Esposito G, van den Heuvel R, Scholten A, Heck AJ (2008) cGMP-binding prepares PKG for substrate binding by disclosing the C-terminal domain. J Mol Biol 375:1380–1393

Amieux PS, Cummings DE, Motamed K, Brandon EP, Wailes LA, Le K, Idzerda RL, McKnight GS (1997) Compensatory regulation of RIalpha protein levels in protein kinase A mutant mice. J Biol Chem 272:3993–3998

Anthonsen MW, Ronnstrand L, Wernstedt C, Degerman E, Holm C (1998) Identification of novel phosphorylation sites in hormone-sensitive lipase that are phosphorylated in response to isoproterenol and govern activation properties in vitro. J Biol Chem 273:215–221

Ashman DF, Lipton R, Melicow MM, Price TD (1963) Isolation of adenosine 3′, 5′-monophosphate and guanosine 3′, 5′-monophosphate from rat urine. Biochem Biophys Res Commun 11:330–334

Ayala JE, Bracy DP, Julien BM, Rottman JN, Fueger PT, Wasserman DH (2007) Chronic treatment with sildenafil improves energy balance and insulin action in high fat–fed conscious mice. Diabetes 56:1025–1033

Bartness TJ, Vaughan CH, Song CK (2010) Sympathetic and sensory innervation of brown adipose tissue. Int J Obes 34:S36

Bechtel PJ, Beavo JA, Krebs EG (1977) Purification and characterization of catalytic subunit of skeletal muscle adenosine 3′:5′-monophosphate-dependent protein kinase. J Biol Chem 252:2691–2697

Bender AT, Beavo JA (2006) Cyclic nucleotide phosphodiesterases: molecular regulation to clinical use. Pharmacol Rev 58:488–520

Bengtsson T, Cannon B, Nedergaard J (2000) Differential adrenergic regulation of the gene expression of the beta-adrenoceptor subtypes beta1, beta2 and beta3 in brown adipocytes. Biochem J 347(Pt 3):643–651

Berthet J, Rall TW, Sutherland EW (1957a) The relationship of epinephrine and glucagon to liver phosphorylase. IV. Effect of epinephrine and glucagon on the reactivation of phosphorylase in liver homogenates. J Biol Chem 224:463–475

Berthet J, Sutherland EW, Rall TW (1957b) The assay of glucagon and epinephrine with use of liver homogenates. J Biol Chem 229:351–361

Billington CJ, Briggs JE, Link JG, Levine AS (1991) Glucagon in physiological concentrations stimulates brown fat thermogenesis in vivo. Am J Phys 261:R501–R507

Bordicchia M, Liu D, Amri EZ, Ailhaud G, Dessì-Fulgheri P, Zhang C, Takahashi N, Sarzani R, Collins S (2012) Cardiac natriuretic peptides act via p38 MAPK to induce the brown fat thermogenic program in mouse and human adipocytes. J Clin Invest 122:1022–1036

Bradley J, Reisert J, Frings S (2005) Regulation of cyclic nucleotide-gated channels. Curr Opin Neurobiol 15:343–349

Brandon EP, Logue SF, Adams MR, Qi M, Sullivan SP, Matsumoto AM, Dorsa DM, Wehner JM, McKnight GS, Idzerda RL (1998) Defective motor behavior and neural gene expression in RIIbeta-protein kinase A mutant mice. J Neurosci 18:3639–3649

Bronnikov G, Houstek J, Nedergaard J (1992) Beta-adrenergic, cAMP-mediated stimulation of proliferation of brown fat cells in primary culture. Mediation via beta 1 but not via beta 3 adrenoceptors. J Biol Chem 267:2006–2013

Bronnikov G, Bengtsson T, Kramarova L, Golozoubova V, Cannon B, Nedergaard J (1999) Beta1 to beta3 switch in control of cyclic adenosine monophosphate during brown adipocyte development explains distinct beta-adrenoceptor subtype mediation of proliferation and differentiation. Endocrinology 140:4185–4197

Bryant NJ, Govers R, James DE (2002) Regulated transport of the glucose transporter GLUT4. Nat Rev Mol Cell Biol 3:267–277

Cao Z, Umek RM, McKnight SL (1991) Regulated expression of three C/EBP isoforms during adipose conversion of 3T3-L1 cells. Genes Dev 5:1538–1552

Cao W, Medvedev AV, Daniel KW, Collins S (2001) beta-Adrenergic activation of p38 MAP kinase in adipocytes: cAMP induction of the uncoupling protein 1 (UCP1) gene requires p38 MAP kinase. J Biol Chem 276:27077–27082

Cao W, Collins QF, Becker TC, Robidoux J, Lupo EG Jr, Xiong Y, Daniel KW, Floering L, Collins S (2005) p38 Mitogen-activated protein kinase plays a stimulatory role in hepatic gluconeogenesis. J Biol Chem 280:42731–42737

Carmona MC, Hondares E, Rodriguez de la Concepcion ML, Rodriguez-Sureda V, Peinado-Onsurbe J, Poli V, Iglesias R, Villarroya F, Giralt M (2005) Defective thermoregulation, impaired lipid metabolism, but preserved adrenergic induction of gene expression in brown fat of mice lacking C/EBPbeta. Biochem J 389:47–56

Chernick SS, Spooner PM, Garrison MM, Scow RO (1986) Effect of epinephrine and other lipolytic agents on intracellular lipolysis and lipoprotein lipase activity in 3T3-L1 adipocytes. J Lipid Res 27:286–294

Chernogubova E, Cannon B, Bengtsson T (2004) Norepinephrine increases glucose transport in brown adipocytes via beta3-adrenoceptors through a cAMP, PKA, and PI3-kinase-dependent pathway stimulating conventional and novel PKCs. Endocrinology 145:269–280

Choi SM, Tucker DF, Gross DN, Easton RM, Dipilato LM, Dean AS, Monks BR, Birnbaum MJ (2010) Insulin regulates adipocyte lipolysis via an Akt-independent signaling pathway. Mol Cell Biol 30:5009–5020

Corbin JD, Sugden PH, Lincoln TM, Keely SL (1977) Compartmentalization of adenosine 3′:5′-monophosphate and adenosine 3′:5′-monophosphate-dependent protein kinase in heart tissue. J Biol Chem 252:3854–3861

Cummings DE, Brandon EP, Planas JV, Motamed K, Idzerda RL, McKnight GS (1996) Genetically lean mice result from targeted disruption of the RII beta subunit of protein kinase A. Nature 382:622–626

Cypess AM, Weiner LS, Roberts-Toler C, Franquet Elia E, Kessler SH, Kahn PA, English J, Chatman K, Trauger SA, Doria A, Kolodny GM (2015) Activation of human brown adipose tissue by a beta3-adrenergic receptor agonist. Cell Metab 21:33–38

Dao KK, Teigen K, Kopperud R, Hodneland E, Schwede F, Christensen AE, Martinez A, Doskeland SO (2006) Epac1 and cAMP-dependent protein kinase holoenzyme have similar cAMP affinity, but their cAMP domains have distinct structural features and cyclic nucleotide recognition. J Biol Chem 281:21500–21511

Darlington GJ, Ross SE, MacDougald OA (1998) The role of C/EBP genes in adipocyte differentiation. J Biol Chem 273:30057–30060

Daval M, Diot-Dupuy F, Bazin R, Hainault I, Viollet B, Vaulont S, Hajduch E, Ferre P, Foufelle F (2005) Anti-lipolytic action of AMP-activated protein kinase in rodent adipocytes. J Biol Chem 280:25250–25257

Deng C, Paoloni-Giacobino A, Kuehne F, Boss O, Revelli JP, Moinat M, Cawthorne MA, Muzzin P, Giacobino JP (1996) Respective degree of expression of beta 1-, beta 2- and beta 3-adrenoceptors in human brown and white adipose tissues. Br J Pharmacol 118:929–934

Denninger JW, Marletta MA (1999) Guanylate cyclase and the·NO/cGMP signaling pathway. Biochim Biophys Acta 1411:334–350

Derbyshire ER, Marletta MA (2012) Structure and regulation of soluble guanylate cyclase. Annu Rev Biochem 81:533–559

Dickson LM, Gandhi S, Layden BT, Cohen RN, Wicksteed B (2016) Protein kinase A induces UCP1 expression in specific adipose depots to increase energy expenditure and improve metabolic health. Am J Physiol Regul Integr Comp Physiol 311:R79–R88

Dipilato LM, Ahmad F, Harms M, Seale P, Manganiello V, Birnbaum MJ (2015) The role of PDE3B phosphorylation in the inhibition of lipolysis by insulin. Mol Cell Biol 35:2752–2760

Djouder N, Tuerk RD, Suter M, Salvioni P, Thali RF, Scholz R, Vaahtomeri K, Auchli Y, Rechsteiner H, Brunisholz RA, Viollet B, Makela TP, Wallimann T, Neumann D, Krek W (2010) PKA phosphorylates and inactivates AMPKalpha to promote efficient lipolysis. EMBO J 29:469–481

Duncan RE, Ahmadian M, Jaworski K, Sarkadi-Nagy E, Sul HS (2007) Regulation of lipolysis in adipocytes. Annu Rev Nutr 27:79–101

Eide T, Tasken KA, Carlson C, Williams G, Jahnsen T, Tasken K, Collas P (2003) Protein kinase A-anchoring protein AKAP95 interacts with MCM2, a regulator of DNA replication. J Biol Chem 278:26750–26756

Enns LC, Morton JF, Treuting PR, Emond MJ, Wolf NS, Dai DF, McKnight GS, Rabinovitch PS, Ladiges WC (2009) Disruption of protein kinase A in mice enhances healthy aging. PLoS One 4:e5963

Fedorenko A, Lishko PV, Kirichok Y (2012) Mechanism of fatty-acid-dependent UCP1 uncoupling in brown fat mitochondria. Cell 151:400–413

Förstermann U, Sessa WC (2012) Nitric oxide synthases: regulation and function. Eur Heart J 33:829–837
Francis SH, Busch JL, Corbin JD (2010) cGMP-dependent protein kinases and cGMP phosphodiesterases in nitric oxide and cGMP action. Pharmacol Rev 62:525–563
Fredriksson JM, Nedergaard J (2002) Norepinephrine specifically stimulates ribonucleotide reductase subunit R2 gene expression in proliferating brown adipocytes: mediation via a cAMP/PKA pathway involving Src and Erk1/2 kinases. Exp Cell Res 274:207–215
Fukuda M, Williams KW, Gautron L, Elmquist JK (2011) Induction of leptin resistance by activation of cAMP-Epac signaling. Cell Metab 13:331–339
Gauthier MS, Miyoshi H, Souza SC, Cacicedo JM, Saha AK, Greenberg AS, Ruderman NB (2008) AMP-activated protein kinase is activated as a consequence of lipolysis in the adipocyte: potential mechanism and physiological relevance. J Biol Chem 283:16514–16524
Glöde A, Naumann J, Gnad T, Cannone V, Kilic A, Burnett JC Jr, Pfeifer A (2017) Divergent effects of a designer natriuretic peptide CD-NP in the regulation of adipose tissue and metabolism. Mol Metab 6:276–287
Gnad T, Scheibler S, von Kugelgen I, Scheele C, Kilic A, Glode A, Hoffmann LS, Reverte-Salisa L, Horn P, Mutlu S, El-Tayeb A, Kranz M, Deuther-Conrad W, Brust P, Lidell ME, Betz MJ, Enerback S, Schrader J, Yegutkin GG, Muller CE, Pfeifer A (2014) Adenosine activates brown adipose tissue and recruits beige adipocytes via A2A receptors. Nature 516:395–399
Gold MG, Lygren B, Dokurno P, Hoshi N, McConnachie G, Tasken K, Carlson CR, Scott JD, Barford D (2006) Molecular basis of AKAP specificity for PKA regulatory subunits. Mol Cell 24:383–395
Granneman JG, Lahners KN (1992) Differential adrenergic regulation of beta 1- and beta 3-adrenoreceptor messenger ribonucleic acids in adipose tissues. Endocrinology 130:109–114
Greco-Perotto R, Zaninetti D, Assimacopoulos-Jeannet F, Bobbioni E, Jeanrenaud B (1987) Stimulatory effect of cold adaptation on glucose utilization by brown adipose tissue. Relationship with changes in the glucose transporter system. J Biol Chem 262:7732–7736
Gruden G, Landi A, Bruno G (2014) Natriuretic peptides, heart, and adipose tissue: new findings and future developments for diabetes research. Diabetes Care 37:2899–2908
Haas B, Mayer P, Jennissen K, Scholz D, Diaz MB, Bloch W, Herzig S, Fässler R, Pfeifer A (2009) Protein kinase G controls brown fat cell differentiation and mitochondrial biogenesis. Sci Signal 2:ra78
Handa P, Tateya S, Rizzo NO, Cheng AM, Morgan-Stevenson V, Han C-Y, Clowes AW, Daum G, O'Brien KD, Schwartz MW, Chait A, Kim F (2011) Reduced vascular nitric oxide–cGMP signaling contributes to adipose tissue inflammation during high-fat feeding. Arterioscler Thromb Vasc Biol 31:2827–2835
Hankir MK, Kranz M, Gnad T, Weiner J, Wagner S, Deuther-Conrad W, Bronisch F, Steinhoff K, Luthardt J, Kloting N, Hesse S, Seibyl JP, Sabri O, Heiker JT, Bluher M, Pfeifer A, Brust P, Fenske WK (2016) A novel thermoregulatory role for PDE10A in mouse and human adipocytes. EMBO Mol Med 8:796–812
Hanoune J, Defer N (2001) Regulation and role of adenylyl cyclase isoforms. Annu Rev Pharmacol Toxicol 41:145–174
Heckemeyer CM, Barker J, Duckworth WC, Solomon SS (1983) Studies of the biological effect and degradation of glucagon in the rat perifused isolated adipose cell. Endocrinology 113:270–276
Himms-Hagen J, Melnyk A, Zingaretti MC, Ceresi E, Barbatelli G, Cinti S (2000) Multilocular fat cells in WAT of CL-316243-treated rats derive directly from white adipocytes. Am J Physiol Cell Physiol 279:C670–C681
Hoffmann LS, Etzrodt J, Willkomm L, Sanyal A, Scheja L, Fischer AWC, Stasch J-P, Bloch W, Friebe A, Heeren J, Pfeifer A (2015) Stimulation of soluble guanylyl cyclase protects against obesity by recruiting brown adipose tissue. Nat Commun 6:7235
Hofmann F, Feil R, Kleppisch T, Schlossmann J (2006) Function of cGMP-dependent protein kinases as revealed by gene deletion. Physiol Rev 86:1–23

Hom GJ, Forrest MJ, Bach TJ, Brady E, Candelore MR, Cascieri MA, Fletcher DJ, Fisher MH, Iliff SA, Mathvink R, Metzger J, Pecore V, Saperstein R, Shih T, Weber AE, Wyvratt M, Zafian P, Macintyre DE (2001) Beta(3)-adrenoceptor agonist-induced increases in lipolysis, metabolic rate, facial flushing, and reflex tachycardia in anesthetized rhesus monkeys. J Pharmacol Exp Ther 297:299–307

Hondares E, Rosell M, Gonzalez FJ, Giralt M, Iglesias R, Villarroya F (2010) Hepatic FGF21 expression is induced at birth via PPARalpha in response to milk intake and contributes to thermogenic activation of neonatal brown fat. Cell Metab 11:206–212

Hondares E, Iglesias R, Giralt A, Gonzalez FJ, Giralt M, Mampel T, Villarroya F (2011) Thermogenic activation induces FGF21 expression and release in brown adipose tissue. J Biol Chem 286:12983–12990

Howland RJ, Benning AD (1986) Differential effects of noradrenaline and glucagon on lipolysis and fatty-acid utilization in brown adipose tissue. FEBS Lett 208:128–132

Hu Y, Robichaux WG 3rd, Mei FC, Kim ER, Wang H, Tong Q, Jin J, Xu M, Chen J, Cheng X (2016) Role of exchange protein directly activated by cyclic AMP isoform 1 in energy homeostasis: regulation of leptin expression and secretion in white adipose tissue. Mol Cell Biol 36:2440–2450

Hutchinson DS, Chernogubova E, Dallner OS, Cannon B, Bengtsson T (2005) Beta-adrenoceptors, but not alpha-adrenoceptors, stimulate AMP-activated protein kinase in brown adipocytes independently of uncoupling protein-1. Diabetologia 48:2386–2395

Hwang M, Go Y, Park JH, Shin SK, Song SE, Oh BC, Im SS, Hwang I, Jeon YH, Lee IK, Seino S, Song DK (2017) Epac2a-null mice exhibit obesity-prone nature more susceptible to leptin resistance. Int J Obes (Lond) 41:279–288

Jahnsen T, Hedin L, Kidd VJ, Schulz T, Richards JS (1988) Molecular cloning of cDNA for a hormone-regulated isoform of the regulatory subunit of type II cAMP-dependent protein kinase from rat ovaries. Methods Enzymol 159:318–324

Jaumann M, Dettling J, Gubelt M, Zimmermann U, Gerling A, Paquet-Durand F, Feil S, Wolpert S, Franz C, Varakina K, Xiong H, Brandt N, Kuhn S, Geisler H-S, Rohbock K, Ruth P, Schlossmann J, Hütter J, Sandner P, Feil R, Engel J, Knipper M, Rüttiger L (2012) cGMP-Prkg1 signaling and Pde5 inhibition shelter cochlear hair cells and hearing function. Nat Med 18:252

Jennissen K, Siegel F, Liebig-Gonglach M, Hermann MR, Kipschull S, van Dooren S, Kunz WS, Fassler R, Pfeifer A (2012) A VASP-Rac-soluble guanylyl cyclase pathway controls cGMP production in adipocytes. Sci Signal 5:ra62

Jennissen K, Haas B, Mitschke MM, Siegel F, Pfeifer A (2013) Analysis of cGMP signaling in adipocytes. Methods Mol Biol 1020:175–192

Jiang C, Zhai M, Yan D, Li D, Li C, Zhang Y, Xiao L, Xiong D, Deng Q, Sun W (2017) Dietary menthol-induced TRPM8 activation enhances WAT "browning" and ameliorates diet-induced obesity. Oncotarget 8:75114–75126

Kasai K, Kon S, Sato N, Muraishi K, Yoshida H, Nakai N, Hamakawa H, Itoh C, Yamaoka S (1999) Case report of lymphoepithelioma-like carcinoma of the lung – lymphoid population consisting of cytotoxic T cells in resting state. Pathol Res Pract 195:773–779

Kaupp UB, Seifert R (2001) Molecular diversity of pacemaker ion channels. Annu Rev Physiol 63:235–257

Kaupp UB, Seifert R (2002) Cyclic nucleotide-gated ion channels. Physiol Rev 82:769–824

Kawasaki H, Springett GM, Mochizuki N, Toki S, Nakaya M, Matsuda M, Housman DE, Graybiel AM (1998) A family of cAMP-binding proteins that directly activate Rap1. Science 282:2275–2279

Kelley GG, Chepurny OG, Schwede F, Genieser HG, Leech CA, Roe MW, Li X, Dzhura I, Dzhura E, Afshari P, Holz GG (2009) Glucose-dependent potentiation of mouse islet insulin secretion by Epac activator 8-pCPT-2′-O-Me-cAMP-AM. Islets 1:260–265

Kim SH, Plutzky J (2016) Brown fat and browning for the treatment of obesity and related metabolic disorders. Diabetes Metab J 40:12–21

Kim SP, Ha JM, Yun SJ, Kim EK, Chung SW, Hong KW, Kim CD, Bae SS (2010) Transcriptional activation of peroxisome proliferator-activated receptor-gamma requires activation of both protein kinase A and Akt during adipocyte differentiation. Biochem Biophys Res Commun 399:55–59

Kinderman FS, Kim C, von Daake S, Ma Y, Pham BQ, Spraggon G, Xuong NH, Jennings PA, Taylor SS (2006) A dynamic mechanism for AKAP binding to RII isoforms of cAMP-dependent protein kinase. Mol Cell 24:397–408

Kinoshita K, Ozaki N, Takagi Y, Murata Y, Oshida Y, Hayashi Y (2014) Glucagon is essential for adaptive thermogenesis in brown adipose tissue. Endocrinology 155:3484–3492

Kitamura T, Kitamura Y, Kuroda S, Hino Y, Ando M, Kotani K, Konishi H, Matsuzaki H, Kikkawa U, Ogawa W, Kasuga M (1999) Insulin-induced phosphorylation and activation of cyclic nucleotide phosphodiesterase 3B by the serine-threonine kinase Akt. Mol Cell Biol 19:6286–6296

Koesling D, Böhme E, Schultz G (1991) Guanylyl cyclases, a growing family of signal-transducing enzymes. FASEB J 5:2785–2791

Koh HJ, Hirshman MF, He H, Li Y, Manabe Y, Balschi JA, Goodyear LJ (2007) Adrenaline is a critical mediator of acute exercise-induced AMP-activated protein kinase activation in adipocytes. Biochem J 403:473–481

Kong X, Banks A, Liu T, Kazak L, Rao RR, Cohen P, Wang X, Yu S, Lo JC, Tseng YH, Cypess AM, Xue R, Kleiner S, Kang S, Spiegelman BM, Rosen ED (2014) IRF4 is a key thermogenic transcriptional partner of PGC-1alpha. Cell 158:69–83

Kots AY, Martin E, Sharina IG, Murad F (2009) A short history of cGMP, guanylyl cyclases, and cGMP-dependent protein kinases. Handb Exp Pharmacol:1–14

Kozak UC, Kozak LP (1994) Norepinephrine-dependent selection of brown adipocyte cell lines. Endocrinology 134:906–913

Kraynik SM, Miyaoka RS, Beavo JA (2013) PDE3 and PDE4 isozyme-selective inhibitors are both required for synergistic activation of brown adipose tissue. Mol Pharmacol 83:1155–1165

Kumar A, Shiloach J, Betenbaugh MJ, Gallagher EJ (2015) The beta-3 adrenergic agonist (CL-316,243) restores the expression of down-regulated fatty acid oxidation genes in type 2 diabetic mice. Nutr Metab (Lond) 12:8

Kusminski CM, Bickel PE, Scherer PE (2016) Targeting adipose tissue in the treatment of obesity-associated diabetes. Nat Rev Drug Discov 15:639–660

Langeberg LK, Scott JD (2015) Signalling scaffolds and local organization of cellular behaviour. Nat Rev Mol Cell Biol 16:232–244

Lee DC, Carmichael DF, Krebs EG, McKnight GS (1983) Isolation of a cDNA clone for the type I regulatory subunit of bovine cAMP-dependent protein kinase. Proc Natl Acad Sci U S A 80:3608–3612

Lee MY, Kong HJ, Cheong J (2001) Regulation of activating transcription factor-2 in early stage of the adipocyte differentiation program. Biochem Biophys Res Commun 281:1241–1247

Lee CYW, Chen HH, Lisy O, Swan S, Cannon C, Lieu HD, Burnett JC (2009) Pharmacodynamics pharmacodynamics of a novel designer natriuretic peptide, CD-NP, in a first-in-human clinical trial in healthy subjects. J Clin Pharmacol 49:668–673

Lee YH, Kim SN, Kwon HJ, Maddipati KR, Granneman JG (2016) Adipogenic role of alternatively activated macrophages in beta-adrenergic remodeling of white adipose tissue. Am J Physiol Regul Integr Comp Physiol 310:R55–R65

Li S, Li Y, Xiang L, Dong J, Liu M, Xiang G (2018) Sildenafil induces browning of subcutaneous white adipose tissue in overweight adults. Metabolism 78:106–117

Loncar D (1991) Convertible adipose tissue in mice. Cell Tissue Res 266:149–161

Louis SN, Jackman GP, Nero TL, Iakovidis D, Louis WJ (2000) Role of beta-adrenergic receptor subtypes in lipolysis. Cardiovasc Drugs Ther 14:565–577

MacDougald OA, Lane MD (1995) Transcriptional regulation of gene expression during adipocyte differentiation. Annu Rev Biochem 64:345–373
McDonald LJ, Murad F (1996) Nitric oxide and cyclic GMP signaling. Exp Biol Med 211:1–6
Mitschke MM, Hoffmann LS, Gnad T, Scholz D, Kruithoff K, Mayer P, Haas B, Sassmann A, Pfeifer A, Kilić A (2013) Increased cGMP promotes healthy expansion and browning of white adipose tissue. FASEB J 27:1621–1630
Miyashita K, Itoh H, Tsujimoto H, Tamura N, Fukunaga Y, Sone M, Yamahara K, Taura D, Inuzuka M, Sonoyama T, Nakao K (2009) Natriuretic peptides/cGMP/cGMP-dependent protein kinase cascades promote muscle mitochondrial biogenesis and prevent obesity. Diabetes 58:2880–2892
Miyoshi H, Souza SC, Zhang HH, Strissel KJ, Christoffolete MA, Kovsan J, Rudich A, Kraemer FB, Bianco AC, Obin MS, Greenberg AS (2006) Perilipin promotes hormone-sensitive lipase-mediated adipocyte lipolysis via phosphorylation-dependent and -independent mechanisms. J Biol Chem 281:15837–15844
Miyoshi H, Perfield JW 2nd, Souza SC, Shen WJ, Zhang HH, Stancheva ZS, Kraemer FB, Obin MS, Greenberg AS (2007) Control of adipose triglyceride lipase action by serine 517 of perilipin A globally regulates protein kinase A-stimulated lipolysis in adipocytes. J Biol Chem 282:996–1002
Moreland RB, Goldstein I, Traish A (1998) Sildenafil, a novel inhibitor of phosphodiesterase type 5 in human corpus cavernosum smooth muscle cells. Life Sci 62:PL309–PL318
Moro C, Lafontan M (2012) Natriuretic peptides and cGMP signaling control of energy homeostasis. Am J Phys Heart Circ Phys 304:H358–H368
Nádvorník R, Vomastek T, Janeček J, Techniková Z, Branny P (1999) Pkg2, a novel transmembrane protein Ser/Thr kinase of Streptomyces granaticolor. J Bacteriol 181:15–23
Newhall KJ, Cummings DE, Nolan MA, McKnight GS (2005) Deletion of the RIIbeta-subunit of protein kinase A decreases body weight and increases energy expenditure in the obese, leptin-deficient ob/ob mouse. Mol Endocrinol 19:982–991
Newlon MG, Roy M, Morikis D, Carr DW, Westphal R, Scott JD, Jennings PA (2001) A novel mechanism of PKA anchoring revealed by solution structures of anchoring complexes. EMBO J 20:1651–1662
Nishikimi T, Maeda N, Matsuoka H (2006) The role of natriuretic peptides in cardioprotection. Cardiovasc Res 69:318–328
Nisoli E, Clementi E, Tonello C, Sciorati C, Briscini L, Carruba MO (1998) Effects of nitric oxide on proliferation and differentiation of rat brown adipocytes in primary cultures. Br J Pharmacol 125:888–894
Nisoli E, Clementi E, Paolucci C, Cozzi V, Tonello C, Sciorati C, Bracale R, Valerio A, Francolini M, Moncada S, Carruba MO (2003) Mitochondrial biogenesis in mammals: the role of endogenous nitric oxide. Science 299:896–899
Nolan MA, Sikorski MA, McKnight GS (2004) The role of uncoupling protein 1 in the metabolism and adiposity of RII beta-protein kinase A-deficient mice. Mol Endocrinol 18:2302–2311
Olsen JM, Sato M, Dallner OS, Sandstrom AL, Pisani DF, Chambard JC, Amri EZ, Hutchinson DS, Bengtsson T (2014) Glucose uptake in brown fat cells is dependent on mTOR complex 2-promoted GLUT1 translocation. J Cell Biol 207:365–374
Olsen JM, Csikasz RI, Dehvari N, Lu L, Sandstrom A, Oberg AI, Nedergaard J, Stone-Elander S, Bengtsson T (2017) beta3-Adrenergically induced glucose uptake in brown adipose tissue is independent of UCP1 presence or activity: mediation through the mTOR pathway. Mol Metab 6:611–619
Omar B, Zmuda-Trzebiatowska E, Manganiello V, Goransson O, Degerman E (2009) Regulation of AMP-activated protein kinase by cAMP in adipocytes: roles for phosphodiesterases, protein kinase B, protein kinase A, Epac and lipolysis. Cell Signal 21:760–766
Pandit K, Mukhopadhyay P, Ghosh S, Chowdhury S (2011) Natriuretic peptides: diagnostic and therapeutic use. Indian J Endocrinol Metab 15:S345–S353

Perea A, Clemente F, Martinell J, Villanueva-Penacarrillo ML, Valverde I (1995) Physiological effect of glucagon in human isolated adipocytes. Horm Metab Res 27:372–375

Perino A, Ghigo A, Ferrero E, Morello F, Santulli G, Baillie GS, Damilano F, Dunlop AJ, Pawson C, Walser R, Levi R, Altruda F, Silengo L, Langeberg LK, Neubauer G, Heymans S, Lembo G, Wymann MP, Wetzker R, Houslay MD, Iaccarino G, Scott JD, Hirsch E (2011) Integrating cardiac PIP3 and cAMP signaling through a PKA anchoring function of p110gamma. Mol Cell 42:84–95

Pfeifer A, Aszodi A, Seidler U, Ruth P, Hofmann F, Fassler R (1996) Intestinal secretory defects and dwarfism in mice lacking cGMP-dependent protein kinase II. Science 274:2082–2086

Pidoux G, Tasken K (2010) Specificity and spatial dynamics of protein kinase A signaling organized by A-kinase-anchoring proteins. J Mol Endocrinol 44:271–284

Pifferi S, Boccaccio A, Menini A (2006) Cyclic nucleotide-gated ion channels in sensory transduction. FEBS Lett 580:2853–2859

Poulos TL (2006) Soluble guanylate cyclase. Curr Opin Struct Biol 16:736–743

Puigserver P, Spiegelman BM (2003) Peroxisome proliferator-activated receptor-gamma coactivator 1 alpha (PGC-1 alpha): transcriptional coactivator and metabolic regulator. Endocr Rev 24:78–90

Rehmann H, Das J, Knipscheer P, Wittinghofer A, Bos JL (2006) Structure of the cyclic-AMP-responsive exchange factor Epac2 in its auto-inhibited state. Nature 439:625–628

Ringheim GE, Taylor SS (1990) Effects of cAMP-binding site mutations on intradomain cross-communication in the regulatory subunit of cAMP-dependent protein kinase I. J Biol Chem 265:19472–19478

Robidoux J, Cao W, Quan H, Daniel KW, Moukdar F, Bai X, Floering LM, Collins S (2005) Selective activation of mitogen-activated protein (MAP) kinase kinase 3 and p38alpha MAP kinase is essential for cyclic AMP-dependent UCP1 expression in adipocytes. Mol Cell Biol 25:5466–5479

Rosen ED, Hsu CH, Wang X, Sakai S, Freeman MW, Gonzalez FJ, Spiegelman BM (2002) C/EBPalpha induces adipogenesis through PPARgamma: a unified pathway. Genes Dev 16:22–26

Schimmel RJ, McCarthy L (1984) Role of adenosine as an endogenous regulator of respiration in hamster brown adipocytes. Am J Physiol 246:C301–C307

Schlueter N, de Sterke A, Willmes DM, Spranger J, Jordan J, Birkenfeld AL (2014) Metabolic actions of natriuretic peptides and therapeutic potential in the metabolic syndrome. Pharmacol Ther 144:12–27

Schreiber R, Diwoky C, Schoiswohl G, Feiler U, Wongsiriroj N, Abdellatif M, Kolb D, Hoeks J, Kershaw EE, Sedej S, Schrauwen P, Haemmerle G, Zechner R (2017) Cold-induced thermogenesis depends on ATGL-mediated lipolysis in cardiac muscle, but not brown adipose tissue. Cell Metab 26:753–763 e7

Scott JD, Glaccum MB, Fischer EH, Krebs EG (1986) Primary-structure requirements for inhibition by the heat-stable inhibitor of the cAMP-dependent protein kinase. Proc Natl Acad Sci U S A 83:1613–1616

Scott JD, Glaccum MB, Zoller MJ, Uhler MD, Helfman DM, McKnight GS, Krebs EG (1987) The molecular cloning of a type II regulatory subunit of the cAMP-dependent protein kinase from rat skeletal muscle and mouse brain. Proc Natl Acad Sci U S A 84:5192–5196

Seale P, Kajimura S, Yang W, Chin S, Rohas LM, Uldry M, Tavernier G, Langin D, Spiegelman BM (2007) Transcriptional control of brown fat determination by PRDM16. Cell Metab 6:38–54

Seitz HJ, Krone W, Wilke H, Tarnowski W (1981) Rapid rise in plasma glucagon induced by acute cold exposure in man and rat. Pflugers Arch 389:115–120

Shabb JB (2001) Physiological substrates of cAMP-dependent protein kinase. Chem Rev 101:2381–2411

Sheth S, Brito R, Mukherjea D, Rybak LP, Ramkumar V (2014) Adenosine receptors: expression, function and regulation. Int J Mol Sci 15:2024–2052

Sheyn D, Pelled G, Tawackoli W, Su S, Ben-David S, Gazit D, Gazit Z (2013) Transient overexpression of Ppargamma2 and C/ebpalpha in mesenchymal stem cells induces brown adipose tissue formation. Regen Med 8:295–308
Shimizu Y, Satoh S, Yano H, Minokoshi Y, Cushman SW, Shimazu T (1998) Effects of noradrenaline on the cell-surface glucose transporters in cultured brown adipocytes: novel mechanism for selective activation of GLUT1 glucose transporters. Biochem J 330(Pt 1):397–403
Shin H, Ma Y, Chanturiya T, Cao Q, Wang Y, Kadegowda AKG, Jackson R, Rumore D, Xue B, Shi H, Gavrilova O, Yu L (2017) Lipolysis in brown adipocytes is not essential for cold-induced thermogenesis in mice. Cell Metab 26:764–777 e5
Sim AT, Scott JD (1999) Targeting of PKA, PKC and protein phosphatases to cellular microdomains. Cell Calcium 26:209–217
Snyder PB, Esselstyn JM, Loughney K, Wolda SL, Florio VA (2005) The role of cyclic nucleotide phosphodiesterases in the regulation of adipocyte lipolysis. J Lipid Res 46:494–503
Su CL, Sztalryd C, Contreras JA, Holm C, Kimmel AR, Londos C (2003) Mutational analysis of the hormone-sensitive lipase translocation reaction in adipocytes. J Biol Chem 278:43615–43619
Svensson KJ, Long JZ, Jedrychowski MP, Cohen P, Lo JC, Serag S, Kir S, Shinoda K, Tartaglia JA, Rao RR, Chedotal A, Kajimura S, Gygi SP, Spiegelman BM (2016) A secreted slit2 fragment regulates adipose tissue thermogenesis and metabolic function. Cell Metab 23:454–466
Szillat D, Bukowiecki LJ (1983) Control of brown adipose tissue lipolysis and respiration by adenosine. Am J Physiol 245:E555–E559
Sztalryd C, Xu G, Dorward H, Tansey JT, Contreras JA, Kimmel AR, Londos C (2003) Perilipin A is essential for the translocation of hormone-sensitive lipase during lipolytic activation. J Cell Biol 161:1093–1103
Tanaka T, Yoshida N, Kishimoto T, Akira S (1997) Defective adipocyte differentiation in mice lacking the C/EBPbeta and/or C/EBPdelta gene. EMBO J 16:7432–7443
Tasken K, Aandahl EM (2004) Localized effects of cAMP mediated by distinct routes of protein kinase A. Physiol Rev 84:137–167
Tiraby C, Langin D (2003) Conversion from white to brown adipocytes: a strategy for the control of fat mass? Trends Endocrinol Metab 14:439–441
Tiraby C, Tavernier G, Lefort C, Larrouy D, Bouillaud F, Ricquier D, Langin D (2003) Acquirement of brown fat cell features by human white adipocytes. J Biol Chem 278:33370–33376
Tsai EJ, Kass DA (2009) Cyclic GMP signaling in cardiovascular pathophysiology and therapeutics. Pharmacol Ther 122:216–238
Uhler MD, Carmichael DF, Lee DC, Chrivia JC, Krebs EG, McKnight GS (1986a) Isolation of cDNA clones coding for the catalytic subunit of mouse cAMP-dependent protein kinase. Proc Natl Acad Sci U S A 83:1300–1304
Uhler MD, Chrivia JC, McKnight GS (1986b) Evidence for a second isoform of the catalytic subunit of cAMP-dependent protein kinase. J Biol Chem 261:15360–15363
Unelius L, Mohell N, Nedergaard J (1990) Cold acclimation induces desensitization to adenosine in brown fat cells without changing receptor binding. Am J Physiol 258:C818–C826
Vaandrager AB, de Jonge HR (1996) Signalling by cGMP-dependent protein kinases. Mol Cell Biochem 157:23–30
Valladares A, Porras A, Alvarez AM, Roncero C, Benito M (2000) Noradrenaline induces brown adipocytes cell growth via beta-receptors by a mechanism dependent on ERKs but independent of cAMP and PKA. J Cell Physiol 185:324–330
Waldman SA, Rapoport RM, Murad F (1984) Atrial natriuretic factor selectively activates particulate guanylate cyclase and elevates cyclic GMP in rat tissues. J Biol Chem 259:14332–14334
Wernet W, Flockerzi V, Hofmann F (1989) The cDNA of the two isoforms of bovine cGMP-dependent protein kinase. FEBS Lett 251:191–196
Yin W, Mu J, Birnbaum MJ (2003) Role of AMP-activated protein kinase in cyclic AMP-dependent lipolysis in 3T3-L1 adipocytes. J Biol Chem 278:43074–43080

Zhang JW, Klemm DJ, Vinson C, Lane MD (2004) Role of CREB in transcriptional regulation of CCAAT/enhancer-binding protein beta gene during adipogenesis. J Biol Chem 279:4471–4478
Zhang X, Ji J, Yan G, Wu J, Sun X, Shen J, Jiang H, Wang H (2010) Sildenafil promotes adipogenesis through a PKG pathway. Biochem Biophys Res Commun 396:1054–1059
Zoller MJ, Kerlavage AR, Taylor SS (1979) Structural comparisons of cAMP-dependent protein kinases I and II from porcine skeletal muscle. J Biol Chem 254:2408–2412

Fatty Acid Metabolites as Novel Regulators of Non-shivering Thermogenesis

Stefanie F. Maurer, Sebastian Dieckmann, Karin Kleigrewe, Cécilia Colson, Ez-Zoubir Amri, and Martin Klingenspor

Contents

1 Introduction 184
2 Origin and Formation of Oxylipins and Cannabinoids 185
  2.1 Oxylipins 185
  2.2 Endocannabinoids and Endocannabinoid-Like Compounds 188
3 Extraction and Analysis of Oxylipins and Endocannabinoids in Adipose Tissues 188
4 Mechanisms of Action 190
5 Effects of Individual Compounds on Brown/Brite Adipocytes 192
  5.1 Prostacyclin 192
  5.2 Prostaglandin E_2 194
  5.3 Prostaglandin $F_{2\alpha}$ 196
  5.4 Lipoxin A_4 197
  5.5 12,13-Dihydroxy-9Z-Octadecenoic Acid 198
  5.6 Hydroxyoctadecadienoic Acids 199
  5.7 10-Oxo-12(Z)-Octadecenoic Acid 200
  5.8 Endocannabinoids 201
6 Nutritional Interventions to Modulate Oxylipin Formation and the Thermogenic Capacity of Adipose Tissues 202
7 Conclusion 204
References 205

S. F. Maurer (✉) · S. Dieckmann · M. Klingenspor
Molecular Nutritional Medicine, Else Kröner-Fresenius Center for Nutritional Medicine, Technical University of Munich, Freising, Germany

ZIEL Institute for Food and Health, TUM School of Life Sciences, Technical University of Munich, Freising, Germany
e-mail: stefanie.maurer@tum.de

K. Kleigrewe
Bavarian Center for Biomolecular Mass Spectrometry (BayBioMS), Technical University of Munich, Freising, Germany

C. Colson · E.-Z. Amri
Université Côte d'Azur, CNRS, Inserm, iBV, Nice, France

A. Pfeifer et al. (eds.), *Brown Adipose Tissue*,
Handbook of Experimental Pharmacology 251, https://doi.org/10.1007/164_2018_150

Abstract

Fatty acids are essential contributors to adipocyte-based non-shivering thermogenesis by acting as activators of uncoupling protein 1 and serving as fuel for mitochondrial heat production. Novel evidence suggests a contribution to this thermogenic mechanism by their conversion to bioactive compounds. Mammalian cells produce a plethora of oxylipins and endocannabinoids, some of which have been identified to affect the abundance or thermogenic activity of brown and brite adipocytes. These effectors are produced locally or at distant sites and signal toward thermogenic adipocytes via a direct interaction with these cells or indirectly via secondary mechanisms. These interactions are evoked by the activation of receptor-mediated pathways. The endogenous production of these compounds is prone to modulation by the dietary intake of the respective precursor fatty acids. The effect of nutritional interventions on uncoupling protein 1-derived thermogenesis may thus at least in part be conferred by the production of a supportive oxylipin and endocannabinoid profile. The manipulation of this system in future studies will help to elucidate the physiological potential of these compounds as novel, endogenous regulators of non-shivering thermogenesis.

Keywords

Brite adipocytes · Brown adipocytes · Endocannabinoids · Oxylipins · PUFAs · Thermogenesis · Ucp1 · ω-3 · ω-6

1 Introduction

The storage of dietary lipids is a hallmark common to brown (BAT) and white adipose tissue (WAT). Based on the primary functions of these tissues, deposited lipids are predominantly released as fatty acids from triglycerides via lipolytic pathways to fuel energy-consuming processes in place or throughout the body. Besides their function as potent energy source, fatty acids are stored as structural components within biological membranes and act as metabolic regulators. The latter quality is significantly influenced by their conversion into bioactive metabolites within manifold cell types and tissues. Oxylipins and endocannabinoids represent two major metabolite classes. Both classes encompass compounds derived from polyunsaturated fatty acids (PUFAs) via designated pathways following their release from membrane phospholipids. Altogether, these metabolites are well-known for a long time to affect countless physiological processes including various adipocyte-related functions. Evidence pointing toward a novel role of these metabolites in mammalian brown/brite adipogenesis and thermogenic adipocyte function has only recently accumulated. Facing the global burden of obesity, an increase in the abundance or activity of thermogenic cells is considered a possibility to increase energy expenditure, thus reducing fat deposition and related metabolic impairment [for a recent review, see Betz and Enerback (2018)]. The potential of novel effectors and pathways targeting brown or brite adipocyte-based thermogenesis therefore seems highly valuable to explore and is currently one focus of research related to energy balance regulation (Giordano et al. 2016; Vargas-Castillo et al. 2017; Pradhan et al. 2017).

We here provide a concise summary of the potential of individual fatty acid-derived candidate compounds to affect the abundance or activity of murine and human thermogenic adipocytes, give an overview of their biosynthesis pathways and mechanisms of action, describe state-of-the-art methods for their extraction from adipose tissues and subsequent quantification, and discuss the potency of nutritional intervention studies to modulate their abundance and consequently the capacity for non-shivering thermogenesis.

2 Origin and Formation of Oxylipins and Cannabinoids

2.1 Oxylipins

Oxylipins are oxygenated metabolites primarily derived from the enzymatic oxidation of ω-3 and ω-6 PUFAs, but they can also be generated through nonenzymatic pathways such as autoxidation. Fatty acids serving as oxylipin progenitors include the ω-3 PUFAs eicosapentaenoic acid (C20:5, EPA), docosahexaenoic acid (C22:6, DHA), and α-linolenic acid (C18:3, ALA) as well as the ω-6 PUFAs arachidonic acid (C20:4, ARA) and linoleic acid (18:2, LA) (Fig. 1). The 18-carbon molecules LA and ALA are essential fatty acids, as mammals in contrast to plants lack the enzyme ω-3-desaturase and are therefore unable to endogenously produce these PUFAs. Consequently, plant oils constitute rich sources for LA and ALA. Long-chained PUFAs such as ARA, EPA, and DHA can be produced by mammalian cells through elongation and desaturation of either LA or ALA, respectively. However, as both LA and ALA compete for the same desaturase enzymes, endogenous production of DHA, EPA, and ARA is limited (Simopoulos 2016). Consequently, their individual production may be insufficient to cover needs, which makes their nutritional uptake essential (Plourde and Cunnane 2007). The composition of PUFAs in tissues and the production of their bioactive oxylipin-derivatives can be altered by dietary lipid composition and targeted supplementation of certain lipids such as ω-3 PUFAs (Balvers et al. 2012; Ostermann and Schebb 2017; Schebb et al. 2014). Accordingly, diet is a major factor influencing oxylipin levels in the body. This is important as dietary patterns have changed toward an increased uptake of ω-6 PUFAs at the expense of ω-3 PUFAs, resulting in a dietary ω-6/ω-3 ratio of 20:1 in modern days compared to a ratio of 1:1 during the Paleolithic period (Simopoulos 2001). Furthermore, both a high ω-6 PUFA intake and a high ω-6/ω-3 ratio are associated with obesity and inflammation [reviewed by (Simopoulos 2016)]. After being resorbed from the diet, PUFAs are either esterified and stored as triglycerides or incorporated into biological membranes as phospholipids. Prior to the generation of oxylipins, PUFAs must be released from these compartments upon cell activation. The cytosolic phospholipase A_2 ($cPLA_2$) enzyme is involved in the release of PUFAs from membrane phospholipids, which is traditionally perceived as the

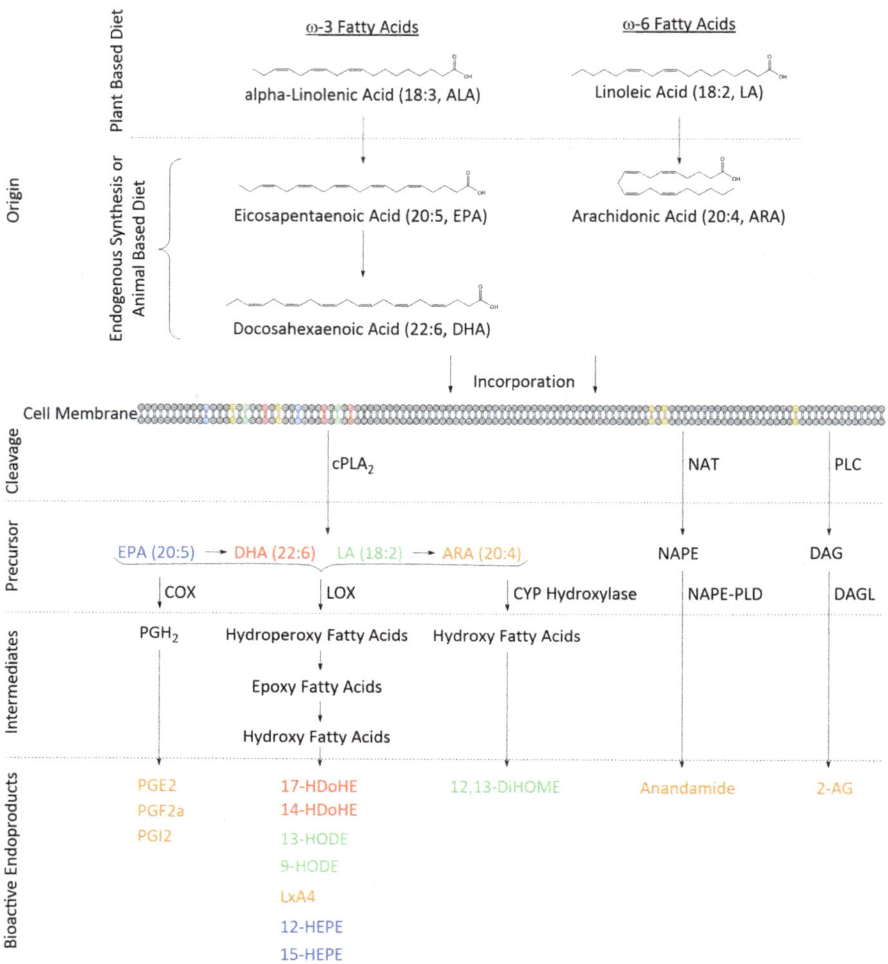

Fig. 1 Synthesis of oxylipins and endocannabinoids from membrane phospholipids. Diet-derived ω-3 and ω-6 fatty acids are incorporated into biological membranes. Through the action of various enzymes, membrane phospholipids are cleaved to release free fatty acids, *N*-arachidonoyl-phosphatidylethanolamine (NAPE) or diacylglycerols (DAG), into the cytosol. These precursors serve as substrates for different enzymatic pathways involved in the synthesis of oxylipins or endocannabinoids. Individual compounds derived from EPA, DHA, LA, and ARA are highlighted in blue, red, green, and orange, respectively. *cPLA*$_2$ cytosolic phospholipase A2, *NAT* *N*-acyltransferase, *PLC* phospholipase C, *COX* cyclooxygenase, *LOX* lipoxygenase, *CYP* cytochrome P450, *NAPE-PLD* NAPE-phospholipase D, *DAG* diacylglycerol, *DAGL* DAG lipase

major origin of oxylipin precursor PUFAs. Deficiency of cPLA$_2$ is, however, not sufficient to entirely attenuate oxylipin production in mice and humans, suggesting the existence of alternative origins (Uozumi and Shimizu 2002; Adler et al. 2008). In fact, the adipose triglyceride lipase-mediated release of PUFAs from triglyceride-rich lipid droplets appears to constitute a novel route for the release of oxylipin precursor fatty acids (Dichlberger et al. 2014; Schlager et al. 2015; Riederer et al. 2017). Subsequently, a vast diversity of oxylipins can be generated depending on the progenitor fatty acid released and on the metabolizing enzymes present in the cells. Based on the precursor PUFA, there are three classes of oxylipins to be distinguished: C18-derived octadecanoids, eicosanoids derived from C20 PUFAs, and C22-derived docosanoids. Fatty acids are converted to oxylipins by three types of enzymes, thus generating distinct oxylipin subclasses (Fig. 1).

The cyclooxygenase (COX) enzyme is responsible for the production of prostanoids and therefore an important mediator of inflammatory processes. Up to date, two COX isoforms have been identified. Although it is still a matter of debate, COX-1 is generally believed to be constitutively expressed in mammalian tissues, while COX-2 is regarded the inducible isoform being upregulated during states of inflammation. The COX-enzyme mainly facilitates the cyclization of its C20-PUFA substrate between C8 and C12, leading to the generation of the intermediary product prostaglandin H (PGH). Therefore, COX-1 and COX-2 are also known as prostaglandin H synthases (PGHS) 1 and 2, respectively. Depending on the substrate involved, PGH_1, PGH_2, or PGH_3 are generated from the C20 PUFAs dihomo-γ-linolenic acid (an intermediate in the synthesis of ARA from LA), ARA, or EPA, respectively. These intermediates serve as precursors for the formation of individual prostanoids of the 1-, 2-, and 3-series upon action of subsequent enzymes such as prostaglandin synthases, for instance. Among the different COX-substrates, ARA serves as the main substrate and precursor of 2-series prostanoids comprising various prostaglandins and thromboxanes [see Gabbs et al. (2015) for a comprehensive review about the generation of COX-derived oxylipins].

The lipoxygenase (LOX) isozymes represent a second oxylipin-producing enzyme class. There are six different human LOX genes (ALOX15, ALOX15B, ALOX12, ALOX12B, ALOXE3, and ALOX5), encoding for six functional isoforms. In contrast, ALOX15B is not found within the murine genome, which comprises the genes ALOX8 and ALOX12E instead. The common property of these enzymes is the introduction of dioxygen into PUFAs, thereby generating hydroperoxy, epoxy, and hydroxy fatty acids (Fig. 1). This includes the formation of hydroxyoctadecadienoic acids (HODEs), hydroxyeicosatetraenoic acids (HETEs), hydroxyeicosapentaenoic acids (HEPEs), and hydroxydocosahexaenoic acids (HDoHEs) from LA, ARA, EPA, and DHA, respectively. Additional products of the LOX pathway are leukotrienes, resolvins, and lipoxins (Kuhn et al. 2015; Gabbs et al. 2015).

Members of the cytochrome P450 (CYP450) system constitute the third class of enzymes able to generate oxylipins. In humans, there are 57 functional genes encoding for enzymes with hydroxylase or epoxygenase activity, while 102 functional genes have been identified in mice (Nelson et al. 2004). The hydroxylase activity of the CYP450 enzymes produces similar metabolite classes as LOX such as HODEs and HETEs, while epoxy fatty acids like ARA-derived epoxyeicosatetraenoic acids (EETs) are generated via epoxygenase activity (Konkel and Schunck 2011).

2.2 Endocannabinoids and Endocannabinoid-Like Compounds

Like oxylipins, endocannabinoids and endocannabinoid-like compounds are derivatives of PUFAs. The endocannabinoids *N*-arachidonoyl-ethanolamide (anandamide, AEA) and 2-arachidonoylglycerol (2-AG) are derived from ARA via two distinct enzymatic pathways (Fig. 1). The synthesis of 2-AG is initiated via the transformation of the membrane phospholipid 2-arachidonoyl-phosphatidylinositol into diacylglycerol by phospholipase C. Subsequently, diacylglycerol is converted by diacylglycerol lipase into 2-AG, which is a short-lived molecule that is either isomerized to 1-arachidonoylglycerol or degraded to ARA by monoacylglycerol lipase. Anandamide is generated by the transfer of ARA from 1-arachidonoyl-phosphatidylcholine to phosphatidylethanolamine, a process catalyzed by the enzyme *N*-acyltransferase. *N*-arachidonoyl-phosphatidylethanolamine (NAPE) is the product of this reaction, which is further converted to anandamide via NAPE-phospholipase D (NAPE-PLD). The degradation of AEA into ARA is subsequently mediated by the action of the fatty acid amide hydrolase. Both AEA and 2-AG are ligands of the cannabinoid receptors 1 and 2 (CB1 and CB2) and therefore considered endocannabinoids.

Endocannabinoid-like compounds represent *N*-acylated-ethanolamides (NAEs) that are generated by the same enzymatic pathways as endocannabinoids. They are, however, not derived from ARA but from other PUFAs or saturated fatty acids. As such, docosahexaenoylethanolamide, eicosapentaenoylethanolamide, or oleoylethanolamide (OEA) are derived from DHA, EPA, or oleic acid, respectively. Their structural similarities with the endocannabinoid AEA constitute their denomination as endocannabinoid-like compounds (although these metabolites barely bind to CB receptors).

3 Extraction and Analysis of Oxylipins and Endocannabinoids in Adipose Tissues

Over the years, several methods were developed for the quantitation of oxylipins and endocannabinoids in biological samples. The first analytical methods were based on gas chromatography coupled to mass spectrometry. Due to laborious sample preparation and the alteration of the lipid pattern based on degradation of specific compounds, these methods were mostly replaced by liquid chromatography coupled to tandem mass spectrometry (LC-MS/MS). Advancements of analytical instruments allow the detection of hundreds of oxylipins and endocannabinoids simultaneously from the low nanomolar to the high millimolar range by mass spectrometry-based methods nowadays. Nevertheless, analyzing these metabolite classes, sample preparation, chromatographic separation, and detection of the desired analytes needs to be validated and optimized. For example, the oxylipin pattern can be altered through autoxidation mechanisms during sample storage. Therefore, antioxidants like butylated hydroxytoluene or ethylenediaminetetraacetic acid are added to the biological samples to minimize the change of the oxylipin pattern. The stable isotope dilution assay is the method of choice for mass spectrometric methods; thus deuterated or 13C-labeled oxylipins or endocannabinoids are added to the samples as internal standards to correct for losses during sample preparation and matrix effects during measurements. One bottleneck is the limited availability of these standards and their high price. Therefore, a compromise often is to use one isotopically labeled standard per lipid class. The lipids are extracted by liquid-liquid extraction, solid-phase extraction, or a combination of both. For example, frozen adipose tissues can be crushed in balanced salt solution, extracted with cold methanol, and subjected to polymeric solid-phase extraction to quantify arachidonic acid metabolites in adipose tissue with LC-MS/MS (Pisani et al. 2014; Le Faouder et al. 2013). Most methods nowadays, where around 100 oxylipins and endocannabinoids can be quantified simultaneously, use LC-MS/MS with negative electro-spray ionization or atmospheric pressure chemical ionization (Strassburg et al. 2012; Dumlao et al. 2011; Krott et al. 2016). Usually, quantitative mass spectrometry is performed on a triple quadrupole mass spectrometer, which is operated in the multiple reaction monitoring mode (MRM) (Schuchardt et al. 2013). Here, the parent ion, often [M-H]-, is fragmented and a specific fragment is detected. The biggest challenge of oxylipins and endocannabinoids analysis is the differentiation of stereoisomers. Chiral chromatography methods have been developed to differentiate, for example, between (R)- and (S)-isomers (Mesaros and Blair 2012). Another example, where good chromatographic separation is needed, is the quantitation of the prostaglandins PGE_2 and PGD_2. These prostaglandins have the same chemical formula ($C_{20}H_{32}O_5$) and therefore the same mass-to-charge ratio and also the same MS/MS fragmentation pattern. In some cases, stereoisomers can be differentiated by a clever choice of mass transitions. For example, product ions of HETEs are formed through α-cleavage adjacent to a double bond, and therefore the individual oxylipins 5-HETE and 8-HETE can be distinguished by their fragment ions (Murphy et al. 2005). In general, MRM methods are sensitive, selective and the method of choice for quantitation of oxylipins and endocannabinoids in adipose tissues (Willenberg et al. 2015). One restriction of these LC-MRM-MS methods is that only compounds can be detected where reference material is available. To identify new lipids and to resolve additional structural information, liquid chromatography coupled to high-resolution mass spectrometry (LC-HRMS/MS) with data-dependent fragmentation is used (Lynes et al. 2017). These methods are less sensitive than the MRM methods but bear the advantage to obtain more structural information due to the recording of complete MS/MS spectra. MS/MS spectra are compared to lipid databases, and the lipid classes are tentatively annotated (Fahy et al. 2007). To verify the correct annotation and to quantify the candidates, reference standards are necessary (Masoodi et al. 2010). To use benefits of the targeted (LC-MRM-MS) and the untargeted (LC-HRMS/MS) approaches, the SWATH (sequential window acquisition of all theoretical mass windows) technology was recently applied to lipidomics and will in the future generate deeper insight into the complex regulation of lipids (Prasain et al. 2015).

Independent of the analysis technique, the interpretation of quantitative metabolite data requires their normalization to a reference quantity. A normalization to the applied mass is easily applicable when tissues are used as starting material but presumes the determination of the dissected tissue weight prior to the analysis. Metabolites expressed per milligram tissue can be extrapolated to depot size, thus providing information on the contribution of a tissue to the physiological metabolite production ("per animal") under the prerequisite that entire depots were dissected and weighed. Similarly, cell number may serve as reference quantity when cultured cells are subject to analysis. Additionally, the protein or lipid content of the material may be quantified to serve as alternative reference. However, the percentage of proteins and lipids may not only vary between different types of tissues but may also be influenced by treatment, which poses limitations to the interpretation of the results. This becomes particularly obvious in an exemplary, comparative analysis of oxylipins between WAT and BAT depots, which are well-known to differ in tissue mass (e.g., 250 mg vs. 65 mg depot mass for the inguinal WAT and interscapular BAT, respectively) and protein content (e.g., 2.5% vs. 7.25% of the respective, dissected tissue masses). Assuming that an oxylipin of interest is quantified with around 100 pg/mg protein in both tissues, a similar production rate would be a straightforward interpretation. However, when this production rate is extrapolated to the indicated tissue masses, the WAT depot is interpreted as a stronger contributor to the systemic synthesis of this oxylipin. In contrast, when the production rate is expressed per milligram tissue, BAT has an approx. threefold higher potency to produce this oxylipin. All normalization options provide valuable information but may result in over- or underestimation of metabolite production, which is particularly crucial when the analysis is applied to tissues of unknown mass or an undefined number of cells. Thus, oxylipin and endocannabinoid data should be analyzed in the light of all available normalization options to select the most reasonable option with regard to the scientific question.

4 Mechanisms of Action

Oxylipins and endocannabinoids are thought to act locally in autocrine or paracrine manner, which is based on their short half-life resulting from rapid degradation of the free, active compounds. However, their occurrence within the blood circulation suggests a possible, endocrine function. Blood oxylipins predominantly occur as esterified or albumin-bound forms (Raz 1972a, b; Maclouf et al. 1980; Shearer and Newman 2008; Shearer et al. 2018). Lipoprotein particles comprise a large quantity of esterified oxylipins, which can be released by lipoprotein lipase (Shearer and Newman 2008; Schebb et al. 2014). Thus, blood lipoproteins may represent a storage compartment for oxylipins to facilitate their delivery and uptake by distant tissues. In contrast, the majority of blood endocannabinoids and endocannabinoid-like compounds does not seem to require lipoproteins as transport vehicle but may bind to serum albumin (Bilgin et al. 2015; Bojesen and Hansen 2003).

The transcellular exchange of oxylipins is likely mediated by plasma membrane transporters rather than diffusion. Several transporters have been identified, which mediate the exchange of prostaglandins and leukotrienes. These proteins regulate the secretion of these compounds into the extracellular space as well as their uptake, which serves the activation of intracellular signaling pathways or their degradation (Schuster et al. 2015; Zhou et al. 2008). The influx and efflux of endocannabinoids is discussed to involve different mechanisms such as specialized transporters, passive diffusion, or endocytosis (Fowler 2013).

Effects of oxylipins and endocannabinoids within the target cell are evoked by activation of receptor-mediated pathways. These compounds interact with different types of receptors located intracellularly or at the cell surface. The latter class includes several members of the transient receptor potential (TRP) ion channel family [for a comprehensive summary, see Kaneko and Szallasi (2014)]. Moreover, many oxylipin receptors as well as the endocannabinoid receptors CB1 and CB2 represent G-protein-coupled receptors of the plasma membrane that signal via intracellular calcium (G_q-protein) or cyclic adenosine monophosphate (cAMP). The cAMP response is triggered by inhibition (G_i-protein) or activation (G_s-protein) of the adenylyl cyclase. Elevation of cAMP levels activates protein kinase A, which subsequently triggers the p38 mitogen-activated protein kinases pathway to stimulate the expression of selected genes. Signaling via this pathway stimulates expression of the uncoupling protein 1 (Ucp1) gene, which classically occurs upon activation of β-adrenoreceptors under conditions of non-shivering thermogenesis (Cao et al. 2001). The same pathway is repressed upon inhibition of the adenylyl cyclase, e.g., upon activation of the CB1 and CB2 receptors (Howlett et al. 2002).

The occurrence of cannabinoid receptors is not restricted to the plasma membrane. Following their synthesis, cannabinoid receptors can be incorporated into endolysosomes of human and rodent cells. These receptors are functional, couple to G_i and/or G_q, and elicit calcium release from intracellular stores (Brailoiu et al. 2011, 2014; Rozenfeld and Devi 2008; Boon et al. 2014). The significance of intracellular vs. plasma membrane cannabinoid receptors for the mediation of physiological effects has not yet been elucidated.

Peroxisome proliferator-activated receptors (PPARs) are a family of ligand-activated nuclear receptors occurring as subtypes α, β/δ, and γ. Oxylipins and endocannabinoids serve as endogenous ligands for the γ-subtype (Bouaboula et al. 2005; Barquissau et al. 2017), which is the central regulator of adipose tissue development, thus regulating white and brown adipogenesis and browning of WAT [reviewed by Harms and Seale (2013) and Seale et al. (2009)]. After its heterodimerization with the retinoic x receptor, PPARγ is able to bind to the PPAR response element representing a DNA-binding domain. Such a PPAR response element is located in the promoter region of the Ucp1 gene, whose product essentially enables non-shivering thermogenesis in brown and brite adipocytes (Enerback et al. 1997; Li et al. 2014). Consequently, treatment of mice or cultured pre-adipocytes with the PPARγ agonist rosiglitazone induces the development of brown and brite adipocytes with marked Ucp1 expression (Carmona et al. 2007; Petrovic et al. 2010).

5 Effects of Individual Compounds on Brown/Brite Adipocytes

Brown and brite adipocytes are functionally thermogenic and provide Ucp1-dependent non-shivering thermogenesis within BAT and WAT (Li et al. 2014). The role of fatty acids for adipocyte-derived thermogenesis is not restricted to their quality as fuel for heat production, but comprises their function as important signaling molecules. Thermogenic signaling cascades may be initiated by fatty acids acting as primary effectors (Quesada-Lopez et al. 2016; Hu et al. 2016; Kim et al. 2016) or by their property to function as precursors for the formation of secondary thermogenic metabolites. The latter is contributed by their conversion into oxylipins and endocannabinoids representing a novel function of these fatty acid derivatives. Here we review the candidate compounds reported to influence the capacity for Ucp1-dependent non-shivering thermogenesis in BAT and WAT.

5.1 Prostacyclin

The two COX enzymes are an important hub of oxylipin synthesis in numerous cell types and tissues. Their abundance in murine WAT is under adrenergic control, rendering these enzymes crucial regulators of brite adipocyte formation (Madsen et al. 2010; Vegiopoulos et al. 2010). This function has been ascribed to the production of prostacyclin (PGI_2), which is a derivative of ARA obtained by the specific action of prostacyclin synthase on COX-derived PGH_2. As a ligand for its G_s-protein-coupled cell surface receptor (the IP receptor) expressed by numerous tissues and cell types, this prostaglandin is well-known to potently act on vasodilation and platelet aggregation. Browning of WAT constitutes a novel function affected by this oxylipin.

The production of PGI_2 in WAT under conditions of non-shivering thermogenesis is likely mediated by mature adipocytes as these cells show increased expression of COX-2 upon adrenergic stimulation (Vegiopoulos et al. 2010). As part of a straightforward signal transduction, the released PGI_2 likely acts in paracrine manner to stimulate Ucp1 expression and the sensitivity toward noradrenergic stimulation during the differentiation of adipocyte precursors (Vegiopoulos et al. 2010). This brite adipogenesis seems to specifically originate from the action of PGI_2 on committed precursor cells of the stromal vascular fraction as the effect is in vitro most efficiently inducible by treatment of purified Lin^-Sca1^+ primary cultures of the posterior subcutaneous fat depot with carbaprostacyclin ($cPGI_2$) (Babaei et al. 2017). The latter represents a chemically stable PGI_2-analogue as PGI_2 itself is quickly converted into the inactive metabolite 6-keto-$PGF_{1\alpha}$.

The $cPGI_2$-induced conversion of committed precursors into brite adipocytes in vitro is most efficient when the compound is added to the adipogenic culture medium during eight consecutive days (Babaei et al. 2017; Bayindir et al. 2015). Thermogenic maturation of these cells in response to $cPGI_2$ treatment is initiated during an advanced stage of differentiation (Bayindir et al. 2015). This effect likewise seems to originate from an interaction of $cPGI_2$ with the cell surface IP receptor as well as with the nuclear receptor PPARγ since $cPGI_2$-induced Ucp1 expression is blunted during the differentiation of primary progenitors obtained from WAT of mice with heterozygous PPARγ ablation or in the presence of an IP receptor antagonist (Vegiopoulos et al. 2010). In fact, expression of the IP receptor is highest in pre-adipocytes and follows a reduction during the adipogenic differentiation of murine and human cells (Vassaux et al. 1992; Borglum et al. 1999). Thus, $cPGI_2$ may orchestrate brite adipogenesis via an initial activation of the IP receptor of committed pre-adipocytes to promote their thermogenic maturation in the further progress of differentiation via a predominant interaction with PPARγ.

Intriguingly, $cPGI_2$-induced brite adipogenesis is not restricted to murine adipocytes but reproducible in human mesenchymal stem cells involving similar signal transduction, either during a continuous treatment or upon 3 days of exposure in the final stage of the adipogenic differentiation (Vegiopoulos et al. 2010; Ghandour et al. 2016). The $cPGI_2$-induced expression of Ucp1 in human and murine cells is rather moderate compared to the potent effect of the PPARγ agonist rosiglitazone (Ghandour et al. 2016; Babaei et al. 2017). This relative difference in potency may be counterbalanced under physiological conditions since the naturally occurring PGI_2 is considered ten times more potent than its synthetic analogue (Aiken and Shebuski 1980; Whittle et al. 1980). However, the physiologic contribution of this system to the recruitment of brite adipocytes in vivo is to date virtually unexplored. The production of PGI_2 in WAT of COX-2 overexpressing mice correlates well with their increased abundance of brite adipocytes. This phenotype is associated with thermogenesis and resistance to diet-induced obesity (Vegiopoulos et al. 2010). Moreover, adrenergically induced browning of WAT is impaired in IP receptor knockout mice and in mice with heterozygous ablation of PPARγ, indicating PGI_2 signaling to be a modulator of systemic energy expenditure (Vegiopoulos et al. 2010).

The effect of PGI_2 on browning of murine adipocytes remains to be corroborated by more studies. The author's laboratories were not able to fully reproduce these findings using our own proven protocols. Prostacyclin-mediated browning of murine cells appears to require very precise culture conditions and may thus not constitute a common, robust effect. Moreover, IP receptor knockout mice appear to be protected from the adipogenic effect of a diet rich in ARA precursors, indicating PGI_2 signaling to promote energy storage and fat mass expansion (Massiera et al. 2003). This observation is well in line with enhanced adipogenic differentiation of cultured cells in the presence of PGI_2 (Negrel et al. 1989; Catalioto et al. 1991). Altogether, these findings indicate a bifunctional role of PGI_2 for adipocyte identity with an effect on both brite and white adipogenesis. The underlying reasons remain to be clarified but may involve, for instance, the PGI_2 target cell itself, the abundance of committed precursors within a tissue or a culture, or the presence of other adipogenic factors in blood or cell culture sera, respectively. Further studies are needed to corroborate the browning potential of PGI_2 and to further explore its specific effect on energy balance regulation under in vivo conditions.

5.2 Prostaglandin E_2

As member of the various prostaglandins produced via the COX pathway, prostaglandin E_2 (PGE_2) specifically arises from the action of the PGE synthase isozymes on ARA-derived PGH_2. As for other prostaglandins, the physiological effects of PGE_2 originate from its interaction with at least four cognate cell surface PGE receptors (EP1-EP4). The individual physiological responses evoked by PGE_2 are not only the result of cell type-dependent variations in receptor expression but also influenced by the specific signaling cascade of the receptor subtypes. While the EP1 subtype is G_q-coupled, the EP2 and EP4 subtypes couple to G_s, whereas the EP3 subtype couples to G_i. The latter is involved in mammalian febrile response (Ushikubi et al. 1998; Lazarus et al. 2007), which is an indirect consequence of central PGE_2 signaling on BAT function. In a healthy organism, the hypothalamic preoptic area (POA) translates the environmental temperature sensed by cutaneous thermoreceptors into an appropriate physiological response to maintain normothermia. In a cold environment, this involves an inhibition of efferent POA neurons culminating in the excitation of sympathetic nerves to activate BAT thermogenesis. The activity of these efferent POA neurons is likely influenced by the EP3 receptor [for a comprehensive review, see Morrison and Madden (2014)]. During febrile response (e.g., to infection), elevated levels of central PGE_2 act on the same neuronal pathways mediating an increase in body temperature and energy expenditure via the stimulation of BAT activity (Morrison 2016; Nakamura et al. 2002; Scammell et al. 1996).

Reduced systemic levels of PGE_2 have been hypothesized to participate in the afebrile regulation of body temperature (Foster et al. 2015). Besides centrally mediated mechanisms, there is evidence for a direct interaction of PGE_2 with adipocytes affecting thermogenesis. In fact, exogenous PGE_2 acutely increases the oxygen consumption of brown adipocytes (Nagai et al. 1996). The administration of 16,16-dimethyl-PGE_2 (a stable analogue of PGE_2) to C57BL/6J mice enhances the expression of Ucp1 mRNA in WAT suggesting an effect on brite adipocyte recruitment (Madsen et al. 2010; Xue et al. 2016). This PGE_2-mediated browning appears to involve a coordinated interaction of distinct molecular effectors (Garcia-Alonso and Claria 2014; Garcia-Alonso et al. 2013). Mice with an adipocyte-specific knockout of PPARγ have elevated adipose tissue levels of PGE_2 besides an increased expression of its synthesizing enzymes COX and microsomal PGE synthase 1 (mPGES1). This indicates a repressive effect of PPARγ on the PGE_2-synthesis pathway, confining the physiological origin of adipose tissue PGE_2 production to a cell type with low or absent PPARγ expression. In line with this, cells of the stromal vascular fraction from murine subcutaneous adipose tissue show higher PGE_2 secretion compared to mature adipocytes (Ying et al. 2017). Exogenous PGE_2 is able to suppress PPARγ expression in murine cells and to stimulate COX-2 and mPGES1 expression of WAT explants, indicating the existence of an autocrine positive feedback loop within the secreting cell to further elevate its own PGE_2 production (Garcia-Alonso et al. 2013). Whereas the presence of PGE_2 during the adipogenic differentiation of primary epididymal pre-adipocytes enhances Ucp1 expression over that of vehicle-treated wild-type cells, Ucp1 expression is unaltered in PPARγ-ablated cells (Garcia-Alonso et al. 2013). Thus, the browning effect of PGE_2 appears to require the presence PPARγ within the target cell, although the underlying mechanism is not yet conclusive in the light of its suppressive effect on PPARγ expression. Moreover, the specific cell type responding to PGE_2 has not yet been determined and may differ between species. Adipocyte progenitors seem to be a plausible target for the murine tissue as an acute treatment with PGE_2 is capable of increasing Ucp1 mRNA levels in cells of the stromal vascular fraction (Vegiopoulos et al. 2010). Conversely, primary mature adipocytes derived from human omental fat respond to acute PGE_2 treatment with an increase in Ucp1 mRNA expression, suggesting PGE_2-mediated browning to arise from trans-differentiation of white adipocytes (Garcia-Alonso et al. 2016). Such hypothesis is, however, not supported by experiments with human multipotent adipose-derived stem (hMADS) cells that serve as an established model system for human brite adipogenesis in culture (Elabd et al. 2009; Pisani et al. 2011). The white-to-brite adipocyte conversion of these cells is dose-dependently inhibited when 16,16-dimethyl-PGE_2 is present in the culture medium during an advanced stage of the adipogenic differentiation (Pisani et al. 2014).

The physiological effects of PGE_2 are mediated via its interaction with the four EP receptors, all of which are expressed in murine adipose tissue. The EP4 receptor has been suggested to be responsible for the browning effect (Madsen et al. 2010), which was further explored in EP4 receptor knockout mice recently (Ying et al. 2017). In fact, these mice show a more pronounced delipidation of their subcutaneous WAT depot than wild-type littermates after 10 days of repeated β_3-adrenergic stimulation. Interestingly, this phenotype is not related to the abundance of Ucp1 but rather originates from mitochondrial biogenesis and a disequilibrium between lipid droplet breakdown and synthesis. Thus, ablation of the EP4 receptor enhances murine WAT remodeling, suggesting this receptor as a negative regulator of adrenergically stimulated WAT browning. In contrast, the inhibitory effect of 16,16-dimethyl-PGE_2 on brite adipogenesis of hMADS cells is even more pronounced in the presence of an EP4 receptor inhibitor, indicating a supportive effect on the trans-differentiation of human adipocytes (Pisani et al. 2014).

A question of profound interest is whether PGE_2 is of any physiological relevance for the recruitment of brite adipocytes in vivo. Pertaining to mice, reported effect sizes on Ucp1 mRNA expression are rather small, and none of the aforementioned studies provides information on Ucp1 protein expression, PGE_2-induced oxygen consumption of brite cultures or tissues, or physiological estimates of thermogenic capacity depending on the size of the affected depot. Studies in COX-2 transgenic mice or in mice with genetically or pharmacologically induced deficiency of COX-2 only provide limited information as physiological effects associated with brown/brite adipocyte-derived thermogenesis (e.g., body temperature, whole-body energy expenditure, body mass development) may be influenced by altered production of various COX-metabolites (Vegiopoulos et al. 2010; Madsen et al. 2010). Altogether, the observations currently available suggest PGE_2 to constitute a candidate compound that may mediate the recruitment of non-shivering thermogenesis in WAT, whose potential requires comprehensive characterization in vitro and in vivo with particular focus on species-specific effects.

5.3 Prostaglandin $F_{2\alpha}$

Together with PGI_2 and PGE_2, prostaglandin $F_{2\alpha}$ ($PGF_{2\alpha}$) represents another COX-derived prostaglandin produced by adipose tissues. The synthesis originates from the metabolization of ARA into PGH_2, which serves as a direct or indirect (via the intermediate production of PGE_2 and PGD_2) substrate for the downstream production of $PGF_{2\alpha}$ by members of the aldo-keto reductase enzyme family. Its cognate receptor (the FP receptor) is abundant in adipose tissues (Volat et al. 2012), located at the cell surface, and couples to G_q, thus mediating the physiological effects of $PGF_{2\alpha}$ via an increase in intracellular calcium. This signaling pathway attenuates Ucp1 expression during brite adipogenesis, thus rendering this compound a negative regulator of thermogenic adipocyte function (Pisani et al. 2014). Intriguingly, the effect of $PGF_{2\alpha}$ on adipocyte browning has not yet been investigated in murine cells but is evident in hMADS cells. These cells can be differentiated into white or brite adipocytes under serum-free conditions depending on the presence (brite) or absence (white) of the PPARγ-activator rosiglitazone during the last days of the differentiation phase. Co-administration of the stable $PGF_{2\alpha}$-analogue fluprostenol during that phase dose-dependently blunts the levels of Ucp1 mRNA and protein in brite hMADS adipocytes (Pisani et al. 2014). The effect is mediated via activation of the FP receptor resulting in intracellular calcium oscillations with subsequent phosphorylation of extracellular signal-regulated kinases (ERK). In line with this, differentiation of brite hMADS in the presence of an ERK inhibitor or a calcium-chelator restores fluprostenol-induced effects on Ucp1 expression. As $PGF_{2\alpha}$ signaling in adipocytes inhibits the activity of PPARγ (Reginato et al. 1998), the repressive effect of this compound on Ucp1 expression may be explained by an antagonization of rosiglitazone during the conversion of white into brite hMADS adipocytes.

Physiologic effects of $PGF_{2\alpha}$ in adipose tissue may comprise both mature and pre-adipocytes as responsive cell types. In fact, $PGF_{2\alpha}$ signaling in cultured 3T3-L1 pre-adipocytes results in an inhibition of adipogenesis (Annamalai and Clipstone 2014; Liu and Clipstone 2007; Ueno and Fujimori 2011), which could be hypothesized to limit the de novo maturation of thermogenic cells in vivo in the long term. Genetically induced reduction of $PGF_{2\alpha}$ production in mice, however, results in the excessive accumulation of adipose tissue and adipocyte hypertrophy (Volat et al. 2012) and thus appears to primarily promote general adipogenesis. As brite adipogenesis in hMADS cells can be interpreted as a model system for trans-differentiation, the thermogenic function of adipocytes may rather be affected by reduced conversion of mature white into brite adipocytes in the presence of $PGF_{2\alpha}$ (Pisani et al. 2014). The increased production of this compound by adipose tissues may antagonize the browning effect of other oxylipins and therefore limit adaptive heat production under conditions of non-shivering thermogenesis.

5.4 Lipoxin A_4

The ARA-derivative lipoxin A_4 (LXA_4) is an eicosanoid belonging to the subclass of lipoxins, whose biosynthesis is mediated via alternative routes within the LOX pathway. The formation of LXA_4 via the LOX-15 isozyme involves the intermediate formation of 15-hydroperoxyeicosatetraenoic acid, whereas the metabolization of ARA via LOX-5 results in the intermediate synthesis of leukotriene A_4 serving as precursor for the subsequent formation of both leukotriene B_4 and LXA_4. The activity of LOX-5 is tightly controlled by interaction with its activating protein ALOX5AP (Dixon et al. 1990; Miller et al. 1990). Overexpression of this protein in adipose tissue of mice increases the levels of LXA_4 but not leukotriene B_4 in BAT, subcutaneous WAT, and within the circulation (Elias et al. 2016). The subcutaneous WAT of these mice is characterized by an increased abundance of multilocular cells and elevated expression of Ucp1 protein under obesogenic and non-obesogenic conditions. Although elevations in WAT Ucp1 expression are reproducible on mRNA level during a 2-day injection of LXA_4 in wild-type mice, it remains to be determined whether adipose tissue browning in ALOX5AP transgenic mice is entirely related to elevated LXA_4 levels. In fact, LOX-5 is involved in the synthesis of a number of oxylipins, some of which may complement the effect of LXA_4 via a direct or indirect interaction with cells of white adipose tissue.

A possible, direct action of LXA_4 on cultured adipocytes has not been investigated, but this oxylipin appears to influence the number and activity of Ucp1-positive cells in ALOX5AP transgenic mice via a secondary mechanism. Cultivation of the liver cell line HepG2 in the presence of LXA_4 results in higher mRNA levels of bile acid synthesizing enzymes. In line with this, ALOX5AP transgenic mice have higher levels of circulating bile acids as well as a higher fecal bile acid excretion (Elias et al. 2016). Intriguingly, systemic bile acid levels have been associated with elevated energy expenditure and reduced propensity to diet-induced obesity (Watanabe et al. 2006, 2011, 2012; Teodoro et al. 2014; da-Silva et al. 2011; Zietak and Kozak 2016). These effects have been hypothesized to originate from Ucp1-dependent thermogenesis based on the ability of bile acids to activate brown adipocytes and to increase the abundance of brite adipocytes (Velazquez-Villegas et al. 2018; Zietak and Kozak 2016; Teodoro et al. 2014; Watanabe et al. 2006). In line with these studies, ALOX5AP transgenic mice have a higher BAT expression of the G-protein-coupled bile acid receptor 1 and display higher energy expenditure that translates into attenuated body mass accumulation upon high-fat diet feeding (Elias et al. 2016). However, the physiological relevance of Ucp1-dependent thermogenesis, activated or recruited via such LXA_4-bile acid-mediated mechanism, requires careful elucidation in further studies. Bile acid-mediated effects on energy expenditure seem to require the presence of Ucp1 (Zietak and Kozak 2016), while such relationship was not reproduced in our own laboratory (manuscript submitted). Thus, the currently available data do not support a direct thermogenic function of LXA_4 for brown and brite adipocytes, while its possible secondary effect on Ucp1-dependent energy expenditure may be of minor physiological relevance.

5.5 12,13-Dihydroxy-9Z-Octadecenoic Acid

Besides the eicosanoids described above, there are several compounds belonging to the class of octadecanoids recently associated with an impact on thermogenic adipocyte function. Among them, 12,13-dihydroxy-9Z-octadecenoic acid (12,13-DiHOME) represents a potent modulator of non-shivering thermogenesis as recently demonstrated by Lynes and coworkers (Lynes et al. 2017). This compound occupies a unique role among the oxylipins described herein as it exclusively influences the thermogenic function of brown but not brite adipocytes. The metabolite is generated in a two-step reaction from LA firstly forming its epoxide-precursor 12,13-epoxy-9Z-octadecenoic acid via the CYP450 pathway, which is subsequently converted into the diol 12,13-DiHOME via four epoxide hydrolase isozymes (Ephx1–4). Transcript levels of the Ephx1 and 2 genes are upregulated in murine BAT in a cold (4°C) environment, rendering BAT an important source of circulating 12,13-DiHOME levels under conditions of non-shivering thermogenesis. In vivo, the cold-induced release of 12,13-DiHOME appears to initiate an autocrine/paracrine feedback reaction to enhance the thermogenic function of BAT. The causative signal transducer has not yet been addressed but may involve PPARγ since such an interaction has been described for its isomer 9,10-DiHOME (Lecka-Czernik et al. 2002). Interestingly, the thermogenic effect of 12,13-DiHOME does not result from a direct effect on Ucp1 expression but rather originates from an immediately augmented membrane translocation of adipocyte fatty acid transporters facilitating the lipid absorption of mature brown adipocytes. Consequently, repeated administration of 12,13-DiHOME to diet-induced obese mice results in enhanced BAT-specific uptake of triglycerides and free fatty acids, thus affecting oral lipid tolerance and serum triglyceride levels. This facilitated lipid uptake appears to improve adaptive heat production as 12,13-DiHOME-treated cold-exposed mice show an attenuated reduction of body temperature, an increased oxygen consumption, and a reduction of the respiratory exchange ratio indicative of lipid oxidation. Thus, 12,13-DiHOME supports non-shivering thermogenesis by increasing the supply of BAT with circulating lipids. The molecular fate of internalized lipids, however, remains subject to further experiments since fatty acids do not exclusively serve as fuel for thermogenesis but also as direct activators of Ucp1. Moreover, further studies are required not only to clarify whether this thermogenic mechanism is dependent on the presence of Ucp1 but also to determine the most effective therapeutic dosage of the oxylipin. In fact, supraphysiological concentrations of LA-diols interfere with the mitochondrial function of mammalian cells (Moran et al. 1997; Sisemore et al. 2001). Although the chronic treatment of diet-induced obese mice with 12,13-DiHOME does not affect body mass development (Lynes et al. 2017), this compound may have therapeutic relevance in the context of dyslipidemia. Intriguingly, circulating 12,13-DiHOME levels in humans correlate with BAT activity in response to acute cold exposure (14°C), emphasizing the translational significance of this novel effector (Lynes et al. 2017).

5.6 Hydroxyoctadecadienoic Acids

Recruitment of brite adipocytes in response to cold exposure involves interleukin-4 (IL-4) dependent, adipose-directed migration of alternatively activated M2 macrophages. This brite adipogenic mechanism has been ascribed to the local production of macrophage-derived catecholamines (Nguyen et al. 2011; Qiu et al. 2014), but this was questioned lately (Fischer et al. 2017). A recent publication suggests the octadecanoids 9-hydroxyoctadecadienoic acid (9-HODE) and 13-HODE as alternative mediators of macrophage-dependent adipose tissue browning (Lee et al. 2016). In their study, Lee and coworkers observed a transiently increased number of lipid-laden, cluster of differentiation 44 positive ($CD44^+$) F4/80 macrophages in the stromal vascular fraction of murine gonadal WAT upon 3 days of β_3-adrenergic treatment. This particular macrophage fraction is characterized by an increased production of 9-HODE and 13-HODE. In vivo, this macrophage-derived secretion of 9-HODE and 13-HODE could be supported by adipocyte-derived fatty acids as $CD44^+$ macrophages accumulate their triglycerides during the clearance of dying adipocytes. This hypothesis is supported by the observation that β_3-adrenergic treatment causes F4/80 macrophages to form crown-like structures in adipose tissue with elevated expression of the genes cluster of differentiation 36 (Cd36) and arachidonate 15-lipoxygenase (Alox15) in the $CD44^+$ fraction. Whereas the first encodes a fatty acid transporter that may catalyze the uptake of adipocyte-derived fatty acids, the second encodes a LOX-enzyme involved in the biosynthesis of HODE from LA. The expression of macrophage Alox15 protein increases during co-culture with dying adipocytes along with the accumulation of markers typical for alternative macrophage activation. As these events occur in the presence but not absence of IL-4, the production of 9-HODE and 13-HODE likely represents a downstream event mediated by $CD44^+$ macrophages following their targeted M2 polarization.

Alternatively activated M2 macrophages not only form crown-like structures to clear the tissue from dying adipocytes but also recruit adipocyte progenitors expressing the platelet-derived growth factor receptor A (Lee et al. 2013, 2014). This particular fraction is capable of undergoing brite adipogenesis (Lee et al. 2012). Indeed, adipogenic differentiation of these precursor cells in the presence of 9-HODE, 13-HODE, or rosiglitazone for 3 days results in an increased sensitivity toward β-adrenergic stimulation characterized by elevated Ucp1 mRNA levels, lipolysis, and Ucp1-dependent oxygen consumption following isoproterenol treatment (Lee et al. 2016). It remains to be determined whether these effects originate from an interaction with either PPARγ, the cation channel TRP vanilloid 1 (TRPV1), or the G-protein-coupled receptor 132, all of which serving as receptors for these oxylipins (Nagy et al. 1998; Obinata et al. 2005; Patwardhan et al. 2009; Huang et al. 1999).

Thus, 9-HODE and 13-HODE represent candidate compounds for the browning of WAT likely acting in a paracrine manner following their orchestrated production from immune cells. The latter fraction complements mature and pre-adipocytes as source of adipose-derived oxylipins with brite adipogenic potential.

5.7 10-Oxo-12(Z)-Octadecenoic Acid

The intestinal microbiome has been repeatedly hypothesized to be a significant contributor to systemic energy balance regulation [see Moran-Ramos et al. (2017) and Zhang et al. (2017) for recent reviews]. This relationship appears to be contributed by the bacterial turnover of dietary LA affecting energy expenditure. Lactic acid bacteria are capable of producing the octadecanoid 10-oxo-12(Z)-octadecenoic acid (KetoA) and further structurally related LA metabolites (Kishino et al. 2013). Considerably higher levels in colon, small intestine, and plasma of conventional vs. germ-free mice render the intestinal microbiome a significant source of these compounds (Kishino et al. 2013). In their recent study, Kim and coworkers show that KetoA represents an activator of TRPV1 (Kim et al. 2017). Physiological effects of oral KetoA administration investigated in the same study rely on the TRPV1-dependent initiation of an afferent signaling cascade within the gastrointestinal tract as demonstrated by the use of TRPV1 knockout mice and mice subjected to gastric vagotomy. Afferent signaling of KetoA via the nervus vagus results in the excitation of efferent sympathetic pathways. KetoA-induced, sympathetic signaling may selectively target BAT and WAT but not the cardiovascular system as the turnover of norepinephrine is elevated locally in these tissues but not in the heart. Moreover, this sympathetic norepinephrine release translates acutely and chronically into elevated Ucp1 expression in BAT and WAT under obesogenic and non-obesogenic conditions. In line with an elevated capacity for non-shivering thermogenesis, KetoA-supplemented mice of the C57BL/6 and KK-Ay strains but not TRPV1 knockout mice show reduced propensity to diet-induced obesity, elevated rectal body temperature and energy expenditure.

The TRPV1 receptor is expressed by adipose tissue, and treatment of cultured cells with the TRPV1 agonist capsaicin affects brown and brite adipogenesis (Kida et al. 2016; Baboota et al. 2014). KetoA may thus be able to complement Ucp1-dependent non-shivering thermogenesis via a direct interaction with TRPV1 of adipocytes upon spillover into the bloodstream. Such direct effect may be supported by its property to interact with PPARγ (Goto et al. 2015), although the knockout of TRPV1 is sufficient to explain the majority of physiologic effects induced by oral KetoA (Kim et al. 2017). In fact, Kim and coworkers report KetoA to be undetectable in plasma of KetoA-supplemented mice (Kim et al. 2017), indicating the physiologic effects induced by this compound to entirely rely on secondary signaling upon activation of TRPV1 in the intestine. The effect of KetoA, however, has not been investigated upon administration via other peripheral routes (e.g., intraperitoneal or subcutaneous) or with cultured adipocytes.

Although the actual relevance of Ucp1-dependent thermogenesis remains to be assessed, the oxylipin KetoA appears to be a potent modulator of systemic energy expenditure of microbial origin. The physiological relevance of this microbiome-host-interaction should be further addressed in studies using KetoA-supplemented conventional and germ-free mice under conditions (e.g., cold exposure) that promote Ucp1-dependent non-shivering thermogenesis.

5.8 Endocannabinoids

The endocannabinoid system is well-known for its implication in energy homeostasis [reviewed by Matias and Di Marzo (2007), Silvestri and Di Marzo (2013), Richard et al. (2009)]. The best studied entity in this regard is the CB1 receptor expressed by various brain areas associated with the regulation of food intake and energy turnover. Its activation increases body mass by elevated food intake as well as reduced energy expenditure and thus centrally signals a positive energy balance (Kirkham et al. 2002). In line with this, treatment of mice with the synthetic CB1 antagonist rimonabant decreases food intake and body mass (Jbilo et al. 2005) and increases glucose and lipoprotein clearance by BAT (Bajzer et al. 2011; Boon et al. 2014).

While the importance of the endocannabinoid system for the regulation of energy homeostasis is strongly supported by the aforementioned studies, it is still a matter of debate (1) whether effects on energy expenditure are exclusively or only partially mediated by central cannabinoid receptors and (2) whether effects on energy expenditure originate from brown and brite adipocyte-derived thermogenesis. On the one hand, exclusive stimulation of peripheral CB1 in diet-induced obese mice is sufficient to reduce their body mass and to increase their energy expenditure, suggesting the occurrence of cannabinoid-dependent thermogenesis in the absence of central signaling (Boon et al. 2014). On the other hand, a genetically induced decrease of 2-AG levels (an endogenous CB1 agonist) in the murine forebrain results in elevated mitochondrial abundance and Ucp1 protein expression in BAT, increased body temperature upon adrenergic stimulation, and reduced body mass accumulation under obesogenic and non-obesogenic conditions (Jung et al. 2012), thus arguing for a central regulation of peripheral energy expenditure by the cannabinoid system. This is further supported by studies in mice with forebrain-specific CB1 deletion, which are resistant to diet-induced obesity and show higher body temperature under cold stimulation (Quarta et al. 2010). Elevated Ucp1 mRNA expression and glucose uptake in BAT of these mice indicates the recruitment of BAT-derived thermogenesis via central cannabinoid signaling (Quarta et al. 2010). Moreover, treatment of cultured murine white and brown adipocytes with a synthetic CB1 agonist decreases Ucp1 protein levels, suggesting a direct regulation of adipocyte-derived thermogenesis by the peripheral endocannabinoid system (Perwitz et al. 2006; Boon et al. 2014). In line with this, the adipose-specific knockout of the CB1 receptor is sufficient to induce a lower body mass in mice and to reduce the propensity to diet-induced obesity (Ruiz de Azua et al. 2017). Rimonabant-treated mice show increased Ucp1 mRNA in BAT independent of sympathetic innervation (Bajzer et al. 2011), providing additional evidence for the importance of peripherally expressed CB1. Hypothesizing the adipocyte CB1 receptor as direct repressor of Ucp1-dependent thermogenesis, ablation of its ligands should be expected to mimic the effect of the CB1 knockout. Such relationship was further investigated in mice with an adipocyte-specific knockout of NAPE-PLD, representing the enzyme that synthesizes the endocannabinoid AEA and endocannabinoid-like compounds (Geurts et al. 2015). As expected, the knockout translates into reduced levels of the respective metabolites. Surprisingly, these mice have reduced Ucp1 mRNA expression and elevated body mass under obesogenic and non-obesogenic conditions, suggesting the endocannabinoid system as a positive regulator of adipocyte-derived thermogenesis. Thus, the significance of the endocannabinoid system as a direct or indirect regulator of Ucp1-dependent thermogenesis requires further deciphering as it appears to involve a complex regulation of the implicated effectors.

The potential of individual endocannabinoids or endocannabinoid-like compounds to activate or recruit Ucp1-dependent non-shivering thermogenesis is largely unexplored. Stimulation of mice by cold exposure or with a β_3-adrenergic agonist increases the levels of endocannabinoids and NAEs like OEA in BAT and especially in WAT (Krott et al. 2016). Krott and coworkers hypothesize this regulation of the endocannabinoid system to mediate an autocrine negative feedback mechanism limiting the capacity for Ucp1-dependent thermogenesis during the recruitment of brown and brite adipocytes. However, mice supplemented with OEA exhibit increased thermogenic capacity in BAT, energy expenditure, and BAT temperature after β_3-adrenergic treatment, suggesting this compound to potentiate the recruitment of Ucp1-dependent heat production (Suarez et al. 2014). As OEA is unable to bind cannabinoid receptors (Fonseca et al. 2013), this effect may be mediated by an interaction with PPARα of brown and brite adipocytes (Fu et al. 2003). Further studies are needed to manifest the positive regulatory role of OEA on Ucp1-dependent non-shivering thermogenesis and to elucidate the potential of other endocannabinoids and endocannabinoid-like compounds.

6 Nutritional Interventions to Modulate Oxylipin Formation and the Thermogenic Capacity of Adipose Tissues

Dietary PUFAs constitute an important part of the human nutrition. The dietary ω-6/ω-3 ratio has shifted from approx. 1:1 during the Paleolithic period to 15:1 and higher in modern societies, which is associated with the occurrence of obesity related to adipose tissue dysfunction (Simopoulos 2016). Concomitantly, the activity of BAT among adult humans is highest in lean and lowest in obese subjects (Saito et al. 2009; Cypess et al. 2009; van Marken Lichtenbelt et al. 2009; Vijgen et al. 2011). Treatment of hMADS cells with ARA blunts Ucp1 expression during brite adipogenesis, which is prevented upon co-treatment with EPA (Ghandour et al. 2018; Pisani et al. 2014). Thus, a causal relationship between dietary PUFA-uptake, the abundance and activity of thermogenic adipocytes, and energy expenditure may be hypothesized.

Evidence supporting a role of fatty acids and their metabolites in the activation of BAT and the recruitment of brite adipocytes has accumulated during recent years (Okla et al. 2017). Several studies have particularly investigated the effects of ω-3 fatty acids and show that supplementation of mice with fish oil comprising diets rich in EPA and DHA results in an elevated expression of Ucp1 in BAT and/or WAT (Kim et al. 2015, 2016; Bargut et al. 2016a, b). These in vivo findings are further supported by in vitro experiments. Treatment of murine primary white and brown pre-adipocytes with EPA increases thermogenic gene expression and concomitantly enhances markers of mitochondrial abundance and function (Kim et al. 2016; Zhao and Chen 2014). Similarly, EPA elevates Ucp1 expression of human primary adipose tissue-derived stem cells, but the same effect is not observed in the presence of DHA (Fleckenstein-Elsen et al. 2016). Thus, ω-3 fatty acids and particularly EPA appear to beneficially affect the recruitment of Ucp1-positive cells in vivo and in vitro. Mechanistically, such effect may be mediated by a direct interaction of EPA with free fatty acid receptors (Quesada-Lopez et al. 2016; Kim et al. 2016), for instance, or via their conversion to bioactive metabolites. In fact, the fatty acid composition of membranes and thus the endogenous oxylipin profile changes upon nutritional interventions. Accordingly, the systemic production of ω-3-derived metabolites is enhanced upon dietary supplementation of ω-3 PUFAs [for a review, see Ostermann et al. (2017)].

The potential of nutritional interventions to modulate the abundance or activity of Ucp1-positive adipocytes as a direct consequence of altered oxylipin production is virtually unexplored. Recent studies support the hypothesized role of the dietary ω-6/ω-3 ratio in this context (Pisani et al. 2014; Ghandour et al. 2018). Mice were subjected to chow diets supplemented with ω-6 or ω-3 PUFAs to achieve a dietary ω-6/ω-3 ratio of 30 or 3.7, respectively. After 11 weeks of feeding, mice were treated with a β_3-adrenergic agonist for another week to stimulate the recruitment of non-shivering thermogenesis in BAT and WAT. The expression of Ucp1 is more pronounced in BAT and WAT of ω-3-supplemented mice. This effect is hypothesized to originate from $PGF_{2\alpha}$, whose production is reduced in both tissues of ω-3- vs. ω-6-supplemented mice upon β_3-stimulation. This reduced production is reproducible in ARA-treated brite hMADS cells upon co-incubation with EPA accompanied by a restoration of Ucp1 expression. As described above (Sect. 5.3), $PGF_{2\alpha}$ is a negative regulator of Ucp1-dependent thermogenesis in hMADS cells. In line with this, the capacity for non-shivering thermogenesis is impaired under conditions that favor the production of this compound. Mice on a chow diet supplemented with ARA have a higher WAT-production of $PGF_{2\alpha}$ compared to mice receiving oleic acid supplementation. This production rate of $PGF_{2\alpha}$ remains elevated following β_3-adrenergic treatment. This regulation is well in line with the impaired recruitment of Ucp1 protein and the reduced abundance of multilocular adipocytes in WAT of ARA-supplemented vs. oleic acid-supplemented mice. Thus, these studies support the hypothesis that non-shivering thermogenesis is modifiable by the oxylipin pattern produced in response to a dietary intervention. These studies suggest this effect to primarily result from an antagonistic mechanism that reduces the production of individual ω-6-derived compounds rather than originating from an elevated abundance of ω-3-derived compounds acting as thermogenic effectors. This regulation may constitute one possible mechanism to explain the beneficial effect of dietary ω-3 supplementation on energy expenditure and cold tolerance (Kim et al. 2015, 2016). A possible relationship between dietary PUFA supplementation, changes in the endocannabinoid profile of adipose tissues, and the induction or activation of thermogenic adipocytes remains to be assessed.

7 Conclusion

Oxylipins and endocannabinoids represent two metabolite classes originating from the metabolic conversion of ω-3 and ω-6 PUFAs. These bioactive compounds affect a plethora of physiological processes constituting their role as essential regulators of metabolic control. Since only a few years, there is accumulating evidence for a novel role for these effectors as regulators of Ucp1-dependent non-shivering thermogenesis. Several individual compounds with the potential to affect brown and brite adipocytes have been identified (Fig. 2). These compounds appear to modulate thermogenic adipocytes on multiple levels by acting as (1) upstream effectors influencing non-shivering thermogenesis via a secondary mechanism, (2) direct effectors of adipocyte identity and function, or (3) sensitizers to enhance Ucp1-derived heat production in response to a thermogenic stimulus. Thus, they can act directly in adipose tissues or indirectly via central or peripheral circuitries. Accordingly, the production of thermogenic effector compounds is not restricted to local entities such as mature adipocytes, pre-adipocytes, or immune cells, but involves distant sites such as the central nervous system or the intestinal microbiota, which may in turn facilitate the delivery of effectors via the blood circulation. Their production and their biological effect are regulated in a complex manner affected

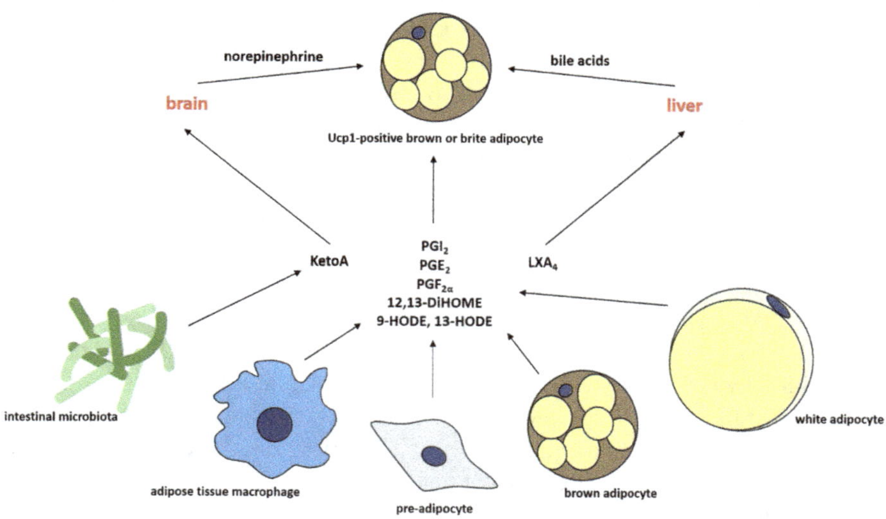

Fig. 2 Schematic summary of oxylipins affecting the recruitment or function of Ucp1-positive brown and brite adipocytes. The compounds PGI_2, PGE_2, $PGF_{2\alpha}$, 12,13-DiHOME, 9-HODE, and 13-HODE likely act in autocrine or paracrine manner upon their local production in adipose tissues by mature adipocytes, pre-adipocytes, or immune cells. The compound LXA_4 produced in BAT and WAT is thought to affect hepatic bile acid production, which is in turn hypothesized to signal Ucp1-dependent thermogenesis. The microbial-derived compound KetoA stimulates gastrointestinal afferences to signal thermogenesis via sympathetic pathways. Endocannabinoids and endocannabinoid-like compounds (not depicted) may complement oxylipin-mediated effects on brown and brite adipocytes via direct or centrally mediated pathways.

by multiple variables such as the abundance of the precursor fatty acid, the enzymatic pathway involved, the responding cell type, and the implicated receptor and signaling pathway. Future studies may not only focus on the physiological potential of such compounds to affect parameters associated with Ucp1-dependent energy expenditure (such as adrenergically induced heat production, cold tolerance, or the regulation of body mass) but also on the manipulation of their production and signaling pathways. Nutritional interventions seem to represent a useful tool, although their effect is not restricted to the modulation of a single compound but may trigger the systemic generation of a profile that favors the abundance or activity of thermogenic adipocytes. Sophisticated analysis techniques with a high sensitivity will not only help to decipher effects of oxylipins and endocannabinoids on brown and brite adipocyte function in the future but also facilitate the identification of further compounds with thermogenic potential.

Acknowledgments This work was supported by French Agence Nationale de la Recherche and Deutsche Forschungsgemeinschaft (ANR/DFG-15-CE14-0033 "Nutribrite"). We thank Tobias Fromme for proofreading the manuscript.

References

Adler DH, Cogan JD, Phillips JA 3rd, Schnetz-Boutaud N, Milne GL, Iverson T, Stein JA, Brenner DA, Morrow JD, Boutaud O, Oates JA (2008) Inherited human cPLA(2alpha) deficiency is associated with impaired eicosanoid biosynthesis, small intestinal ulceration, and platelet dysfunction. J Clin Invest 118(6):2121–2131. https://doi.org/10.1172/JCI30473

Aiken JW, Shebuski RJ (1980) Comparison in anesthetized dogs of the anti-aggregatory and hemodynamic effects of prostacyclin and a chemically stable prostacyclin analog, 6a-carba-PGI2 (carbacyclin). Prostaglandins 19(4):629–643

Annamalai D, Clipstone NA (2014) Prostaglandin F2alpha inhibits adipogenesis via an autocrine-mediated interleukin-11/glycoprotein 130/STAT1-dependent signaling cascade. J Cell Biochem 115(7):1308–1321. https://doi.org/10.1002/jcb.24785

Babaei R, Bayindir-Buchhalter I, Meln I, Vegiopoulos A (2017) Immuno-magnetic isolation and thermogenic differentiation of white adipose tissue progenitor cells. Methods Mol Biol 1566:37–48. https://doi.org/10.1007/978-1-4939-6820-6_5

Baboota RK, Singh DP, Sarma SM, Kaur J, Sandhir R, Boparai RK, Kondepudi KK, Bishnoi M (2014) Capsaicin induces "brite" phenotype in differentiating 3T3-L1 preadipocytes. PLoS One 9(7):e103093. https://doi.org/10.1371/journal.pone.0103093

Bajzer M, Olivieri M, Haas MK, Pfluger PT, Magrisso IJ, Foster MT, Tschop MH, Krawczewski-Carhuatanta KA, Cota D, Obici S (2011) Cannabinoid receptor 1 (CB1) antagonism enhances glucose utilisation and activates brown adipose tissue in diet-induced obese mice. Diabetologia 54(12):3121–3131. https://doi.org/10.1007/s00125-011-2302-6

Balvers MG, Verhoeckx KC, Bijlsma S, Rubingh CM, Meijerink J, Wortelboer HM, Witkamp RF (2012) Fish oil and inflammatory status alter the n-3 to n-6 balance of the endocannabinoid and oxylipin metabolomes in mouse plasma and tissues. Metabolomics 8(6):1130–1147. https://doi.org/10.1007/s11306-012-0421-9

Bargut TC, Silva-e-Silva AC, Souza-Mello V, Mandarim-de-Lacerda CA, Aguila MB (2016a) Mice fed fish oil diet and upregulation of brown adipose tissue thermogenic markers. Eur J Nutr 55(1):159–169. https://doi.org/10.1007/s00394-015-0834-0

Bargut TC, Souza-Mello V, Mandarim-de-Lacerda CA, Aguila MB (2016b) Fish oil diet modulates epididymal and inguinal adipocyte metabolism in mice. Food Funct 7(3):1468–1476. https://doi.org/10.1039/c5fo00909j

Barquissau V, Ghandour RA, Ailhaud G, Klingenspor M, Langin D, Amri EZ, Pisani DF (2017) Control of adipogenesis by oxylipins, GPCRs and PPARs. Biochimie 136:3–11. https://doi.org/10.1016/j.biochi.2016.12.012

Bayindir I, Babaeikelishomi R, Kocanova S, Sousa IS, Lerch S, Hardt O, Wild S, Bosio A, Bystricky K, Herzig S, Vegiopoulos A (2015) Transcriptional pathways in cPGI2-induced adipocyte progenitor activation for browning. Front Endocrinol (Lausanne) 6:129. https://doi.org/10.3389/fendo.2015.00129

Betz MJ, Enerback S (2018) Targeting thermogenesis in brown fat and muscle to treat obesity and metabolic disease. Nat Rev Endocrinol 14(2):77–87. https://doi.org/10.1038/nrendo.2017.132

Bilgin M, Bindila L, Graessler J, Shevchenko A (2015) Quantitative profiling of endocannabinoids in lipoproteins by LC-MS/MS. Anal Bioanal Chem 407(17):5125–5131. https://doi.org/10.1007/s00216-015-8559-8

Bojesen IN, Hansen HS (2003) Binding of anandamide to bovine serum albumin. J Lipid Res 44 (9):1790–1794. https://doi.org/10.1194/jlr.M300170-JLR200

Boon MR, Kooijman S, van Dam AD, Pelgrom LR, Berbee JF, Visseren CA, van Aggele RC, van den Hoek AM, Sips HC, Lombes M, Havekes LM, Tamsma JT, Guigas B, Meijer OC, Jukema JW, Rensen PC (2014) Peripheral cannabinoid 1 receptor blockade activates brown adipose tissue and diminishes dyslipidemia and obesity. FASEB J 28(12):5361–5375. https://doi.org/10.1096/fj.13-247643

Borglum JD, Pedersen SB, Ailhaud G, Negrel R, Richelsen B (1999) Differential expression of prostaglandin receptor mRNAs during adipose cell differentiation. Prostaglandins Other Lipid Mediat 57(5–6):305–317

Bouaboula M, Hilairet S, Marchand J, Fajas L, Le Fur G, Casellas P (2005) Anandamide induced PPARgamma transcriptional activation and 3T3-L1 preadipocyte differentiation. Eur J Pharmacol 517(3):174–181. https://doi.org/10.1016/j.ejphar.2005.05.032

Brailoiu GC, Oprea TI, Zhao P, Abood ME, Brailoiu E (2011) Intracellular cannabinoid type 1 (CB1) receptors are activated by anandamide. J Biol Chem 286(33):29166–29174. https://doi.org/10.1074/jbc.M110.217463

Brailoiu GC, Deliu E, Marcu J, Hoffman NE, Console-Bram L, Zhao P, Madesh M, Abood ME, Brailoiu E (2014) Differential activation of intracellular versus plasmalemmal CB2 cannabinoid receptors. Biochemistry 53(30):4990–4999. https://doi.org/10.1021/bi500632a

Cao W, Medvedev AV, Daniel KW, Collins S (2001) Beta-adrenergic activation of p38 MAP kinase in adipocytes: cAMP induction of the uncoupling protein 1 (UCP1) gene requires p38 MAP kinase. J Biol Chem 276(29):27077–27082. https://doi.org/10.1074/jbc.M101049200

Carmona MC, Louche K, Lefebvre B, Pilon A, Hennuyer N, Audinot-Bouchez V, Fievet C, Torpier G, Formstecher P, Renard P, Lefebvre P, Dacquet C, Staels B, Casteilla L, Penicaud L, Consortium of the French Ministry of Research Technology (2007) S 26948: a new specific peroxisome proliferator activated receptor gamma modulator with potent antidiabetes and antiatherogenic effects. Diabetes 56(11):2797–2808. https://doi.org/10.2337/db06-1734

Catalioto RM, Gaillard D, Maclouf J, Ailhaud G, Negrel R (1991) Autocrine control of adipose cell differentiation by prostacyclin and PGF2 alpha. Biochim Biophys Acta 1091(3):364–369

Cypess AM, Lehman S, Williams G, Tal I, Rodman D, Goldfine AB, Kuo FC, Palmer EL, Tseng YH, Doria A, Kolodny GM, Kahn CR (2009) Identification and importance of brown adipose tissue in adult humans. N Engl J Med 360(15):1509–1517. https://doi.org/10.1056/NEJMoa0810780

Dichlberger A, Schlager S, Maaninka K, Schneider WJ, Kovanen PT (2014) Adipose triglyceride lipase regulates eicosanoid production in activated human mast cells. J Lipid Res 55 (12):2471–2478. https://doi.org/10.1194/jlr.M048553

Dixon RA, Diehl RE, Opas E, Rands E, Vickers PJ, Evans JF, Gillard JW, Miller DK (1990) Requirement of a 5-lipoxygenase-activating protein for leukotriene synthesis. Nature 343 (6255):282–284. https://doi.org/10.1038/343282a0

Dumlao DS, Buczynski MW, Norris PC, Harkewicz R, Dennis EA (2011) High-throughput lipidomic analysis of fatty acid derived eicosanoids and N-acylethanolamines. Biochim Biophys Acta 1811(11):724–736. https://doi.org/10.1016/j.bbalip.2011.06.005

Elabd C, Chiellini C, Carmona M, Galitzky J, Cochet O, Petersen R, Penicaud L, Kristiansen K, Bouloumie A, Casteilla L, Dani C, Ailhaud G, Amri EZ (2009) Human multipotent adipose-derived stem cells differentiate into functional brown adipocytes. Stem Cells 27 (11):2753–2760. https://doi.org/10.1002/stem.200

Elias I, Ferre T, Vila L, Munoz S, Casellas A, Garcia M, Molas M, Agudo J, Roca C, Ruberte J, Bosch F, Franckhauser S (2016) ALOX5AP overexpression in adipose tissue leads to LXA4 production and protection against diet-induced obesity and insulin resistance. Diabetes 65 (8):2139–2150. https://doi.org/10.2337/db16-0040

Enerback S, Jacobsson A, Simpson EM, Guerra C, Yamashita H, Harper ME, Kozak LP (1997) Mice lacking mitochondrial uncoupling protein are cold-sensitive but not obese. Nature 387 (6628):90–94. https://doi.org/10.1038/387090a0

Fahy E, Sud M, Cotter D, Subramaniam S (2007) LIPID MAPS online tools for lipid research. Nucleic Acids Res 35(Web Server):W606–W612. https://doi.org/10.1093/nar/gkm324

Fischer K, Ruiz HH, Jhun K, Finan B, Oberlin DJ, van der Heide V, Kalinovich AV, Petrovic N, Wolf Y, Clemmensen C, Shin AC, Divanovic S, Brombacher F, Glasmacher E, Keipert S, Jastroch M, Nagler J, Schramm KW, Medrikova D, Collden G, Woods SC, Herzig S, Homann D, Jung S, Nedergaard J, Cannon B, Tschop MH, Muller TD, Buettner C (2017) Alternatively activated macrophages do not synthesize catecholamines or contribute to adipose tissue adaptive thermogenesis. Nat Med 23(5):623–630. https://doi.org/10.1038/nm.4316

Fleckenstein-Elsen M, Dinnies D, Jelenik T, Roden M, Romacho T, Eckel J (2016) Eicosapentaenoic acid and arachidonic acid differentially regulate adipogenesis, acquisition of a brite phenotype and mitochondrial function in primary human adipocytes. Mol Nutr Food Res 60(9):2065–2075. https://doi.org/10.1002/mnfr.201500892

Fonseca BM, Costa MA, Almada M, Correia-da-Silva G, Teixeira NA (2013) Endogenous cannabinoids revisited: a biochemistry perspective. Prostaglandins Other Lipid Mediat 102–103:13–30. https://doi.org/10.1016/j.prostaglandins.2013.02.002

Foster J, Mauger AR, Chrismas BC, Thomasson K, Taylor L (2015) Is prostaglandin E2 (PGE2) involved in the thermogenic response to environmental cooling in healthy humans? Med Hypotheses 85(5):607–611. https://doi.org/10.1016/j.mehy.2015.07.022

Fowler CJ (2013) Transport of endocannabinoids across the plasma membrane and within the cell. FEBS J 280(9):1895–1904. https://doi.org/10.1111/febs.12212

Fu J, Gaetani S, Oveisi F, Lo Verme J, Serrano A, Rodriguez De Fonseca F, Rosengarth A, Luecke H, Di Giacomo B, Tarzia G, Piomelli D (2003) Oleylethanolamide regulates feeding and body weight through activation of the nuclear receptor PPAR-alpha. Nature 425 (6953):90–93. https://doi.org/10.1038/nature01921

Gabbs M, Leng S, Devassy JG, Monirujjaman M, Aukema HM (2015) Advances in our understanding of oxylipins derived from dietary PUFAs. Adv Nutr 6(5):513–540. https://doi.org/10.3945/an.114.007732

Garcia-Alonso V, Claria J (2014) Prostaglandin E2 signals white-to-brown adipogenic differentiation. Adipocytes 3(4):290–296. https://doi.org/10.4161/adip.29993

Garcia-Alonso V, Lopez-Vicario C, Titos E, Moran-Salvador E, Gonzalez-Periz A, Rius B, Parrizas M, Werz O, Arroyo V, Claria J (2013) Coordinate functional regulation between microsomal prostaglandin E synthase-1 (mPGES-1) and peroxisome proliferator-activated receptor gamma (PPARgamma) in the conversion of white-to-brown adipocytes. J Biol Chem 288(39):28230–28242. https://doi.org/10.1074/jbc.M113.468603

Garcia-Alonso V, Titos E, Alcaraz-Quiles J, Rius B, Lopategi A, Lopez-Vicario C, Jakobsson PJ, Delgado S, Lozano J, Claria J (2016) Prostaglandin E2 exerts multiple regulatory actions on human obese adipose tissue remodeling, inflammation, adaptive thermogenesis and lipolysis. PLoS One 11(4):e0153751. https://doi.org/10.1371/journal.pone.0153751

Geurts L, Everard A, Van Hul M, Essaghir A, Duparc T, Matamoros S, Plovier H, Castel J, Denis RG, Bergiers M, Druart C, Alhouayek M, Delzenne NM, Muccioli GG, Demoulin JB, Luquet S, Cani PD (2015) Adipose tissue NAPE-PLD controls fat mass development by altering the browning process and gut microbiota. Nat Commun 6:6495. https://doi.org/10.1038/ncomms7495

Ghandour RA, Giroud M, Vegiopoulos A, Herzig S, Ailhaud G, Amri EZ, Pisani DF (2016) IP-receptor and PPARs trigger the conversion of human white to brite adipocyte induced by carbaprostacyclin. Biochim Biophys Acta 1861(4):285–293. https://doi.org/10.1016/j.bbalip.2016.01.007

Ghandour RA, Colson C, Giroud M, Maurer S, Rekima S, Ailhaud GP, Klingenspor M, Amri EZ, Pisani DF (2018) Impact of dietary omega3 polyunsaturated fatty acid supplementation on brown and brite adipocyte function. J Lipid Res. https://doi.org/10.1194/jlr.M081091

Giordano A, Frontini A, Cinti S (2016) Convertible visceral fat as a therapeutic target to curb obesity. Nat Rev Drug Discov 15(6):405–424. https://doi.org/10.1038/nrd.2016.31

Goto T, Kim YI, Furuzono T, Takahashi N, Yamakuni K, Yang HE, Li Y, Ohue R, Nomura W, Sugawara T, Yu R, Kitamura N, Park SB, Kishino S, Ogawa J, Kawada T (2015) 10-Oxo-12(Z)-octadecenoic acid, a linoleic acid metabolite produced by gut lactic acid bacteria, potently activates PPARgamma and stimulates adipogenesis. Biochem Biophys Res Commun 459 (4):597–603. https://doi.org/10.1016/j.bbrc.2015.02.154

Harms M, Seale P (2013) Brown and beige fat: development, function and therapeutic potential. Nat Med 19(10):1252–1263. https://doi.org/10.1038/nm.3361

Howlett AC, Barth F, Bonner TI, Cabral G, Casellas P, Devane WA, Felder CC, Herkenham M, Mackie K, Martin BR, Mechoulam R, Pertwee RG (2002) International Union of Pharmacology. XXVII. Classification of cannabinoid receptors. Pharmacol Rev 54(2):161–202

Hu J, Kyrou I, Tan BK, Dimitriadis GK, Ramanjaneya M, Tripathi G, Patel V, James S, Kawan M, Chen J, Randeva HS (2016) Short-chain fatty acid acetate stimulates Adipogenesis and mitochondrial biogenesis via GPR43 in Brown adipocytes. Endocrinology 157(5):1881–1894. https://doi.org/10.1210/en.2015-1944

Huang JT, Welch JS, Ricote M, Binder CJ, Willson TM, Kelly C, Witztum JL, Funk CD, Conrad D, Glass CK (1999) Interleukin-4-dependent production of PPAR-gamma ligands in macrophages by 12/15-lipoxygenase. Nature 400(6742):378–382. https://doi.org/10.1038/22572

Jbilo O, Ravinet-Trillou C, Arnone M, Buisson I, Bribes E, Peleraux A, Penarier G, Soubrie P, Le Fur G, Galiegue S, Casellas P (2005) The CB1 receptor antagonist rimonabant reverses the diet-induced obesity phenotype through the regulation of lipolysis and energy balance. FASEB J 19 (11):1567–1569. https://doi.org/10.1096/fj.04-3177fje

Jung KM, Clapper JR, Fu J, D'Agostino G, Guijarro A, Thongkham D, Avanesian A, Astarita G, DiPatrizio NV, Frontini A, Cinti S, Diano S, Piomelli D (2012) 2-Arachidonoylglycerol signaling in forebrain regulates systemic energy metabolism. Cell Metab 15(3):299–310. https://doi.org/10.1016/j.cmet.2012.01.021

Kaneko Y, Szallasi A (2014) Transient receptor potential (TRP) channels: a clinical perspective. Br J Pharmacol 171(10):2474–2507. https://doi.org/10.1111/bph.12414

Kida R, Yoshida H, Murakami M, Shirai M, Hashimoto O, Kawada T, Matsui T, Funaba M (2016) Direct action of capsaicin in brown adipogenesis and activation of brown adipocytes. Cell Biochem Funct 34(1):34–41. https://doi.org/10.1002/cbf.3162

Kim M, Goto T, Yu R, Uchida K, Tominaga M, Kano Y, Takahashi N, Kawada T (2015) Fish oil intake induces UCP1 upregulation in brown and white adipose tissue via the sympathetic nervous system. Sci Rep 5:18013. https://doi.org/10.1038/srep18013

Kim J, Okla M, Erickson A, Carr T, Natarajan SK, Chung S (2016) Eicosapentaenoic acid potentiates Brown thermogenesis through FFAR4-dependent up-regulation of miR-30b and miR-378. J Biol Chem 291(39):20551–20562. https://doi.org/10.1074/jbc.M116.721480

Kim M, Furuzono T, Yamakuni K, Li Y, Kim YI, Takahashi H, Ohue-Kitano R, Jheng HF, Takahashi N, Kano Y, Yu R, Kishino S, Ogawa J, Uchida K, Yamazaki J, Tominaga M, Kawada T, Goto T (2017) 10-Oxo-12(Z)-octadecenoic acid, a linoleic acid metabolite produced by gut lactic acid bacteria, enhances energy metabolism by activation of TRPV1. FASEB J 31 (11):5036–5048. https://doi.org/10.1096/fj.201700151R

Kirkham TC, Williams CM, Fezza F, Di Marzo V (2002) Endocannabinoid levels in rat limbic forebrain and hypothalamus in relation to fasting, feeding and satiation: stimulation of eating by 2-arachidonoyl glycerol. Br J Pharmacol 136(4):550–557. https://doi.org/10.1038/sj.bjp.0704767

Kishino S, Takeuchi M, Park SB, Hirata A, Kitamura N, Kunisawa J, Kiyono H, Iwamoto R, Isobe Y, Arita M, Arai H, Ueda K, Shima J, Takahashi S, Yokozeki K, Shimizu S, Ogawa J (2013) Polyunsaturated fatty acid saturation by gut lactic acid bacteria affecting host lipid composition. Proc Natl Acad Sci U S A 110(44):17808–17813. https://doi.org/10.1073/pnas.1312937110

Konkel A, Schunck WH (2011) Role of cytochrome P450 enzymes in the bioactivation of polyunsaturated fatty acids. Biochim Biophys Acta 1814(1):210–222. https://doi.org/10.1016/j.bbapap.2010.09.009

Krott LM, Piscitelli F, Heine M, Borrino S, Scheja L, Silvestri C, Heeren J, Di Marzo V (2016) Endocannabinoid regulation in white and brown adipose tissue following thermogenic activation. J Lipid Res 57(3):464–473. https://doi.org/10.1194/jlr.M065227

Kuhn H, Banthiya S, van Leyen K (2015) Mammalian lipoxygenases and their biological relevance. Biochim Biophys Acta 1851(4):308–330. https://doi.org/10.1016/j.bbalip.2014.10.002

Lazarus M, Yoshida K, Coppari R, Bass CE, Mochizuki T, Lowell BB, Saper CB (2007) EP3 prostaglandin receptors in the median preoptic nucleus are critical for fever responses. Nat Neurosci 10(9):1131–1133. https://doi.org/10.1038/nn1949

Le Faouder P, Baillif V, Spreadbury I, Motta JP, Rousset P, Chene G, Guigne C, Terce F, Vanner S, Vergnolle N, Bertrand-Michel J, Dubourdeau M, Cenac N (2013) LC-MS/MS method for rapid and concomitant quantification of pro-inflammatory and pro-resolving polyunsaturated fatty acid metabolites. J Chromatogr B Analyt Technol Biomed Life Sci 932:123–133. https://doi.org/10.1016/j.jchromb.2013.06.014

Lecka-Czernik B, Moerman EJ, Grant DF, Lehmann JM, Manolagas SC, Jilka RL (2002) Divergent effects of selective peroxisome proliferator-activated receptor-gamma 2 ligands on adipocyte versus osteoblast differentiation. Endocrinology 143(6):2376–2384. https://doi.org/10.1210/endo.143.6.8834

Lee YH, Petkova AP, Mottillo EP, Granneman JG (2012) In vivo identification of bipotential adipocyte progenitors recruited by beta3-adrenoceptor activation and high-fat feeding. Cell Metab 15(4):480–491. https://doi.org/10.1016/j.cmet.2012.03.009

Lee YH, Petkova AP, Granneman JG (2013) Identification of an adipogenic niche for adipose tissue remodeling and restoration. Cell Metab 18(3):355–367. https://doi.org/10.1016/j.cmet.2013.08.003

Lee YH, Thacker RI, Hall BE, Kong R, Granneman JG (2014) Exploring the activated adipogenic niche: interactions of macrophages and adipocyte progenitors. Cell Cycle 13(2):184–190. https://doi.org/10.4161/cc.27647

Lee YH, Kim SN, Kwon HJ, Maddipati KR, Granneman JG (2016) Adipogenic role of alternatively activated macrophages in beta-adrenergic remodeling of white adipose tissue. Am J Physiol Regul Integr Comp Physiol 310(1):R55–R65. https://doi.org/10.1152/ajpregu.00355.2015

Li Y, Fromme T, Schweizer S, Schottl T, Klingenspor M (2014) Taking control over intracellular fatty acid levels is essential for the analysis of thermogenic function in cultured primary brown and brite/beige adipocytes. EMBO Rep 15(10):1069–1076. https://doi.org/10.15252/embr.201438775

Liu L, Clipstone NA (2007) Prostaglandin F2alpha inhibits adipocyte differentiation via a G alpha q-calcium-calcineurin-dependent signaling pathway. J Cell Biochem 100(1):161–173. https://doi.org/10.1002/jcb.21044

Lynes MD, Leiria LO, Lundh M, Bartelt A, Shamsi F, Huang TL, Takahashi H, Hirshman MF, Schlein C, Lee A, Baer LA, May FJ, Gao F, Narain NR, Chen EY, Kiebish MA, Cypess AM, Bluher M, Goodyear LJ, Hotamisligil GS, Stanford KI, Tseng YH (2017) The cold-induced lipokine 12,13-diHOME promotes fatty acid transport into brown adipose tissue. Nat Med 23 (5):631–637. https://doi.org/10.1038/nm.4297

Maclouf J, Kindahl H, Granstrom E, Samuelsson B (1980) Interactions of prostaglandin H2 and thromboxane A2 with human serum albumin. Eur J Biochem 109(2):561–566

Madsen L, Pedersen LM, Lillefosse HH, Fjaere E, Bronstad I, Hao Q, Petersen RK, Hallenborg P, Ma T, De Matteis R, Araujo P, Mercader J, Bonet ML, Hansen JB, Cannon B, Nedergaard J, Wang J, Cinti S, Voshol P, Doskeland SO, Kristiansen K (2010) UCP1 induction during recruitment of brown adipocytes in white adipose tissue is dependent on cyclooxygenase activity. PLoS One 5(6):e11391. https://doi.org/10.1371/journal.pone.0011391

van Marken Lichtenbelt WD, Vanhommerig JW, Smulders NM, Drossaerts JM, Kemerink GJ, Bouvy ND, Schrauwen P, Teule GJ (2009) Cold-activated brown adipose tissue in healthy men. N Engl J Med 360(15):1500–1508. https://doi.org/10.1056/NEJMoa0808718

Masoodi M, Eiden M, Koulman A, Spaner D, Volmer DA (2010) Comprehensive lipidomics analysis of bioactive lipids in complex regulatory networks. Anal Chem 82(19):8176–8185. https://doi.org/10.1021/ac1015563

Massiera F, Saint-Marc P, Seydoux J, Murata T, Kobayashi T, Narumiya S, Guesnet P, Amri EZ, Negrel R, Ailhaud G (2003) Arachidonic acid and prostacyclin signaling promote adipose tissue development: a human health concern? J Lipid Res 44(2):271–279. https://doi.org/10.1194/jlr.M200346-JLR200

Matias I, Di Marzo V (2007) Endocannabinoids and the control of energy balance. Trends Endocrinol Metab 18(1):27–37. https://doi.org/10.1016/j.tem.2006.11.006

Mesaros C, Blair IA (2012) Targeted chiral analysis of bioactive arachidonic acid metabolites using liquid-chromatography-mass spectrometry. Meta 2(2):337–365. https://doi.org/10.3390/metabo2020337

Miller DK, Gillard JW, Vickers PJ, Sadowski S, Leveille C, Mancini JA, Charleson P, Dixon RA, Ford-Hutchinson AW, Fortin R et al (1990) Identification and isolation of a membrane protein necessary for leukotriene production. Nature 343(6255):278–281. https://doi.org/10.1038/343278a0

Moran JH, Weise R, Schnellmann RG, Freeman JP, Grant DF (1997) Cytotoxicity of linoleic acid diols to renal proximal tubular cells. Toxicol Appl Pharmacol 146(1):53–59. https://doi.org/10.1006/taap.1997.8197

Moran-Ramos S, Lopez-Contreras BE, Canizales-Quinteros S (2017) Gut microbiota in obesity and metabolic abnormalities: a matter of composition or functionality? Arch Med Res. https://doi.org/10.1016/j.arcmed.2017.11.003

Morrison SF (2016) Central control of body temperature. F1000Res 5. doi:https://doi.org/10.12688/f1000research.7958.1

Morrison SF, Madden CJ (2014) Central nervous system regulation of brown adipose tissue. Compr Physiol 4(4):1677–1713. https://doi.org/10.1002/cphy.c140013

Murphy RC, Barkley RM, Zemski Berry K, Hankin J, Harrison K, Johnson C, Krank J, McAnoy A, Uhlson C, Zarini S (2005) Electrospray ionization and tandem mass spectrometry of eicosanoids. Anal Biochem 346(1):1–42. https://doi.org/10.1016/j.ab.2005.04.042

Nagai M, Tuchiya K, Kojima H (1996) Prostaglandin E2 increases the calcium concentration in rat brown adipocytes and their consumption of oxygen. Prostaglandins 51(6):377–386

Nagy L, Tontonoz P, Alvarez JG, Chen H, Evans RM (1998) Oxidized LDL regulates macrophage gene expression through ligand activation of PPARgamma. Cell 93(2):229–240

Nakamura K, Matsumura K, Kaneko T, Kobayashi S, Katoh H, Negishi M (2002) The rostral raphe pallidus nucleus mediates pyrogenic transmission from the preoptic area. J Neurosci 22 (11):4600–4610

Negrel R, Gaillard D, Ailhaud G (1989) Prostacyclin as a potent effector of adipose-cell differentiation. Biochem J 257(2):399–405

Nelson DR, Zeldin DC, Hoffman SM, Maltais LJ, Wain HM, Nebert DW (2004) Comparison of cytochrome P450 (CYP) genes from the mouse and human genomes, including nomenclature recommendations for genes, pseudogenes and alternative-splice variants. Pharmacogenetics 14 (1):1–18

Nguyen KD, Qiu Y, Cui X, Goh YP, Mwangi J, David T, Mukundan L, Brombacher F, Locksley RM, Chawla A (2011) Alternatively activated macrophages produce catecholamines to sustain adaptive thermogenesis. Nature 480(7375):104–108. https://doi.org/10.1038/nature10653

Obinata H, Hattori T, Nakane S, Tatei K, Izumi T (2005) Identification of 9-hydroxyoctadecadienoic acid and other oxidized free fatty acids as ligands of the G protein-coupled receptor G2A. J Biol Chem 280(49):40676–40683. https://doi.org/10.1074/jbc.M507787200

Okla M, Kim J, Koehler K, Chung S (2017) Dietary factors promoting brown and beige fat development and thermogenesis. Adv Nutr 8(3):473–483. https://doi.org/10.3945/an.116.014332

Ostermann AI, Schebb NH (2017) Effects of omega-3 fatty acid supplementation on the pattern of oxylipins: a short review about the modulation of hydroxy-, dihydroxy-, and epoxy-fatty acids. Food Funct 8(7):2355–2367. https://doi.org/10.1039/c7fo00403f

Ostermann AI, Waindok P, Schmidt MJ, Chiu CY, Smyl C, Rohwer N, Weylandt KH, Schebb NH (2017) Modulation of the endogenous omega-3 fatty acid and oxylipin profile in vivo – a comparison of the fat-1 transgenic mouse with C57BL/6 wildtype mice on an omega-3 fatty acid enriched diet. PLoS One 12(9):e0184470. https://doi.org/10.1371/journal.pone.0184470

Patwardhan AM, Scotland PE, Akopian AN, Hargreaves KM (2009) Activation of TRPV1 in the spinal cord by oxidized linoleic acid metabolites contributes to inflammatory hyperalgesia. Proc Natl Acad Sci U S A 106(44):18820–18824. https://doi.org/10.1073/pnas.0905415106

Perwitz N, Fasshauer M, Klein J (2006) Cannabinoid receptor signaling directly inhibits thermogenesis and alters expression of adiponectin and visfatin. Horm Metab Res 38(5):356–358. https://doi.org/10.1055/s-2006-925401

Petrovic N, Walden TB, Shabalina IG, Timmons JA, Cannon B, Nedergaard J (2010) Chronic peroxisome proliferator-activated receptor gamma (PPARgamma) activation of epididymally derived white adipocyte cultures reveals a population of thermogenically competent, UCP1-containing adipocytes molecularly distinct from classic brown adipocytes. J Biol Chem 285 (10):7153–7164. https://doi.org/10.1074/jbc.M109.053942

Pisani DF, Djedaini M, Beranger GE, Elabd C, Scheideler M, Ailhaud G, Amri EZ (2011) Differentiation of human adipose-derived stem cells into "Brite" (Brown-in-White) adipocytes. Front Endocrinol (Lausanne) 2:87. https://doi.org/10.3389/fendo.2011.00087

Pisani DF, Ghandour RA, Beranger GE, Le Faouder P, Chambard JC, Giroud M, Vegiopoulos A, Djedaini M, Bertrand-Michel J, Tauc M, Herzig S, Langin D, Ailhaud G, Duranton C, Amri EZ (2014) The omega6-fatty acid, arachidonic acid, regulates the conversion of white to brite adipocyte through a prostaglandin/calcium mediated pathway. Mol Metab 3(9):834–847. https://doi.org/10.1016/j.molmet.2014.09.003

Plourde M, Cunnane SC (2007) Extremely limited synthesis of long chain polyunsaturates in adults: implications for their dietary essentiality and use as supplements. Appl Physiol Nutr Metab 32 (4):619–634. https://doi.org/10.1139/H07-034

Pradhan RN, Zachara M, Deplancke B (2017) A systems perspective on brown adipogenesis and metabolic activation. Obes Rev 18(Suppl 1):65–81. https://doi.org/10.1111/obr.12512

Prasain JK, Wilson L, Hoang HD, Moore R, Miller MA (2015) Comparative lipidomics of Caenorhabditis elegans metabolic disease models by SWATH non-targeted tandem mass spectrometry. Metabolites 5(4):677–696. https://doi.org/10.3390/metabo5040677

Qiu Y, Nguyen KD, Odegaard JI, Cui X, Tian X, Locksley RM, Palmiter RD, Chawla A (2014) Eosinophils and type 2 cytokine signaling in macrophages orchestrate development of functional beige fat. Cell 157(6):1292–1308. https://doi.org/10.1016/j.cell.2014.03.066

Quarta C, Bellocchio L, Mancini G, Mazza R, Cervino C, Braulke LJ, Fekete C, Latorre R, Nanni C, Bucci M, Clemens LE, Heldmaier G, Watanabe M, Leste-Lassere T, Maitre M, Tedesco L, Fanelli F, Reuss S, Klaus S, Srivastava RK, Monory K, Valerio A, Grandis A, De

Giorgio R, Pasquali R, Nisoli E, Cota D, Lutz B, Marsicano G, Pagotto U (2010) CB(1) signaling in forebrain and sympathetic neurons is a key determinant of endocannabinoid actions on energy balance. Cell Metab 11(4):273–285. https://doi.org/10.1016/j.cmet.2010.02.015

Quesada-Lopez T, Cereijo R, Turatsinze JV, Planavila A, Cairo M, Gavalda-Navarro A, Peyrou M, Moure R, Iglesias R, Giralt M, Eizirik DL, Villarroya F (2016) The lipid sensor GPR120 promotes brown fat activation and FGF21 release from adipocytes. Nat Commun 7:13479. https://doi.org/10.1038/ncomms13479

Raz A (1972a) Interaction of prostaglandins with blood plasma proteins. Comparative binding of prostaglandins A 2, F 2 and E 2 to human plasma proteins. Biochem J 130(2):631–636

Raz A (1972b) Interaction of prostaglandins with blood plasma proteins. I. Binding of prostaglandin E 2 to human plasma proteins and its effect on the physiological activity of prostaglandin E 2 in vitro and in vivo. Biochim Biophys Acta 280(4):602–613

Reginato MJ, Krakow SL, Bailey ST, Lazar MA (1998) Prostaglandins promote and block adipogenesis through opposing effects on peroxisome proliferator-activated receptor gamma. J Biol Chem 273(4):1855–1858

Richard D, Guesdon B, Timofeeva E (2009) The brain endocannabinoid system in the regulation of energy balance. Best Pract Res Clin Endocrinol Metab 23(1):17–32. https://doi.org/10.1016/j.beem.2008.10.007

Riederer M, Lechleitner M, Kofeler H, Frank S (2017) Reduced expression of adipose triglyceride lipase decreases arachidonic acid release and prostacyclin secretion in human aortic endothelial cells. Arch Physiol Biochem 123(4):249–253. https://doi.org/10.1080/13813455.2017.1309052

Rozenfeld R, Devi LA (2008) Regulation of CB1 cannabinoid receptor trafficking by the adaptor protein AP-3. FASEB J 22(7):2311–2322. https://doi.org/10.1096/fj.07-102731

Ruiz de Azua I, Mancini G, Srivastava RK, Rey AA, Cardinal P, Tedesco L, Zingaretti CM, Sassmann A, Quarta C, Schwitter C, Conrad A, Wettschureck N, Vemuri VK, Makriyannis A, Hartwig J, Mendez-Lago M, Bindila L, Monory K, Giordano A, Cinti S, Marsicano G, Offermanns S, Nisoli E, Pagotto U, Cota D, Lutz B (2017) Adipocyte cannabinoid receptor CB1 regulates energy homeostasis and alternatively activated macrophages. J Clin Invest 127(11):4148–4162. https://doi.org/10.1172/JCI83626

Saito M, Okamatsu-Ogura Y, Matsushita M, Watanabe K, Yoneshiro T, Nio-Kobayashi J, Iwanaga T, Miyagawa M, Kameya T, Nakada K, Kawai Y, Tsujisaki M (2009) High incidence of metabolically active brown adipose tissue in healthy adult humans: effects of cold exposure and adiposity. Diabetes 58(7):1526–1531. https://doi.org/10.2337/db09-0530

Scammell TE, Elmquist JK, Griffin JD, Saper CB (1996) Ventromedial preoptic prostaglandin E2 activates fever-producing autonomic pathways. J Neurosci 16(19):6246–6254

Schebb NH, Ostermann AI, Yang J, Hammock BD, Hahn A, Schuchardt JP (2014) Comparison of the effects of long-chain omega-3 fatty acid supplementation on plasma levels of free and esterified oxylipins. Prostaglandins Other Lipid Mediat 113-115:21–29. https://doi.org/10.1016/j.prostaglandins.2014.05.002

Schlager S, Goeritzer M, Jandl K, Frei R, Vujic N, Kolb D, Strohmaier H, Dorow J, Eichmann TO, Rosenberger A, Wolfler A, Lass A, Kershaw EE, Ceglarek U, Dichlberger A, Heinemann A, Kratky D (2015) Adipose triglyceride lipase acts on neutrophil lipid droplets to regulate substrate availability for lipid mediator synthesis. J Leukoc Biol 98(5):837–850. https://doi.org/10.1189/jlb.3A0515-206R

Schuchardt JP, Schmidt S, Kressel G, Dong H, Willenberg I, Hammock BD, Hahn A, Schebb NH (2013) Comparison of free serum oxylipin concentrations in hyper- vs. normolipidemic men. Prostaglandins Leukot Essent Fatty Acids 89(1):19–29. https://doi.org/10.1016/j.plefa.2013.04.001

Schuster VL, Chi Y, Lu R (2015) The prostaglandin transporter: eicosanoid reuptake, control of signaling, and development of high-affinity inhibitors as drug candidates. Trans Am Clin Climatol Assoc 126:248–257

Seale P, Kajimura S, Spiegelman BM (2009) Transcriptional control of brown adipocyte development and physiological function – of mice and men. Genes Dev 23(7):788–797. https://doi.org/10.1101/gad.1779209

Shearer GC, Newman JW (2008) Lipoprotein lipase releases esterified oxylipins from very low-density lipoproteins. Prostaglandins Leukot Essent Fatty Acids 79(6):215–222. https://doi.org/10.1016/j.plefa.2008.09.023

Shearer GC, Borkowski K, Puumala SL, Harris WS, Pedersen TL, Newman JW (2018) Abnormal lipoprotein oxylipins in metabolic syndrome and partial correction by omega-3 fatty acids. Prostaglandins Leukot Essent Fatty Acids 128:1–10. https://doi.org/10.1016/j.plefa.2017.10.006

da-Silva WS, Ribich S, Arrojo e Drigo R, Castillo M, Patti ME, Bianco AC (2011) The chemical chaperones tauroursodeoxycholic and 4-phenylbutyric acid accelerate thyroid hormone activation and energy expenditure. FEBS Lett 585(3):539–544. https://doi.org/10.1016/j.febslet.2010.12.044

Silvestri C, Di Marzo V (2013) The endocannabinoid system in energy homeostasis and the etiopathology of metabolic disorders. Cell Metab 17(4):475–490. https://doi.org/10.1016/j.cmet.2013.03.001

Simopoulos AP (2001) Evolutionary aspects of diet and essential fatty acids. World Rev Nutr Diet 88:18–27

Simopoulos AP (2016) An increase in the Omega-6/Omega-3 fatty acid ratio increases the risk for obesity. Nutrients 8(3):128. https://doi.org/10.3390/nu8030128

Sisemore MF, Zheng J, Yang JC, Thompson DA, Plopper CG, Cortopassi GA, Hammock BD (2001) Cellular characterization of leukotoxin diol-induced mitochondrial dysfunction. Arch Biochem Biophys 392(1):32–37. https://doi.org/10.1006/abbi.2001.2434

Strassburg K, Huijbrechts AM, Kortekaas KA, Lindeman JH, Pedersen TL, Dane A, Berger R, Brenkman A, Hankemeier T, van Duynhoven J, Kalkhoven E, Newman JW, Vreeken RJ (2012) Quantitative profiling of oxylipins through comprehensive LC-MS/MS analysis: application in cardiac surgery. Anal Bioanal Chem 404(5):1413–1426. https://doi.org/10.1007/s00216-012-6226-x

Suarez J, Rivera P, Arrabal S, Crespillo A, Serrano A, Baixeras E, Pavon FJ, Cifuentes M, Nogueiras R, Ballesteros J, Dieguez C, Rodriguez de Fonseca F (2014) Oleoylethanolamide enhances beta-adrenergic-mediated thermogenesis and white-to-brown adipocyte phenotype in epididymal white adipose tissue in rat. Dis Model Mech 7(1):129–141. https://doi.org/10.1242/dmm.013110

Teodoro JS, Zouhar P, Flachs P, Bardova K, Janovska P, Gomes AP, Duarte FV, Varela AT, Rolo AP, Palmeira CM, Kopecky J (2014) Enhancement of brown fat thermogenesis using chenodeoxycholic acid in mice. Int J Obes (Lond) 38(8):1027–1034. https://doi.org/10.1038/ijo.2013.230

Ueno T, Fujimori K (2011) Novel suppression mechanism operating in early phase of adipogenesis by positive feedback loop for enhancement of cyclooxygenase-2 expression through prostaglandin F2alpha receptor mediated activation of MEK/ERK-CREB cascade. FEBS J 278(16):2901–2912. https://doi.org/10.1111/j.1742-4658.2011.08213.x

Uozumi N, Shimizu T (2002) Roles for cytosolic phospholipase A2alpha as revealed by gene-targeted mice. Prostaglandins Other Lipid Mediat 68-69:59–69

Ushikubi F, Segi E, Sugimoto Y, Murata T, Matsuoka T, Kobayashi T, Hizaki H, Tuboi K, Katsuyama M, Ichikawa A, Tanaka T, Yoshida N, Narumiya S (1998) Impaired febrile response in mice lacking the prostaglandin E receptor subtype EP3. Nature 395(6699):281–284. https://doi.org/10.1038/26233

Vargas-Castillo A, Fuentes-Romero R, Rodriguez-Lopez LA, Torres N, Tovar AR (2017) Understanding the biology of thermogenic fat: is browning a new approach to the treatment of obesity? Arch Med Res 48(5):401–413. https://doi.org/10.1016/j.arcmed.2017.10.002

Vassaux G, Gaillard D, Darimont C, Ailhaud G, Negrel R (1992) Differential response of preadipocytes and adipocytes to prostacyclin and prostaglandin E2: physiological implications. Endocrinology 131(5):2393–2398. https://doi.org/10.1210/endo.131.5.1330499

Vegiopoulos A, Muller-Decker K, Strzoda D, Schmitt I, Chichelnitskiy E, Ostertag A, Berriel Diaz M, Rozman J, Hrabe de Angelis M, Nusing RM, Meyer CW, Wahli W, Klingenspor M, Herzig S (2010) Cyclooxygenase-2 controls energy homeostasis in mice by de novo recruitment of brown adipocytes. Science 328(5982):1158–1161. https://doi.org/10.1126/science.1186034

Velazquez-Villegas LA, Perino A, Lemos V, Zietak M, Nomura M, Pols TWH, Schoonjans K (2018) TGR5 signalling promotes mitochondrial fission and beige remodelling of white adipose tissue. Nat Commun 9(1):245. https://doi.org/10.1038/s41467-017-02068-0

Vijgen GH, Bouvy ND, Teule GJ, Brans B, Schrauwen P, van Marken Lichtenbelt WD (2011) Brown adipose tissue in morbidly obese subjects. PLoS One 6(2):e17247. https://doi.org/10.1371/journal.pone.0017247

Volat FE, Pointud JC, Pastel E, Morio B, Sion B, Hamard G, Guichardant M, Colas R, Lefrancois-Martinez AM, Martinez A (2012) Depressed levels of prostaglandin F2alpha in mice lacking Akr1b7 increase basal adiposity and predispose to diet-induced obesity. Diabetes 61 (11):2796–2806. https://doi.org/10.2337/db11-1297

Watanabe M, Houten SM, Mataki C, Christoffolete MA, Kim BW, Sato H, Messaddeq N, Harney JW, Ezaki O, Kodama T, Schoonjans K, Bianco AC, Auwerx J (2006) Bile acids induce energy expenditure by promoting intracellular thyroid hormone activation. Nature 439(7075):484–489. https://doi.org/10.1038/nature04330

Watanabe M, Horai Y, Houten SM, Morimoto K, Sugizaki T, Arita E, Mataki C, Sato H, Tanigawara Y, Schoonjans K, Itoh H, Auwerx J (2011) Lowering bile acid pool size with a synthetic farnesoid X receptor (FXR) agonist induces obesity and diabetes through reduced energy expenditure. J Biol Chem 286(30):26913–26920. https://doi.org/10.1074/jbc.M111.248203

Watanabe M, Morimoto K, Houten SM, Kaneko-Iwasaki N, Sugizaki T, Horai Y, Mataki C, Sato H, Murahashi K, Arita E, Schoonjans K, Suzuki T, Itoh H, Auwerx J (2012) Bile acid binding resin improves metabolic control through the induction of energy expenditure. PLoS One 7(8): e38286. https://doi.org/10.1371/journal.pone.0038286

Whittle BJ, Moncada S, Whiting F, Vane JR (1980) Carbacyclin – a potent stable prostacyclin analogue for the inhibition of platelet aggregation. Prostaglandins 19(4):605–627

Willenberg I, Ostermann AI, Schebb NH (2015) Targeted metabolomics of the arachidonic acid cascade: current state and challenges of LC-MS analysis of oxylipins. Anal Bioanal Chem 407 (10):2675–2683. https://doi.org/10.1007/s00216-014-8369-4

Xue Y, Xu X, Zhang XQ, Farokhzad OC, Langer R (2016) Preventing diet-induced obesity in mice by adipose tissue transformation and angiogenesis using targeted nanoparticles. Proc Natl Acad Sci U S A 113(20):5552–5557. https://doi.org/10.1073/pnas.1603840113

Ying F, Cai Y, Cai Y, Wang Y, Tang EHC (2017) Prostaglandin E receptor subtype 4 regulates lipid droplet size and mitochondrial activity in murine subcutaneous white adipose tissue. FASEB J 31(9):4023–4036. https://doi.org/10.1096/fj.201700191R

Zhang P, Meng X, Li D, Calderone R, Mao D, Sui B (2017) Commensal homeostasis of gut microbiota-host for the impact of obesity. Front Physiol 8:1122. https://doi.org/10.3389/fphys.2017.01122

Zhao M, Chen X (2014) Eicosapentaenoic acid promotes thermogenic and fatty acid storage capacity in mouse subcutaneous adipocytes. Biochem Biophys Res Commun 450 (4):1446–1451. https://doi.org/10.1016/j.bbrc.2014.07.010

Zhou SF, Wang LL, Di YM, Xue CC, Duan W, Li CG, Li Y (2008) Substrates and inhibitors of human multidrug resistance associated proteins and the implications in drug development. Curr Med Chem 15(20):1981–2039

Zietak M, Kozak LP (2016) Bile acids induce uncoupling protein 1-dependent thermogenesis and stimulate energy expenditure at thermoneutrality in mice. Am J Physiol Endocrinol Metab 310 (5):E346–E354. https://doi.org/10.1152/ajpendo.00485.2015

Regulatory Small and Long Noncoding RNAs in Brite/Brown Adipose Tissue

Marcel Scheideler

Contents

1 Introduction 216
2 Noncoding RNAs 217
3 MicroRNAs in Brite/Brown Adipose Tissue 219
3.1 miRNAs Targeting PPARγ and PRDM16 in Brite/Brown Adipogenesis 219
3.2 miRNAs Targeting the cAMP-PKA-CREB and p38-ATF2 Signaling Pathways in BAT 221
3.3 miRNAs Ultimately Modulating C/EBPβ Activity 222
3.4 miRNAs Modulating the Transcriptional Coactivator PGC-1α 223
3.5 miRNA Targeting ADAM17 and PTEN 223
3.6 miRNA Targeting the β-Secretase BACE1 224
3.7 miRNAs with Impact on Brite/Brown Adipocyte Function and Formation but with Unknown Direct Targets 224
3.8 miRNAs as Serum Biomarker 225
4 Long ncRNAs in Brite/Brown Adipose Tissue 225
4.1 Long ncRNA BLNC1 225
4.2 Long ncRNA BATE1 227
4.3 LncRNA BATE10 227
4.4 Long ncRNAs in Circulation as Potential Biomarkers 228
5 Outlook 228
References 233

M. Scheideler (✉)
Institute for Diabetes and Cancer (IDC), Helmholtz Zentrum München, German Research Center for Environmental Health, Neuherberg, Germany

Joint Heidelberg-IDC Translational Diabetes Program, Heidelberg University Hospital, Heidelberg, Germany

German Center for Diabetes Research (DZD), Neuherberg, Germany
e-mail: marcel.scheideler@helmholtz-muenchen.de

A. Pfeifer et al. (eds.), *Brown Adipose Tissue*,
Handbook of Experimental Pharmacology 251, https://doi.org/10.1007/164_2018_123

Abstract

Brite/brown adipose tissue (BAT) is a thermogenic tissue able to dissipate energy via non-shivering thermogenesis. It is naturally activated by cold and has been demonstrated to increase thermogenic capacity, elevate energy expenditure, and to ultimately contribute to fat mass reduction. Thus, it emerges as novel therapeutic concept for pharmacological intervention in obesity and other metabolic disorders. Therefore, the comprehensive understanding of the regulatory network in thermogenic adipocytes is in demand.

The surprising findings that (1) all human protein-coding genes make up not more than 2% of our genome, (2) organismal complexity goes well along with the percentage of nonprotein-coding sequences, and that (3) three quarters of our genome are pervasively transcribed, provide evidence that noncoding RNAs (ncRNAs) are not junk, but a significant and even predominant part of our transcriptome representing a treasure chest worth retrieving regulatory determinants in biological processes and diseases.

In this chapter, the impact of regulatory small and long ncRNAs (lncRNAs) in particular microRNAs and lncRNAs on BAT formation and metabolic function and their involvement in physiological and pathological conditions has been reviewed.

Keywords

Brite/brown thermogenic adipocyte · Brown adipose tissue · Long noncoding RNA · Metabolism · microRNA · Noncoding RNA · Obesity · Regulatory RNA

1 Introduction

The adipose organ is an important player in the regulation of whole-body energy homeostasis, including fatty acid and glucose metabolism, and contributes to many of an organism's pivotal requirements of survival: fuel for metabolism, immune responses, lactation, and thermogenesis (Cinti 2012). The adipose organ can be divided into two distinct types of adipose tissues, white (WAT) and brown (BAT) adipose tissue: WAT is specialized for the storage and release of chemical energy (Cinti 2012; Cohen and Spiegelman 2016), while BAT is able to dissipate energy in the form of heat (thermogenesis). Interestingly, WAT and BAT do not display clear anatomical boundaries; as in rodents and humans, islands of brown-like adipocytes emerge within WAT depots after cold or β-adrenergic receptor stimulation. These adipocytes, termed "brite" (brown-in-white) or "beige" adipocytes, differ by embryonic origin from genuine brown adipocytes but are functional, i.e., thermogenically active (Petrovic et al. 2010; Wu et al. 2012). Brite/brown adipocytes are endowed with high capacity of glucose and lipid oxidation thus making the brite/brown adipose tissue a promising target for lowering plasma levels of glucose and fatty acids thus diminishing the risks of overweight, obesity, and follow-up complications (Nedergaard et al. 2011). Indeed, brite/brown adipocyte formation and activation emerge as a novel therapeutic concept for pharmacological intervention in obesity and other metabolic disorders (Nedergaard and Cannon 2010), and the identification of regulatory factors and drugs able to initiate the formation and activation of thermogenic adipocytes, particularly in humans, is in demand and constitutes a highly active research field.

In this review, I tempt to summarize recent findings on a novel class of regulatory determinants in BAT biology with impact on metabolism and disease, noncoding RNAs (ncRNAs).

2 Noncoding RNAs

For a long time, research on molecular mechanisms has been centered on protein-coding genes. However, efforts investigating our entire genome brought up surprising discoveries. Firstly, the Human Genome Project revealed much less protein-coding genes than previously expected, and all 20,500 protein-coding genes known in human make up not more than 2% of our genome, raising questions for the majority of 98% of our genome (Lander et al. 2001; Venter et al. 2001). Secondly, even single cell organisms such as *Tetrahymena thermophila* exceed the number of human protein-coding genes, indicating that this is not the determinant of organismal complexity (Taft et al. 2007). However, organismal complexity goes well along with the increasing relative amount of nonprotein-coding sequences, suggesting indeed a function for the noncoding part of the genome in higher organisms (Taft et al. 2007). Thirdly, the ENCODE project intriguingly demonstrated transcriptional activity for 74.7% of the human genome, with many novel nonprotein-coding transcripts of small and long ncRNA (Birney et al. 2007; Djebali et al. 2012; Dogini et al. 2014). Altogether, these findings indicate that the noncoding part of the genome is not "junk" DNA, but a significant and even predominant part of our genome representing a promising treasure chest worth retrieving regulatory determinants in both biological processes and diseases.

ncRNAs are defined as RNA transcripts that do not encode a protein and are divided into two primary categories: small ncRNAs (<200 nt) and long ncRNAs (lncRNAs; >200 nt). Some small ncRNAs are housekeeping RNAs, such as tRNA, snRNA, and snoRNA, which are crucial for cell physiology, while others, such as microRNAs (miRNAs) and piRNAs, are associated with protein-coding gene regulation (Fig. 1). Also lncRNAs comprise housekeeping RNAs, such as ribosomal RNAs (rRNAs), and regulatory elements such as lncRNAs, including antisense RNAs (AS-RNAs) and enhancer RNAs (eRNAs) (Ponting et al. 2009; Wilusz et al. 2009; Vance and Ponting 2014; Sun and Kraus 2015; Iyer et al. 2015).

miRNAs were first discovered in 1993 and are to date the most extensively studied class of ncRNAs, with more than 2,500 candidates in human and 1,900 candidates in mouse (Lee et al. 1993). MiRNAs are endogenous, single-stranded, noncoding, small RNAs with a length of 21–22 nucleotides that are involved in regulating gene expression in the cytoplasm by incorporating into the RNA-induced silencing complex (RISC), and preferential binding to specific sequences in the 3′-UTR of their target mRNAs suppress translation or induce mRNA degradation (Filipowicz et al. 2008).

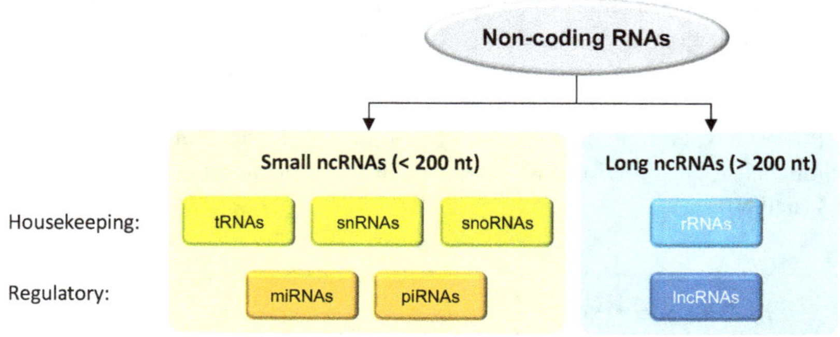

Fig. 1 Classification of noncoding RNAs (ncRNAs): ncRNAs with less than 200 nucleotides (nt) in length are small ncRNAs, while noncoding transcripts with a length of more than 200 nt are defined as long ncRNAs (lncRNAs). *tRNAs* transfer RNAs, *snRNAs* small nuclear RNAs, *snoRNAs* small nucleolar RNAs, *miRNAs* microRNAs, *piRNAs* piwi-associated RNAs, *rRNAs* ribosomal RNAs, *lncRNAs* long non-coding RNAs

On the other hand, regulatory lncRNAs were already discovered in 1990, with now more than 58,000 loci found in human (Brannan et al. 1990; Iyer et al. 2015). They are found in the cytoplasm as well in the nucleus where they bind to enhancer regions, promoter sequences, 5′-UTRs, exons, introns, intragenic regions, intergenic sequences, antisense sequences, and 3′-UTRs. The regulatory role of lncRNAs is directly dependent on their cellular localization. In the cytoplasm, lncRNAs can act as molecular decoys for proteins and microRNAs, while in the nucleus, lncRNAs have been shown to perform as transcriptional activators or inhibitors in *cis*, i.e., regulating neighboring genes, or in *trans*, i.e., regulating genes from other regions or chromosomes (Zhang et al. 2014). However, in contrast to miRNAs, lncRNAs are poorly conserved between species and are highly tissue-specific, which makes them specifically and tightly regulated, even though they are found at lower abundance compared to mRNAs (Babak et al. 2005; Mercer et al. 2008; Guttman et al. 2009; Ramsköld et al. 2009; Derrien et al. 2012).

Over the last decades, it has been unraveled that small and lncRNAs govern the formation and function of tissues and organs, including the adipose organ. For the characterization of miRNAs in the adipose organ, several mouse models with adipose-specific knockout of key regulators of miRNA biogenesis were generated. Importantly, fat-selective inactivation of Dicer, an essential factor in miRNA biogenesis, resulted in mice which were almost devoid of WAT (Mudhasani et al. 2010, 2011). Moreover, adipose-specific ablation of Dicer or DGCR8 in mice, another crucial determinant in miRNA biogenesis, displayed enlarged but pale interscapular BAT, decreased expression of genes characteristic of brown fat, and intolerance to cold exposure (Mori et al. 2014; Kim et al. 2014). These findings suggest a pivotal role of miRNAs in the formation of white, brite, and brown adipocytes. In recent years, an explosion in the identification of ncRNAs and their functions was observed, yet one only began to understand the complexity of this new regulatory RNA world, in particular how ncRNAs control various aspects of gene expression and their involvement in diseases (Prasanth and Spector 2007; de Almeida et al. 2016). The impact of miRNAs on diseases is acknowledged by their deployment as biomarkers, drugs, and/or drug targets, with first candidates in clinical trials phase 1 and 2 (Wahid et al. 2010, 2014; van Rooij et al. 2012; Hydbring and Badalian-Very 2013; Christopher et al. 2016), while the impact of lncRNAs as potential diagnostic markers and/or valuable therapeutic targets for diseases is just emerging. Here, we review the regulatory impact of miRNAs and lncRNAs on BAT biology and thermogenic capacity and their involvement in metabolic disease.

3 MicroRNAs in Brite/Brown Adipose Tissue

Brite/brown adipocyte differentiation is tightly governed in a coordinated manner using various regulatory pathways which finally activate several transcription factors and coactivators (Xue et al. 2005; Kajimura et al. 2010), such as peroxisome proliferator-activated receptor α (PPARα) and γ (PPARγ), PPARγ-coactivators-1 [PGC-1α (Puigserver et al. 1998; Rohas et al. 2007) and PGC-1β (Villena 2015)], cAMP responsive element binding protein (CREB), CCAAT/enhancer-binding proteins (C/EBPs) (Tanaka et al. 1997; Rosen et al. 2002), PR domain containing 16 (PRDM16) (Seale et al. 2008, 2011), and activating transcription factor 2 (ATF2) (Cao et al. 2004), but can also repress transcriptional repressors, such as the corepressor receptor-interacting protein 140 (RIP140) (Kiskinis et al. 2014) in order to induce the expression of browning genes, with UCP1 as the hallmark of brite/brown adipocytes. These transcriptional factors drive the brite/brown adipogenic program, which ultimately leads to the remodeling of the adipocyte including increased mitochondrial density and size, altered mitochondrial morphology with laminar cristae, lipid droplets of smaller size but of higher number per adipocyte, and increased oxidative capacity of carbohydrates and fatty acids. The pathways which are known to be targeted by miRNAs are illustrated in Fig. 2.

3.1 miRNAs Targeting PPARγ and PRDM16 in Brite/Brown Adipogenesis

In this context, miR-27 was found to be endogenously downregulated during white and brite/brown adipocyte differentiation, as well as in BAT and WAT. Moreover, miR-27 was identified as central upstream inhibitor of PPARγ in white adipocyte differentiation (Karbiener et al. 2009) and was shown to directly target and repress a number of essential factors of the brite/brown transcriptional network, e.g., PRDM16, CREB, PGC-1α/β, PPARα, and PPARγ, during brite/brown adipogenesis (Fig. 2) (Sun and Trajkovski 2014; Zhu et al. 2014). Under pathophysiological conditions, miR-27 was less abundant in mature adipocytes of obese compared to lean mice (Kim et al. 2010), thus probably allowing adipose tissue hyperplasia, while miR-27 was found to be more abundant in the adipose tissue of hyperglycemic Goto-Kakizaki rats compared to normoglycemic Brown Norway rats, which could

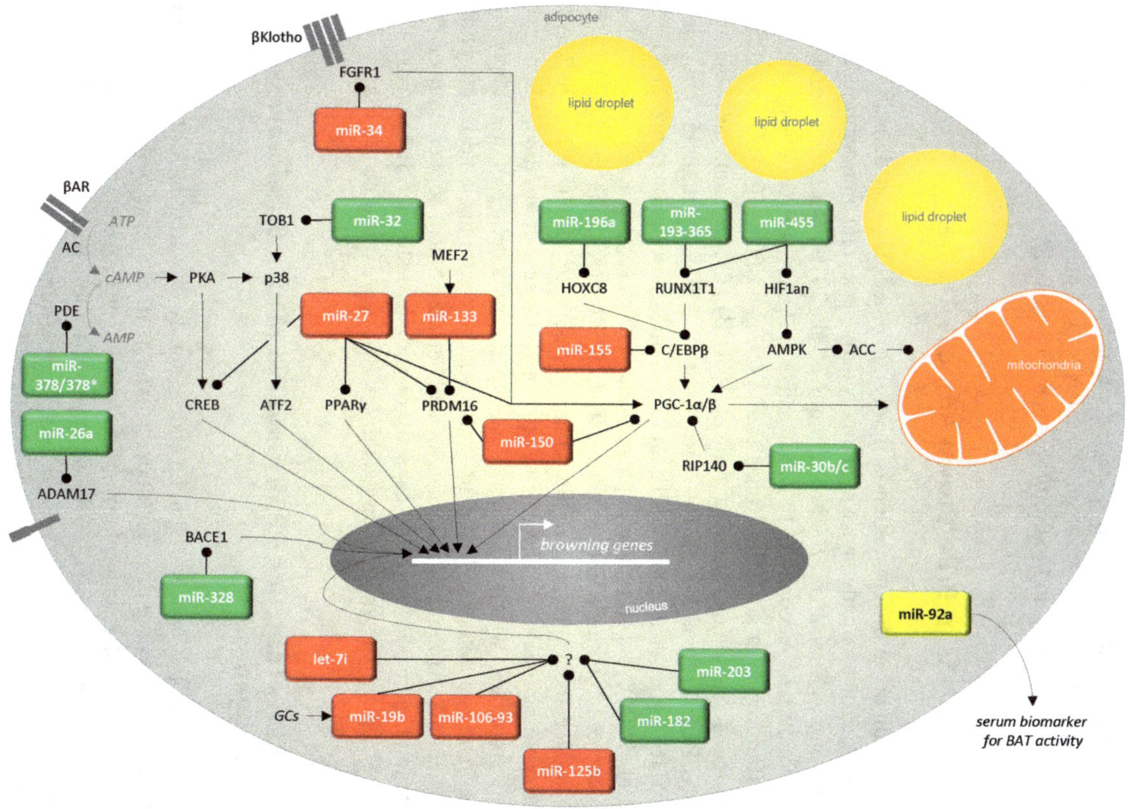

Fig. 2 miRNAs in brite/brown adipose tissue (BAT). miRNAs with a positive impact on brite/brown adipocyte formation and function are displayed in green, while miRNAs with a repressive role are displayed in red. miRNAs which are known to be secreted are indicated in yellow

also be corroborated by elevated miR-27 levels in 3T3-L1 adipocytes, a murine adipogenesis model, upon exposure to increased glucose concentrations (Herrera et al. 2010).

The muscle-enriched miR-133 was found to be markedly downregulated in BAT and subcutaneous WAT. This repression is a result of decreased expression of the myocyte enhancer factor 2 (MEF2), its transcriptional regulator, which is also repressed upon elevated cyclic AMP (cAMP) levels after cold exposure. In line with this, inhibition of miR-133 or Mef2 promotes brite and brown adipogenesis. Mechanistically, miR-133 directly targets and represses PRDM16 (Trajkovski et al. 2012).

The targeted deletion of the RNA-binding protein KSRP, that regulates gene expression at several levels, caused a reduction in adiposity, with elevated expression of brite/brown marker genes in subcutaneous WAT, and in reduced expression of miR-150. In this context, forced expression of miR-150 indeed attenuated the browning program. Mechanistically, miR-150 directly targets and represses PRDM16 and PGC-1α (Chou et al. 2014).

3.2 miRNAs Targeting the cAMP-PKA-CREB and p38-ATF2 Signaling Pathways in BAT

β-adrenergic receptor signaling leads to elevated levels of cAMP, followed by p38/MAPK signaling, which all play a vital role in BAT thermogenic response leading to ATF2 activation, thus promoting transcription of downstream targets such as PGC-1α, UCP1, and FGF21 (Cao et al. 2004; Robidoux et al. 2005). Conversely, repressors of p38/MAPK signaling, such as TOB1, are able to silence the pathway under normal conditions (Sun et al. 2013; Wu et al. 2015).

In this context, miR-378/378* is a miRNA encoded within the PGC-1β gene and was the first miRNA found to increase BAT mass and is sufficient to prevent both genetic and high fat diet (HFD)-induced obesity. Mechanistic studies at the molecular level revealed that miR-378/378* directly targets the phosphodiesterase (PDE) (Fig. 2), which then leads to diminished degradation of cAMP to AMP, thus leading to elevated cAMP levels which activate PKA and downstream signaling pathways (Pan et al. 2014). Moreover, another study demonstrated that the ω-3 fatty acid eicosapentaenoic acid (EPA) binds and activates the free fatty acid receptor (FFAR4), a functional receptor for n-3 polyunsaturated fatty acids (PUFA), which positively modulates miR-378 leading to elevated cAMP levels and ultimately UCP1 expression (Kim et al. 2016).

Another miRNA, miR-32, was identified to be expressed selectively in BAT, and its levels were elevated by cold exposure. Mechanistically, miR-32 directly targets the p38/MAPK signaling repressor TOB1, thus diminishing the repressive effect of TOB1 on p38/MAPK signaling which leads to phosphorylation and activation of ATF2 with enhanced BAT thermogenesis (Fig. 2). Interestingly, this also drives FGF21 expression and secretion from BAT, thereby also trans-activating the browning of WAT (Ng et al. 2017).

3.3 miRNAs Ultimately Modulating C/EBPβ Activity

C/EBPβ is a critical transcription factor that activates transcription of C/EBPα and PPARγ, two important transcriptional inducers of the adipogenic brite/brown transcriptional program. C/EBPβ cooperates with PRDM16 in a complex that initiates both brown adipocyte differentiation from myoblastic precursors and brite adipocyte formation in subcutaneous WAT (Kajimura et al. 2009; Seale et al. 2011; Jimenez-Preitner et al. 2011).

In this context, miR-155 has been demonstrated to be enriched in BAT, highly enriched in proliferating brown preadipocytes, and declines after induction of brown adipogenesis. Thus inhibition of miR-155 enhanced brite and brown adipogenesis and increased BAT thermogenesis and browning of WAT in mice, while mice transgenically overexpressing miR-155 exhibited reduced BAT mass and function. Interestingly, as direct target of miR-155, C/EBPβ was identified, with C/EBPβ repressing again miR-155 expression, thus forming a self-inhibitory feedback loop that tightly governs brite/brown adipogenesis (Fig. 2) (Chen et al. 2013).

miR-196a was found to be specifically required for the induction of the browning program of WAT, not BAT, progenitor cells. Mechanistically, HOXC8 was identified as direct miR-196a target and was repressed post-transcriptionally (Fig. 2). HOXC8 is a white-fat gene, which represses C/EBPβ and UCP1. In line with that, transgenic mice with elevated miR-196a levels exhibited enhanced energy expenditure and resistance to diet-induced obesity. Thus, these data indicate that the induced brite adipocytes in the inguinal WAT are indeed metabolically functional (Mori et al. 2012).

The first described miRNAs in murine brown adipogenesis were the miRNA cluster miR-193b-365 which was enriched in BAT. Blocking miR-193b and/or miR-365 in brown preadipocytes impaired brown adipogenesis and promoted the expression of myogenic markers, while forced expression in C2C12 myoblasts blocked the entire program of myogenesis and promoted brown adipogenesis (Sun et al. 2011). The runt-related transcription factor 1 (RUNX1T1) was identified as direct miR-193b-365 target, which is known to act as inhibitor of C/EBPβ and consequently of white adipogenesis (Fig. 2) (Rochford et al. 2004). However, another study challenged these in vitro results by demonstrating that in mice with an inactivated miR-193b-365 locus the development, differentiation, and function of BAT was unaffected, indicating that BAT does not require the presence of miR-193b and miR-365 (Feuermann et al. 2013).

Another miRNA, which also targets RUNX1T1, besides other direct targets such as Necdin, and promotes brown and brite adipocyte differentiation, is miR-455. This miRNA exhibited a BAT-specific expression pattern and is induced by cold and the browning inducer bone morphogenetic protein 7 (BMP7). In adipose-specific transgenic mice, elevated miR-455 levels led to marked browning of subcutaneous WAT upon cold exposure by activating the hypoxia inducible factor 1α inhibitor (HIF1an), that further activates AMPK, which then promotes the browning program including PGC-1α expression and mitochondrial biogenesis (Zhang et al. 2015).

3.4 miRNAs Modulating the Transcriptional Coactivator PGC-1α

The thermogenic program of BAT includes mitochondrial biogenesis, and PGC-1α is a key regulator of mitochondrial biogenesis, oxidative metabolism, and UCP1 (Puigserver et al. 1998; Cannon and Nedergaard 2004). However, the transcriptional repressor RIP140 is able to block PGC-1α effects, as mice devoid of RIP140 are lean, show resistance to HFD-induced obesity and have increased oxygen consumption, with a marked increase in expression of genes involved in energy dissipation and mitochondrial uncoupling, including UCP1 (Leonardsson et al. 2004).

MiRNA miR-34a has been shown to be elevated in WAT and BAT upon obesity, associated with inhibited browning of WAT and pale BAT. Mechanistically, miR-34a directly targets the fibroblast growth factor receptor 1 (FGFR1), reduces expression of βklotho and SIRT1, which results in reduced FGF21/SIRT1-dependent deacetylation of PGC-1α, finally repressing the browning program (Fig. 2). Thus, lentiviral-mediated repression of miR-34a levels in adipose depots of mice with diet-induced obesity to levels which were detected in lean mice reduced adiposity and improved mitochondrial biogenesis and oxidative metabolism (Fu et al. 2014). However, global miR-34a knockout mice are again susceptible to diet-induced obesity (Lavery et al. 2016).

Members of the miR-30 family, miR-30b and miR-30c, were greatly elevated in expression levels during adipocyte differentiation and are stimulated by cold or β-adrenergic receptor stimulation. Interestingly, the corepressor RIP140 was identified as direct target of miR-30b/c (Fig. 2). Consequently, overexpression of miR-30b/c induced the browning program, including UCP1 and mitochondrial respiration, in the development of white and brown adipocytes. Moreover, miR-30b/c was able to potentiate β-adrenergic receptor stimulation-induced browning, suggesting a positive feedback loop of miR-30 family members on the β-adrenergic receptor signaling and action (Hu et al. 2015).

3.5 miRNA Targeting ADAM17 and PTEN

The miR-26 family members miR-26a and miR-26b have been identified to be upregulated in murine WAT upon cold exposure. So far, they are the first in-depth characterized miRNAs able to shift human adipocyte differentiation from white to brite via inducing UCP1 expression, increasing mitochondrial density, changing mitochondrial morphology towards brown adipocyte characteristics, and elevating coupled and uncoupled respiration (Karbiener et al. 2014). The identified and validated target that at least partially mediates the miR-26 effects on both adipocyte differentiation and browning is ADAM17, also known as TNFα converting enzyme (TACE), which upon knockdown causes a lean, hypermetabolic phenotype in mice (Gelling et al. 2008) (Fig. 2). However, how ADAM17 mediates mechanistically the browning of WAT yet needs to be elucidated. On the other hand, miR-26b has been identified to directly target the phosphatase and tensin homolog (PTEN), thereby improving insulin sensitivity in human mature adipocytes (Xu et al. 2015), which is in line with the results from in vivo studies where transgenic mice, which globally or liver-specifically overexpress miR-26a, also exhibited increased insulin sensitivity (Fu et al. 2015). Interestingly, in these mice HFD-induced obesity is not ameliorated upon liver-specific overexpression, but upon global miR-26a overexpression. This indicates that this obesity resistant phenotype of miR-26 action is dependent on its function in another organ than the liver.

3.6 miRNA Targeting the β-Secretase BACE1

miR-328 has recently been identified to promote the shift in cell commitment from muscle to BAT. Repressed miR-328 function blocked adipogenesis, and miR-328 overexpression promoted brown adipogenesis while diminishing myogenesis. Mechanistically, the β-secretase BACE1 was identified as direct target of miR-328 (Oliverio et al. 2016) (Fig. 2). Reduced BACE1 levels are known to decrease body weight, to protect against diet-induced obesity, at least partially via UCP1 induction, and to enhance insulin sensitivity in mice (Meakin et al. 2012), and to control the G-protein coupled receptor 5b (GPRC5b), a known link between diet-induced obesity and type 2 diabetes (Kim et al. 2012).

3.7 miRNAs with Impact on Brite/Brown Adipocyte Function and Formation but with Unknown Direct Targets

There are a number of further miRNAs with impact on brite/brown adipocyte formation and function; however, they still lack a known direct target by which the miRNA effects are mediated (Fig. 2).

For example, let-7i is a repressor of brite adipocyte function, as inhibition was able to promote the conversion of adipocytes from white to brite in mouse and human, while let-7i mimic injection in murine subcutaneous WAT partially blocked β-adrenergic activation of the browning process (Giroud et al. 2016a). Another miRNA that was characterized in humans and rodents is miR-125b, which was found to be downregulated upon β-adrenergic receptor stimulation in WAT and BAT and lower expressed in BAT than in WAT. While miR-125b overexpression led to decreased mitochondrial biogenesis and respiration, miR-125b inhibition promoted both in human adipocytes (Giroud et al. 2016b). However, a direct target has not been validated yet.

miR-19b has recently been identified to be transcriptionally upregulated by glucocorticoids (GC) which are known to inhibit the function of BAT and browning of WAT (Kong et al. 2015). While miR-19b overexpression had the same effect as GC treatment, miR-19b inhibition blocked dexamethasone-mediated suppression of the browning program, placing miR-19b as an essential target for GC-mediated control of adipose tissue browning (Lv et al. 2018).

Expression levels of the miRNA cluster miR-106b-93 were found to be elevated in BAT of HFD-fed mice compared to mice that fed a low fat diet, and knockdown of miR-106b and miR-93 significantly induced the adipogenic browning program, while ectopic expression of both miRNAs suppressed brown marker genes, such as UCP1 (Wu et al. 2013).

The miRNAs miR-182 and miR-203 were found in brown adipocytes of DGCR8 KO mice to be among the ten most downregulated miRNAs. Inhibition of both miRNAs in brown adipogenesis led to a reduction in brown adipogenic marker genes, including UCP1, PGC-1α, CIDEA, and PPARα, but not common adipogenic marker genes. Thus, these two miRNAs are required for brown adipocyte differentiation (Kim et al. 2014).

3.8 miRNAs as Serum Biomarker

Beyond functioning as energy buffer, the adipose organ vitally cross-talks with other organs as adipocytes are endowed with secretory abilities of different bioactive compounds which can act in a paracrine, autocrine, and/or endocrine manner (Ailhaud 2000; Villarroya et al. 2017). This also includes the brown adipose depots contributing 87% of the total amount of exosomes under cold exposure (Chen et al. 2016). Profiling of miRNAs in these exosomes revealed miR-92a to be inversely correlated with human BAT activity which is usually measured by ^{18}F-fluorodeoxyglucose (^{18}F-FDG) positron emission tomography coupled with computer tomography (PET/CT). Thus, exosomal miR-92a represents a potential serum biomarker for BAT activity in mice and humans (Chen et al. 2016).

4 Long ncRNAs in Brite/Brown Adipose Tissue

Increasing the thermogenic capacity of adipose tissue has been proposed as a strategy for combating obesity and its associated metabolic disorders. As various small ncRNAs have been implied in the formation and function of brite/brown adipocytes, it is also worth to pay attention to lncRNAs and their regulatory functions in thermogenic adipocytes. And indeed, lncRNAs have been demonstrated to have a regulatory impact on BAT biology; however, only a few candidates have been functionally characterized so far which we review in the following (Fig. 3).

4.1 Long ncRNA BLNC1

PPARγ cooperates with C/EBPα to control the expression of a large number of adipogenic genes, and PRDM16, PGC-1α, and the early B-cell factor 2 (EBF2) have been identified as transcription factors that work with PPARγ to selectively promote the formation of adipocytes with thermogenic capacity. In this context, the first lncRNA named brown fat lncRNA 1 (BLNC1) was identified, which has a length of 965 nucleotides, does not associate with ribosomes, is not translated into a protein, and is highly conserved between mice and humans. Overexpression of BLNC1

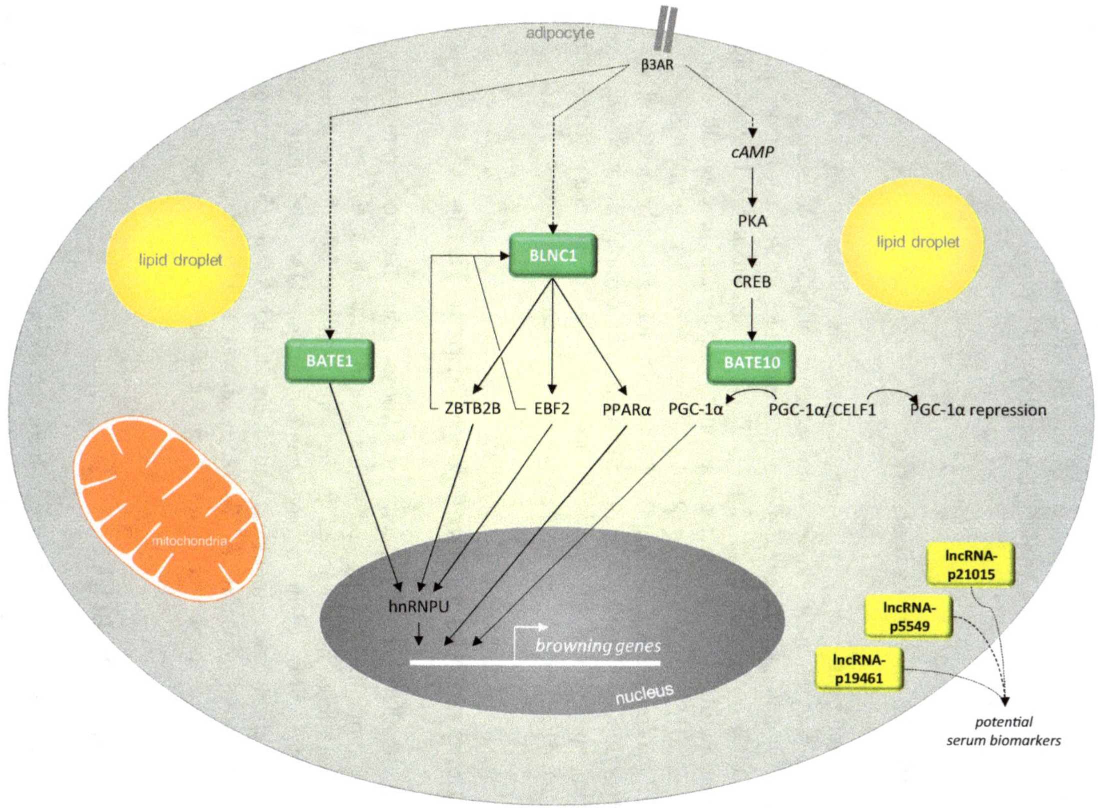

Fig. 3 LncRNAs in BAT. LncRNAs with a positive impact on brite/brown adipocyte formation and function are displayed in green. LncRNAs which are known to be secreted are indicated in yellow

promoted the browning program in white and brown adipocytes, including elevated UCP1 expression, mitochondrial content, total respiratory capacity, and uncoupled respiration, while BLNC1 inhibition impaired these characteristics. Mechanistically, BLNC1 has been found to be strictly dependent on EBF2, forming a ribonucleoprotein complex with EBF2, which on the one hand promotes expression of BLINC1 itself, and on the other hand facilitates binding to the UCP1 promoter leading there to higher activity (Jones and Tontonoz 2014; Zhao et al. 2014) (Fig. 3). This BLNC1/EBF2 complex is formed by support of the heterogeneous nuclear ribonucleoprotein U (hnRNPU) (Mi et al. 2017). Moreover, the zinc finger and BTB domain-containing protein 7b (ZBTB7B) has been identified as potent transcription factor of brite and brown adipocyte differentiation and thermogenic capacity. Interestingly, ZBTB7B is able to recruit the BLNC1/hnRNPU ribonucleoprotein complex to drive the browning program (Li et al. 2017).

4.2 Long ncRNA BATE1

Another transcriptomics study comparing three different murine adipose depots (BAT, subcutaneous WAT, and visceral WAT) revealed 127 lncRNAs with an expression pattern restricted to BAT that are often targeted in their promoter region by the transcriptional regulators C/EBPα, C/EBPβ, and PPARγ. One of them is lnc-BATE1, which has been found to be enriched 10–20-fold during brown adipogenesis. Functional studies elucidated that lnc-BATE1 is required for the establishment and maintenance of BAT identity and thermogenic capacity. Interestingly, also lnc-BATE1 interacts with hnRNPU to form a functional ribonucleoprotein complex to regulate brown adipogenesis (Alvarez-Dominguez et al. 2015) (Fig. 3).

4.3 LncRNA BATE10

Another BAT-enriched lncRNA is lnc-BATE10, which was found to be highly upregulated during brown adipogenesis, is higher expressed in brown compared to white adipocytes, and is induced upon BAT activation in cold exposed mice as well as in subcutaneous WAT by β-adrenergic receptor stimulation. Inhibition of lnc-BATE10 depleted the response to norepinephrine and significantly impaired the expression of BAT-selective genes such as UCP1 and PGC-1α. Thus, lnc-BATE10 is required for BAT-selective gene expression in white and brown adipocytes. Mechanistically, lnc-BATE10 has been shown to be regulated by the cAMP-CREB signaling pathway and interacts with the CUG-binding protein and ELAV-like family member 1 (CELF1) to finally compete with PGC-1α for CELF1 binding (Bai et al. 2017). CELF1 is known to bind the 3′-UTR of its target mRNAs to promote RNA degradation and to repress translation. By competing with CELF1, lnc-BATE10 blocked its inhibitory function on PGC-1α mRNA thus promoting brite/brown adipogenesis.

4.4 Long ncRNAs in Circulation as Potential Biomarkers

Moreover, also circulating lncRNAs have been studied in lean and obese human subjects, as well as in obese patients submitted to diet for 12 weeks. It appeared that three lncRNAs, lncRNA-p5549, lncRNA-p21015, and lncRNA-p19461, are inversely correlated with body mass index (BMI), waist circumference, waist-to-hip ratio, and fasting insulin levels (Sun et al. 2016).

5 Outlook

The worldwide epidemic of obesity is inexorably progressing and thus demands the development of novel and more effective therapeutic approaches. Adipose tissue is the core unit in energy metabolism which can cope with a positive energy balance either by energy storage in white adipocytes or an increase in energy expenditure via non-shivering thermogenesis in brite/brown adipocytes. The latter has ameliorating impact on blood glucose and triglyceride levels as well as on insulin sensitivity. However, the comprehensive regulatory network of brite/brown adipocyte formation remains to be elucidated.

In this context, small and lncRNAs are an emerging novel regulatory layer in energy metabolism, involved in physiological and pathological conditions. So far, numerous miRNAs have been identified and characterized to govern BAT biology, while for lncRNAs only three candidates have been revealed. Interestingly, while miRNAs have been shown to promote or impair brite/brown adipocyte formation and function, the so far all identified lncRNAs promote brite/brown adipogenesis and thermogenic capacity (Table 1). While most endogenous miRNAs and all lncRNAs which are known so far to be involved in BAT biology are characterized in mouse models, only three miRNAs, miR-26, let-7i, and miR-125b, have also been functionally characterized in human. ncRNA candidates that also have a protective role in diet-induced obesity are still rare and will further shrink when criteria for therapeutic applications are applied, such as cross-species conserved function and a comprehensive list of validated direct targets and mediators in order to allow and extrapolate animal studies to humans and to minimize adverse side effects in the long run. Thus, the list of ncRNA candidates is far from being exhaustive, with plenty of space for further research in that field.

At the moment, anti-RNA treatments are currently being developed to expand the options available to clinicians (Wahid et al. 2010, 2014; Kole et al. 2012; Slaby et al. 2017). However, from the current perspective, lncRNAs seem to be more challenging than miRNAs, as lncRNAs can have high turnover rates, lower transcriptional expression, and less cross-species conservation, and a lack of mechanistic understanding hinders further investigation into the application of targeted therapeutics. Nevertheless, miRNAs and lncRNAs as drug targets can be targeted by RNA interference technology, while both classes of ncRNAs as drugs have, in contrast to small molecules and antibodies, very similar physicochemical properties which will emerge as advantage in targeted drug delivery. The reason is that once a targeted

Table 1 miRNAs and lncRNAs with impact on brite/brown adipose tissue (BAT) biology

MicroRNA	Organism	Models used for functional characterization	Function	Validated direct target(s)	Reference
miRNAs targeting PPAR γ and PRDM16 in brite/brown adipogenesis					
miR-27	Mouse	Adipocyte precursors	Directly represses components of the brown transcriptional network and decreases brown differentiation	PRDM16, PPARα, PGC-1β, CREB1	Sun and Trajkovski (2014) and Zhu et al. (2014)
miR-133	Mouse	Adipocyte precursors	Prevents differentiation to brown adipocytes in both BAT and SAT precursors	PRDM16	Trajkovski et al. (2012)
miR-150	Mouse	Ksrp−/− mice; adipocyte precursors	Forced expression attenuates the elevated expression of brown fat genes caused by KSRP deletion	PRDM16, PGC-1α	Chou et al. (2014)
miRNAs targeting the cAMP-PKA-CREB and p38-ATF2 signaling pathways in BAT					
miR-32	Mouse	Brown preadipocytes; AAV-bassed inhibition and overexpression of miR-32 in mice	Overexpression promotes increased BAT themogenesis and serum FGF21 levels, which further promotes white fat browning	TOB1	Ng et al. (2017)
miR-378/378*	Mouse	Mice overexpressing miR-378/378* fat-enriched under the control of the aP2 promoter, adipocyte precursors; mesenchymal stem cells	Forced expression increases classical BAT mass, suppreseses formation of brite adipocytes in subcutaneous WAT, and prevents high fat diet-induced obesity	PDE1 in BAT	Pan et al. (2014) and Kim et al. (2016)
miRNAs ultimately modulating C/EBP β activity					
miR-155	Mouse	Adipocyte precursors; mice overexpressing miR-155 globally under the control of the PGK promoter; mice overexpressing miR-155 BAT-specific under the control of the UCP1 promoter	Inhibition enhances brown and brite adipocyte differentiation, while forced expression reduces brown adipocyte recruitment and function	C/EBPβ	Chen et al. (2013)

(continued)

Table 1 (continued)

MicroRNA	Organism	Models used for functional characterization	Function	Validated direct target(s)	Reference
miR-193b-365	Mouse	Adipocyte precursors; murine myoblasts	Inhibition impairs brown adipocyte adipogenesis, while forced expression blocks myogenesis in myoblasts and induces myoblasts to differentiate into brown adipocytes, while BAT of mice with the incactivated miR-193b-365 cluster is unaffected	RUNX1T1	Sun et al. (2011) and Feuermann et al. (2013)
miR-196a	Mouse	Adipocyte precursors; mice overexpressing miR-196a fat-enriched under the control of the aP2 promoter	Inhibition prevented the induction of thermogenic markers during white adipocyte differentiation, while forced expression induces the recruitment of brite adipocytes and prevents diet-induced obesity	HOXC8	Mori et al. (2012)
miR-455	Mouse	Adipocyte precursors; mesenchymal stem cells	Forced expression induces the recruitment of brown adipocytes incl. mitochondrial biogenesis and prevents diet-induced obesity	RUNX1T1, Necdin, HIF1an	Zhang et al. (2015)
miRNAs modulating the transcriptional coactivator PGC-1 α					
miR-30b/c	Mouse	Adipocyte precursors	Forced expression increases brite and brown adipocyte differentiation, while miR-30b/c inhibition impairs brite/brown characteristics	RIP140	Hu et al. (2015)
miR-34	Mouse	Adipocyte precursors; wild-type mice injected with lentiviral-mediated miR-34a repression upon high fat diet	Inhibition promotes brite and brown marker expression and decreases adiposity, while whole-body miR-34 KO mice are again susceptible to diet-induced obesity	FGFR1	Fu et al. (2014) and Lavery et al. (2016)

miRNA targeting ADAM17 and PTEN					
miR-26a/b	Human and mouse	Mesenchymal stem cells; adipocyte precursors, transgenic mice with liver-specific and global miR-26a overexpression	Promotes brite adipocyte formation and function, improves insulin sensitivity, and diminishes weight gain upon high fat diet-induced obesity	ADAM17	Karbiener et al. (2014) and Fu et al. (2015)
miRNA targeting the β-secretase BACE1					
miR-328	Mouse	Primary immortalized brown adipocytes (PIBA)	Inhibition blocks preadipocyte commitment, while overexpression instigates BAT differentiation	BACE1	Oliverio et al. (2016)
miRNAs with impact on brite/brown adipocyte function and formation but with unknown direct targets					
let-7i	Human and mouse	Mature adipocytes, wild-type mice injected with miRNA mimics	Prevents conversion of white to brite adipocyte formation and brite adipocyte function	–	Giroud et al. (2016a)
miR-125b	Human and mouse	Mature adipocytes, wild-type mice injected with miRNA mimics or inhibitors	Responsive to beta-adrenergic stimulation, mimic injection inhibits conversion of white to brite adipocytes, while inhibition promotes brite adipocyte formation	–	Giroud et al. (2016b)
miR-19b	Mouse	Primary white and brown preadipocytes	Inhibition promotes expression of browning marker genes, while overexpression blocked brown adipogenesis	–	Lv et al. (2018)
miR-106b-93	Mouse	Adipocyte presursors; wild-type mice upon high fat diet	Impairs brown adipocyte differentiation, while inhibition promotes brown adipogenesis	–	Wu et al. (2013)
miR-182	Mouse	Murine brown preadipocytes	Inhibition caused a reduction of brown fat but not common adipogenic markers	–	Kim et al. (2014)
miR-203	Mouse	Murine brown preadipocytes	Inhibition caused a reduction of brown fat but not common adipogenic markers	–	Kim et al. (2014)

(continued)

Table 1 (continued)

MicroRNA	Organism	Models used for functional characterization	Function	Validated direct target(s)	Reference
Circulating miRNAs as potential serum biomarkers					
miR-92a	Human and mouse	Supernatant-derived exosomes of primary murine white and brown preadipocytes and 3T3-L1 murine preadipocytes, serum-derived exosomes of mice and humans	Reduced secretion via exosomes upon BAT activation	–	Sun et al. (2016)
lncRNAs with impact on brite/brown adipocyte function and formation					
BLNC1	Mouse	Immortalized murine brown preadipocytes and mature adipocytes, C3H10T1/2 cells, 3T3-L1 preadipocytes, NU/J Foxn1^{nu} (nude), and C57BL/6 wt mice	Promotes brite and brown adipogenic differentiation	EBF2, hnRNPU, ZBTB7B	Zhao et al. (2014), Li et al. (2017), and Mi et al. (2017)
lnc-BATE1	Mouse	Primary white and brown preadipocytes, C2C12 cells	Brown adipogenic differentiation	hnRNPU	Alvarez-Dominguez et al. (2015)
lnc-BATE10	Mouse	Primary white and brown preadipocytes, 3T3-L1 preadipocytes, and C57BL/6 mice	Brown adipogenic differentiation	CELF1	Bai et al. (2017)
Circulating lncRNAs as potential serum biomarkers					
lncRNA-p5549	Human	Blood samples of obese and nonobese subjects	Reduced secretion upon increasing BMI, waist circumference, waist-to-hip ratio, and fasting insulin	–	Sun et al. (2016)
lncRNA-p21015	Human	Blood samples of obese and nonobese subjects	Reduced secretion upon increasing BMI, waist circumference, waist-to-hip ratio, and fasting insulin	–	Sun et al. (2016)
lncRNA-p19461	Human	Blood samples of obese and nonobese subjects	Reduced secretion upon increasing BMI, waist circumference, waist-to-hip ratio, and fasting insulin	–	Sun et al. (2016)

ncRNA delivery system has been developed for one specific cell type, it could be easily loaded with any other member of this ncRNA class, thus changing the paradigm from one delivery system per drug to one delivery system per class of drugs. Moreover, due to the substantial number of lncRNA loci in the mammalian genome, lncRNAs are a treasure chest yet to be discovered and applied for therapeutic applications. To conclude, despite the currently existing obstacles, we have reached the point where modulating ncRNA expression and function has become a viable option for the modulation of energy metabolism and metabolic diseases. Moreover, it will be interesting to determine whether miRNA/lncRNA-targeting therapeutics could be combined with other chemical or biological drugs for multi-drug therapy.

References

Ailhaud G (2000) Adipose tissue as an endocrine organ. Int J Obes Relat Metab Disord 24(Suppl 2): S1–S3

Alvarez-Dominguez JR, Bai Z, Xu D et al (2015) De Novo reconstruction of adipose tissue transcriptomes reveals long non-coding RNA regulators of brown adipocyte development. Cell Metab 21:764–776. https://doi.org/10.1016/j.cmet.2015.04.003

Babak T, Blencowe BJ, Hughes TR (2005) A systematic search for new mammalian noncoding RNAs indicates little conserved intergenic transcription. BMC Genomics 6:104. https://doi.org/10.1186/1471-2164-6-104

Bai Z, Chai X-R, Yoon MJ et al (2017) Dynamic transcriptome changes during adipose tissue energy expenditure reveal critical roles for long noncoding RNA regulators. PLoS Biol 15: e2002176. https://doi.org/10.1371/journal.pbio.2002176

Birney E, Stamatoyannopoulos JA, Dutta A et al (2007) Identification and analysis of functional elements in 1% of the human genome by the ENCODE pilot project. Nature 447:799–816. https://doi.org/10.1038/nature05874

Brannan CI, Dees EC, Ingram RS, Tilghman SM (1990) The product of the H19 gene may function as an RNA. Mol Cell Biol 10:28–36

Cannon B, Nedergaard J (2004) Brown adipose tissue: function and physiological significance. Physiol Rev 84:277–359. https://doi.org/10.1152/physrev.00015.2003

Cao W, Daniel KW, Robidoux J et al (2004) p38 mitogen-activated protein kinase is the central regulator of cyclic AMP-dependent transcription of the brown fat uncoupling protein 1 gene. Mol Cell Biol 24:3057–3067

Chen Y, Siegel F, Kipschull S et al (2013) miR-155 regulates differentiation of brown and beige adipocytes via a bistable circuit. Nat Commun 4:1769. https://doi.org/10.1038/ncomms2742

Chen Y, Buyel JJ, Hanssen MJW et al (2016) Exosomal microRNA miR-92a concentration in serum reflects human brown fat activity. Nat Commun 7:11420. https://doi.org/10.1038/ncomms11420

Chou C-F, Lin Y-Y, Wang H-K et al (2014) KSRP ablation enhances brown fat gene program in white adipose tissue through reduced miR-150 expression. Diabetes 63:2949–2961. https://doi.org/10.2337/db13-1901

Christopher AF, Kaur RP, Kaur G et al (2016) MicroRNA therapeutics: discovering novel targets and developing specific therapy. Perspect Clin Res 7:68–74. https://doi.org/10.4103/2229-3485.179431

Cinti S (2012) The adipose organ at a glance. Dis Model Mech 5:588–594. https://doi.org/10.1242/dmm.009662

Cohen P, Spiegelman BM (2016) Cell biology of fat storage. Mol Biol Cell 27:2523–2527. https://doi.org/10.1091/mbc.E15-10-0749

de Almeida RA, Fraczek MG, Parker S et al (2016) Non-coding RNAs and disease: the classical ncRNAs make a comeback. Biochem Soc Trans 44:1073–1078. https://doi.org/10.1042/BST20160089

Derrien T, Johnson R, Bussotti G et al (2012) The GENCODE v7 catalog of human long noncoding RNAs: analysis of their gene structure, evolution, and expression. Genome Res 22:1775–1789. https://doi.org/10.1101/gr.132159.111

Djebali S, Davis CA, Merkel A et al (2012) Landscape of transcription in human cells. Nature 489:101–108. https://doi.org/10.1038/nature11233

Dogini DB, Pascoal VDB, Avansini SH et al (2014) The new world of RNAs. Genet Mol Biol 37:285–293

Feuermann Y, Kang K, Gavrilova O et al (2013) MiR-193b and miR-365-1 are not required for the development and function of brown fat in the mouse. RNA Biol 10:1807–1814. https://doi.org/10.4161/rna.27239

Filipowicz W, Bhattacharyya SN, Sonenberg N (2008) Mechanisms of post-transcriptional regulation by microRNAs: are the answers in sight? Nat Rev Genet 9:102–114. https://doi.org/10.1038/nrg2290

Fu T, Seok S, Choi S et al (2014) MicroRNA 34a inhibits beige and brown fat formation in obesity in part by suppressing adipocyte fibroblast growth factor 21 signaling and SIRT1 function. Mol Cell Biol 34:4130–4142. https://doi.org/10.1128/MCB.00596-14

Fu X, Dong B, Tian Y et al (2015) MicroRNA-26a regulates insulin sensitivity and metabolism of glucose and lipids. J Clin Investig 125:2497–2509. https://doi.org/10.1172/JCI75438

Gelling RW, Yan W, Al-Noori S et al (2008) Deficiency of TNFalpha converting enzyme (TACE/ADAM17) causes a lean, hypermetabolic phenotype in mice. Endocrinology 149:6053–6064. https://doi.org/10.1210/en.2008-0775

Giroud M, Karbiener M, Pisani DF et al (2016a) Let-7i-5p represses brite adipocyte function in mice and humans. Sci Rep 6:28613. https://doi.org/10.1038/srep28613

Giroud M, Pisani DF, Karbiener M et al (2016b) miR-125b affects mitochondrial biogenesis and impairs brite adipocyte formation and function. Mol Metab 5:615–625. https://doi.org/10.1016/j.molmet.2016.06.005

Guttman M, Amit I, Garber M et al (2009) Chromatin signature reveals over a thousand highly conserved large non-coding RNAs in mammals. Nature 458:223–227. https://doi.org/10.1038/nature07672

Herrera BM, Lockstone HE, Taylor JM et al (2010) Global microRNA expression profiles in insulin target tissues in a spontaneous rat model of type 2 diabetes. Diabetologia 53:1099–1109. https://doi.org/10.1007/s00125-010-1667-2

Hu F, Wang M, Xiao T et al (2015) miR-30 promotes thermogenesis and the development of beige fat by targeting RIP140. Diabetes 64:2056–2068. https://doi.org/10.2337/db14-1117

Hydbring P, Badalian-Very G (2013) Clinical applications of microRNAs. F1000Res 2:136. https://doi.org/10.12688/f1000research.2-136.v3

Iyer MK, Niknafs YS, Malik R et al (2015) The landscape of long noncoding RNAs in the human transcriptome. Nat Genet 47:199–208. https://doi.org/10.1038/ng.3192

Jimenez-Preitner M, Berney X, Uldry M et al (2011) Plac8 is an inducer of C/EBPβ required for brown fat differentiation, thermoregulation, and control of body weight. Cell Metab 14:658–670. https://doi.org/10.1016/j.cmet.2011.08.008

Jones M, Tontonoz P (2014) Enhanced thermogenesis in the blinc of an eye. Mol Cell 55:343–344. https://doi.org/10.1016/j.molcel.2014.07.015

Kajimura S, Seale P, Kubota K et al (2009) Initiation of myoblast to brown fat switch by a PRDM16-C/EBP-beta transcriptional complex. Nature 460:1154–1158. https://doi.org/10.1038/nature08262

Kajimura S, Seale P, Spiegelman BM (2010) Transcriptional control of brown fat development. Cell Metab 11:257–262. https://doi.org/10.1016/j.cmet.2010.03.005

Karbiener M, Fischer C, Nowitsch S et al (2009) microRNA miR-27b impairs human adipocyte differentiation and targets PPARgamma. Biochem Biophys Res Commun 390:247–251. https://doi.org/10.1016/j.bbrc.2009.09.098
Karbiener M, Pisani DF, Frontini A et al (2014) MicroRNA-26 family is required for human adipogenesis and drives characteristics of brown adipocytes. Stem Cells 32:1578–1590. https://doi.org/10.1002/stem.1603
Kim SY, Kim AY, Lee HW et al (2010) miR-27a is a negative regulator of adipocyte differentiation via suppressing PPARgamma expression. Biochem Biophys Res Commun 392:323–328. https://doi.org/10.1016/j.bbrc.2010.01.012
Kim Y-J, Sano T, Nabetani T et al (2012) GPRC5B activates obesity-associated inflammatory signaling in adipocytes. Sci Signal 5:ra85. https://doi.org/10.1126/scisignal.2003149
Kim H-J, Cho H, Alexander R et al (2014) MicroRNAs are required for the feature maintenance and differentiation of brown adipocytes. Diabetes 63:4045–4056. https://doi.org/10.2337/db14-0466
Kim J, Okla M, Erickson A et al (2016) Eicosapentaenoic acid potentiates brown thermogenesis through FFAR4-dependent up-regulation of miR-30b and miR-378. J Biol Chem 291:20551–20562. https://doi.org/10.1074/jbc.M116.721480
Kiskinis E, Chatzeli L, Curry E et al (2014) RIP140 represses the "brown-in-white" adipocyte program including a futile cycle of triacylglycerol breakdown and synthesis. Mol Endocrinol 28:344–356. https://doi.org/10.1210/me.2013-1254
Kole R, Krainer AR, Altman S (2012) RNA therapeutics: beyond RNA interference and antisense oligonucleotides. Nat Rev Drug Discov 11:125–140. https://doi.org/10.1038/nrd3625
Kong X, Yu J, Bi J et al (2015) Glucocorticoids transcriptionally regulate miR-27b expression promoting body fat accumulation via suppressing the browning of white adipose tissue. Diabetes 64:393–404. https://doi.org/10.2337/db14-0395
Lander ES, Linton LM, Birren B et al (2001) Initial sequencing and analysis of the human genome. Nature 409:860–921. https://doi.org/10.1038/35057062
Lavery CA, Kurowska-Stolarska M, Holmes WM et al (2016) miR-34a(-/-) mice are susceptible to diet-induced obesity. Obesity (Silver Spring) 24:1741–1751. https://doi.org/10.1002/oby.21561
Lee RC, Feinbaum RL, Ambros V (1993) The C. elegans heterochronic gene lin-4 encodes small RNAs with antisense complementarity to lin-14. Cell 75:843–854
Leonardsson G, Steel JH, Christian M et al (2004) Nuclear receptor corepressor RIP140 regulates fat accumulation. Proc Natl Acad Sci U S A 101:8437–8442. https://doi.org/10.1073/pnas.0401013101
Li S, Mi L, Yu L et al (2017) Zbtb7b engages the long noncoding RNA Blnc1 to drive brown and beige fat development and thermogenesis. Proc Natl Acad Sci U S A 114:E7111–E7120. https://doi.org/10.1073/pnas.1703494114
Lv Y-F, Yu J, Sheng Y-L et al (2018) Glucocorticoids suppress the browning of adipose tissue via miR-19b in male mice. Endocrinology 159:310–322. https://doi.org/10.1210/en.2017-00566
Meakin PJ, Harper AJ, Hamilton DL et al (2012) Reduction in BACE1 decreases body weight, protects against diet-induced obesity and enhances insulin sensitivity in mice. Biochem J 441:285–296. https://doi.org/10.1042/BJ20110512
Mercer TR, Dinger ME, Sunkin SM et al (2008) Specific expression of long noncoding RNAs in the mouse brain. Proc Natl Acad Sci U S A 105:716–721. https://doi.org/10.1073/pnas.0706729105
Mi L, Zhao X-Y, Li S et al (2017) Conserved function of the long noncoding RNA Blnc1 in brown adipocyte differentiation. Mol Metab 6:101–110. https://doi.org/10.1016/j.molmet.2016.10.010
Mori M, Nakagami H, Rodriguez-Araujo G et al (2012) Essential role for miR-196a in brown adipogenesis of white fat progenitor cells. PLoS Biol 10:e1001314. https://doi.org/10.1371/journal.pbio.1001314
Mori MA, Thomou T, Boucher J et al (2014) Altered miRNA processing disrupts brown/white adipocyte determination and associates with lipodystrophy. J Clin Invest 124:3339–3351. https://doi.org/10.1172/JCI73468

Mudhasani R, Imbalzano AN, Jones SN (2010) An essential role for Dicer in adipocyte differentiation. J Cell Biochem 110:812–816. https://doi.org/10.1002/jcb.22625
Mudhasani R, Puri V, Hoover K et al (2011) Dicer is required for the formation of white but not brown adipose tissue. J Cell Physiol 226:1399–1406. https://doi.org/10.1002/jcp.22475
Nedergaard J, Cannon B (2010) The changed metabolic world with human brown adipose tissue: therapeutic visions. Cell Metab 11:268–272. https://doi.org/10.1016/j.cmet.2010.03.007
Nedergaard J, Bengtsson T, Cannon B (2011) New powers of brown fat: fighting the metabolic syndrome. Cell Metab 13(3):238–240. https://doi.org/10.1016/j.cmet.2011.02.009
Ng R, Hussain NA, Zhang Q et al (2017) miRNA-32 drives brown fat thermogenesis and trans-activates subcutaneous white fat browning in mice. Cell Rep 19:1229–1246. https://doi.org/10.1016/j.celrep.2017.04.035
Oliverio M, Schmidt E, Mauer J et al (2016) Dicer1-miR-328-Bace1 signalling controls brown adipose tissue differentiation and function. Nat Cell Biol 18:328–336. https://doi.org/10.1038/ncb3316
Pan D, Mao C, Quattrochi B et al (2014) MicroRNA-378 controls classical brown fat expansion to counteract obesity. Nat Commun 5:4725. https://doi.org/10.1038/ncomms5725
Petrovic N, Walden TB, Shabalina IG et al (2010) Chronic peroxisome proliferator-activated receptor gamma (PPARgamma) activation of epididymally derived white adipocyte cultures reveals a population of thermogenically competent, UCP1-containing adipocytes molecularly distinct from classic brown adipocytes. J Biol Chem 285:7153–7164. https://doi.org/10.1074/jbc.M109.053942
Ponting CP, Oliver PL, Reik W (2009) Evolution and functions of long noncoding RNAs. Cell 136:629–641. https://doi.org/10.1016/j.cell.2009.02.006
Prasanth KV, Spector DL (2007) Eukaryotic regulatory RNAs: an answer to the "genome complexity" conundrum. Genes Dev 21:11–42. https://doi.org/10.1101/gad.1484207
Puigserver P, Wu Z, Park CW et al (1998) A cold-inducible coactivator of nuclear receptors linked to adaptive thermogenesis. Cell 92:829–839
Ramsköld D, Wang ET, Burge CB, Sandberg R (2009) An abundance of ubiquitously expressed genes revealed by tissue transcriptome sequence data. PLoS Comput Biol 5:e1000598. https://doi.org/10.1371/journal.pcbi.1000598
Robidoux J, Cao W, Quan H et al (2005) Selective activation of mitogen-activated protein (MAP) kinase kinase 3 and p38alpha MAP kinase is essential for cyclic AMP-dependent UCP1 expression in adipocytes. Mol Cell Biol 25:5466–5479. https://doi.org/10.1128/MCB.25.13.5466-5479.2005
Rochford JJ, Semple RK, Laudes M et al (2004) ETO/MTG8 is an inhibitor of C/EBPbeta activity and a regulator of early adipogenesis. Mol Cell Biol 24:9863–9872. https://doi.org/10.1128/MCB.24.22.9863-9872.2004
Rohas LM, St-Pierre J, Uldry M et al (2007) A fundamental system of cellular energy homeostasis regulated by PGC-1alpha. Proc Natl Acad Sci U S A 104:7933–7938. https://doi.org/10.1073/pnas.0702683104
Rosen ED, Hsu C-H, Wang X et al (2002) C/EBPalpha induces adipogenesis through PPARgamma: a unified pathway. Genes Dev 16:22–26. https://doi.org/10.1101/gad.948702
Seale P, Bjork B, Yang W et al (2008) PRDM16 controls a brown fat/skeletal muscle switch. Nature 454:961–967. https://doi.org/10.1038/nature07182
Seale P, Conroe HM, Estall J et al (2011) Prdm16 determines the thermogenic program of subcutaneous white adipose tissue in mice. J Clin Invest 121:96–105. https://doi.org/10.1172/JCI44271
Slaby O, Laga R, Sedlacek O (2017) Therapeutic targeting of non-coding RNAs in cancer. Biochem J 474:4219–4251. https://doi.org/10.1042/BCJ20170079
Sun M, Kraus WL (2015) From discovery to function: the expanding roles of long noncoding RNAs in physiology and disease. Endocr Rev 36:25–64. https://doi.org/10.1210/er.2014-1034
Sun L, Trajkovski M (2014) MiR-27 orchestrates the transcriptional regulation of brown adipogenesis. Metab Clin Exp 63:272–282. https://doi.org/10.1016/j.metabol.2013.10.004

Sun L, Xie H, Mori MA et al (2011) Mir193b-365 is essential for brown fat differentiation. Nat Cell Biol 13:958–965. https://doi.org/10.1038/ncb2286

Sun K-K, Zhong N, Yang Y et al (2013) Enhanced radiosensitivity of NSCLC cells by transducer of erbB2.1 (TOB1) through modulation of the MAPK/ERK pathway. Oncol Rep 29:2385–2391. https://doi.org/10.3892/or.2013.2403

Sun J, Ruan Y, Wang M et al (2016) Differentially expressed circulating LncRNAs and mRNA identified by microarray analysis in obese patients. Sci Rep 6:35421. https://doi.org/10.1038/srep35421

Taft RJ, Pheasant M, Mattick JS (2007) The relationship between non-protein-coding DNA and eukaryotic complexity. BioEssays 29:288–299. https://doi.org/10.1002/bies.20544

Tanaka T, Yoshida N, Kishimoto T, Akira S (1997) Defective adipocyte differentiation in mice lacking the C/EBPbeta and/or C/EBPdelta gene. EMBO J 16:7432–7443. https://doi.org/10.1093/emboj/16.24.7432

Trajkovski M, Ahmed K, Esau CC, Stoffel M (2012) MyomiR-133 regulates brown fat differentiation through Prdm16. Nat Cell Biol 14:1330–1335. https://doi.org/10.1038/ncb2612

van Rooij E, Purcell AL, Levin AA (2012) Developing microRNA therapeutics. Circ Res 110:496–507. https://doi.org/10.1161/CIRCRESAHA.111.247916

Vance KW, Ponting CP (2014) Transcriptional regulatory functions of nuclear long noncoding RNAs. Trends Genet 30:348–355. https://doi.org/10.1016/j.tig.2014.06.001

Venter JC, Adams MD, Myers EW et al (2001) The sequence of the human genome. Science 291:1304–1351. https://doi.org/10.1126/science.1058040

Villarroya F, Cereijo R, Villarroya J, Giralt M (2017) Brown adipose tissue as a secretory organ. Nat Rev Endocrinol 13:26–35. https://doi.org/10.1038/nrendo.2016.136

Villena JA (2015) New insights into PGC-1 coactivators: redefining their role in the regulation of mitochondrial function and beyond. FEBS J 282:647–672. https://doi.org/10.1111/febs.13175

Wahid F, Shehzad A, Khan T, Kim YY (2010) MicroRNAs: synthesis, mechanism, function, and recent clinical trials. Biochim Biophys Acta 1803:1231–1243. https://doi.org/10.1016/j.bbamcr.2010.06.013

Wahid F, Khan T, Kim YY (2014) MicroRNA and diseases: therapeutic potential as new generation of drugs. Biochimie 104:12–26. https://doi.org/10.1016/j.biochi.2014.05.004

Wilusz JE, Sunwoo H, Spector DL (2009) Long noncoding RNAs: functional surprises from the RNA world. Genes Dev 23:1494–1504. https://doi.org/10.1101/gad.1800909

Wu J, Boström P, Sparks LM et al (2012) Beige adipocytes are a distinct type of thermogenic fat cell in mouse and human. Cell 150:366–376. https://doi.org/10.1016/j.cell.2012.05.016

Wu Y, Zuo J, Zhang Y et al (2013) Identification of miR-106b-93 as a negative regulator of brown adipocyte differentiation. Biochem Biophys Res Commun 438:575–580. https://doi.org/10.1016/j.bbrc.2013.08.016

Wu D, Zhou W, Wang S et al (2015) Tob1 enhances radiosensitivity of breast cancer cells involving the JNK and p38 pathways. Cell Biol Int 39:1425–1430. https://doi.org/10.1002/cbin.10545

Xu G, Ji C, Song G et al (2015) MiR-26b modulates insulin sensitivity in adipocytes by interrupting the PTEN/PI3K/AKT pathway. Int J Obes 39:1523–1530. https://doi.org/10.1038/ijo.2015.95

Xue B, Coulter A, Rim JS et al (2005) Transcriptional synergy and the regulation of Ucp1 during brown adipocyte induction in white fat depots. Mol Cell Biol 25:8311–8322. https://doi.org/10.1128/MCB.25.18.8311-8322.2005

Zhang K, Shi Z-M, Chang Y-N et al (2014) The ways of action of long non-coding RNAs in cytoplasm and nucleus. Gene 547:1–9. https://doi.org/10.1016/j.gene.2014.06.043

Zhang H, Guan M, Townsend KL et al (2015) MicroRNA-455 regulates brown adipogenesis via a novel HIF1an-AMPK-PGC1α signaling network. EMBO Rep 16:1378–1393. https://doi.org/10.15252/embr.201540837

Zhao X-Y, Li S, Wang G-X et al (2014) A long noncoding RNA transcriptional regulatory circuit drives thermogenic adipocyte differentiation. Mol Cell 55:372–382. https://doi.org/10.1016/j.molcel.2014.06.004

Zhu Y, Zhang X, Ding X et al (2014) miR-27 inhibits adipocyte differentiation via suppressing CREB expression. Acta Biochim Biophys Sin Shanghai 46:590–596. https://doi.org/10.1093/abbs/gmu036

Brown Adipokines

Francesc Villarroya, Aleix Gavaldà-Navarro, Marion Peyrou,
Joan Villarroya, and Marta Giralt

Contents

1 Brown and White Adipose Tissues: Distinct Secretory Functions 240
2 BAT-Secreted Factors with Autocrine Action 241
2.1 Non-peptidic Autocrine Factors 241
2.2 Peptidic Autocrine Factors 243
3 BAT-Secreted Factors with Paracrine Action 245
4 BAT-Secreted Factors with Endocrine Action 246
5 Brown Adipokines: Circulating Biomarkers of BAT Activity? 249
6 Conclusions: Challenges and Opportunities to a Comprehensive Understanding of the BAT Secretome 250
6.1 Experimental Approaches to Elucidating the BAT Secretome 250
6.2 Human BAT Secretome 251
6.3 Brown Versus Beige Secretome 251
6.4 Brown/Beige Secretome at Distinct Anatomical Sites 251
References 252

Abstract

Brown adipokines are regulatory factors secreted by brown and beige adipocytes that exhibit endocrine, paracrine, and autocrine actions. Peptidic and non-peptidic molecules, including miRNAs and lipids, are constituents of brown adipokines.

F. Villarroya (✉) · A. Gavaldà-Navarro · M. Peyrou · M. Giralt
Departament de Bioquímica i Biomedicina Molecular, Institut de Biomedicina, Universitat de Barcelona, Barcelona, Catalonia, Spain

CIBER Fisiopatología de la Obesidad y Nutrición, Barcelona, Spain
e-mail: fvillarroya@ub.edu

J. Villarroya
Departament de Bioquímica i Biomedicina Molecular, Institut de Biomedicina, Universitat de Barcelona, Barcelona, Catalonia, Spain

Hospital de la Santa Creu I Sant Pau, Barcelona, Spain

A. Pfeifer et al. (eds.), *Brown Adipose Tissue*,
Handbook of Experimental Pharmacology 251, https://doi.org/10.1007/164_2018_119

Brown adipose tissue remodeling to meet thermogenic needs is dependent on the secretory properties of brown/beige adipocytes. The association between brown fat activity and a healthy metabolic profile, in relation to energy balance and glucose and lipid homeostasis, is influenced by the endocrine actions of brown adipokines. A comprehensive knowledge of the brown adipocyte secretome is still lacking. Advancements in the identification and characterization of brown adipokines will facilitate therapeutic interventions for metabolic diseases, as these molecules are obvious candidates to therapeutic agents. Moreover, identification of brown adipokines as circulating biomarkers of brown adipose tissue activity may be particularly useful for noninvasive assessment of brown adipose tissue alterations in human pathologies.

Keywords
Adipokine · Beige adipocyte · Brown adipocyte · Brown adipose tissue

1 Brown and White Adipose Tissues: Distinct Secretory Functions

Examination of the physiology of adipose tissues has traditionally focused on their crucial role in energy metabolism. In white adipose tissue (WAT), white adipocytes are responsible for storing fat that will be used when food is not available. However, in brown adipose tissue (BAT), the characteristic cells are brown adipocytes, which oxidize metabolites to produce heat (Cannon and Nedergaard 2004). Anatomically defined BAT depots are located in the interscapular region in rodents and in the supraclavicular region and other upper trunk areas in adult humans (Gesta et al. 2007). In conditions of high thermogenic activity of the organism, "beige" or "brite" adipocytes, which are brown adipocyte-like cells, appear in WAT depots (Petrovic et al. 2010; Wu et al. 2012). These cells share a thermogenic role with "classical" brown adipocytes in BAT depots (Shabalina et al. 2013; Kazak et al. 2015). "Browning" of WAT is the commonly used term for the process of enrichment of WAT in beige cells. Subcutaneous WAT is more prone to experience browning than visceral WAT depots. Both WAT browning and BAT activation in response to thermogenic challenges, such as exposure to a cold environment, are mainly induced by activation of the sympathetic nervous system. β-adrenergic receptors on brown adipocyte surfaces mediate most of the intracellular events leading to thermogenic activation (Cannon and Nedergaard 2004), although non-sympathetic mechanisms can also lead to activation (Villarroya and Vidal-Puig 2013).

Other adipose depots contain brown or beige adipocytes. In rodents, perivascular adipose tissue around the thoracic aorta exhibits a strong BAT-like phenotype (Fitzgibbons et al. 2011), whereas distinct proportions of pure white and BAT-like adipocytes are present at other locations of perivascular adipose tissue (e.g., mesenteric tissue surrounding the abdominal aorta) (Friederich-Persson et al. 2017; Fitzgibbons et al. 2011). BAT-like molecular features, reminiscent of a beige phenotype, have been reported in adipose tissue around human coronary vessels (Chatterjee et al. 2009). Moreover, epicardial adipose tissue, a fat depot closely associated with the myocardium and present in humans and other mammalian species (Sacks et al. 2013), has BAT/beige features. Brown adipocyte-like features have also been attributed to bone marrow adipocytes, based on intermediate gene expression pattern between BAT and WAT (Krings et al. 2012).

WAT was determined to be an endocrine organ a few decades ago, mainly after the discovery of leptin. As an endocrine organ, WAT is capable of secreting the so-called adipokines which are bioactive molecules that act on multiple biological processes and help maintain overall energy homeostasis (Blüher and Mantzoros 2015). In fact, almost all the current knowledge of adipokines is derived from studies on WAT. The potential secretory role and specific secretory properties of BAT are just beginning to be recognized (Villarroya et al. 2017), perhaps because the expression and secretion of leptin and pro-inflammatory cytokines are less in BAT than in WAT (Cannon and Nedergaard 2004).

Brown adipokines (also named "batokines") released by brown and/or beige adipocytes may act on different targets, including distant cells and organs, after being released into circulation (endocrine function) or locally to affect brown/beige adipocytes (autocrine function) or neighboring cells in BAT/beige depots (paracrine function). It is important to note that, similar to the general term "adipokine" used for WAT-released factors, the term "brown adipokines" may be applied generally to factors released by any type of BAT cell or may refer to factors released specifically by brown adipocytes. In this chapter, the term "brown adipokine" is applied only to factors secreted by brown and beige adipocytes and/or preferentially released when BAT thermogenesis is activated.

2 BAT-Secreted Factors with Autocrine Action

Most factors that show autocrine action and are secreted by brown and beige cells enhance thermogenic activity. However, secretion of inhibitory factors has also been documented. Brown adipokines with autocrine actions may be peptidic and non-peptidic molecules (Fig. 1).

2.1 Non-peptidic Autocrine Factors

Nitric Oxide (NO) Brown adipocytes produce NO through the noradrenergic induction of NO synthase activities, which inhibit proliferation and induce differentiation of brown fat cells in culture (Nisoli et al. 1997, 1998). Inorganic nitrate also induces the synthesis of NO, leading to the browning of WAT (Roberts et al. 2015). Serial reduction of nitrate to nitrite and then to NO, thus contributing to the release of NO by brown adipocytes, has been proposed to occur in BAT (Roberts et al. 2015). In addition to its autocrine role, brown adipocyte-released NO is believed to play a paracrine role by targeting vascular cells in BAT (see below).

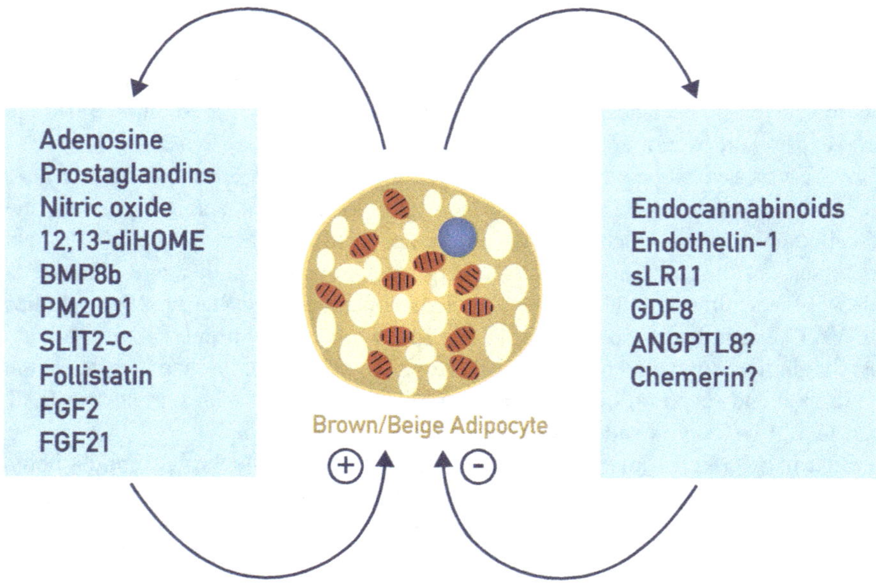

Fig. 1 Autocrine factors released by brown/beige adipocytes. Brown and beige adipocytes secrete factors with autocrine action that lead to positive (+) or negative (−) action on thermogenic activity

Prostaglandins and Prostaglandin-Related Molecules The importance of prostaglandins as local activators of WAT browning was first reported by Vegiopoulos et al. (2010) who described the importance of cyclooxygenase (COX)-2, a rate-limiting enzyme in prostaglandin synthesis, to elicit the recruitment of beige adipocytes in WAT adipose depots. Further studies confirmed that prostaglandins synthetized and released by brown and beige adipocytes induce BAT activity and WAT browning in mice (García-Alonso et al. 2013, 2016). This was concluded based on the finding that invalidation of the genes encoding cyclooxygenase-2 and lipocalin prostaglandin D synthase, which are required for prostaglandin synthesis, impairs WAT browning (Vegiopoulos et al. 2010; Virtue et al. 2012). Similarly, prostaglandin E synthase-1 downregulation impairs the appearance of beige cells in WAT, whereas prostaglandin E2 induces the browning of WAT (García-Alonso and Clària 2014).

Adenosine Sympathetic-activated brown adipocytes release adenosine, which elicits an autocrine enhancement of thermogenic activation (Gnad et al. 2014). The autocrine action of BAT-secreted adenosine occurs thanks to the presence of adenosine A2A receptors in human and murine BAT. In addition to autocrine effects on brown adipocytes themselves, adenosine promotes the induction of browning in WAT.

Endocannabinoids With activation of BAT and WAT browning, an autocrine negative feedback mechanism resulting from thermogenically induced endocannabinoid release by brown/beige adipocytes is triggered (Krott et al. 2016).

Endocannabinoids appear to be negative regulators of brown adipocyte thermogenic activity (Boon et al. 2014) through interference of the β3-adrenergic pathway for thermogenic activation (Krott et al. 2016).

12,13-Dihydroxy-9Z-Octadecenoic Acid (12,13-diHOME) BAT was recently shown to release lipids with signaling properties ("lipokines"). Among them, 12,13-diHOME has been identified as a stimulator of BAT activity through the promotion of fatty acid uptake into brown adipocytes (Lynes et al. 2017). 12,13-diHOME is produced in BAT in response to cold following activation of their biosynthesis enzymes.

2.2 Peptidic Autocrine Factors

Bone Morphogenetic Proteins (BMPs) White/beige/brown adipogenesis is affected by different BMPs. BMP7 is expressed early in brown fat development as a secreted factor necessary for the formation of classical BAT depots (Tseng et al. 2008). BMP7 is mainly produced by stromal vascular cells within adipose tissue and promotes both brown and beige adipocyte commitment and differentiation (Schulz et al. 2013). Until recently, BMP4 was thought to specifically regulate white adipogenesis. However, BMP4 is now also recognized as an important endogenous promoter of beige adipocyte differentiation (Qian et al. 2013).

In response to thermogenic and nutritional factors, mature brown adipocytes secrete BMP8b (Whittle et al. 2012). This secreted protein acts locally by sensitizing brown adipocytes to the β-adrenergic stimulus. Moreover, BMP8b also targets the hypothalamus to specifically induce sympathetic nervous system activation of BAT thermogenesis (Whittle et al. 2012).

Peptidase M20 Domain Containing 1 (PM20D1) PM20D1 is a secreted enzyme that converts fatty acids and amino acids into *N*-acyl amino acids, which increase cellular respiration by acting as endogenous, UCP1-independent, mitochondrial uncouplers (Long et al. 2016).

SLIT2-C Fragment (SLIT2-C) SLIT2-C is a beige adipocyte-secreted peptide cleaved from the full-length SLIT2 protein. SLIT2-C promotes adipose tissue thermogenesis by inducing protein kinase A signaling activity downstream the β-adrenergic pathway (Svensson et al. 2016). Recent studies in human patients indicate that circulating SLIT2 levels negatively correlate with metabolic markers of diabetes mellitus (Kang et al. 2017), and an endocrine role of SLIT2-C, in addition to its autocrine action, cannot be ruled out.

Follistatin This is a soluble glycoprotein that is upregulated in cold-exposed BAT and exerts positive effects on brown thermogenic activity through distinct mechanisms (Singh et al. 2014). Among these are antagonism of the inhibitory effects of GDF8/myostatin on thermogenic activation (see below) and activation of p38 MAP kinase (Singh et al. 2014, 2017).

Other peptidic autocrine factors with inhibitory effects on BAT activity include the following:

Soluble Form of the Low-Density Lipoprotein Receptor Relative LR11 (sLR11) This is a secreted form of LR11 (also called SorLA) which is induced in BAT after cold-induced thermogenic activation. However, this factor inhibits thermogenesis (Whittle et al. 2015) via the BMP/transforming growth factor-β signaling pathway, which has been interpreted as a mechanism to prevent excessive energy wastage when thermogenesis is activated.

Growth Differentiation Factor-8 (GDF8/Myostatin) GDF8/myostatin secretion by brown adipocytes (Fournier et al. 2012; Stanford et al. 2016) following hunger-related neural circuit stimulation acts as a negative autocrine factor in the regulation of WAT browning and metabolic activity and of BAT thermogenic activity (Braga et al. 2014).

Angiopoietin-Like 8 (ANGPTL8) This protein, also called lipasin, betatrophin, or RIFL ("refeeding-induced fat and liver"), is a secreted regulatory factor that increases in BAT in response to cold (Fu et al. 2013). However, ANGPTL8 represses lipoprotein lipase activity (Zhang and Abou-Samra 2013). The role of ANGPTL8 in the context of thermogenic regulation is unclear. It may counteract a potential overload of fatty acids in BAT associated with enhanced lipoprotein lipase activity and subsequent hydrolysis of circulating triglycerides that occurs in association with BAT thermogenic activation.

Chemerin This is a secreted protein whose expression is reduced in BAT in response to cold exposure but increased in response to a high-fat diet in mice (Hansen et al. 2014). Expression of chemerin receptors in brown adipocytes (Rourke et al. 2014) and the effects of chemerin on triglyceride accumulation suggest that chemerin affects lipid metabolism in an autocrine manner (Rourke et al. 2014).

Endothelin-1 Endothelin-1 secretion by brown/beige adipocytes is inhibited by adrenergic activation. The thermogenic activity of brown/beige adipocytes induced via Gq signaling is repressed by endothelin-1, which acts as an autocrine factor (Klepac et al. 2016).

Basic Fibroblast Growth Factor (FGF2) Brown adipocytes secrete FGF2 which induces the proliferation of preadipocytes in BAT and contributes to BAT recruitment in response to sustained thermogenic activation (Yamashita et al. 1994).

3 BAT-Secreted Factors with Paracrine Action

Many cell types have been identified in BAT, such as vascular endothelial cells, smooth muscular cells, neurons, and immune cells, in addition to precursor/preadipocyte cells and mature brown adipocytes (Cannon and Nedergaard 2004). These cells conform to the tissue architecture and support blood flow or innervation, which is essential for BAT function (Pellegrinelli et al. 2016). There is convincing evidence that regulatory factors released by brown and beige adipocytes and acting on numerous cell types inside adipose depots are crucial for WAT browning and BAT recruitment (Fig. 2).

Vascular Endothelial Growth Factor A (VEGF-A) An increase in tissue blood perfusion to sustain the supply of metabolic substrates and oxygen required to fuel thermogenesis is necessary for WAT browning, and BAT recruitment in response to continued thermogenic activation (Cannon and Nedergaard 2004; Orava et al. 2011). In response to sympathetic system-mediated recruitment of BAT, brown adipocytes produce the angiogenic factor VEGF-A (Xue et al. 2009; Sun et al. 2014). In an obese mice model, decreased vessel density and increased brown adipocyte mitophagy, which leads to a loss of thermogenic potentiality of the tissue, have been reported to occur in association with reduced levels of VEGF-A in BAT (Shimizu et al. 2014). VEGF-A can also target brown adipocytes themselves to induce BAT activity in an autocrine manner (Mahdaviani et al. 2016).

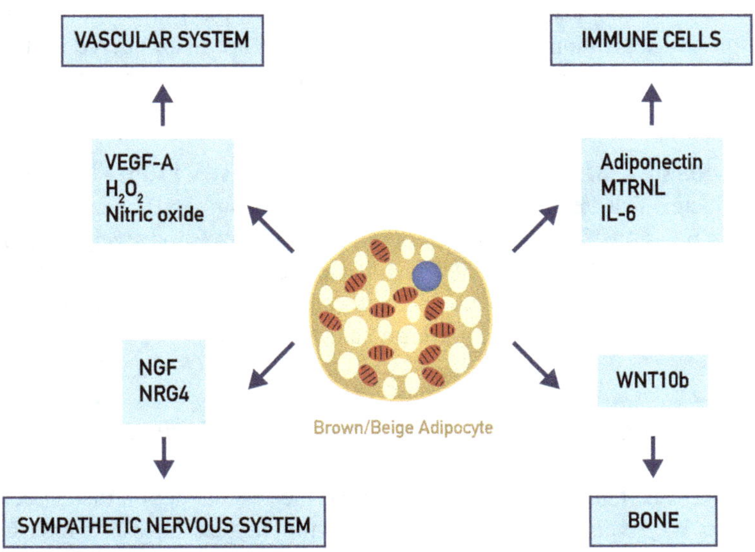

Fig. 2 Paracrine factors released by brown/beige adipocytes. Brown and beige adipocytes secrete factors with paracrine actions on the bone and vascular, nervous, and immune cells present in brown and beige adipose depots

Hydrogen Peroxide (H_2O_2) Brown adipocytes have recently been found to release H_2O_2, which acts on vascular cell-suppression contractility (Friederich-Persson et al. 2017). H_2O_2 is particularly relevant for the paracrine action of brown/beige adipocytes present in perivascular adipose tissue. The release of H_2O_2 largely mediates the anticontractile effect of mesenteric perivascular adipose tissue on vascular cells. The main responsible for H_2O_2 production by brown adipocytes is the enzyme NADPH oxidase-4 (Nox4), which is highly expressed in BAT (Friederich-Persson et al. 2017).

Nerve Growth Factor (NGF) BAT development and thermogenic function are strongly correlated with sympathetic innervation of BAT, as norepinephrine released by nerve endings is the main regulator of BAT thermogenesis (Cannon and Nedergaard 2004). An increased density of noradrenergic nerve fibers correlates with browning of WAT depots (Murano et al. 2009). NGF is released by brown adipocytes and induces nerve growth and differentiation (Néchad et al. 1994; Nisoli et al. 1996), promoting the remodeling of sympathetic innervation in brown and beige adipose tissues.

Meteorin-Like (MTRNL) In response to thermogenic activation, MTRNL is produced by beige adipocytes that reside in WAT depots, but not by brown adipocytes in BAT (Rao et al. 2014). Increased recruitment of anti-inflammatory, alternatively activated, M2-type macrophages, as well as eosinophils (Nguyen et al. 2011), leads to enhanced BAT activity and promotion of WAT browning. MTRNL targets immune cells locally and contributes to increase M2 macrophages via enhanced infiltration and activation of eosinophils, which are the main source of IL-4 and IL-13 (cytokines that promote alternative activation of macrophages) (Rao et al. 2014).

Wingless-Related MMTV Integration Site 10b (WNT10b) WNT10b is a secreted factor that regulates both energy metabolism and bone homeostasis. Targeted expression of WNT10b in adipose tissue improves body weight and insulin sensitivity in mice (Wright et al. 2007) and also increases bone mass (Bennett et al. 2005). WNT10b is released by brown/beige cells residing in bone marrow and has anabolic effects on bone formation (Rahman et al. 2013).

4 BAT-Secreted Factors with Endocrine Action

There is evidence that BAT secretes endocrine factors that act on distant cells and organs (Fig. 3). Experimental BAT transplantation studies have provided indirect evidence for the endocrine actions of brown adipokines (see below).

Triiodothyronine (T3) T3 is possibly the first BAT-released endocrine factor identified. In contrast to WAT, BAT expresses thyroxine deiodinase type II (Dio2), which is an enzyme capable of converting thyroxine into the thyroid active

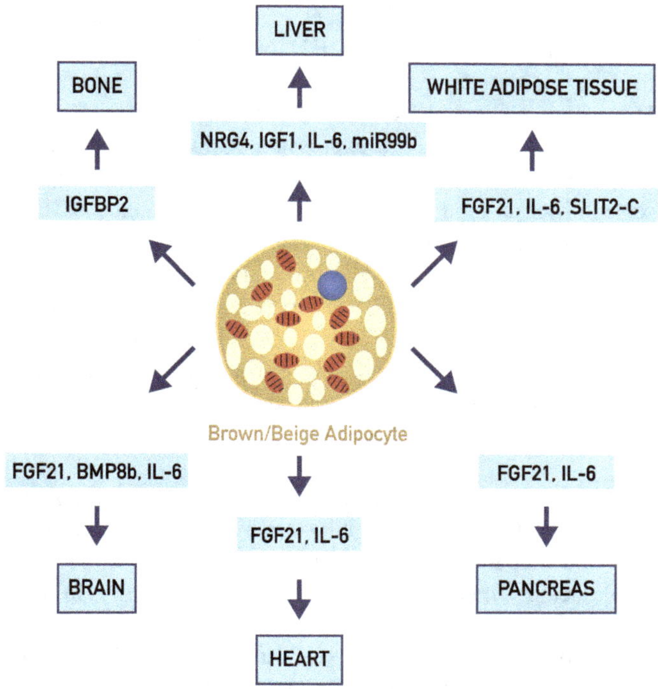

Fig. 3 Endocrine factors released by brown/beige adipocytes. Brown and beige adipocytes secrete factors that might signal to distinct tissues, including the brain, pancreas, liver, heart, bone, and white adipose tissue

hormone T3. In response to cold exposure via noradrenergic stimulation, brown adipocytes strongly induce Dio2 activity. Locally produced T3 is important for the induction of thermogenic machinery in brown adipocytes (Rabelo et al. 1995), and, in these conditions, BAT also becomes a relevant source of systemic T3. In studies of rats acclimated to cold, BAT was estimated to account for about half of the total daily production of T3 (Silva and Larsen 1985; Fernandez et al. 1987).

Fibroblast Growth Factor 21 (FGF21) FGF21 is a member of the so-called "endocrine" subfamily of FGFs. It is also a BAT-secreted factor with a clear endocrine role, in addition to its autocrine actions. In rodent models of obesity and insulin resistance (Giralt et al. 2015) and in humans (Gaich et al. 2013), FGF21 has been shown to exert healthy effects on blood glucose and lipid levels. FGF21 acts on the brain and peripheral tissues such as adipose, pancreas, and heart tissues, among others. Most circulating FGF21 is liver derived, although release of FGF21 by brown adipocytes increases in response to thermogenic activation of BAT (e.g., by cold exposure) (Hondares et al. 2011; Chartoumpekis et al. 2011). FGF21 gene transcription is induced by norepinephrine acting through β3-adrenergic receptors on brown adipocytes (Hondares et al. 2011). In cold-exposed rodents, despite downregulation of hepatic FGF21 gene expression, blood levels of FGF21 tend to increase.

Assessment of the arteriovenous differences of FGF21 levels across interscapular BAT in rats confirmed the release of FGF21 by BAT (Hondares et al. 2011). In contrast with rodents, the expression of FGF21 in WAT in humans is almost negligible. However, substantial expression and release of FGF21 in cultures of human brown adipocytes, as well as expression of FGF21 by human brown and beige adipose tissues, have been demonstrated (Hondares et al. 2014; Di Franco et al. 2016). The recent observation that BAT from mice with targeted UCP1 invalidation (and, therefore, mice that were thermogenically deficient) released large amounts of FGF21, leading to increased blood FGF21 levels, is consistent with the potential for BAT to alter systemic levels of FGF21 (Keipert et al. 2015). The mechanisms by which such large amounts of FGF21 are released by UCP1-deficient brown adipocytes are still unknown. Although it appears that FGF21 can target multiple tissues, evidence of tissues and processes that are specifically influenced by BAT-originated FGF21 is still lacking.

Neuregulin-4 (NRG4) BAT-released NRG4 has endocrine activity that targets the liver and represses hepatic lipogenesis. NRG4 may contribute to protection against hyperlipidemia and hepatic steatosis, and NRG4 systemic levels have been associated with a healthy metabolic profile (Wang et al. 2014). In addition to this endocrine role, NRG4 has been proposed to have a paracrine action inside BAT depots through the promotion terminal nerve branching (Rosell et al. 2014). Recent studies have confirmed the pleiotropic beneficial effects of NRG4 on energy balance and glucose and lipid metabolism (Chen et al. 2017) and impaired NRG4 levels in gestational diabetes (Kralisch et al. 2017).

Insulin Growth Factor-1 (IGF-1) Exposure of rats to cold temperatures increases the expression of IGF-1 in BAT (Duchamp et al. 1997). In pharmacologically and genetically altered rodent models of type I diabetes, transplantation of small amounts of embryonic BAT was found to reverse diabetes, which the authors speculated was associated with a transplant-associated increase in IGF-1 (Gunawardana and Piston 2012, 2015). IGF-1 causes increased proliferation and differentiation of brown adipocytes (Lorenzo et al. 1993). Thus, brown adipocyte-released IGF-1 appears to have autocrine, as well as endocrine, actions.

Interleukin-6 (IL-6) IL-6 gene expression and IL-6 release are induced by cold exposure in vivo or noradrenergic action on brown adipocytes (Burysek and Houstek 1997). When diet-induced obese mice underwent transplantation of small amounts of adult BAT, obesity and insulin resistance was reversed; however, when the source of transplanted BAT was IL-6-null mice, this effect was abolished (Stanford et al. 2013). IL-6 targets multiple cell types including those in the brain, pancreas, heart, and liver (Pal et al. 2014). BAT IL-6-dependent effects on metabolism have been suggested to occur indirectly via FGF21 action, because BAT transplantation results in increased FGF21 levels, and such augmentation is blunted when IL-6-null BAT was used for transplantation. In addition to its endocrine actions, IL-6 may also target local immune cells, mainly to promote M2 macrophage activation (Mauer et al. 2014), a process associated with BAT and beige thermogenic activation.

Retinol-Binding Protein-4 (RBP4) This is a protein involved in intracellular transport of vitamin A and its derivatives. RBP4 behaves as an adipokine when released by WAT (Blüher and Mantzoros 2015). However, when exposed to thermogenic or noradrenergic stimuli, brown adipocytes release RBP4 (Rosell et al. 2012). The physiological significance of BAT-released RBP4 is unclear. WAT-released RBP4 has been found to impair insulin sensitivity, which may appear contradictory with the induction of RBP4 release by BAT following thermogenic activation. However, RBP4 is also a retinoic acid transporter, and there are indications that noradrenergic-induced lipolysis in BAT may affect retinyl esters leading to the release of retinoic acid, which may be actively exported through binding to RBP4. Research is still needed for a comprehensive understanding of the role of RBP4 as a brown adipokine.

Insulin-Like Growth Factor-Binding Protein-2 (IGFBP2) Beige adipocytes secrete IGFBP2, which is involved in the control of bone turnover and contributes to bone gain. The cells producing IGFBP2 may either be beige cells in adipose depots or beige-like cells inside the bone marrow (Rahman et al. 2013; Lidell and Enerbäck 2015).

Exosomal miRNAs It has been recently found that endosomes containing specific microRNAs (miRNAs) are released into circulation by BAT, target the liver, and have regulatory functions. Exosomes are microvesicles secreted by cells. They contain proteins and RNAs, including miRNAs. These short 20–22-nucleotide RNA molecules have been shown to reduce the stability of mRNAs and/or their translation for a gene silencing effect. Exosomes have distinct miRNA composition and are released by BAT in relation to the activation status of the tissue. Thermogenic activation of BAT determines a reduction in the release of miR-92a-containing exosomes (Chen et al. 2016). Conversely, miR-99b-containing exosomes are released by activated BAT and exhibit hepatic effects (Thomou et al. 2017). miR-99b arising from BAT represses hepatic expression of FGF21, which results in systemic consequences on BAT activation by determining FGF21 levels. It has been shown that conditions of BAT activation such as cold exposure result in repressed FGF21 expression in the liver (Hondares et al. 2011), and, in this case, exosomes carrying miR-99b may be mediators of BAT-to-liver communication. New areas of research include the identification of BAT-secreted miRNAs other than miR-99b and the potential hepatic and non-hepatic targets of BAT-secreted miRNAs.

5 Brown Adipokines: Circulating Biomarkers of BAT Activity?

Identification of circulating biomarkers of BAT activity appears to be a very relevant area of research in the field of BAT-secreted factors. Because the extent of BAT activity can only be ascertained using complex radioimaging system (Chen et al. 2014), recognition of biomarkers of BAT activity is a currently challenging and non-accomplished goal. Despite extensive research in this area, a brown adipokine exclusively expressed in BAT has not been identified, which indicates the difficulties in identifying highly reliable circulating biomarkers of BAT activity. In some conditions, levels of circulating FGF21 (Lee et al. 2014; Hanssen et al. 2015) or ANGPTL8 (Martinez-Perez et al. 2016) may reflect BAT activity, but definitive conclusions cannot be drawn, because these molecules are expressed and released by other tissues. A negative association between BAT activity and miR-92a plasma levels has also been proposed (Chen et al. 2016).

6 Conclusions: Challenges and Opportunities to a Comprehensive Understanding of the BAT Secretome

At present, comprehensive knowledge of the BAT secretome is lacking. Moreover, the roles of specific brown adipokines that contribute to the systemic hormonal pool and relative targeting of these adipokines to distinct tissues and organs remain unclear. However, a healthy metabolic profile is associated with BAT activation, and the secretory profile of BAT may contribute to this profile. Close to 15 studies from different laboratories have shown that experimental transplantation of small amounts of healthy BAT to rodent models of obesity and insulin resistance ameliorates metabolic profiles, body weight, and even the fertility of recipients (Villarroya and Giralt 2015; Villarroya et al. 2017). The consensual conclusion from these experiments is that intrinsic thermogenic activity of the transplanted BAT cannot account for the observed effects; rather, factors secreted by BAT may account for the healthy systemic effects. These findings highlight the importance of the identification of brown adipokines for potential future metabolic health therapies.

6.1 Experimental Approaches to Elucidating the BAT Secretome

To date, three methods have been used to identify brown adipokines: direct measurement of components of the proteome of brown adipocyte cell culture medium, sequence trap technology, and bioinformatics prediction of "secretability" based on genome-wide transcriptomic data from BAT or brown adipocytes following exposure to thermogenic stimuli. To strengthen the identification of novel brown adipokines, proteomics and transcriptomic technologies should be improved, and more powerful bioinformatic tools must be developed. On the other hand, assessment of the actual contribution of BAT to the secretion of a given brown adipokine not expressed exclusively in BAT is a challenge. This contribution may be ascertained using existing technologies for targeted invalidation of brown adipokine-encoding genes, specifically in BAT (e.g., UCP1 promoter-driven invalidation). However, conclusions may be limited by compensatory processes often occurring in these experiments when other tissues supplant secretion by BAT.

6.2 Human BAT Secretome

Another question is to what extent data on the brown adipokines obtained from experimental studies in rodents can be extrapolated to humans. Similar systemic effects of the BAT secretome in humans and rodents have been suggested based on the healthy metabolic effects elicited by transplantation of human beige cells grown in culture into mice; these effects are similar to those observed when rodent BAT is transplanted (Min et al. 2016). However, an assessment of the roles of distinct brown adipokines in humans has not yet been conducted. For example, FGF21 is expressed similarly by rodent WAT and BAT under basal conditions, whereas FGF21 is almost undetectable in human WAT but significantly expressed in human BAT (Hondares et al. 2014). However, Dio2, the enzyme responsible for generation of T3 by BAT in rodents, is also preferentially expressed in human BAT versus WAT (Virtanen et al. 2009). Transcriptome analysis of gene expression of adipose tissue from pheochromocytoma patients (a model of WAT browning) revealed the induction of Dio2 and some (e.g., NRG4, PM20D1, follistatin), but not all, batokines identified in rodents (Villarroya et al., unpublished observations). Recently, enhanced expression of IL-6 in human BAT has been reported to be associated with metabolic parameter improvements (Jorge et al. 2017). Several studies have reported changes in the levels of batokine candidates such as FGF21, ANGPTL8, SLIT2-C, or NRG4 (see above) in the blood from human patients under distinct pathophysiological conditions, consistent with potential release by active BAT/beige adipose tissue. However, direct evidence of human BAT/beige tissue as responsible for changes in blood levels of these molecules is still lacking.

6.3 Brown Versus Beige Secretome

Another area of research is aimed at determining whether beige adipocytes exhibit the same pattern of brown adipokine secretion as classical BAT. In recent years, attention has been focused on the process of WAT "browning" because of the high abundance of beige cells in adult human BAT and the strong association between the abundance/activity of beige cells and healthy metabolic parameters. No distinction has been made between brown and beige adipocyte secretomes in the present chapter, because very little information on this distinction is available. MTRNL appears to be released almost exclusively by beige adipose tissue (Rao et al. 2014), and FGF21 may be more intensely expressed by beige adipocytes than by brown adipocytes (Wu et al. 2012). Assessment of a potentially distinct beige secretome will undoubtedly be an area of future research.

6.4 Brown/Beige Secretome at Distinct Anatomical Sites

Beyond the distinction between "classical" brown adipocytes present in BAT depots and beige adipocytes in WAT depots that experience browning, there is a growing awareness of the presence of thermogenic brown adipocyte type of cells in the perivascular and epicardial adipose tissues, as well as in the bone marrow. Is there a specific pattern of secretion from brown/beige-type adipose cells present at these anatomic sites separate from major adipose depots? What role do such secreted factors play locally on vessels or bone? Further research will be needed to ascertain the answers to these critical questions.

Acknowledgments This work has been supported by Grants SAF2017-85722-R from the Ministerio de Ciencia e Innovación (MINECO) and PI17/00420 from the Instituto de Salud Carlos III, Spain, cofinanced by the European Regional Development Fund (ERDF), and Fundació Marató de TV3 (grant 201612-30). M.P. and J.V. are "Juan de la Cierva" researchers (MINECO, Spain).

References

Bennett CN, Longo KA, Wright WS, Suva LJ, Lane TF, Hankenson KD et al (2005) Regulation of osteoblastogenesis and bone mass by Wnt10b. Proc Natl Acad Sci U S A 102:3324–3329

Blüher M, Mantzoros CS (2015) From leptin to other adipokines in health and disease: facts and expectations at the beginning of the 21st century. Metabolism 64:131–145

Boon MR, Kooijman S, van Dam AD, Pelgrom LR, Berbée JF, Visseren CA et al (2014) Peripheral cannabinoid 1 receptor blockade activates brown adipose tissue and diminishes dyslipidemia and obesity. FASEB J 28:5361–5375

Braga M, Reddy ST, Vergnes L, Pervin S, Grijalva V, Stout D et al (2014) Follistatin promotes adipocyte differentiation, browning, and energy metabolism. J Lipid Res 55:375–384

Burysek L, Houstek J (1997) β-Adrenergic stimulation of interleukin-1α and interleukin-6 expression in mouse brown adipocytes. FEBS Lett 411:83–86

Cannon B, Nedergaard J (2004) Brown adipose tissue: function and physiological significance. Physiol Rev 84:277–259

Chartoumpekis DV, Habeos IG, Ziros PG, Psyrogiannis AI, Kyriazopoulou VE, Papavassiliou AG (2011) Brown adipose tissue responds to cold and adrenergic stimulation by induction of FGF21. Mol Med 17:736–740

Chatterjee TK, Stoll LL, Denning GM, Harrelson A, Blomkalns AL, Idelman G et al (2009) Proinflammatory phenotype of perivascular adipocytes: influence of high-fat feeding. Circ Res 104:541–549

Chen KY, Cypess AM, Laughlin MR, Haft CR, Hu HH, Bredella MA et al (2014) Brown adipose reporting criteria in imaging studies (BARCIST 1.0): recommendations for standardized FDG-PET/CT experiments in humans. Cell Metab 24:210–222

Chen Y, Buyel JJ, Hanssen MJ, Siegel F, Pan R, Naumann J et al (2016) Exosomal microRNA miR-92a concentration in serum reflects human brown fat activity. Nat Commun 7:11420

Chen Z, Wang GX, Ma SL, Jung DY, Ha H, Altamimi T et al (2017) Nrg4 promotes fuel oxidation and a healthy adipokine profile to ameliorate diet-induced metabolic disorders. Mol Metab 6:863–872

Di Franco A, Guasti D, Squecco R, Mazzanti B, Rossi F, Idrizaj E et al (2016) Searching for classical brown fat in humans: development of a novel human fetal brown stem cell model. Stem Cells 34:1679–1691

Duchamp C, Burton KA, Géloën A, Dauncey MJ (1997) Transient upregulation of IGF-I gene expression in brown adipose tissue of cold-exposed rats. Am J Physiol 272:E453–EE60

Fernandez JA, Mampel T, Villarroya F, Iglesias R (1987) Direct assessment of brown adipose tissue as a site of systemic tri-iodothyronine production in the rat. Biochem J 243:281–284

Fitzgibbons TP, Kogan S, Aouadi M, Hendricks GM, Straubhaar J, Czech MP (2011) Similarity of mouse perivascular and brown adipose tissues and their resistance to diet-induced inflammation. Am J Physiol Heart Circ Physiol 301:H1425–H1437

Fournier B, Murray B, Gutzwiller S, Marcaletti S, Marcellin D, Bergling S et al (2012) Blockade of the activin receptor IIb activates functional brown adipogenesis and thermogenesis by inducing mitochondrial oxidative metabolism. Mol Cell Biol 32:2871–2879

Friederich-Persson M, Nguyen Dinh Cat A, Persson P, Montezano AC, Touyz RM (2017) Brown adipose tissue regulates small artery function through NADPH oxidase 4-derived hydrogen peroxide and redox-sensitive protein kinase G-1α. Arterioscler Thromb Vasc Biol 37:455–465

Fu Z, Yao F, Abou-Samra AB, Zhang R (2013) Lipasin, thermoregulated in brown fat, is a novel but atypical member of the angiopoietin-like protein family. Biochem Biophys Res Commun 430:1126–1131

Gaich G, Chien JY, Fu H, Glass LC, Deeg MA, Holland WL et al (2013) The effects of LY2405319, an FGF21 analog, in obese human subjects with type 2 diabetes. Cell Metab 18:333–340

García-Alonso V, Clària J (2014) Prostaglandin E2 signals white-to-brown adipogenic differentiation. Adipocytes 3:290–296

García-Alonso V, López-Vicario C, Titos E, Morán-Salvador E, González-Périz A, Rius B et al (2013) Coordinate functional regulation between microsomal prostaglandin E synthase-1 (mPGES-1) and peroxisome proliferator activated receptor γ (PPARγ) in the conversion of white-to-brown adipocytes. J Biol Chem 288:28230–28242

García-Alonso V, Titos E, Alcaraz-Quiles J, Rius B, Lopategi A, López-Vicario C et al (2016) Prostaglandin E2 exerts multiple regulatory actions on human obese adipose tissue remodeling, inflammation, adaptive thermogenesis and lipolysis. PLoS One 11:e0153751

Gesta S, Tseng YH, Kahn CR (2007) Developmental origin of fat: tracking obesity to its source. Cell 131:242–256

Giralt M, Gavaldà-Navarro A, Villarroya F (2015) Fibroblast growth factor-21, energy balance and obesity. Mol Cell Endocrinol 418:66–63

Gnad T, Scheibler S, von Kügelgen I, Scheele C, Kilić A, Glöde A et al (2014) Adenosine activates brown adipose tissue and recruits beige adipocytes via A2A receptors. Nature 516:395–399

Gunawardana SC, Piston DW (2012) Reversal of type 1 diabetes in mica by brown adipose tissue transplant. Diabetes 61:674–682

Gunawardana SC, Piston DW (2015) Insulin-independent reversal of type 1 diabetes in nonobese diabetic mice with brown adipose tissue transplant. Am J Physiol Endocrinol Metab 308: E1043–E1055

Hansen IR, Jansson KM, Cannon B, Nedergaard J (2014) Contrasting effects of cold acclimation versus obesogenic diets on chemerin gene expression in brown and brite adipose tissues. Biochim Biophys Acta 1841:1691–1699

Hanssen MJ, Broeders E, Samms RJ, Vosselman MJ, van der Lans AA, Cheng CC et al (2015) Serum FGF21 levels are associated with brown adipose tissue activity in humans. Sci Rep 5:10275

Hondares E, Iglesias R, Giralt A, Gonzalez FJ, Giralt M, Mampel T et al (2011) Thermogenic activation induces FGF21 expression and release in brown adipose tissue. J Biol Chem 286:12983–12990

Hondares E, Gallego-Escuredo JM, Flachs P, Frontini A, Cereijo R, Goday A et al (2014) Fibroblast growth factor-21 is expressed in neonatal and pheochromocytoma-induced adult human brown adipose tissue. Metabolism 63:312–317

Jorge AS, Jorge GC, Paraíso AF, Franco RM, Vieira LJ, Hilzenderger AM et al (2017) Brown and White adipose tissue expression of IL6, UCP1 and SIRT1 are associated with alterations in clinical, metabolic and anthropometric parameters in obese humans. Exp Clin Endocrinol Diabetes 125:163–170

Kang YE, Choung S, Lee JH, Kim HJ, Ku BJ (2017) The role of circulating Slit2, the one of the newly Batokines, in human diabetes mellitus. Endocrinol Metab (Seoul) 32:383–388

Kazak L, Chouchani ET, Jedrychowski MP, Erickson BK, Shinoda K, Cohen P et al (2015) A creatine-driven substrate cycle enhances energy expenditure and thermogenesis in beige fat. Cell 163:643–655

Keipert S, Kutschke M, Lamp D, Brachthäuser L, Neff F, Meyer CW et al (2015) Genetic disruption of uncoupling protein 1 in mice renders brown adipose tissue a significant source of FGF21 secretion. Mol Metab 4:537–542

Klepac K, Kilić A, Gnad T, Brown LM, Herrmann B, Wilderman A et al (2016) The Gq signalling pathway inhibits brown and beige adipose tissue. Nat Commun 7:10895

Kralisch S, Hoffmann A, Kratzsch J, Blüher M, Stumvoll M, Fasshauer M et al (2017) The brown-fat-secreted adipokine neuregulin 4 is decreased in gestational diabetes mellitus. Diabetes Metab. pii: S1262-3636(17)30465-2

Krings A, Rahman S, Huang S, Lu Y, Czernik PJ, Lecka-Czernik B (2012) Bone marrow fat has brown adipose tissue characteristics, which are attenuated with aging and diabetes. Bone 50:546–552

Krott LM, Piscitelli F, Heine M, Borrino S, Scheja L, Silvestri C et al (2016) Endocannabinoid regulation in white and brown adipose tissue following thermogenic activation. J Lipid Res 57:464–473

Lee P, Linderman JD, Smith S, Brychta RJ, Wang J, Idelson C et al (2014) Irisin and FGF21 are cold-induced endocrine activators of brown fat function in humans. Cell Metab 19:302–309

Lidell ME, Enerbäck S (2015) Brown adipose tissue and bone. Int J Obes Suppl 5:S23–S27

Long JZ, Svensson KJ, Bateman LA, Lin H, Kamenecka T, Lokurkar IA et al (2016) The secreted enzyme PM20D1 regulates lipidated amino acid uncouplers of mitochondria. Cell 166:424–435

Lorenzo M, Valverde AM, Teruel T, Benito M (1993) IGF-I is a mitogen also involved in differentiation related gene expression in fetal brown adipocytes. J Cell Biol 123:1567–1575

Lynes MD, Leiria LO, Lundh M, Bartelt A, Shamsi F, Huang TL et al (2017) The cold-induced lipokine 12,13-diHOME promotes fatty acid transport into brown adipose tissue. Nat Med 23:631–637

Mahdaviani K, Chess D, Wu Y, Shirihai O, Aprahamian TR (2016) Autocrine effect of vascular endothelial growth factor-A is essential for mitochondrial function in brown adipocytes. Metabolism 65:26–35

Martinez-Perez B, Ejarque M, Gutierrez C, Nuñez-Roa C, Roche K, Vila-Bedmar R et al (2016) Angiopoietin-like protein 8 (ANGPTL8) in pregnancy: a brown adipose tissue-derived endocrine factor with a potential role in fetal growth. Transl Res 178:1–12

Mauer J, Chaurasia B, Goldau J, Vogt MC, Ruud J, Nguyen KD et al (2014) Signaling by IL-6 promotes alternative activation of macrophages to limit endotoxemia and obesity-associated resistance to insulin. Nat Immunol 15:423–430

Min SY, Kady J, Nam M, Rojas-Rodriguez R, Berkenwald A, Kim JH et al (2016) Human "brite/beige" adipocytes develop from capillary networks, and their implantation improves metabolic homeostasis in mice. Nat Med 22:312–318

Murano I, Barbatelli G, Giordano A, Cinti S (2009) Noradrenergic parenchymal nerve fiber branching after cold acclimatisation correlates with brown adipocyte density in mouse adipose organ. J Anat 214:171–178

Néchad M, Ruka E, Thibault J (1994) Production of nerve growth factor by brown fat in culture: relation with the in vivo developmental stage of the tissue. Comp Biochem Physiol Comp Physiol 107:381–388

Nguyen KD, Qiu Y, Cui X, Goh YP, Mwangi J, David T et al (2011) Alternatively activated macrophages produce catecholamines to sustain adaptive thermogenesis. Nature 480:104–108

Nisoli E, Tonello C, Benarese M, Liberini P, Carruba MO (1996) Expression of nerve growth factor in brown adipose tissue: implications for thermogenesis and obesity. Endocrinology 137:495–503

Nisoli E, Tonello C, Briscini L, Carruba MO (1997) Inducible nitric oxide synthase in rat brown adipocytes: implications for blood flow to brown adipose tissue. Endocrinology 138:676–682

Nisoli E, Clementi E, Tonello C, Sciorati C, Briscini L, Carruba MO (1998) Effects of nitric oxide on proliferation and differentiation of rat brown adipocytes in primary cultures. Br J Pharmacol 125:888–894

Orava J, Nuutila P, Lidell ME, Oikonen V, Noponen T, Viljanen T et al (2011) Different metabolic responses of human brown adipose tissue to activation by cold and insulin. Cell Metab 14:272–279

Pal M, Febbraio MA, Whitham M (2014) From cytokine to myokine: the emerging role of interleukin-6 in metabolic regulation. Immunol Cell Biol 92:331–339

Pellegrinelli V, Carobbio S, Vidal-Puig A (2016) Adipose tissue plasticity: how fat depots respond differently to pathophysiological cues. Diabetologia 59:1075–1088

Petrovic N, Walden TB, Shabalina IG, Timmons JA, Cannon B, Nedergaard J (2010) Chronic peroxisome proliferator-activated receptor gamma (PPARgamma) activation of epididymally derived white adipocyte cultures reveals a population of thermogenically competent, UCP1-containing adipocytes molecularly distinct from classic brown adipocytes. J Biol Chem 285:7153–7164

Qian SW, Tang Y, Li X, Liu Y, Zhang YY, Huang HY et al (2013) BMP4-mediated brown fat-like changes in white adipose tissue alter glucose and energy homeostasis. Proc Natl Acad Sci U S A 110:E798–E807

Rabelo R, Schifman A, Rubio A, Sheng X, Silva JE (1995) Delineation of thyroid hormone-responsive sequences within a critical enhancer in the rat uncoupling protein gene. Endocrinology 136:1003–1013

Rahman S, Lu Y, Czernik PJ, Rosen CJ, Enerback S, Lecka-Czernik B (2013) Inducible brown adipose tissue, or beige fat, is anabolic for the skeleton. Endocrinology 154:2687–2701

Rao RR, Long JZ, White JP, Svensson KJ, Lou J, Lokurkar I et al (2014) Meteorin-like is a hormone that regulates immune adipose interactions to increase beige fat thermogenesis. Cell 157:1279–1291

Roberts LD, Ashmore T, Kotwica AO, Murfitt SA, Fernandez BO, Feelisch M et al (2015) Inorganic nitrate promotes the browning of white adipose tissue through the nitrate-nitrite-nitric oxide pathway. Diabetes 64:471–484

Rosell M, Hondares E, Iwamoto S, Gonzalez FJ, Wabitsch M, Staels B et al (2012) Peroxisome proliferator-activated receptors-alpha and -gamma, and cAMP-mediated pathways, control retinol-binding protein-4 gene expression in brown adipose tissue. Endocrinology 153:1162–1173

Rosell M, Kaforou M, Frontini A, Okolo A, Chan YW, Nikolopoulou E et al (2014) Brown and white adipose tissues: intrinsic differences in gene expression and response to cold exposure in mice. Am J Physiol Endocrinol Metab 306:E945–E964

Rourke JL, Muruganandan S, Dranse HJ, McMullen NM, Sinal CJ (2014) Gpr1 is an active chemerin receptor influencing glucose homeostasis in obese mice. J Endocrinol 222:201–215

Sacks HS, Fain JN, Bahouth SW, Ojha S, Frontini A, Budge H et al (2013) Adult epicardial fat exhibits beige features. J Clin Endocrinol Metab 98:E1448–E1455

Schulz TJ, Huang P, Huang TL, Xue R, McDougall LE, Townsend KL et al (2013) Brown-fat paucity due to impaired BMP signalling induces compensatory browning of white fat. Nature 495:379–383

Shabalina IG, Petrovic N, de Jong JM, Kalinovich AV, Cannon B, Nedergaard J (2013) UCP1 in brite/beige adipose tissue mitochondria is functionally thermogenic. Cell Rep 5:1196–1203

Shimizu I, Aprahamian T, Kikuchi R, Shimizu A, Papanicolaou KN, MacLauchlan S et al (2014) Vascular rarefaction mediates whitening of brown fat in obesity. J Clin Invest 124:2099–2112

Silva JE, Larsen PR (1985) Potential 520 of brown adipose tissue type II thyroxine 5′-deiodinase as a local and systemic source of triiodothyronine in rats. J Clin Invest 76:2296–2305

Singh R, Braga M, Pervin S (2014) Regulation of brown adipocyte metabolism by myostatin/follistatin signaling. Front Cell Dev Biol 2:60

Singh R, Braga M, Reddy ST, Lee SJ, Parveen M, Grijalva V et al (2017) Follistatin targets distinct pathways to promote brown adipocyte characteristics in brown and white adipose tissues. Endocrinology 158:1217–1230

Stanford KI, Middelbeek RJ, Townsend KL, An D, Nygaard EB, Hitchcox KM et al (2013) Brown adipose tissue regulates glucose homeostasis and insulin sensitivity. J Clin Invest 123:215–223

Stanford KI, Middelbeek RJ, Townsend KL, An D, Nygaard EB, Hitchcox KM et al (2016) AgRP neurons control systemic insulin sensitivity via myostatin expression in brown adipose tissue. Cell 165:125–138

Sun K, Kusminski CM, Luby-Phelps K, Spurgin SB, An YA, Wang QA et al (2014) Brown adipose tissue derived VEGF-A modulates cold tolerance and energy expenditure. Mol Metab 3:474–483

Svensson KJ, Long JZ, Jedrychowski MP, Cohen P, Lo JC, Serag S et al (2016) A secreted Slit2 fragment regulates adipose tissue thermogenesis and metabolic function. Cell Metab 23:454–466

Thomou T, Mori MA, Dreyfuss JM, Konishi M, Sakaguchi M, Wolfrum C et al (2017) Adipose-derived circulating miRNAs regulate gene expression in other tissues. Nature 542:450–455

Tseng YH, Kokkotou E, Schulz TJ, Huang TL, Winnay JN, Taniguchi CM et al (2008) New role of bone morphogenetic protein 7 in brown adipogenesis and energy expenditure. Nature 454:1000–1004

Vegiopoulos A, Müller-Decker K, Strzoda D, Schmitt I, Chichelnitskiy E, Ostertag A et al (2010) Cyclooxygenase-2 controls energy homeostasis in mice by de novo recruitment of brown adipocytes. Science 328:1158–1161

Villarroya F, Giralt M (2015) The beneficial effects of brown fat transplantation: further evidence of an endocrine role of brown adipose tissue. Endocrinology 156:2368–2370

Villarroya F, Vidal-Puig A (2013) Beyond the sympathetic tone: the new brown fat activators. Cell Metab 17:638–643

Villarroya F, Cereijo R, Villarroya J, Giralt M (2017) Brown adipose tissue as a secretory organ. Nat Rev Endocrinol 13:26–35

Virtanen KA, Lidell ME, Orava J, Heglind M, Westergren R, Niemi T et al (2009) Functional brown adipose tissue in healthy adults. N Engl J Med 360:1518–1525

Virtue S, Feldmann H, Christian M, Tan CY, Masoodi M, Dale M et al (2012) A new role for lipocalin prostaglandin D synthase in the regulation of brown adipose tissue substrate utilization. Diabetes 61:3139–3147

Wang GX, Zhao XY, Meng ZX, Kern M, Dietrich A, Chen Z et al (2014) The brown fat-enriched secreted factor Nrg4 preserves metabolic homeostasis through attenuation of hepatic lipogenesis. Nat Med 20:1436–1443

Whittle AJ, Carobbio S, Martins L, Slawik M, Hondares E, Vázquez MJ et al (2012) BMP8B increases brown adipose tissue thermogenesis through both central and peripheral actions. Cell 149:871–885

Whittle AJ, Jiang M, Peirce V, Relat J, Virtue S, Ebinuma H et al (2015) Soluble LR11/SorLA represses thermogenesis in adipose tissue and correlates with BMI in humans. Nat Commun 6:8951

Wright WS, Longo KA, Dolinsky VW, Gerin I, Kang S, Bennett CN et al (2007) Wnt10b inhibits obesity in ob/ob and agouti mice. Diabetes 56:295–303

Wu J, Boström P, Sparks LM, Ye L, Choi JH, Giang AH et al (2012) Beige adipocytes are a distinct type of thermogenic fat cell in mouse and human. Cell 150:366–376

Xue Y, Petrovic N, Cao R, Larsson O, Lim S, Chen S et al (2009) Hypoxia-independent angiogenesis in adipose tissues during cold acclimation. Cell Metab 9:99–109

Yamashita H, Sato Y, Kizaki T, Oh S, Nagasawa J, Ohno H (1994) Basic fibroblast growth factor (bFGF) contributes to the enlargement of brown adipose tissue during cold acclimation. Pflugers Arch 428:352–356

Zhang R, Abou-Samra AB (2013) Emerging roles of lipasin as a critical lipid regulator. Biochem Biophys Res Commun 432:401–405

Part III

Detection of BAT In Vivo

Infrared Thermography

James Law, David E. Morris, Helen Budge, and Michael E. Symonds

Contents

1 Introduction 260
2 Infrared Radiation and Temperature 261
3 Infrared Thermography 263
4 Use of IRT in Studies of BAT 266
5 Acquiring IR Thermographs 267
5.1 Environment 267
5.2 Participants 269
5.3 Positioning 269
5.4 Stimulation 270
5.5 Imaging Duration 272
5.6 Additional Measurements 273
6 Image Analysis 273
6.1 Conversion 273
6.2 Region of Interest Identification 274
6.3 Averaging 274
6.4 Measures 276
7 Summary 278
References 279

Abstract

Historically, brown adipose tissue has been elusive and not easy to detect, hence its relative obscurity in human physiology until its rediscovery in 2009. At that point, it was proven that the symmetrical artefacts frequently detected on positron

J. Law · H. Budge · M. E. Symonds (✉)
School of Medicine, Faculty of Medicine and Health Sciences, University of Nottingham, Nottingham, UK
e-mail: Michael.Symonds@nottingham.ac.uk

D. E. Morris
Bioengineering Research Group, Faculty of Engineering, University of Nottingham, Nottingham, UK

A. Pfeifer et al. (eds.), *Brown Adipose Tissue*,
Handbook of Experimental Pharmacology 251, https://doi.org/10.1007/164_2018_137

emission tomography-computed tomography (PET-CT), which resolved if the environment was kept warm, were brown adipose tissue deposits. PET-CT has remained the stalwart of human brown adipose tissue research and is still considered the gold standard. However, PET-CT exposes the participant to ionising radiation, limiting studies to large, but retrospective, review of clinical imaging or a small-scale, but prospective, design. Within this context, alternative imaging modalities have been sought. Due to the heat-generating properties of brown adipose tissue, infrared thermography is a natural candidate for measuring its activity and the supraclavicular depot is relatively superficial, allowing detection of the heat signature. Infrared thermography is a non-invasive, non-contact technique for measuring temperature remotely. Recent developments in image analysis techniques have facilitated the use of infrared thermography to study brown adipose tissue activation in populations, and in ways, not previously feasible.

Keywords
Brown adipose tissue · Brown fat · Image analysis · Infrared radiation · Infrared thermography · Region of interest identification · Thermal imaging

1 Introduction

Historically, brown adipose tissue (BAT) has been elusive and not easy to detect, hence its relative obscurity in human physiology until its rediscovery in 2009 (Celi 2009; Cypess et al. 2009; van Marken Lichtenbelt et al. 2009; Virtanen et al. 2009). At that point, it was proven that the symmetrical artefacts frequently detected on 18F-fluorodeoxyglucose (18F-FDG) positron emission tomography-computed tomography (PET-CT), which resolved if the environment was kept warm, were BAT deposits. PET-CT has remained the stalwart of human BAT research and is still considered the gold standard. However, PET-CT exposes the participant to a high dose of ionising radiation (~8 mSv). Such studies either rely on retrospective review of clinical imaging (Cypess et al. 2009) or are small-scale prospective studies (van Marken Lichtenbelt et al. 2009; Virtanen et al. 2009). Retrospective studies are able to review results from large numbers of patients, but clinical imaging adopts methodologies to minimise BAT detection, resulting in an underestimate of prevalence and activity, and is largely performed on non-healthy populations. Prospective research studies, while able to maximise BAT detection and be undertaken on healthy volunteers, are ethically limited to small numbers of volunteers with no, or limited, repeat imaging and exclude potentially vulnerable groups, such as children. In addition, 18F-FDG PET-CT is not suitable for postprandial imaging as the signal uptake into muscle precludes detection of BAT (Vosselman et al. 2013; Vrieze et al. 2012). Direct measurement of BAT, by biopsy, is similarly limited to small-scale studies due the proximity of BAT to important anatomical structures such as major vessels.

Within this context, alternative imaging modalities have been sought. Due to the heat-generating properties of brown adipose tissue (Cannon and Nedergaard 2004), infrared thermography (IRT) is a natural candidate for measuring its activity.

2 Infrared Radiation and Temperature

Infrared (IR) radiation is part of the electromagnetic spectrum – a continuum of wavelengths (Fig. 1) – and includes radio waves (which have wavelengths of as much as a few hundred metres) and gamma rays (which have wavelengths of less than 0.01 nm). Wavelengths of less than 10 nm are able to ionise atoms with which they interact (i.e. cause the loss of an electron) and are termed ionising radiation. The visible spectrum is in between these extremes, with wavelengths of between 390 and 700 nm (Tattersall 2016), adjacent to the IR region. IR radiation was discovered by Sir William Herschel in 1800 who showed an effect on a thermometer of radiation below red light (Herschel 1800), hence 'infrared' (although this term was not used until later). All objects with a temperature greater than absolute zero (−273.15°C) emit IR radiation dependent on their temperature (Planck 1914). The IR spectrum is subdivided for practical purposes and there are a few different classification systems in use. Thermal imaging utilises the portion commonly referred to as long-wavelength IR which is emitted by objects whose temperature is between −80 and 89°C, a range relevant to humans and other living creatures.

The relationship between IR radiation and temperature is defined by Planck's law (Eq. 1). Planck's law describes the radiation emitted at a given wavelength by a black body in thermal equilibrium, where a black body is an object that absorbs and radiates heat perfectly and does not possess any reflective property. Such a body will, therefore, absorb all radiant energy that reaches it and emit radiation depending on its temperature. The equilibrium temperature is the point at which the energy being absorbed is equal to the energy being emitted: below this temperature, the body will absorb more energy than it emits and will, therefore, warm; above this temperature the body will emit more energy than it absorbs and will, therefore, cool (Fig. 2).

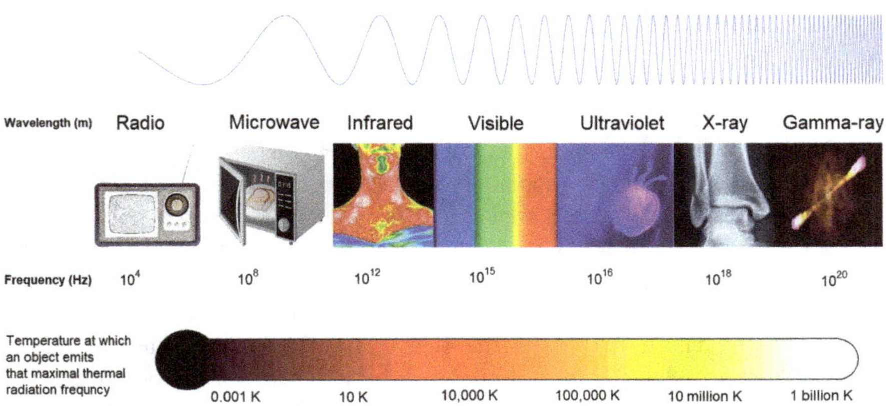

Fig. 1 Electromagnetic spectrum from radio waves to gamma rays. K: Kelvin. Images of radio, microwave, ultraviolet and X-ray examples reproduced under CC0 licences. Image of gamma ray example credited to NASA

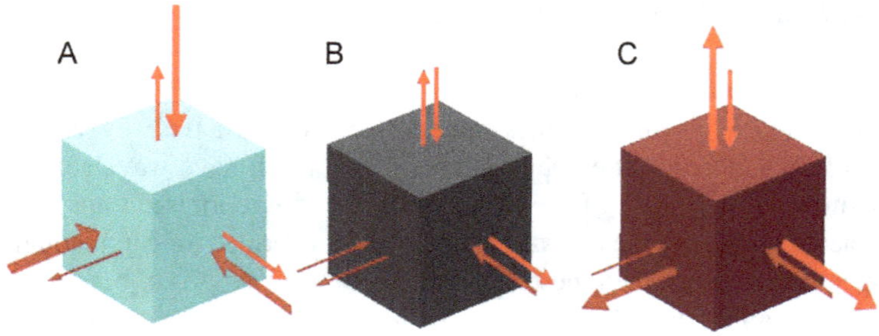

Fig. 2 (**a**) Black body object below thermoneutral point where radiation absorbed (arrows in) is greater than radiation emitted (arrows out) resulting in temperature increasing towards thermoneutral point. (**b**) At thermoneutral point, radiation absorbed is equal to radiation emitted and temperature equilibrium is maintained. (**c**) If the black body object is above the thermal equilibrium temperature, more radiation is emitted than absorbed, and the temperature falls towards the thermoneutral point

Planck's law can be used to solve for an unknown temperature of a black body in thermal equilibrium from the radiometric data measured by IR sensors. A black body at temperature, T, will have a spectral radiance, B_λ, which describes the energy emitted at different wavelengths, λ.

$$B_\lambda(\lambda, T) = \frac{2hc^2}{\lambda^5}\frac{1}{e^{\frac{hc}{\lambda k_B T}} - 1} \tag{1}$$

where h is the Planck constant, k_B is the Boltzmann constant, and c is the speed of light.

A black body is a hypothetical entity. Real objects deviate from being ideal black bodies by a greater or lesser amount and the degree of deviation is denoted by the emissivity (ε). Black bodies have an emissivity of 1 and a perfect reflector would have an emissivity of 0. Real-world objects lie somewhere between these two extremes. To calculate an object's temperature, the conversion equation must allow for its emissivity. For example, given an object with $\varepsilon = 0.7$, 70% of the radiation measured would be radiated from the object and 30% would be reflected from the environment. Therefore, to calculate the unknown temperature of an object, the object's emissivity and the apparent reflected temperature must be known. The apparent reflected temperature is measured using a reflector.

The emissivity of human skin is usually taken to be 0.98 and is constant across the long-wavelength IR range (Steketee 1973). As only 2% of the measured radiation from human skin is reflected, errors in measured environmental and reflected temperature will have a relatively small effect on the calculated skin temperature.

Some radiation emitted by the object will be lost between the object and the receiver. Most notably, water absorbs IR radiation (Roberts et al. 1976) and so the amount of water vapour between the object and the receiver affects measurements. The effect can be mitigated by having knowledge of the object-receiver distance, the relative atmospheric humidity and the atmospheric temperature. Similarly, if the object and receiver are separated by an external 'window' (i.e. an object that is at least partially transparent to IR radiation), the properties of the window must also be known and allowed for. For measurement of brown adipose tissue (BAT), the IR camera is typically less than 1 m from the participant being imaged and the participant and camera are not separated by an IR window. Over such short distances, the absorption of IR radiation by water vapour is minimal.

Finally, maximal radiation is emitted perpendicular to the surface of the object. The plane of the receiver should ideally be placed parallel to the plane of the object along an imaginary line perpendicular from the object with the plane of the detector. Any deviation from this optimal position will reduce the signal being detected and will result in an underestimate of temperature.

Key Points

- Knowledge of emissivity and reflected temperature are necessary to calculate the unknown temperature of an object of interest.
- The emissivity of human skin is 0.98.
- Knowledge of atmospheric temperature and relative humidity are necessary to calculate the loss of signal between the object and sensor.

3 Infrared Thermography

Work during the Second World War accelerated the development of IRT making it a practical reality. It has many beneficial properties that have allowed it to find uses in a wide range of disciplines. IR radiation can be measured remotely and without destruction of the object being studied. This makes its measurement a useful tool both in conservation and heritage work (Candoré et al. 2012) and in structural or electronic engineering (Clark et al. 2003; King et al. 2000); being part of the electromagnetic spectrum, it can travel in a vacuum, allowing the measurement of celestial objects in astronomy (Baldwin et al. 1973). Within medicine, it was used for the identification of breast malignancy in the 1950s. For this particular purpose, alternative modalities of ultrasound and mammography became dominant, but IRT continues to be used in many health-related areas (Ring and Ammer 2012) such as sports medicine (Hildebrandt et al. 2012) and arthritis (Collins and Cosh 1970; Ring and Collins 1970), and its use in malignancy continues to be of research interest (Arora et al. 2008; Kontos et al. 2011).

A modern IR camera measures the radiation arriving at a sensor array that has been calibrated by the manufacturer against a black body and stores the raw signal within the metadata of the image file (see Sect. 6.1). Using the principles outlined above and the calibration constants determined by the manufacturer, the radiometric measurements are converted to temperature values in the preferred units (degrees Celsius, degrees Fahrenheit or Kelvin). Finally, a colourmap can be defined where

A

20476	20525	20576	20626	20676	20727
20178	20227	20277	20326	20376	20426
19885	19933	19982	20031	20080	20129
19595	19643	19691	19740	19788	19836
19310	19358	19405	19452	19500	19548

B

35.79	36.03	36.28	36.52	36.76	37
34.34	34.59	34.83	35.07	35.31	35.55
32.9	33.14	33.38	33.62	33.86	34.1
31.45	31.69	31.93	32.17	32.41	32.66
30	30.24	30.48	30.72	30.97	31.21

C

Fig. 3 Conversion of (**a**) example radiometric data to (**b**) temperature data (degrees Celsius) and displayed as (**c**) a thermal image with associated colour bar and scale alongside

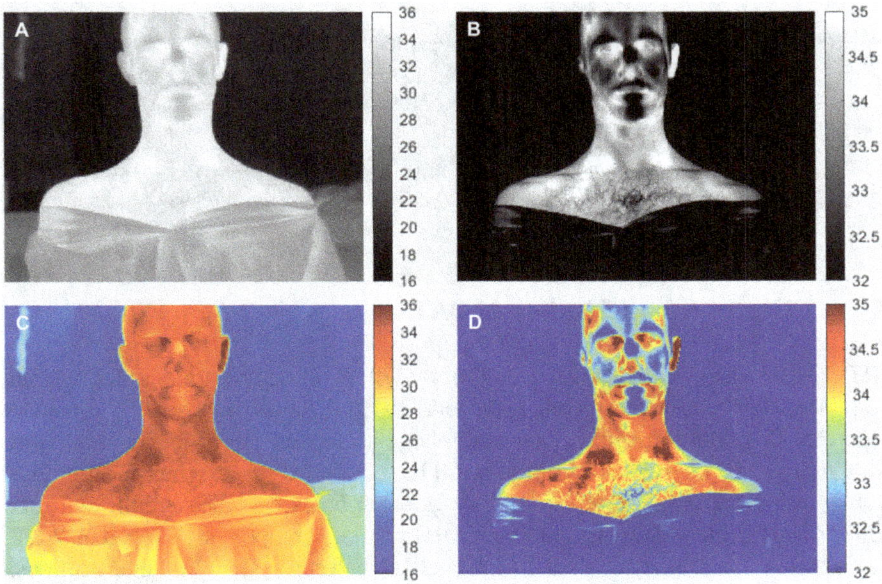

Fig. 4 False-colour images of the same thermal image (**a**, **b**) in greyscale and (**c**, **d**) using the 'jet' palette (MATLAB™) with (**a**, **c**) a wide temperature range and (**b**, **d**) a narrow temperature range. Comparison of **a** and **b** demonstrates the importance of choosing an appropriate temperature range to ascertain detail within the region of interest. The 'jet' colourmap does not have uniform luminance and can cause the false appearance of thermal contrast (e.g. the bottom section of background appears to contrast with the participant's clothing more in **c** than is evident in **a**)

each temperature value is assigned a colour value, allowing the data to be graphically displayed as a false-colour thermal image (Fig. 3).

The colourmap and range can be adapted as required (Fig. 4). The default setting in image viewing software will be to scale the colourmap over the full range of temperature in the image. The background of the image is likely to be cooler than the participant and of little interest. Setting a narrow temperature range is more appropriate for BAT analysis as the dynamic range of the area of interest is limited and because a wide temperature range will reduce the visual thermal resolution of the image.

The choice of colourmap can markedly affect the appearance of the final image and can create the appearance of demarcated regions and features where they do not exist. This can be the case with commonly used colourmaps such as 'jet' (MATLAB™, MathWorks, Natick, MA, USA) and 'rainbow' (FLIR Systems, Wilsonville, OR, USA) and is due to peaks in the luminance of the colourmap. The best colourmaps are sequential ones with a uniform luminance gradient – and the simplest of these is greyscale.

Summary of Stages of Infrared Thermography

1. The radiometric signal arriving at a sensor array is measured and the value stored.
2. The raw signal is converted to temperature values.
3. Temperature data is displayed as a false-colour image.

4 Use of IRT in Studies of BAT

IRT was first used to investigate BAT in the 1970s to demonstrate a thermogenic response to ephedrine (Rothwell and Stock 1979). The non-invasive, non-contact nature of IRT makes it suitable for use in large-scale studies, for repeat imaging in longitudinal and intervention studies, and it has been used for imaging of children as young as 3 years old (Symonds et al. 2012). It can be used to measure the real-time effect of acute interventions including meals or dietary components (Ang et al. 2017; Lee et al. 2011; Scotney et al. 2017), as well as the effect of chronic interventions.

BAT depots in humans are located in multiple anatomical locations (Heaton 1972; Leitner et al. 2017). Of these, the supraclavicular (SCV) BAT depot is most amenable to imaging with IRT due to its relatively superficial location, underneath the subcutaneous adipose tissue in the SCV fossa. IRT measures *surface* temperature and so, for the heat signature of the BAT depot to be detected, at least a proportion of the heat must be transmitted through the tissues to the surface. The more insulating tissue present (for instance, increased subcutaneous adipose tissue), the less heat that will be transmitted to the surface. This can pose a challenge if trying to compare results between individuals directly or if intra-participant changes are sought in a longitudinal study where adiposity has varied through the study duration. Methods to measure local subcutaneous adipose thickness and directly adjust IRT measurements are not yet established, but validated indirect adjustments are discussed below (see Sect. 6.4).

Since the evolutionary purpose of BAT is adaptive thermogenesis (Cannon and Nedergaard 2004), the location of the SCV BAT depot close to the major vessels of the neck suggests a role in maintaining cranial temperature (Smith and Horwitz 1969). Therefore, at least some of the heat must be transmitted into the body and not available to be measured with IRT. In addition, the skin surface temperature changes are affected by other homeostatic mechanisms, most notably blood flow (Braverman 2000), controlled by arteriovenous anastomoses under sympathetic nervous system (SNS) and hormonal regulation (Ootsuka and Tanaka 2015) in response to temperature changes, hypovolaemia and arousal. It is important, therefore, that these factors are controlled when interpreting skin temperature in measurements of BAT activity.

Despite these potential limitations, IRT is reproducible (Haq et al. 2017) and has been shown repeatedly to be able to detect heat changes following cold stimulation (Law et al. 2017a), typically exposure of a limb to cold water or a cooling blanket (see Sect. 5.4). In addition, IRT has been shown to correlate well with PET-CT (Law et al. 2017b). 18F-FDG PET-CT measures glucose uptake, whereas IRT measures heat production, which are not equivalent, and so perfect correlation would not be expected. Optimal methods for detecting BAT are still being refined and, at the moment, comparison between studies can be difficult due to variations in methodologies and the details reported (Law et al. 2017a). Within the general field of physiological measurements using IRT, sports and exercise medicine have led the way in developing consensus opinion (Moreira et al. 2017) and many of the recommendations (Table 1) are equally valid when using IRT for BAT measurements.

5 Acquiring IR Thermographs

In addition to the methodological advantages outlined above, IRT has the benefit of requiring minimal, relatively inexpensive equipment, compared to PET-CT and MRI. However, the same rigour and care needs to be observed when collecting data using IRT as with other imaging techniques. The rapid expansion of the use of IRT, largely driven by industrial uses, has brought down the price and increased the specification of equipment, with a resultant increase in image quality and definition (Fig. 5).

- Modern cameras typically measure between 320 × 240 and 640 × 480 pixels, although true high-definition options are available.
- The option for radiometric video is standard on many cameras allowing 30 frames per second, or greater, to be captured, compared with still images of four to six per minute previously.
- Increased sensitivity means that changes of as little as 0.01°C can be measured.

The camera settings must be checked and adjusted for each session as necessary, including distance from the SCV region to the camera, emissivity (0.98 for human skin) and environmental temperature and humidity (measured using a thermohygrometer).

5.1 Environment

Laboratory-based studies should be undertaken in a temperature-controlled environment at thermoneutrality [approximately 22–24°C for an adult in light cotton shorts and vest, away from airflow (Houdas and Ring 1982)]. The assessment should be undertaken away from both sources of infrared radiation (including lighting and heating) and airflow (such as air-conditioning) (Moreira et al. 2017). The nature of IRT means that, where desired, the equipment is highly portable, allowing real-world studies to be undertaken in the field (Robinson et al. 2014; Symonds et al. 2012), but even here it is important to control as many variables as possible and acknowledge where it has not been possible to do so.

Table 1 Recommendations from a Delphi consensus statement on the measurement of human skin temperature with thermographic imaging in sports and exercise medicine

1) The relevant individual data of the participants must be provided.
Note: These could include, but are not limited to, age, sex, body mass, height, body mass index, ethnicity and whether they are smokers or not. An indication of physical activity profile (e.g. frequency, duration, intensity and activity description) should be reported.
❑ Yes ❑ No ❑ Unclear
2) Participants should be instructed to avoid alcohol beverages, smoking, caffeine, large meals, ointments, cosmetics and showering for 4 h before the assessment. Also, sunbathing (e.g. UV sessions or direct sun without protection) should be avoided before the assessment.
Note: This should be confirmed verbally before the assessment. The use of any medicinal treatments or drugs should be recorded. Any condition that could not be avoided should be reported.
❑ Yes ❑ No ❑ Unclear
3) Extrinsic factors affecting skin temperature (e.g. physical activity prior to the assessment, massage, electrotherapy, ultrasound, heat or cold exposure, cryotherapy) should be clearly described.
❑ Yes ❑ No ❑ Unclear
4) Ambient temperature and relative humidity of the location where the assessment took place must be recorded and reported as mean ± standard deviation.
❑ Yes ❑ No ❑ Unclear
5) The assessment should be completed away from any source of infrared radiation (e.g. electronic devices, lightning) or airflow (e.g. under an air conditioning unit).
Note: Any condition that could not be controlled should be reported.
❑ Yes ❑ No ❑ Unclear
6) The manufacturer, model and accuracy of the camera used should be provided.
Note: When available it is recommended to provide the maintenance information of the equipment (e.g. when and where it completed the last calibration).
❑ Yes ❑ No ❑ Unclear
7) An acclimation period in the examination room should be completed.
Note: This item is only applicable for initial baseline measurements or basal analysis.
❑ Yes ❑ No ❑ Unclear
8) If necessary the camera should be turned on for some time prior to the test to allow sensor stabilisation following the manufacturer's guidelines.
❑ Yes ❑ No ❑ Unclear
9) Conditions of image recording such as mean distance between object and camera and percentage of the region of interest within the image should be detailed.
❑ Yes ❑ No ❑ Unclear
10) The camera should be positioned perpendicular to the region of interest.
❑ Yes ❑ No ❑ Unclear
11) Emissivity settings of the camera must be reported.
Note: 0.98 of emissivity is suggested for a dry clean skin surface.
❑ Yes ❑ No ❑ Unclear
12) The time of day at which the images were taken should be reported.
❑ Yes ❑ No ❑ Unclear

(continued)

Table 1 (continued)

13) The standard body position of the subject and the regions of interest must be well described and appropriately selected. A visual example (with temperature scale presented and scale of colours properly configured) is recommended.
❑ Yes ❑ No ❑ Unclear
14) If the skin is dried (e.g. to remove surface water), the drying method should be clearly described.
❑ Yes ❑ No ❑ Unclear
15) The evaluation of thermograms and collection of temperature from the software should be clearly described.
❑ Yes ❑ No ❑ Unclear

Reprinted from Moreira et al. (2017). ©2017 Reproduced with permission from Elsevier

5.2 Participants

Participants should be screened for factors likely to affect BAT activity or skin temperature and excluded unless explicitly required for the study. These should include smoking (Grassi et al. 1992) and medications such as beta-blockers or beta-agonists (Alexander and Stevens 1980; Law et al. 2014; Thureson-Klein et al. 1976). The region to be imaged should be hair-free, as the emissivity of hair results in a lower measured temperature (Fernández-Cuevas et al. 2015). In addition, they should be asked to refrain from showers; from consuming alcohol, caffeine-containing drinks or large meals; and from strenuous activity for at least 4 h prior to imaging (Moreira et al. 2017) and to avoid use of cosmetics and ointments (Engel 1984; Fernández-Cuevas et al. 2015; Steketee 1976). Where feasible, assessment should be undertaken at a consistent time of day to avoid potential circadian variations in BAT activity (Redlin et al. 1992). A detailed account of the factors influencing skin temperature is provided by Fernández-Cuevas et al. (2015).

Participants should be asked to don standardised study clothing, usually consisting of light cotton shorts and vest, as soon after arrival as practical. This allows the participant to begin to acclimatise to the room and, therefore, they should remain in the study room for the duration of this period. The vest should be of a style and fit to allow complete exposure of the neck and shoulders above the line of the clavicle. Basic anthropometric and characteristics must be measured and recorded including height, weight, age, gender, ethnicity, smoking status and medications.

5.3 Positioning

Following initial acclimation, the participant should be invited to sit in an upright position. The head should be in a near-neutral position, slightly extended if necessary to expose the SCV region. The participant should be positioned in such a way to avoid movement during the imaging period. Firstly, this should be such that they can be expected to remain comfortable for the duration of the imaging period; and,

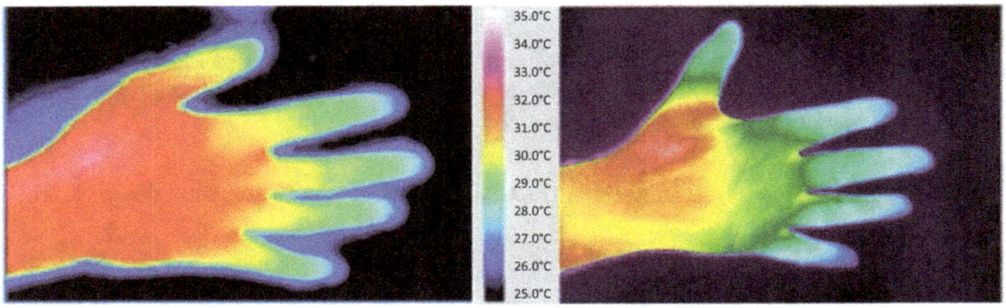

Fig. 5 Infrared thermograph of the dorsum of a hand taken in 1990 (320 × 240 pixels) and at in 2011 (640 × 480 pixels). From Ring and Ammer (2012) © Institute of Physics and Engineering in Medicine. Reproduced by permission of IOP Publishing. All rights reserved

secondly, they should be able to avoid needing to move when the stimulus is introduced (see Sect. 5.4).

They should be directly facing the thermal imaging camera, which should be secured on a tripod able to allow adjustment in height and rotation in all axes. Prior to commencement of imaging, the camera should be positioned such that the receiver is perpendicular to an imaginary line itself perpendicular to the centre of the larynx (Fig. 6). To optimally measure the radiation from the SCV fossa, the camera would ideally be placed along a line perpendicular to that plane, which is externally and cranially rotated relative to the coronal plane. However, the resultant variable and subjective positioning of the camera would reduce reproducibility and would only allow unilateral imaging. The recommended position is a compromise between consistency and maximising detection of radiation. It should, therefore, be recognised that results will underestimate the true SCV temperature.

5.4 Stimulation

BAT stimulation can be altered in several different ways, such as by diet (Kim et al.2015), pharmacological agents (Scotney et al. 2017) or temperature. Stimulation may occur acutely (over minutes) or chronically (over days and weeks). IRT study design can be adapted to measure both acute changes (e.g. increase from a resting baseline) and chronic changes (e.g. measuring change in resting or change in stimulated temperatures at serial study sessions).

Perhaps the most frequently used stimulus is acute cold. This is in the form of either a cold room (Jang et al. 2014; Lee et al. 2011; Ramage et al. 2016) or a cooling stimulus [for instance, cold water (Ang et al. 2017; El Hadi et al. 2016; Peterson et al. 2017; Ramage et al. 2016; Robinson et al. 2014; Symonds et al. 2012) or a cooling blanket (Law et al. 2017b; Salem et al. 2016)] applied directly to the skin but away from the SCV region. An extremity, such as a foot (Ang et al. 2017; Peterson et al. 2017; Symonds et al. 2012; Virtanen et al. 2009) or hand (Ang et al. 2017; El Hadi et al. 2016; Ramage et al. 2016; Robinson et al. 2014; Symonds et al. 2012), is often

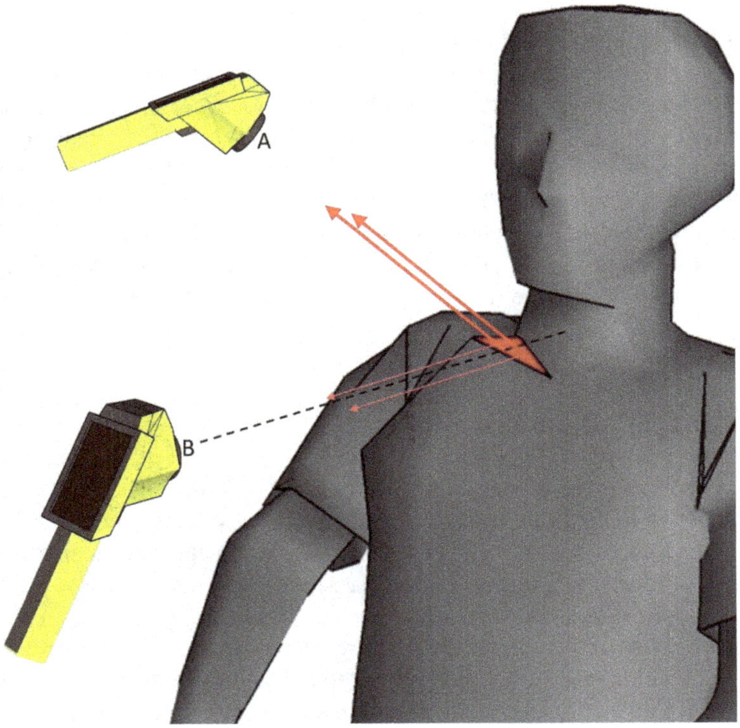

Fig. 6 Representation of infrared radiation emitted by the supraclavicular region (red triangle) which is maximally detected by the camera at position A. While the camera at position B will detect less radiation and therefore underestimate the true skin temperature in the region of interest, the position is reproducible and therefore recommended. From Law et al. (2017a) reproduced by permission of Taylor & Francis Ltd., http://www.tandfonline.com

used as the site of cooling. If a cooling stimulus is to be used, care should be taken to avoid change of body position when the stimulus is introduced. In the case of cold water, the hand (or foot) should therefore be positioned in an empty receptacle prior to image acquisition period. For the same reason, if a cooling blanket is to be used, this should be put in place at the start of imaging, without flow and at a neutral temperature initially.

Cooling should be sufficient to achieve SNS BAT activation, but not so extreme as to induce shivering, an alternative adaptive thermogenesis mechanism, or pain, an alternative SNS activation pathway. Where a standardised temperature is used, the water is typically 15–20°C (Ang et al. 2017; Ramage et al. 2016; Robinson et al. 2014; Symonds et al. 2012).

The alternative is to use an individualised protocol. In this case, the participant is cooled to the point of shivering, then the stimulus temperature is gradually increased until shivering ceases (van der Lans et al. 2013; Vosselman et al. 2012). Shivering should be defined as any one of three measures: participant report, researcher observation and electromyogram (van der Lans et al. 2014). The temperature at which shivering ceases is then used as the participant's individualised stimulus temperature. Although there is much that is appealing about this approach, and it is recommended in PET-CT studies of BAT (van der Lans et al. 2014), caution must be exercised when applying it to IRT studies. If the set point is established on the day, there may be an insufficient washout period and the earlier cooling period may affect the later study session. PET-CT measures substrate uptake at a given point and so is less affected by the effect of earlier cooling than IRT which compares change from baseline (see Sect. 6.4). Shivering points for individuals vary over time, so a set point established prior to the study day may not be valid at a later date (Davis 1961).

5.5 Imaging Duration

There is considerable variation in the length of imaging periods utilised. Acclimatisation to the study room temperature is achieved after an hour (Haq et al. 2017). The change in SCV temperature during acclimatisation is likely to vary depending on the environment that the participant was most recently in. Despite the initial acclimation period, there is frequently a cooling of the skin temperature upon commencement of imaging in our experience, thought to be due to the cessation of voluntary muscle activity in the sedentary state in which imaging is undertaken. For this reason, a prolonged baseline period of imaging is recommended. A baseline period of at least 10 min is necessary to achieve a steady SCV temperature prior to introduction of an acute stimulus. To demonstrate achievement of a steady state, at least three measurements in the last 5 min of the baseline period are required.

The initial response to cold stimulus is rapid. Five minutes of stimulation is sufficient to detect a change from baseline following the introduction of a direct contact cold stimulus (Symonds et al. 2012), but longer periods of up to an hour may show a further increase (Law et al. 2017b). Detailed longitudinal information about the response to a cold environment challenge is not available, but studies have reported results between 30 min (Lee et al. 2011) and 2 h (Jang et al. 2014; Ramage et al. 2016).

Images should be collected at regular intervals throughout the baseline and stimulation periods. Still images should be collected at a rate of at least four to six images per minute, as the camera will allow. Less frequent capture is likely to permit significant movement between images which both reduces image consistency and increases thermogenesis from voluntary muscle movement. Video is captured at typically up to 30 frames per second and may allow more detailed changes to be seen, but the increased data set created may add complexity to the data analysis phase (see Sect. 6).

5.6 Additional Measurements

To interpret fully the data from the infrared thermographs, additional physiological measurements are necessary. These should include measures of SNS activation including heart rate and blood pressure. In addition, core temperature should be measured at the start and end of the session as a minimum, and at regular intervals of longer imaging sessions.

Additional skin temperature measurements are also valuable to use to compare to the changes in BAT temperature (discussed further in Sect. 6.4).

6 Image Analysis

6.1 Conversion

The details of the file structure of thermographs will vary depending on the manufacturer and equipment used, and the researcher will need to be familiar with the particular system in use. The information discussed here is based on experience with FLIR® equipment but will be applicable to many others.

Thermographs are stored in a variety of file formats including, but not limited to, JPEG and TIFF. The radiometric data is stored distinctly from the displayed false-colour image and may be obfuscated to a greater or lesser extent to encourage use of proprietary software for image analysis. The displayed false-colour image cannot be analysed in any meaningful way- and temperature data can only be approximately estimated by comparison to a colour bar. The radiometric data within the file must be accessed and converted to temperature data before progressing. This may be done by utilising proprietary software. Free versions will typically have limited functionality especially of the type required for analysis of BAT. Commercially available versions have a wider range of tools, and it is important to ensure that those required are included.

Alternatively, data can be accessed outside of manufacturers' proprietary software, and this usually requires a conversion step first. Options for conversion include:

- Opening images within proprietary software that allows saving in a variety of formats, such as comma-separated values files (CSV) or MATLAB™ data files (MAT).
- Conversion to an openly accessible format such as portable network graphics (PNG) (Law et al. 2017a, b) using a Python-based script, or equivalent, where the raw radiometric data is stored as the image rather than in the metadata (see Sect. 3).
- FLIR®, for instance, provides a MATLAB™ executable file that allows images taken on FLIR®-manufactured devices, including videos, to be imported into the MATLAB™ programming environment directly, following which data can be analysed in a flexible way.

As well as allowing more flexible analysis, conversion also protects against data loss from proprietary file formats becoming obsolete. This is a significant risk as manufacturers continue to develop new cameras with new file formats and may stop supporting the current ones.

6.2 Region of Interest Identification

Once the image has been opened within a suitable environment, it must be analysed to identify the region of interest (ROI). The shape and border used to define the ROI varies significantly between publications. Early studies used circles or simple polygons to identify the region of interest (Jang et al. 2014; Kim et al. 2014; Lee et al. 2011; Symonds et al. 2012). Where small ROIs over the SCV fossa have been used, the output is usually a summary measure of the whole ROI (e.g. mean or maximum temperature) (Gatidis et al. 2016; Jang et al. 2014). This approach has the problem of requiring subjective placement.

Alternatively, a larger ROI can be identified and a BAT "hotspot" is defined within it: a summary measure of the hotspot being calculated as the output measure (Robinson et al. 2014; Symonds et al. 2012). The latter approach has progressed from simple ROIs to more refined options. Current techniques include the automatic identification of the shoulder contour (Fig. 7) to allow rapid and accurate analysis of large numbers of images (Law et al. 2017b; Scotney et al. 2017). The hotspot is then defined either as an upper percentile of pixels within the ROI (Symonds et al. 2012) or using a watershed method, with a summary measure calculated. Alternatively, a seeded-region growing algorithm approach can be used to define the area of BAT activity, which avoids imposing a fixed size or sharp cut-off (Ang et al. 2017). However, this limits the number of images analysed due to the computational time required and has not yet demonstrated a definite advantage over simpler methods.

6.3 Averaging

As with any other imaging modality, or measurement, each radiometric datum is subject to an error term. While summary measures will reduce the error by averaging across thousands of pixels, time series demonstrate variation around the underlying trend due to physiological rhythms, inherent noise in the measurement or a combination of the two (Fig. 8). Studies that use only single or isolated images (Gatidis et al. 2016; Kim et al. 2014; Peterson et al. 2017), or only acquire one image per minute (Robinson et al. 2014; Symonds et al. 2012), are, therefore, exposed to significantly reduced reproducibility. In contrast, radiometric video, while increasing the noise in each individual frame, allows frames to be flattened in the z-axis (Fig. 9), increasing accuracy. In general, averaging across n frames will reduce the noise (here defined as standard deviation) by $\sqrt{n}$. When considering the noise in data, it is usually useful to compare the noise to the mean as noise is frequently expected to increase proportionately to the mean. This can either be undertaken by calculating a

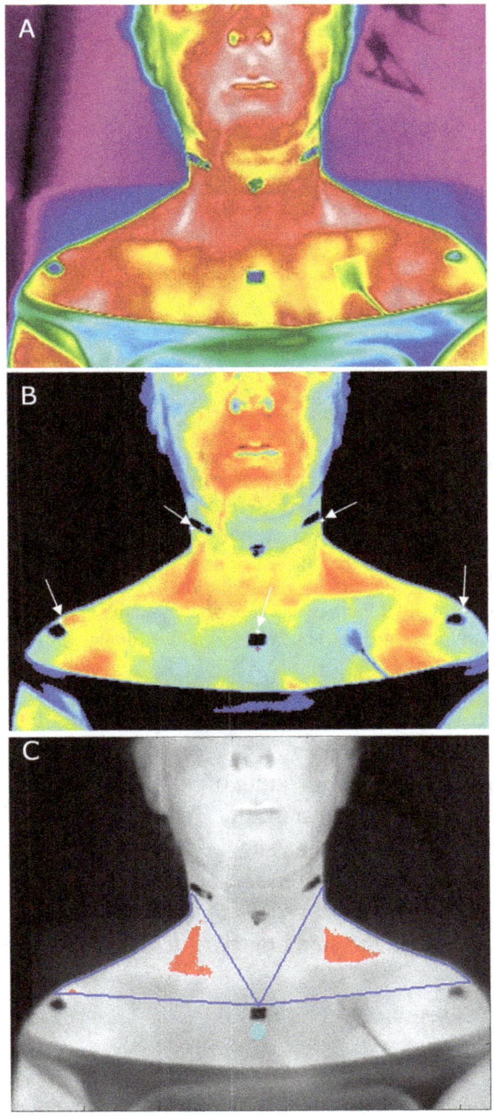

Fig. 7 Contour of region of interest, including border of the shoulder, automatically identified following manual identification of apices. (**a**) Original thermal image (**b**) apices identified (white arrows) (**c**) output image highlighting ROI contour (blue), BAT 'hotspot' (red) and sternal reference region (cyan)

signal-to-noise ratio (where SNR is the mean divided by standard deviation) or a coefficient of variation (CV% is the standard deviation divided by mean, multiplied by 100). As an example, the same absolute accuracy would not be expected from a heavy goods vehicle weighing bridge as would be expected from a set of laboratory scales, but the ratio between the signal (weight) and noise may be similar.

Comparison of this form is less straightforward when dealing with temperature because it is not a linear proportionate scale, i.e. a temperature of 15°C is not half of a

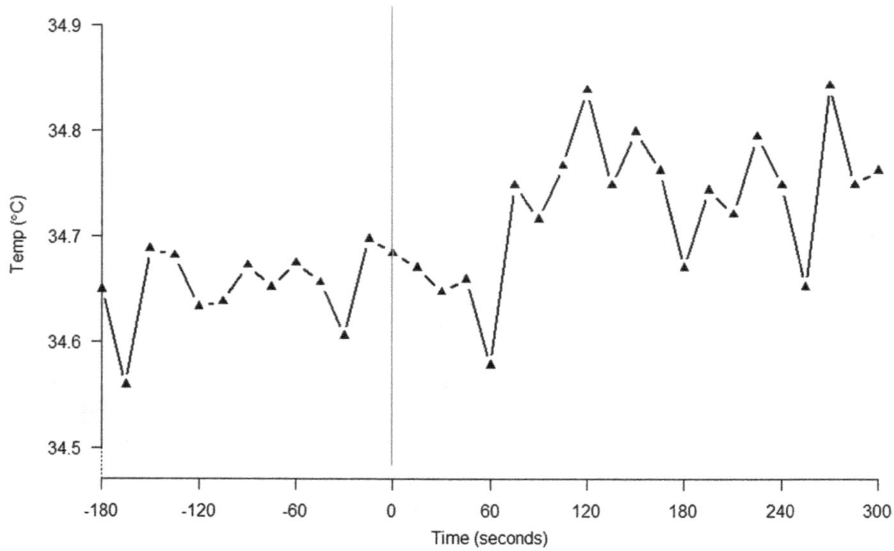

Fig. 8 Time series of infrared thermography measurements from a single participant, with four images per minute over 5 min of stimulation showing a rising trend following introduction of a cold stimulus (blue line) but with significant noise

temperature of 30°C. It follows then that the error in a measurement with a mean temperature of 15°C would not be expected to be half of the error in a measurement with a mean temperature of 30°C. Secondly, human skin temperature is generally in a small dynamic range of ~30–37°C (Law et al. 2017a). There is little change in the mean, and so changes in the SNR or CV are primarily determined by changes in the noise rather than the signal. In theory, a signal could be calculated from the raw radiometric data or from absolute temperature (Kelvin), but this would further emphasise the small range of temperature measurements.

6.4 Measures

Whether or not attempts are made to reduce noise by averaging frames, once the region of interest has been identified, an output measure will need to be calculated. The output measure selected will depend to an extent on the ROI chosen and the software being used. Most proprietary software will allow at least rectangular and elliptical ROIs to be drawn and, from these, provide a minimum, maximum and, possibly, mean temperature. If the ROI has been chosen to represent a small area over a potential BAT deposit, these may not be unreasonable options but are all susceptible to influence from extreme values, especially minimums and maximums. In addition, each image will need to be analysed by hand and the output recorded manually, risking transcribing errors and setting a practical limit on the number of images analysed.

If a larger ROI has been selected, from which the BAT 'hotspot' forms a smaller part, this should be reflected in the outcome measure chosen. The top 10% of temperature pixels within a SCV ROI as previously described (see Sect. 6.2) can be considered to represent the BAT hotspot. Even if the temperature of pixels in the ROI is normally distributed, the temperature of pixels in the hotspot would not be, and so the mean is no longer an appropriate point measure. Instead, the median value is chosen, which has the additional advantage of being less susceptible to influence from extreme values. The median value of the top 10% of values is equivalent to the 95th percentile of the original ROI, which can be easily calculated by any suitable software such as Excel, R (both by exporting the CSV files of the ROI) or MATLAB™.

As discussed above (see Sect. 4), the temperature of the SCV BAT outcome measure (T_{SCV}) can be influenced by factors other than BAT activity, such as obesity (increased insulation) and environmental temperature (skin perfusion). Perfusion can be accounted for by comparing T_{SCV} to a reference skin temperature (T_{Ref}). Most easily, this is undertaken by selecting a reference region within the thermal image over a non-BAT area such as the sternal (Law et al. 2017b) or deltoid regions (Fig. 7), but these are limited to the frame of the image. An alternative is to calculate

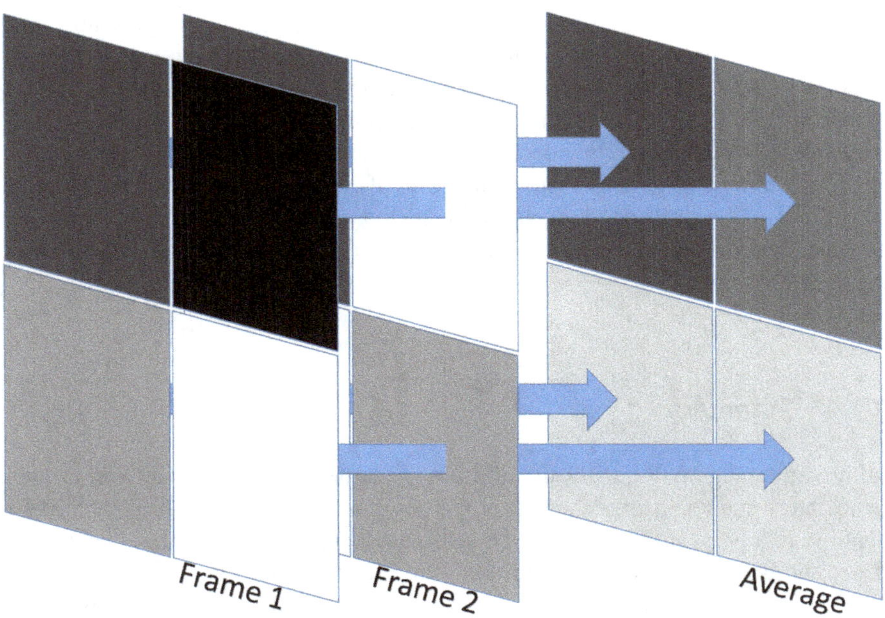

Fig. 9 Simplified example of effect of frame averaging. Frame 1 and Frame 2 are images of a homogenous radiation source and each consist of a 2 × 2 array of 2-bit (0–3) encoded pixels represented using a greyscale. Each pixel in the averaged frame is the mean of the corresponding pixels in Frames 1 and 2, reducing the variation in measurements

a mean skin temperature (Robinson et al. 2016) using contact skin temperature probes (such as iButtons™ or thermocouples attached to real-time display) to measure multiple points simultaneously and continuously. The median temperature of the non-BAT area or the mean skin temperature (T_{Ref}) can then be compared to T_{SCV} to calculate a relative temperature (T_{Rel}) (Eq. 2).

$$T_{Rel} = T_{SCV} - T_{Ref} \tag{2}$$

Relative supraclavicular temperature (T_{Rel}) calculated as the difference between absolute supraclavicular temperature (T_{SCV}) and a reference temperature (T_{Ref}).

Finally, change following stimulation (ΔT) can be calculated from baseline resting temperature (Eq. 3). Note that in some publications, ΔT is defined as the difference between BAT and non-BAT skin temperature, here called T_{Rel}.

$$\Delta T = \text{stimulated } T - \text{resting } T \tag{3}$$

Change in temperature (ΔT) from resting following stimulation.

ΔT further uses each participant as their own control, standardising results and better allowing direct comparison between individuals. ΔT_{Rel} has been shown to correlate closely with BAT activity measured using PET-CT (Law et al. 2017b). ΔT_{Rel} relies on being able to measure both a rested ('off') state and a stimulated ('on') state. A participant who has maximally activated BAT at baseline may show no further increase with stimulation and, therefore, provide indistinguishable data from the participant who has minimal active BAT in both resting and stimulated states (Law et al. 2017a). This is particularly pertinent where individualised cooling protocols (see Sect. 5.4) are being determined on the day of stimulation, potentially resulting in a failure to fully induce a resting state in BAT activity at baseline. For studies of chronic BAT activation, comparison of changes in stimulated temperature over the duration of the study would mitigate these potential pitfalls. Acute studies are able to produce reliable results where the above issues are carefully considered in the design.

7 Summary

Infrared thermography utilises the intrinsic properties of infrared radiation to infer temperature and reconstruct the heat properties of a scene. The supraclavicular BAT depot is relatively superficial and the heat-generating property of this major BAT depot can be detected directly using inexpensive and readily available equipment. Cooling produces a reproducible increase in the temperature of the skin overlying the supraclavicular BAT depot which can be determined by careful identification and analysis of the region of interest. By comparing the results to a non-BAT reference region and calculating change from resting temperature, results from IRT show strong correlation with PET-CT. In this way, the non-invasive, non-contact methods of IRT open the possibility of studying populations and interventions not previously possible.

References

Alexander G, Stevens D (1980) Sympathetic innervation and the development of structure and function of brown adipose tissue: studies on lambs chemically sympathectomized in utero with 6-hydroxydopamine. J Dev Physiol 2:119–137

Ang QY, Goh HJ, Cao Y, Li Y, Chan SP, Swain JL, Henry CJ, Leow MK (2017) A new method of infrared thermography for quantification of brown adipose tissue activation in healthy adults (TACTICAL): a randomized trial. J Physiol Sci 67:395–406. https://doi.org/10.1007/s12576-016-0472-1

Arora N, Martins D, Ruggerio D, Tousimis E, Swistel AJ, Osborne MP, Simmons RM (2008) Effectiveness of a noninvasive digital infrared thermal imaging system in the detection of breast cancer. Am J Surg 196:523–526. https://doi.org/10.1016/j.amjsurg.2008.06.015

Baldwin JE, Harris CS, Ryle M (1973) 5 GHz observations of the infrared star MWC 349, and the H II condensation W3(OH). Nature 241:38–39

Braverman IM (2000) The cutaneous microcirculation. J Investig Dermatol Symp Proc 5:3–9. https://doi.org/10.1046/j.1087-0024.2000.00010.x

Candoré JC, Bodnar JL, Detalle V, Grossel P (2012) Non-destructive testing of works of art by stimulated infrared thermography. Eur Phys J Appl Phys 57:21002. https://doi.org/10.1051/epjap/2011110266

Cannon B, Nedergaard J (2004) Brown adipose tissue: function and physiological significance. Physiol Rev 84:277–359. https://doi.org/10.1152/physrev.00015.2003

Celi FS (2009) Brown adipose tissue--when it pays to be inefficient. N Engl J Med 360:1553–1556. https://doi.org/10.1056/NEJMe0900466

Clark MR, McCann DM, Forde MC (2003) Application of infrared thermography to the non-destructive testing of concrete and masonry bridges. NDT E Int 36:265–275. https://doi.org/10.1016/S0963-8695(02)00060-9

Collins AJ, Cosh JA (1970) Temperature and biochemical studies of joint inflammation. A preliminary investigation. Ann Rheum Dis 29:386

Cypess AM, Lehman S, Williams G, Tal I, Rodman D, Goldfine AB, Kuo FC, Palmer EL, Tseng YH, Doria A, Kolodny GM, Kahn CR (2009) Identification and importance of brown adipose tissue in adult humans. N Engl J Med 360:1509–1517. https://doi.org/10.1056/NEJMoa0810780

Davis TR (1961) Chamber cold acclimatization in man. J Appl Physiol 16:1011–1015

El Hadi H, Frascati A, Granzotto M, Silvestrin V, Ferlini E, Vettor R, Rossato M (2016) Infrared thermography for indirect assessment of activation of brown adipose tissue in lean and obese male subjects. Physiol Meas 37:N118–N128. https://doi.org/10.1088/0967-3334/37/12/n118

Engel JM (1984) Physical and physiological influence of medical ointments on infrared thermography. In: Ring EFJ, Phillips B (eds) Recent advances in medical thermology. Springer, New York, pp 177–183

Fernández-Cuevas I, Bouzas Marins JC, Arnáiz Lastras J, Gómez Carmona PM, Piñonosa Cano S, García-Concepción MÁ, Sillero-Quintana M (2015) Classification of factors influencing the use of infrared thermography in humans: a review. Infrared Phys Technol 71:28–55. https://doi.org/10.1016/j.infrared.2015.02.007

Gatidis S, Schmidt H, Pfannenberg CA, Nikolaou K, Schick F, Schwenzer NF (2016) Is it possible to detect activated brown adipose tissue in humans using single-time-point infrared thermography under thermoneutral conditions? Impact of BMI and subcutaneous adipose tissue thickness. PLoS One 11:e0151152. https://doi.org/10.1371/journal.pone.0151152

Grassi G, Seravalle G, Calhoun DA, Bolla G, Mancia G (1992) Cigarette smoking and the adrenergic nervous system. Clin Exp Hypertens A Theory Pract 14:251–260. https://doi.org/10.3109/10641969209036186

Haq T, Crane JD, Kanji S, Gunn E, Tarnopolsky MA, Gerstein HC, Steinberg GR, Morrison KM (2017) Optimizing the methodology for measuring supraclavicular skin temperature using infrared thermography; implications for measuring brown adipose tissue activity in humans. Sci Rep 7:11934. https://doi.org/10.1038/s41598-017-11537-x

Heaton JM (1972) The distribution of brown adipose tissue in the human. J Anat 112:35–39

Herschel W (1800) Experiments on the refrangibility of the invisible rays of the Sun. By William Herschel, LL. D. F. R. S. Philos Trans R Soc Lond 90:284–292. https://doi.org/10.1098/rstl.1800.0015

Hildebrandt C, Zeilberger K, Ring EFJ, Raschner C (2012) The application of medical infrared thermography in sports medicine. In: Zaslav KR (ed) An international perspective on topics in sports medicine and sports injury. InTech, Rijeka

Houdas Y, Ring EFJ (1982) Human body temperature: its measurement and regulation. Springer, New York

Jang C, Jalapu S, Thuzar M, Law PW, Jeavons S, Barclay JL, Ho KKY (2014) Infrared thermography in the detection of brown adipose tissue in humans. Physiol Rep 2:e12167. https://doi.org/10.14814/phy2.12167

Kim MS, Hu HH, Aggabao PC, Geffner ME, Gilsanz V (2014) Presence of brown adipose tissue in an adolescent with severe primary hypothyroidism. J Clin Endocrinol Metab 99:E1686–E1690. https://doi.org/10.1210/jc.2014-1343

Kim M, Goto T, Yu R, Uchida K, Tominaga M, Kano Y, Takahashi N, Kawada T (2015) Fish oil intake induces UCP1 upregulation in brown and white adipose tissue via the sympathetic nervous system. Sci Rep 5:18013. https://doi.org/10.1038/srep18013

King DL, Kratochvil JA, Quintana MA, McMahon TJ (2000) Applications for infrared imaging equipment in photovoltaic cell, module, and system testing. In: Conference record of the twenty-eighth IEEE photovoltaic specialists conference – 2000 (Cat. No.00CH37036), IEEE, Anchorage, pp 1487–1490

Kontos M, Wilson R, Fentiman I (2011) Digital infrared thermal imaging (DITI) of breast lesions: sensitivity and specificity of detection of primary breast cancers. Clin Radiol 66:536–539. https://doi.org/10.1016/j.crad.2011.01.009

Law J, Bloor I, Budge H, Symonds ME (2014) The influence of sex steroids on adipose tissue growth and function. Horm Mol Biol Clin Invest 19:13–24. https://doi.org/10.1515/hmbci-2014-0015

Law J, Chalmers J, Morris DE, Robinson L, Budge H, Symonds ME (2017a) The use of infrared thermography in the measurement and characterization of brown adipose tissue activation. Temperature:1–15. https://doi.org/10.1080/23328940.2017.1397085

Law JM, Morris DE, Izzi-Engbeaya C, Salem V, Coello C, Robinson L, Jayasinghe M, Scott R, Gunn R, Rabiner E, Tan T, Dhillo W, Bloom S, Budge H, Symonds M (2017b) Thermal imaging is a non-invasive alternative to PET-CT for measurement of brown adipose tissue activity in humans. J Nucl Med. https://doi.org/10.2967/jnumed.117.190546

Lee P, Ho KK, Lee P, Greenfield JR, Ho KK, Greenfield JR (2011) Hot fat in a cool man: infrared thermography and brown adipose tissue. Diabetes Obes Metab 13:92–93. https://doi.org/10.1111/j.1463-1326.2010.01318.x

Leitner BP, Huang S, Brychta RJ, Duckworth CJ, Baskin AS, McGehee S, Tal I, Dieckmann W, Gupta G, Kolodny GM, Pacak K, Herscovitch P, Cypess AM, Chen KY (2017) Mapping of human brown adipose tissue in lean and obese young men. Proc Natl Acad Sci U S A 114:8649–8654. https://doi.org/10.1073/pnas.1705287114

Moreira DG, Costello JT, Brito CJ, Adamczyk JG, Ammer K, Bach AJE, Costa CMA, Eglin C, Fernandes AA, Fernández-Cuevas I, Ferreira JJA, Formenti D, Fournet D, Havenith G, Howell K, Jung A, Kenny GP, Kolosovas-Machuca ES, Maley MJ, Merla A, Pascoe DD, Priego Quesada JI, Schwartz RG, Seixas ARD, Selfe J, Vainer BG, Sillero-Quintana M (2017)

Thermographic imaging in sports and exercise medicine: a Delphi study and consensus statement on the measurement of human skin temperature. J Therm Biol 69:155–162. https://doi.org/10.1016/j.jtherbio.2017.07.006

Ootsuka Y, Tanaka M (2015) Control of cutaneous blood flow by central nervous system. Temperature 2:392–405. https://doi.org/10.1080/23328940.2015.1069437

Peterson CM, Orooji M, Johnson DN, Naraghi-Pour M, Ravussin E (2017) Brown adipose tissue does not seem to mediate metabolic adaptation to overfeeding in men. Obesity 25:502–505. https://doi.org/10.1002/oby.21721

Planck M (1914) The theory of heat radiation. P. Blakiston's Son & Co., Philadelphia

Ramage LE, Akyol M, Fletcher Alison M, Forsythe J, Nixon M, Carter Roderick N, van Beek Edwin JR, Morton Nicholas M, Walker Brian R, Stimson Roland H (2016) Glucocorticoids acutely increase brown adipose tissue activity in humans, revealing species-specific differences in UCP-1 regulation. Cell Metab 24:130–141. https://doi.org/10.1016/j.cmet.2016.06.011

Redlin U, Nuesslein B, Schmidt I (1992) Circadian changes of brown adipose tissue thermogenesis in juvenile rats. Am J Phys Regul Integr Comp Phys 262:R504–R508

Ring EFJ, Ammer K (2012) Infrared thermal imaging in medicine. Physiol Meas 33:R33. https://doi.org/10.1088/0967-3334/33/3/R33

Ring EFJ, Collins AJ (1970) Quantitative thermography. Rheumatology 10:337–341. https://doi.org/10.1093/rheumatology/10.7.337

Roberts RE, Selby JEA, Biberman LM (1976) Infrared continuum absorption by atmospheric water vapor in the 8–12-μm window. Appl Opt 15:2085–2090. https://doi.org/10.1364/AO.15.002085

Robinson L, Ojha S, Symonds ME, Budge H (2014) Body mass index as a determinant of brown adipose tissue function in healthy children. J Pediatr 164:318–322. https://doi.org/10.1016/j.jpeds.2013.10.005

Robinson LJ, Law JM, Symonds ME, Budge H (2016) Brown adipose tissue activation as measured by infrared thermography by mild anticipatory psychological stress in lean healthy females. Exp Physiol 101:549–557. https://doi.org/10.1113/ep085642

Rothwell NJ, Stock MJ (1979) A role for brown adipose tissue in diet-induced thermogenesis. Nature 281:31–35. https://doi.org/10.1038/281031a0

Salem V, Izzi-Engbeaya C, Coello C, Thomas DB, Chambers ES, Comninos AN, Buckley A, Win Z, Al-Nahhas A, Rabiner EA, Gunn RN, Budge H, Symonds ME, Bloom SR, Tan TM, Dhillo WS (2016) Glucagon increases energy expenditure independently of brown adipose tissue activation in humans. Diabetes Obes Metab 18:72–81. https://doi.org/10.1111/dom.12585

Scotney H, Symonds ME, Law J, Budge H, Sharkey D, Manolopoulos KN (2017) Glucocorticoids modulate human brown adipose tissue thermogenesis in vivo. Metab Clin Exp 70:125–132. https://doi.org/10.1016/j.metabol.2017.01.024

Smith RE, Horwitz BA (1969) Brown fat and thermogenesis. Physiol Rev 49:330–425

Steketee J (1973) Spectral emissivity of skin and pericardium. Phys Med Biol 18:686

Steketee J (1976) The influence of cosmetics and ointments on the spectral emissivity of skin (skin temperature measurement). Phys Med Biol 21:920

Symonds ME, Henderson K, Elvidge L, Bosman C, Sharkey D, Perkins AC, Budge H (2012) Thermal imaging to assess age-related changes of skin temperature within the supraclavicular region co-locating with brown adipose tissue in healthy children. J Pediatr 161:892–898. https://doi.org/10.1016/j.jpeds.2012.04.056

Tattersall GJ (2016) Infrared thermography: a non-invasive window into thermal physiology. Comp Biochem Physiol A Mol Integr Physiol 202:78–98. https://doi.org/10.1016/j.cbpa.2016.02.022

Thureson-Klein Å, Lagercrantz H, Barnard T (1976) Chemical sympathectomy of interscapular brown adipose tissue. Acta Physiol Scand 98:8–18. https://doi.org/10.1111/j.1748-1716.1976.tb10296.x

van der Lans AAJJ, Hoeks J, Brans B, Vijgen GHEJ, Visser MGW, Vosselman MJ, Hansen J, Jörgensen JA, Wu J, Mottaghy FM, Schrauwen P, van Marken Lichtenbelt WD (2013) Cold

acclimation recruits human brown fat and increases nonshivering thermogenesis. J Clin Investig 123:3395–3403. https://doi.org/10.1172/JCI68993
van der Lans AAJJ, Wierts R, Vosselman MJ, Schrauwen P, Brans B, van Marken Lichtenbelt WD (2014) Cold-activated brown adipose tissue in human adults: methodological issues. Am J Physiol Regul Integr Comp Physiol 307:R103–R113. https://doi.org/10.1152/ajpregu.00021.2014
van Marken Lichtenbelt WD, Vanhommerig JW, Smulders NM, Drossaerts JM, Kemerink GJ, Bouvy ND, Schrauwen P, Teule GJ (2009) Cold-activated brown adipose tissue in healthy men. N Engl J Med 360:1500–1508. https://doi.org/10.1056/NEJMoa0808718
Virtanen KA, Lidell ME, Orava J, Heglind M, Westergren R, Niemi T, Taittonen M, Laine J, Savisto NJ, Enerback S, Nuutila P (2009) Functional brown adipose tissue in healthy adults. N Engl J Med 360:1518–1525. https://doi.org/10.1056/NEJMoa0808949
Vosselman MJ, van der Lans AAJJ, Brans B, Wierts R, van Baak MA, Schrauwen P, van Marken Lichtenbelt WD (2012) Systemic β-adrenergic stimulation of thermogenesis is not accompanied by brown adipose tissue activity in humans. Diabetes 61:3106–3113. https://doi.org/10.2337/db12-0288
Vosselman MJ, Brans B, van der Lans AA, Wierts R, van Baak MA, Mottaghy FM, Schrauwen P, van Marken Lichtenbelt WD (2013) Brown adipose tissue activity after a high-calorie meal in humans. Am J Clin Nutr 98:57–64. https://doi.org/10.3945/ajcn.113.059022
Vrieze A, Schopman JE, Admiraal WM, Soeters MR, Nieuwdorp M, Verberne HJ, Holleman F (2012) Fasting and postprandial activity of brown adipose tissue in healthy men. J Nucl Med 53:1407–1410. https://doi.org/10.2967/jnumed.111.100701

In Vivo Detection of Human Brown Adipose Tissue During Cold and Exercise by PET/CT

Emmani B. M. Nascimento and Wouter D. van Marken Lichtenbelt

Contents

1 Introduction 284
2 Positron Emission Tomography and Computed Tomography 285
3 Tracers 285
3.1 [^{18}F]-Fluorodeoxyglucose 286
3.2 14-[^{18}F]Fluoro-6-Thiaheptadecanoic Acid 287
4 Human Brown Adipose Tissue Activity 287
4.1 Acute Cold Exposure and Cold Acclimatization 287
4.2 Pharmacological Activation of Brown Adipose Tissue 288
4.3 Exercise Stimulated Brown Adipose Tissue Activity 289
4.3.1 Direct Effects of Exercise on Brown Adipose Tissue Activity 289
4.3.2 Indirect Effects of Exercise on Brown Adipose Tissue Activity 290
5 Critical Evaluation of Commonly Used Tracers to Measure Human Brown Adipose Tissue Activity 292
References 293

Abstract

The role of brown adipose tissue (BAT) in non-shivering thermogenesis is well established in animals. BAT is activated following cold exposure, resulting in non-shivering thermogenesis, to ensure a constant body temperature. In mitochondria of brown adipocytes, glucose and fatty acids are used as substrate for uncoupling resulting in heat production. Activated BAT functions as a sink for glucose and fatty acids and this hallmark has designated BAT a target in the fight against metabolic diseases like type 2 diabetes mellitus and obesity. In order to make valid claims regarding BAT activity in humans, BAT activity needs to be quantified. The combination of positron emission tomography (PET) and computer tomography (CT) analysis is currently the most frequently used imaging

E. B. M. Nascimento · W. D. van Marken Lichtenbelt (✉)
Maastricht University, Maastricht, The Netherlands
e-mail: markenlichtenbelt@maastrichtuniversity.nl

A. Pfeifer et al. (eds.), *Brown Adipose Tissue*,
Handbook of Experimental Pharmacology 251, https://doi.org/10.1007/164_2018_121

technique to determine BAT activity in humans. Here, we will discuss the history of PET/CT and radioisotopes used to determine BAT activity in humans. Moreover, we will assess how PET/CT is used to determine BAT activity following cold and exercise.

Keywords
[^{18}F]FDG · [^{18}F]FTHA · Brown adipose tissue · Cold · Exercise · PET/CT · Radioisotope

1 Introduction

Energy homeostasis in humans is a delicate balance between energy intake and energy expenditure. When the balance tips towards a surplus of energy intake, this leads to storage of the excess of energy in white adipose tissue (WAT). Too much WAT results in obesity, which in turn may lead to the metabolic syndrome.

Brown adipose tissue (BAT), on the other hand, does not store but uses energy when activated. BAT contains many mitochondria. In these mitochondria, glucose and fatty acids are combusted to heat in a process called mitochondrial uncoupling (Cannon and Nedergaard 2004). This uncoupling process is executed by uncoupling protein 1 (UCP1), a unique inner membrane mitochondrial protein for BAT. UCP1 uncouples the electron transport chain, causing increased proton leak over the mitochondrial inner membrane. This results in production of heat instead of ATP. Activated BAT uses huge amounts of energy. In rodents, this comes down to 300 W/kg brown fat (Rothwell and Stock 1983), which is about two orders of magnitude higher than normal metabolic rate of other tissues. Because of this BAT hallmark, BAT has been coined a potential target to combat detrimental effects of increased WAT mass. The most effective way to activate BAT is via cold exposure that results in release of norepinephrine from the sympathetic neuron onto the brown adipocyte.

In 2009, several research groups demonstrated the presence of functionally (i.e., in response to cold) active BAT in adult humans, which reignited the interest in BAT. BAT is mostly appreciated for thermogenesis, so detection of heat production or change in temperature would be the most direct approach to measure BAT activity. However, directly measuring heat production is invasive, and apart from measuring temperature one needs to obtain a measure of heat flow through the tissue. This is not feasible in humans. Measuring infrared radiation is an alternative to measure temperature, but this only provides the temperature of the skin surface overlaying the BAT region, not the temperature or heat production of the tissue itself. At best, this technique can provide an indication of BAT activation. The most widely used technique to detect active BAT in humans is via the combined positron emission tomography and computed tomography (PET/CT) using the glucose tracer [^{18}F]-fluorodeoxyglucose ([^{18}F]FDG). This chapter is dedicated to the origin of PET/CT analysis and how this technique is currently used to measure BAT activity in humans following cold stimulation or exercise.

2 Positron Emission Tomography and Computed Tomography

PET is a method to detect gamma rays from a positron-emitting radioisotope. Positrons released by a beta-plus radioisotope collide with an atomic electron resulting in a release of two gamma photons in opposite directions. The travel path of these photons provides information regarding the source. This way, specific radioisotopes can provide information regarding metabolic status of the organ of interest. However with PET, tissue identification is not possible. Therefore, PET is nowadays combined with computer tomography (CT) for anatomical information. CT uses X-ray scans from different angles to create cross-sectional images thereby allowing spatial resolution to occur. Besides tissue identification, the CT derived Hounsfield scale provides a quantitative scale for describing radiodensity, which can be used to determine the relative water content of BAT. After activation of BAT, several studies (Chondronikola et al. 2016; Ouellet et al. 2012) showed decreased density implying the use of internal fat stores for combustion. The two noninvasive methods combined result in PET/CT. Instead, PET/MRI (magnetic resonance imaging) systems gradually become more available and make it possible to detect new magnetic resonance (MR)-based markers and at the same time use PET tracers. The amount BAT studies using PET/MRI are currently very limited (Deng et al. 2017).

The combination of PET and CT was applied in the 1990s by David Townsend and Ronn Nutt. In 1991, a PET scanner using rotating banks of bismuth germanate detectors was created by David Townsend and collaborators in Geneva. Oncologist Rudi Egeli suggested including an imaging unit, because there was space between the rotating detectors. For the first time, PET and CT were combined in such a way that resulted in one single entity that rotates in order to image human beings. First PET/CT scanner was designed, built by CTI PET system in Knoxville, and used in 1998. The use of this PET/CT in a clinical setting involving cancer patients of PET was a valuable addition for physicians to exactly localize what happened where in their patients (Charron et al. 2000). These positive results generated the need for more commercial devices for other researchers and health practitioners to use as well. In 2001, various commercial PET/CT scanners were released on the market: Discovery LS (GE Healthcare), Biograph (Siemens Medical Solutions), and Gemini (Philips Medical Solutions).

3 Tracers

For PET imaging, a wide range of natural molecules can be radioactively labeled without major change occurring in characteristics of the molecule. Alternatively, the chemical change of the labeled molecule that does occur can be advantageous for imaging, see, for instance, the trapping of [^{18}F]FDG and [^{18}F]FTHA taken up by cells (see below). BAT activity scans have mainly been obtained using glucose or fatty acid tracer, since these are natural sources of energy for BAT. Radioisotope

Fluorine-18 decays by positron emission and electron capture yielding stable Oxygen-18. Fluorine-18 has a half-life of 110 min.

Fluorine-18 is practical in use for PET/CT analysis, because it can be incorporated into organic molecules such as glucose and it has a relatively long half-life. For instance, Oxygen-15 or Carbon-11 have much shorter half-lives of 2 and 20 min, respectively. Therefore, using Oxygen-15 requires a cyclotron in the vicinity of the experimental location. Not many institutes are equipped to carry out metabolic studies with such radioactive isotopes. Nevertheless for detection of BAT, various short-lived positron-emitting radioisotopes have been used for measuring oxygen consumption [^{15}O] (Din et al. 2016; Muzik et al. 2012), blood flow and oxygen metabolism [^{15}O] (Din et al. 2016; Muzik et al. 2012; Orava et al. 2011), oxidative capacity [^{11}C]acetate (Blondin et al. 2014; Ouellet et al. 2012), and mitochondrial membrane potential [^{18}F]fluoro-benzyl triphenyl phosphonium (FBnTP) (Madar et al. 2015). The following general reviews discuss various tracers further (Bauwens et al. 2014; Chondronikola et al. 2018).

In the following part, we will examine two Fluorine-18 derived tracers which are most frequently used to determine human BAT activity: One glucose based – [^{18}F] FDG (Fig. 1) – and another fatty acid based – [^{18}F]FTHA (Fig. 2).

3.1 [^{18}F]-Fluorodeoxyglucose

[^{18}F]FDG is Fluorine-18 labeled 2-deoxyglucose (DG). When the hydroxyl group present at the C2 position in DG is replaced by Fluorine-18, this yields the radioactive molecule [^{18}F]FDG (Fig. 1) which can be used for PET/CT analysis. Glucose uptake and subsequent glycolysis result in the generation of pyruvate and NADH which are necessary for ATP production. The step following glucose uptake is phosphorylation of glucose by hexokinase resulting in glucose-6-phosphate. However, the difference with DG is that phosphorylation of DG by hexokinase yields DG-6-phosphate. This molecule cannot be further metabolized in the glycolytic pathway and therefore the tracer gets trapped. [^{18}F]FDG-PET/CT can therefore be used to image glucose uptake. Radioisotope Fluorine-18 (half-life 110 min) decays to the stable isotope Oxygen-18, which gives rise to a hydroxyl group. This ultimately results in the formation of nonradioactive [^{18}O]-glucose-6-phosphate which can be further metabolized in the cell. This was already observed in the

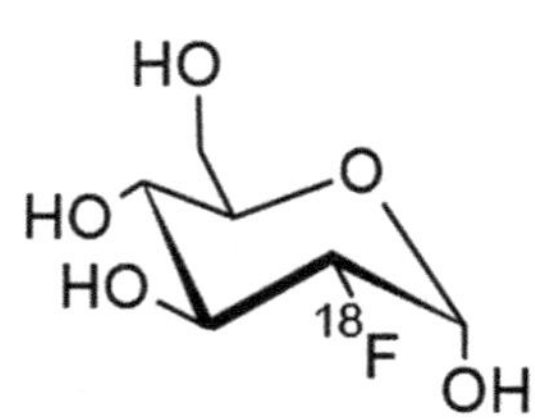

Fig. 1 Chemical structure of [^{18}F]-fluorodeoxyglucose ([^{18}F]FDG)

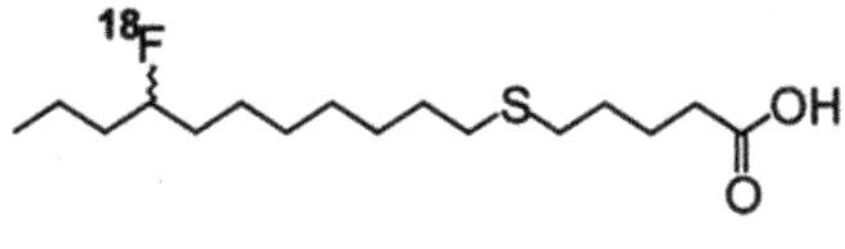

Fig. 2 Chemical structure of 14-[^{18}F]fluoro-6-thiaheptadecanoic acid (FTHA)

1970s when tracking glucose uptake in the rat brain (Sokoloff et al. 1977). Nowadays, [^{18}F]FDG is one of the most frequently used tracers in nuclear medicine, for instance, in cancer diagnostics. Besides uptake in tumors and activated BAT, [^{18}F]FDG is typically taken up by the brain, to a lesser extent by the liver, and skeletal muscle especially when activated. The above described procedure reveals a so-called static scan. Besides static scans, dynamic scans can be performed to define glucose uptake rate in BAT. The latter is used to construct time activity curves of the supraclavicular regions. Glucose uptake rates can then be calculated using Patlak curve fitting (Patlak and Blasberg 1985).

3.2 14-[^{18}F]Fluoro-6-Thiaheptadecanoic Acid

Break down of fatty acids also known as beta-oxidation results in acetyl-CoA (which enters the citric acid cycle) and NADH and $FADH_2$ (which are used in the electron transport chain to generate ATP). 14-Fluoro-6-thiaheptadecanoic acid (FTHA) is a nonesterified long chain fatty acid analogue. FTHA is transported to the mitochondria where it undergoes beta-oxidation. The metabolized FTHA is trapped in the cell after the initial steps of beta-oxidation. These findings originate from the 1990s where fatty acid transport was examined in rat and human heart (DeGrado et al. 1991; Ebert et al. 1994). This tracer is not yet widely used in a clinical setting (Fig. 2).

4 Human Brown Adipose Tissue Activity

Thus far, cold is the most effective way to activate BAT. In the next section, we will discuss methods of cold exposure or pharmacological means that increase BAT activity.

4.1 Acute Cold Exposure and Cold Acclimatization

[^{18}F]FDG was initially used in neuroscience research. The subsequent discovery in 1980 that [^{18}F]FDG accumulates in tumors made it an important tracer to diagnose cancer. In cancer patients who were feeling cold, uptake of [^{18}F]FDG was demonstrated besides the tumor in the supraclavicular area. Initially, it was assumed that this was tense muscle considering symmetrical uptake, but the issue was raised that [^{18}F]FDG uptake in this case could be related to BAT, by that time called USA fat (Cohade et al. 2003b; Hany et al. 2002). It was until 2009 that actual functional cold-stimulated BAT activity using [^{18}F]FDG-PET/CT was demonstrated in adult humans (Cohade et al. 2003a; Cypess et al. 2009; Saito et al. 2009; van Marken Lichtenbelt et al. 2009; Virtanen et al. 2009). BAT presence and activity appeared to be related to seasons and decreased with increased BMI, body fat percentage, and age. Just as in animals, BAT appeared to be a flexible tissue that can be recruited upon demand. BAT activity increased following a 10-day, 4-week, or 6-week cold acclimation in healthy human subjects with obesity, demonstrating plasticity of human BAT (Blondin et al. 2014; Hanssen et al. 2016; van der Lans et al. 2013; Yoneshiro et al. 2013). Finally in line with animal models, (mild) cold is linked to glucose metabolism: whole body insulin sensitivity improved significantly following cold acclimatization (Hanssen et al. 2016) or even following single mild cold exposure (Iwen et al. 2017). However, the quantitative contribution of BAT in human non-shivering thermogenesis, glucose, and lipid metabolism still needs to be established.

In human research, not many studies have been conducted using [^{18}F]FHTA to assess BAT activity. In six healthy male subjects, cold resulted in increased uptake of [^{18}F]FTHA in BAT (Ouellet et al. 2012). Other studies showed increased uptake of [^{18}F]FTHA in BAT following acute cold which was not observed in WAT (Din et al. 2016) thus demonstrating specificity for BAT over WAT. One study showed that cold-stimulated NEFA uptake in BAT was similar between 3 patient groups of healthy young subject, age-matched subject and subjects with type 2 diabetes mellitus (Blondin et al. 2015). FTHA is a fatty acid, so besides infusion into a vein it can be administered orally via a liquid meal. Oral administration resulted in higher uptake of dietary fatty acids in BAT compared to WAT and skeletal muscle; however, a 4-week acclimatization protocol was without effect on boosting uptake of dietary fatty acids in BAT (Blondin et al. 2017).

4.2 Pharmacological Activation of Brown Adipose Tissue

Cold exposure results in the release of norepinephrine, which binds the beta-adrenergic receptors on brown adipocyte resulting in uncoupling. Various strategies have been attempted to enhance BAT activity via targeting of the beta-adrenergic receptors in a pharmacological manner. Nonselective beta-adrenergic receptor agonist isoprenaline increased heart rate and stimulated energy expenditure as expected from stimulation of the beta-adrenergic receptors; however, no change was observed in BAT activity measured via FDG-PET/CT (Vosselman et al. 2012). Similar results were generated with ephedrine, resulting in increased heart rate and energy expenditure; however, BAT activity was also unchanged (Cypess et al. 2012). Mirabegron is a specific beta 3-adrenergic receptor agonist used to treat an overactive bladder. A relatively high dose of mirabegron showed significant increased BAT activity and energy expenditure in healthy subjects, however also increased heart rate and blood pressure (Cypess et al. 2015). Although targeting the beta-adrenergic receptors might be an efficient way to stimulate BAT activity, further research is necessary to bypass effects of beta-adrenergic stimulation of the heart.

Are there any natural compounds that can activate BAT? The bile acid CDCA (Broeders et al. 2015) and glucocorticoid prednisolone (Ramage et al. 2016) are indeed natural compounds that increase BAT activity in humans. Capsinoids, which are less pungent forms of capsaicin that can be found in red pepper, appear to activate BAT. Acute ingestion of 9 mg of capsinoids increased energy expenditure in BAT positive healthy human subjects, while this was not observed in BAT negative human subjects (Yoneshiro et al. 2012). Also, a 1.5-month daily ingestion of capsinoids resulted in increased BAT activity assessed by [^{18}F]FDG-PET/CT (Nirengi et al. 2016). All the [^{18}F]FDG-PET/CT analyses were performed using cooling. However, uncooled administration of capsinoids in healthy human subjects did not affect BAT activity. The combination of cold and capsinoids was needed to show increased BAT activity and increased energy expenditure (Sun et al. 2018). It needs to be determined which other natural compounds can also specifically activate human BAT.

4.3 Exercise Stimulated Brown Adipose Tissue Activity

Physical activity is an effective way to reduce WAT mass. However, it is not fully understood if and how exercise affects BAT. In the next section, we will discuss direct effects of exercise on BAT activity and also look at factors secreted by the exercising body that could alter BAT activity.

4.3.1 Direct Effects of Exercise on Brown Adipose Tissue Activity

Whole body energy expenditure can be roughly split in obligatory (required for basal body functions) and facultative thermogenesis (van Marken Lichtenbelt and Schrauwen 2011). Both cold and physical activity increase the so-called facultative thermogenesis. Cold, as discussed above, can elevate energy expenditure by shivering and NST, where BAT is involved in NST. The magnitude of exercise-induced thermogenesis is highly dependent on the level of exercise. Besides effects on energy expenditure, exercise is known to benefit a variety of organ systems. For instance, 7 weeks of high intensity interval training in obese mice is more favorable on metabolism compared to moderate intensity training resulting in lowered body weight and fat mass while also improving lipid metabolism in the liver (Wang et al. 2017). In 2012, a study indicated the potential of exercise-induced BAT recruitment by a protein called irisin (Bostrom et al. 2012). However, the results of studies on exercise and BAT are not consistent (see below). Comparison between animal and human studies is difficult, partly because [^{18}F]FDG-PET/CT is not included in most animal studies.

In rodents, no direct link is established between [^{18}F]FDG-PET/CT determined BAT activity and exercise. However, in a recent review, UCP1 expression in mouse and rat BAT has been examined following exercise (Flouris et al. 2017). The examined exercise protocols varied in duration of exercise (1 week up to 8 weeks) and type of exercise (running wheel, treadmill, and swimming). Five out of nine publications showed increased UCP1 mRNA expression following exercise while in the other four no change in UCP1 mRNA expression was detected. However, in the two cases where BAT UCP1 protein was examined, only increases in UCP1 following exercise were observed (Seebacher and Glanville 2010; Slocum et al. 2013). The authors do not have an explanation for the change in UCP1 mRNA expression following different types of exercise; however, they do discuss confounding factors (e.g., training protocol, temperature, diet, and time of tissue extraction) that should be studied in future.

In humans, few studies have been conducted where [^{18}F]FDG-PET/CT BAT activity is examined following exercise. Endurance trained athletes (strong exercise) have lower [^{18}F]FDG-PET/CT BAT activity compared to control subjects (Vosselman et al. 2015). Similar trends were obtained in young female athletes; however, this effect did not reach statistical significance (Singhal et al. 2016). These results are interesting considering that obesity is negatively related to BAT activity (van Marken Lichtenbelt et al. 2009), thus indicating that both trained athletes and people with obesity have decreased human BAT activity. However, an athlete would not directly benefit from increased BAT activity/heat production, because that would impede additional heat stress on the exercising body. On the other hand, in a cohort of 40 cancer patients, increasing habitual physical exercise (light exercise) resulted in higher BAT activity (Dinas et al. 2015). These results could indicate that untrained sedentary people could reap benefits of increased BAT activity by starting a light exercise program. A bigger trial has been performed looking at various exercise intensities in young adults to establish effect on BAT activity measured using [^{18}F] FDG-PET/CT (Sanchez-Delgado et al. 2015b). Although the trial ended mid-July of 2017, no results have been published.

4.3.2 Indirect Effects of Exercise on Brown Adipose Tissue Activity

Exercise is known to result in systemic increase of proteins that stimulate BAT (for reviews, see Aldiss et al. 2017; Sanchez-Delgado et al. 2015a). In the following section, we will highlight two factors induced/secreted by the active muscle (FGF21, irisin).

Fibroblast Growth Factor 21

In mammals, the fibroblast growth factor family comprises 18 different ligands (Turner et al. 2012). In mice, fibroblast growth factor 21 (FGF21) is expressed in many tissues including skeletal muscle and liver (Izumiya et al. 2008). FGF21 is activated by norepinephrine (Hondares et al. 2011) thus creating a link towards BAT activation. This link is further strengthened by in vitro treatment of human adipocytes with FGF21, which stimulates a beige/thermogenic profile (Lee et al. 2014). In humans, FGF21 is regulated in various manners following exercise; however, these results are not always consistent. Acute exercise can increase plasma

FGF21 following treadmill exercise (Kim et al. 2013; Tanimura et al. 2016); however, this increase is not observed in obese subjects or subjects with type 2 diabetes (Hansen et al. 2016; Slusher et al. 2015). Participants in cold and physically challenging Yukon Artic Ultra race did not show changes in FGF21 plasma levels during the race (Coker et al. 2017). Long-term exercise increased FGF21 following a 2-week treadmill exercise program in healthy female subjects, but interestingly acute exercise did not increase FGF21 in circulation (Cuevas-Ramos et al. 2012). Three-month combined exercise, composed of a mix of aerobic exercise and resistance training, decreased circulating levels of FGF21 in obese nondiabetic female subjects (Yang et al. 2011) and a 5-week endurance training program in elderly men also decreased circulating FGF21 (Taniguchi et al. 2016). FGF21 regulation after exercise warrants more investigation, especially the link towards thermogenesis in humans is inadequately explored. FGF21 analogue PF-05231023 has been given to obese subject with type 2 diabetes resulting in decreased body weight and improved lipid profile (Talukdar et al. 2016), so studies dedicated to BAT activity and PF-05231023 are interesting to demonstrate increased BAT activity in humans following FGF21 activation.

Irisin

Peroxisome proliferator-activated receptor gamma coactivator 1-alpha (PGC1a) is an important regulator of mitochondrial biogenesis and therefore the step to exercise/muscle contraction is not far, because mitochondria generate ATP that the exercising muscle requires. Irisin (named after the Greek messenger of the Gods Iris) was coined a myokine because it was linked to muscle contraction and PGC1a (Bostrom et al. 2012). Fibronectin type III domain-containing protein 5 (FNDC5) encodes a protein that will be cleaved to irisin, which is subsequently released in the circulation where it can stimulate a brown adipocyte-like thermogenic profile in white adipocytes. Indeed, in vitro treatment of human adipocytes with FNDC5 stimulates this thermogenic profile (Lee et al. 2014). However, controversies arose regarding irisin in humans. The sequences of FNDC5 are identical in humans and mice; however, mutations have occurred in the start codon, resulting in different expression levels in humans compared to mice. The original paper identifying irisin showed increased FNDC5 following 10 weeks of endurance training (Bostrom et al. 2012); however, endurance or strength exercise in another study did not change FNDC5 expression in muscle biopsies (Raschke et al. 2013). Furthermore, determination of irisin levels by ELISA was questioned after examination of the polyclonal antibodies used in the ELISA (Albrecht et al. 2015). Nonetheless, a potential link has been introduced between FGF21 and irisin. Exercise-mediated irisin secretion may be evolutionary derived from shivering to further boost the effects of FGF21 on thermogenesis (Lee et al. 2014). More research on irisin research is needed in order to get a better grasp on human BAT metabolism.

5 Critical Evaluation of Commonly Used Tracers to Measure Human Brown Adipose Tissue Activity

Although [^{18}F]FDG-PET/CT is the most widely used imaging technique for the measurement of human BAT activity, it should be noted that this method only detects uptake (rate) of glucose analog [^{18}F]FDG. It is well established that both glucose and fatty acids serve as fuel for brown adipocytes. Fatty acids derived from internal triglyceride stores or from the circulation are thought to be the main fuel; therefore, [^{18}F]FDG may not reflect the actual brown fat metabolic activity.

Moreover, several factors can influence the results obtained using [^{18}F]FDG-PET/CT to examine BAT activity in humans. For instance, tissue insulin sensitivity determines glucose uptake. BAT activity is reduced in obese diabetic rodents (Mercer and Trayhurn 1984; Yoshioka et al. 1989). Some of these studies indicated that not obesity per se but also insulin resistance contributes significantly to cold intolerance and reduced BAT activity (Burcelin et al. 1993; Marette et al. 1991; Yoshioka et al. 1989). In humans, a negative association between BAT activity and diabetic status, independent of BMI and age, has been reported in a retrospective study (Ouellet et al. 2011). And in a study on fasting-induced insulin resistance using dynamic [^{18}F]FDG-PET/CT, we revealed a lower glucose uptake in brown adipocyte (Hanssen et al. 2015).

As illustrated above, determination of BAT activity by [^{18}F]FDG-PET/CT in humans is subject to variation. In part, this is caused by individual or tissue specific characteristics. Partly, however, this is caused by methodological issues. In November 2014, specialists in the field gathered to discuss minimal requirements needed to perform BAT experiment in human using PET/CT analysis. This resulted in the creation and publication of BARCIST 1.0: Brown adipose reporting criteria in imaging studies (Chen et al. 2016). For BAT activity measurements using PET/CT, two things are important: (1) standardization of PET protocol, method of data acquisition, and experimental criteria for scanner performance. (2) Reporting of participant characteristics, participant preparation, active BAT volume based, and BAT activity within a specified anatomical field of view. Following the suggestions and guidelines present in BARCIST 1.0 will help researchers and the rest of the scientific community to standardize and to compare [^{18}F]FDG-PET/CT studies with regard to BAT activity in humans.

Another issue is the interpretation of BAT activity from retrospective studies. It is well appreciated that under basal conditions/room temperature, no or little uptake of [^{18}F]FDG is detected. Since retrospective studies generally do not use well-controlled cooling protocols, the activity of BAT may depend on indoor temperature in hospitals, clothing insulation, the sensitivity to cold of the patients, and medication. Retrospective studies, for instance, suggested significantly higher BAT activity in women compared to men (Cypess et al. 2009), but well-controlled studies showed no such difference (van der Lans et al. 2013). Nevertheless, retrospective studies may provide interesting insights and suggestions for dedicated studies on human BAT activity. A study involving over 15,000 human subjects showed that high BAT activity is associated with many factors which include: use of diazepam, breast cancer, colder months of the year, low blood glucose and early morning injection of [^{18}F]FDG (Steinberg et al. 2017).

More studies are warranted regarding fatty acid kinetics, with for instance [^{18}F] FTHA, preferably in combination with [^{18}F]FDG tracer. Because the same Fluorine label is used for both tracers, these measurements always need to be performed separately. The main drawback of FTHA provided intravenously is that fatty acids normally are not free in circulation. When FTHA is given orally, the fatty acids will circulate as lipoproteins or chylomicrons, thus requiring a much higher dose for PET/CT. Studies are underway to develop other fatty acid based tracers (Paulus et al. 2017). Secondly, especially after short-term activation of BAT, internal triglyceride stores serve as fuel and FTHA can therefore underestimate BAT activity.

For actual quantification of BAT activity and the determination of the nutrients used, it may be important to combine different tracers when the research protocol allows it (e.g., [^{18}F]FDG, [^{18}F]FTHA, and [^{15}O]). Next, the combination with CT not only provides information on the kind of tissue, but it may also provide information on internal stores used by BAT due to changes in radiodensity. In addition, there are interesting opportunities related to MRI-based techniques in combination with the PET tracers. New PET/MRI systems make it possible to provide a more complete picture on BAT metabolism.

References

Albrecht E, Norheim F, Thiede B, Holen T, Ohashi T, Schering L, Lee S, Brenmoehl J, Thomas S, Drevon CA, Erickson HP, Maak S (2015) Irisin – a myth rather than an exercise-inducible myokine. Sci Rep 5:8889. https://doi.org/10.1038/srep08889

Aldiss P, Betts J, Sale C, Pope M, Symonds ME (2017) Exercise-induced 'browning' of adipose tissues. Metabolism 81:63–70. https://doi.org/10.1016/j.metabol.2017.11.009

Bauwens M, Wierts R, van Royen B, Bucerius J, Backes W, Mottaghy F, Brans B (2014) Molecular imaging of brown adipose tissue in health and disease. Eur J Nucl Med Mol Imaging 41:776–791. https://doi.org/10.1007/s00259-013-2611-8

Blondin DP, Labbe SM, Tingelstad HC, Noll C, Kunach M, Phoenix S, Guerin B, Turcotte EE, Carpentier AC, Richard D, Haman F (2014) Increased brown adipose tissue oxidative capacity in cold-acclimated humans. J Clin Endocrinol Metab 99:E438–E446. https://doi.org/10.1210/jc.2013-3901

Blondin DP, Labbé SM, Noll C, Kunach M, Phoenix S, Guérin B, Turcotte ÉE, Haman F, Richard D, Carpentier AC (2015) Selective impairment of glucose but not fatty acid or oxidative metabolism in brown adipose tissue of subjects with type 2 diabetes. Diabetes 64(7):2388–2397. https://doi.org/10.2337/db14-1651

Blondin DP, Tingelstad HC, Noll C, Frisch F, Phoenix S, Guerin B, Turcotte EE, Richard D, Haman F, Carpentier AC (2017) Dietary fatty acid metabolism of brown adipose tissue in cold-acclimated men. Nat Commun 8:14146. https://doi.org/10.1038/ncomms14146

Bostrom P, Wu J, Jedrychowski MP, Korde A, Ye L, Lo JC, Rasbach KA, Bostrom EA, Choi JH, Long JZ, Kajimura S, Zingaretti MC, Vind BF, Tu H, Cinti S, Hojlund K, Gygi SP, Spiegelman BM (2012) A PGC1-alpha-dependent myokine that drives brown-fat-like development of white fat and thermogenesis. Nature 481:463–468. https://doi.org/10.1038/nature10777

Broeders EP, Nascimento EB, Havekes B, Brans B, Roumans KH, Tailleux A, Schaart G, Kouach M, Charton J, Deprez B, Bouvy ND, Mottaghy F, Staels B, van Marken Lichtenbelt

WD, Schrauwen P (2015) The bile acid chenodeoxycholic acid increases human brown adipose tissue activity. Cell Metab 22:418–426. https://doi.org/10.1016/j.cmet.2015.07.002
Burcelin R, Kande J, Ricquier D, Girard J (1993) Changes in uncoupling protein and GLUT4 glucose transporter expressions in interscapular brown adipose tissue of diabetic rats: relative roles of hyperglycaemia and hypoinsulinaemia. Biochem J 291(Pt 1):109–113
Cannon B, Nedergaard J (2004) Brown adipose tissue: function and physiological significance. Physiol Rev 84:277–359. https://doi.org/10.1152/physrev.00015.2003
Charron M, Beyer T, Bohnen NN, Kinahan PE, Dachille M, Jerin J, Nutt R, Meltzer CC, Villemagne V, Townsend DW (2000) Image analysis in patients with cancer studied with a combined PET and CT scanner. Clin Nucl Med 25:905–910
Chen KY, Cypess AM, Laughlin MR, Haft CR, Hu HH, Bredella MA, Enerback S, Kinahan PE, Lichtenbelt W, Lin FI, Sunderland JJ, Virtanen KA, Wahl RL (2016) Brown Adipose Reporting Criteria in Imaging STudies (BARCIST 1.0): recommendations for standardized FDG-PET/CT experiments in humans. Cell Metab 24:210–222. https://doi.org/10.1016/j.cmet.2016.07.014
Chondronikola M, Volpi E, Borsheim E, Chao T, Porter C, Annamalai P, Yfanti C, Labbe SM, Hurren NM, Malagaris I, Cesani F, Sidossis LS (2016) Brown adipose tissue is linked to a distinct thermoregulatory response to mild cold in people. Front Physiol 7:129. https://doi.org/10.3389/fphys.2016.00129
Chondronikola M, Beeman SC, Wahl RL (2018) Non-invasive methods for the assessment of brown adipose tissue in humans. J Physiol 596:363–378. https://doi.org/10.1113/JP274255
Cohade C, Mourtzikos KA, Wahl RL (2003a) "USA-Fat": prevalence is related to ambient outdoor temperature-evaluation with 18F-FDG PET/CT. J Nucl Med 44:1267–1270
Cohade C, Osman M, Pannu HK, Wahl RL (2003b) Uptake in supraclavicular area fat ("USA-Fat"): description on 18F-FDG PET/CT. J Nucl Med 44:170–176
Coker RH, Weaver AN, Coker MS, Murphy CJ, Gunga HC, Steinach M (2017) Metabolic responses to the Yukon Arctic ultra: longest and coldest in the world. Med Sci Sports Exerc 49:357–362. https://doi.org/10.1249/MSS.0000000000001095
Cuevas-Ramos D, Almeda-Valdes P, Meza-Arana CE, Brito-Cordova G, Gomez-Perez FJ, Mehta R, Oseguera-Moguel J, Aguilar-Salinas CA (2012) Exercise increases serum fibroblast growth factor 21 (FGF21) levels. PLoS One 7:e38022. https://doi.org/10.1371/journal.pone.0038022
Cypess AM, Lehman S, Williams G, Tal I, Rodman D, Goldfine AB, Kuo FC, Palmer EL, Tseng YH, Doria A, Kolodny GM, Kahn CR (2009) Identification and importance of brown adipose tissue in adult humans. N Engl J Med 360:1509–1517. https://doi.org/10.1056/NEJMoa0810780
Cypess AM, Chen YC, Sze C, Wang K, English J, Chan O, Holman AR, Tal I, Palmer MR, Kolodny GM, Kahn CR (2012) Cold but not sympathomimetics activates human brown adipose tissue in vivo. Proc Natl Acad Sci U S A 109:10001–10005. https://doi.org/10.1073/pnas.1207911109
Cypess AM, Weiner LS, Roberts-Toler C, Franquet Elia E, Kessler SH, Kahn PA, English J, Chatman K, Trauger SA, Doria A, Kolodny GM (2015) Activation of human brown adipose tissue by a beta3-adrenergic receptor agonist. Cell Metab 21:33–38. https://doi.org/10.1016/j.cmet.2014.12.009
DeGrado TR, Coenen HH, Stocklin G (1991) 14(R,S)-[18F]fluoro-6-thia-heptadecanoic acid (FTHA): evaluation in mouse of a new probe of myocardial utilization of long chain fatty acids. J Nucl Med 32:1888–1896
Deng J, Neff LM, Rubert NC, Zhang B, Shore RM, Samet JD, Nelson PC, Landsberg L (2017) MRI characterization of brown adipose tissue under thermal challenges in normal weight, overweight, and obese young men. J Magn Reson Imaging 47(4):936–947. https://doi.org/10.1002/jmri.25836
Din M, Raiko J, Saari T, Kudomi N, Tolvanen T, Oikonen V, Teuho J, Sipila HT, Savisto N, Parkkola R, Nuutila P, Virtanen KA (2016) Human brown adipose tissue [(15)O]O2 PET

imaging in the presence and absence of cold stimulus. Eur J Nucl Med Mol Imaging 43:1878–1886. https://doi.org/10.1007/s00259-016-3364-y

Dinas PC, Nikaki A, Jamurtas AZ, Prassopoulos V, Efthymiadou R, Koutedakis Y, Georgoulias P, Flouris AD (2015) Association between habitual physical activity and brown adipose tissue activity in individuals undergoing PET-CT scan. Clin Endocrinol 82:147–154. https://doi.org/10.1111/cen.12620

Ebert A, Herzog H, Stocklin GL, Henrich MM, DeGrado TR, Coenen HH, Feinendegen LE (1994) Kinetics of 14(R,S)-fluorine-18-fluoro-6-thia-heptadecanoic acid in normal human hearts at rest, during exercise and after dipyridamole injection. J Nucl Med 35:51–56

Flouris AD, Dinas PC, Valente A, Andrade CMB, Kawashita NH, Sakellariou P (2017) Exercise-induced effects on UCP1 expression in classical brown adipose tissue: a systematic review. Horm Mol Biol Clin Invest 31. https://doi.org/10.1515/hmbci-2016-0048

Hansen JS, Pedersen BK, Xu G, Lehmann R, Weigert C, Plomgaard P (2016) Exercise-induced secretion of FGF21 and follistatin are blocked by pancreatic clamp and impaired in type 2 diabetes. J Clin Endocrinol Metab 101:2816–2825. https://doi.org/10.1210/jc.2016-1681

Hanssen MJ, Wierts R, Hoeks J, Gemmink A, Brans B, Mottaghy FM, Schrauwen P, van Marken Lichtenbelt WD (2015) Glucose uptake in human brown adipose tissue is impaired upon fasting-induced insulin resistance. Diabetologia 58:586–595. https://doi.org/10.1007/s00125-014-3465-8

Hanssen MJ, van der Lans AA, Brans B, Hoeks J, Jardon KM, Schaart G, Mottaghy FM, Schrauwen P, van Marken Lichtenbelt WD (2016) Short-term cold acclimation recruits brown adipose tissue in obese humans. Diabetes 65:1179–1189. https://doi.org/10.2337/db15-1372

Hany TF, Gharehpapagh E, Kamel EM, Buck A, Himms-Hagen J, von Schulthess GK (2002) Brown adipose tissue: a factor to consider in symmetrical tracer uptake in the neck and upper chest region. Eur J Nucl Med Mol Imaging 29:1393–1398. https://doi.org/10.1007/s00259-002-0902-6

Hondares E, Iglesias R, Giralt A, Gonzalez FJ, Giralt M, Mampel T, Villarroya F (2011) Thermogenic activation induces FGF21 expression and release in brown adipose tissue. J Biol Chem 286:12983–12990. https://doi.org/10.1074/jbc.M110.215889

Iwen KA, Backhaus J, Cassens M, Waltl M, Hedesan OC, Merkel M, Heeren J, Sina C, Rademacher L, Windjager A, Haug AR, Kiefer FW, Lehnert H, Schmid SM (2017) Cold-induced brown adipose tissue activity alters plasma fatty acids and improves glucose metabolism in men. J Clin Endocrinol Metab 102:4226–4234. https://doi.org/10.1210/jc.2017-01250

Izumiya Y, Bina HA, Ouchi N, Akasaki Y, Kharitonenkov A, Walsh K (2008) FGF21 is an Akt-regulated myokine. FEBS Lett 582:3805–3810. https://doi.org/10.1016/j.febslet.2008.10.021

Kim KH, Kim SH, Min YK, Yang HM, Lee JB, Lee MS (2013) Acute exercise induces FGF21 expression in mice and in healthy humans. PLoS One 8:e63517. https://doi.org/10.1371/journal.pone.0063517

Lee P, Linderman JD, Smith S, Brychta RJ, Wang J, Idelson C, Perron RM, Werner CD, Phan GQ, Kammula US, Kebebew E, Pacak K, Chen KY, Celi FS (2014) Irisin and FGF21 are cold-induced endocrine activators of brown fat function in humans. Cell Metab 19:302–309. https://doi.org/10.1016/j.cmet.2013.12.017

Madar I, Naor E, Holt D, Ravert H, Dannals R, Wahl R (2015) Brown adipose tissue response dynamics: in vivo insights with the voltage sensor 18F-fluorobenzyl triphenyl phosphonium. PLoS One 10:e0129627. https://doi.org/10.1371/journal.pone.0129627

Marette A, Deshaies Y, Collet AJ, Tulp O, Bukowiecki LJ (1991) Major thermogenic defect associated with insulin resistance in brown adipose tissue of obese diabetic SHR/N-cp rats. Am J Phys 261:E204–E213. https://doi.org/10.1152/ajpendo.1991.261.2.E204

Mercer SW, Trayhurn P (1984) The development of insulin resistance in brown adipose tissue may impair the acute cold-induced activation of thermogenesis in genetically obese (Ob/Ob) mice. Biosci Rep 4:933–940

Muzik O, Mangner TJ, Granneman JG (2012) Assessment of oxidative metabolism in brown fat using PET imaging. Front Endocrinol 3:15. https://doi.org/10.3389/fendo.2012.00015

Nirengi S, Homma T, Inoue N, Sato H, Yoneshiro T, Matsushita M, Kameya T, Sugie H, Tsuzaki K, Saito M, Sakane N, Kurosawa Y, Hamaoka T (2016) Assessment of human brown adipose tissue density during daily ingestion of thermogenic capsinoids using near-infrared time-resolved spectroscopy. J Biomed Opt 21:091305. https://doi.org/10.1117/1.JBO.21.9.091305

Orava J, Nuutila P, Lidell ME, Oikonen V, Noponen T, Viljanen T, Scheinin M, Taittonen M, Niemi T, Enerback S, Virtanen KA (2011) Different metabolic responses of human brown adipose tissue to activation by cold and insulin. Cell Metab 14:272–279. https://doi.org/10.1016/j.cmet.2011.06.012

Ouellet V, Routhier-Labadie A, Bellemare W, Lakhal-Chaieb L, Turcotte E, Carpentier AC, Richard D (2011) Outdoor temperature, age, sex, body mass index, and diabetic status determine the prevalence, mass, and glucose-uptake activity of 18F-FDG-detected BAT in humans. J Clin Endocrinol Metab 96:192–199. https://doi.org/10.1210/jc.2010-0989

Ouellet V, Labbe SM, Blondin DP, Phoenix S, Guerin B, Haman F, Turcotte EE, Richard D, Carpentier AC (2012) Brown adipose tissue oxidative metabolism contributes to energy expenditure during acute cold exposure in humans. J Clin Invest 122:545–552. https://doi.org/10.1172/JCI60433

Patlak CS, Blasberg RG (1985) Graphical evaluation of blood-to-brain transfer constants from multiple-time uptake data. Generalizations. J Cereb Blood Flow Metab 5:584–590. https://doi.org/10.1038/jcbfm.1985.87

Paulus A, Maenen M, Drude N, Nascimento EBM, van Marken Lichtenbelt WD, Mottaghy FM, Bauwens M (2017) Synthesis, radiosynthesis and in vitro evaluation of 18F-Bodipy-C16/triglyceride as a dual modal imaging agent for brown adipose tissue. PLoS One 12:e0182297. https://doi.org/10.1371/journal.pone.0182297

Ramage LE, Akyol M, Fletcher AM, Forsythe J, Nixon M, Carter RN, van Beek EJ, Morton NM, Walker BR, Stimson RH (2016) Glucocorticoids acutely increase brown adipose tissue activity in humans, revealing species-specific differences in UCP-1 regulation. Cell Metab 24:130–141. https://doi.org/10.1016/j.cmet.2016.06.011

Raschke S, Elsen M, Gassenhuber H, Sommerfeld M, Schwahn U, Brockmann B, Jung R, Wisloff U, Tjonna AE, Raastad T, Hallen J, Norheim F, Drevon CA, Romacho T, Eckardt K, Eckel J (2013) Evidence against a beneficial effect of irisin in humans. PLoS One 8:e73680. https://doi.org/10.1371/journal.pone.0073680

Rothwell NJ, Stock MJ (1983) Luxuskonsumption, diet-induced thermogenesis and brown fat: the case in favour. Clin Sci (Lond) 64:19–23

Saito M, Okamatsu-Ogura Y, Matsushita M, Watanabe K, Yoneshiro T, Nio-Kobayashi J, Iwanaga T, Miyagawa M, Kameya T, Nakada K, Kawai Y, Tsujisaki M (2009) High incidence of metabolically active brown adipose tissue in healthy adult humans: effects of cold exposure and adiposity. Diabetes 58:1526–1531. https://doi.org/10.2337/db09-0530

Sanchez-Delgado G, Martinez-Tellez B, Olza J, Aguilera CM, Gil A, Ruiz JR (2015a) Role of exercise in the activation of brown adipose tissue. Ann Nutr Metab 67:21–32. https://doi.org/10.1159/000437173

Sanchez-Delgado G, Martinez-Tellez B, Olza J, Aguilera CM, Labayen I, Ortega FB, Chillon P, Fernandez-Reguera C, Alcantara JMA, Martinez-Avila WD, Munoz-Hernandez V, Acosta FM, Prados-Ruiz J, Amaro-Gahete FJ, Hidalgo-Garcia L, Rodriguez L, Ruiz YA, Ramirez-Navarro A, Muros-de Fuentes MA, Garcia-Rivero Y, Sanchez-Sanchez R, de Dios Beas Jimenez J, de Teresa C, Navarrete S, Lozano R, Brea-Gomez E, Rubio-Lopez J, Ruiz MR, Cano-Nieto A, Llamas-Elvira JM, Jimenez Rios JA, Gil A, Ruiz JR (2015b) Activating brown adipose tissue through exercise (ACTIBATE) in young adults: rationale, design and methodology. Contemp Clin Trials 45:416–425. https://doi.org/10.1016/j.cct.2015.11.004

Seebacher F, Glanville EJ (2010) Low levels of physical activity increase metabolic responsiveness to cold in a rat (Rattus fuscipes). PLoS One 5:e13022. https://doi.org/10.1371/journal.pone.0013022

Singhal V, Maffazioli GD, Ackerman KE, Lee H, Elia EF, Woolley R, Kolodny G, Cypess AM, Misra M (2016) Effect of chronic athletic activity on brown fat in young women. PLoS One 11: e0156353. https://doi.org/10.1371/journal.pone.0156353

Slocum N, Durrant JR, Bailey D, Yoon L, Jordan H, Barton J, Brown RH, Clifton L, Milliken T, Harrington W, Kimbrough C, Faber CA, Cariello N, Elangbam CS (2013) Responses of brown adipose tissue to diet-induced obesity, exercise, dietary restriction and ephedrine treatment. Exp Toxicol Pathol 65:549–557. https://doi.org/10.1016/j.etp.2012.04.001

Slusher AL, Whitehurst M, Zoeller RF, Mock JT, Maharaj M, Huang CJ (2015) Attenuated fibroblast growth factor 21 response to acute aerobic exercise in obese individuals. Nutr Metab Cardiovasc Dis 25:839–845. https://doi.org/10.1016/j.numecd.2015.06.002

Sokoloff L, Reivich M, Kennedy C, Des Rosiers MH, Patlak CS, Pettigrew KD, Sakurada O, Shinohara M (1977) The [14C]deoxyglucose method for the measurement of local cerebral glucose utilization: theory, procedure, and normal values in the conscious and anesthetized albino rat. J Neurochem 28:897–916

Steinberg JD, Vogel W, Vegt E (2017) Factors influencing brown fat activation in FDG PET/CT: a retrospective analysis of 15,000+ cases. Br J Radiol 90:20170093. https://doi.org/10.1259/bjr.20170093

Sun L, Camps SG, Goh HJ, Govindharajulu P, Schaefferkoetter JD, Townsend DW, Verma SK, Velan SS, Sun L, Sze SK, Lim SC, Boehm BO, Henry CJ, Leow MK (2018) Capsinoids activate brown adipose tissue (BAT) with increased energy expenditure associated with subthreshold 18-fluorine fluorodeoxyglucose uptake in BAT-positive humans confirmed by positron emission tomography scan. Am J Clin Nutr 107:62–70. https://doi.org/10.1093/ajcn/nqx025

Talukdar S, Zhou Y, Li D, Rossulek M, Dong J, Somayaji V, Weng Y, Clark R, Lanba A, Owen BM, Brenner MB, Trimmer JK, Gropp KE, Chabot JR, Erion DM, Rolph TP, Goodwin B, Calle RA (2016) A long-acting FGF21 molecule, PF-05231023, decreases body weight and improves lipid profile in non-human primates and type 2 diabetic subjects. Cell Metab 23:427–440. https://doi.org/10.1016/j.cmet.2016.02.001

Taniguchi H, Tanisawa K, Sun X, Kubo T, Higuchi M (2016) Endurance exercise reduces hepatic fat content and serum fibroblast growth factor 21 levels in elderly men. J Clin Endocrinol Metab 101:191–198. https://doi.org/10.1210/jc.2015-3308

Tanimura Y, Aoi W, Takanami Y, Kawai Y, Mizushima K, Naito Y, Yoshikawa T (2016) Acute exercise increases fibroblast growth factor 21 in metabolic organs and circulation. Physiol Rep 4:e12828. https://doi.org/10.14814/phy2.12828

Turner CA, Watson SJ, Akil H (2012) The fibroblast growth factor family: neuromodulation of affective behavior. Neuron 76:160–174. https://doi.org/10.1016/j.neuron.2012.08.037

van der Lans AA, Hoeks J, Brans B, Vijgen GH, Visser MG, Vosselman MJ, Hansen J, Jorgensen JA, Wu J, Mottaghy FM, Schrauwen P, van Marken Lichtenbelt WD (2013) Cold acclimation recruits human brown fat and increases nonshivering thermogenesis. J Clin Invest 123:3395–3403. https://doi.org/10.1172/JCI68993

van Marken Lichtenbelt WD, Schrauwen P (2011) Implications of nonshivering thermogenesis for energy balance regulation in humans. Am J Physiol Regul Integr Comp Physiol 301:R285–R296. https://doi.org/10.1152/ajpregu.00652.2010

van Marken Lichtenbelt WD, Vanhommerig JW, Smulders NM, Drossaerts JM, Kemerink GJ, Bouvy ND, Schrauwen P, Teule GJ (2009) Cold-activated brown adipose tissue in healthy men. N Engl J Med 360:1500–1508. https://doi.org/10.1056/NEJMoa0808718

Virtanen KA, Lidell ME, Orava J, Heglind M, Westergren R, Niemi T, Taittonen M, Laine J, Savisto NJ, Enerback S, Nuutila P (2009) Functional brown adipose tissue in healthy adults. N Engl J Med 360:1518–1525. https://doi.org/10.1056/NEJMoa0808949

Vosselman MJ, van der Lans AA, Brans B, Wierts R, van Baak MA, Schrauwen P, van Marken Lichtenbelt WD (2012) Systemic beta-adrenergic stimulation of thermogenesis is not accompanied by brown adipose tissue activity in humans. Diabetes 61:3106–3113. https://doi.org/10.2337/db12-0288

Vosselman MJ, Hoeks J, Brans B, Pallubinsky H, Nascimento EB, van der Lans AA, Broeders EP, Mottaghy FM, Schrauwen P, van Marken Lichtenbelt WD (2015) Low brown adipose tissue activity in endurance-trained compared with lean sedentary men. Int J Obes 39:1696–1702. https://doi.org/10.1038/ijo.2015.130

Wang N, Liu Y, Ma Y, Wen D (2017) High-intensity interval versus moderate-intensity continuous training: superior metabolic benefits in diet-induced obesity mice. Life Sci 191:122–131. https://doi.org/10.1016/j.lfs.2017.08.023

Yang SJ, Hong HC, Choi HY, Yoo HJ, Cho GJ, Hwang TG, Baik SH, Choi DS, Kim SM, Choi KM (2011) Effects of a three-month combined exercise programme on fibroblast growth factor 21 and fetuin-A levels and arterial stiffness in obese women. Clin Endocrinol 75:464–469. https://doi.org/10.1111/j.1365-2265.2011.04078.x

Yoneshiro T, Aita S, Kawai Y, Iwanaga T, Saito M (2012) Nonpungent capsaicin analogs (capsinoids) increase energy expenditure through the activation of brown adipose tissue in humans. Am J Clin Nutr 95:845–850. https://doi.org/10.3945/ajcn.111.018606

Yoneshiro T, Aita S, Matsushita M, Kayahara T, Kameya T, Kawai Y, Iwanaga T, Saito M (2013) Recruited brown adipose tissue as an antiobesity agent in humans. J Clin Invest 123:3404–3408. https://doi.org/10.1172/JCI67803

Yoshioka K, Yoshida T, Wakabayashi Y, Nishioka H, Kondo M (1989) The role of insulin in norepinephrine turnover and thermogenesis in brown adipose tissue after acute cold-exposure. Endocrinol Jpn 36:491–499

Techniques and Applications of Magnetic Resonance Imaging for Studying Brown Adipose Tissue Morphometry and Function

Dimitrios C. Karampinos, Dominik Weidlich, Mingming Wu, Houchun H. Hu, and Daniela Franz

Contents

1 Introduction 300
2 Literature Search 301
3 BAT-Containing Anatomical Regions Studied by MR 302
4 BAT MRI in Radiological Reports 302
5 MR Contrast Mechanisms in Characterizing BAT Morphometry 303
  5.1 Water–Fat Environment and Quantitative MR 303
  5.2 MR Measuring BAT Water–Fat Composition 303
  5.3 MR Measuring BAT Microstructure 307
6 MR Contrast Mechanisms in Characterizing BAT Function 309
  6.1 MR Measuring Directly BAT Metabolic Activity 309
  6.2 MR Measuring BAT Blood Flow and Perfusion Effects 309
  6.3 MR Measuring BAT Temperature 312
7 Multimodal Imaging 314
8 MR Image Analysis Aspects 314
9 Applications of BAT MR 315
  9.1 MR BAT Markers Versus Anthropometric Metrics 315
  9.2 MRI of BAT in Disease 316
10 Future Perspectives 317
11 Conclusion 318
References 318

Abstract

The present review reports on the current knowledge and recent findings in magnetic resonance imaging (MRI) and spectroscopy (MRS) of brown adipose

D. C. Karampinos (✉) · D. Weidlich · M. Wu · D. Franz
Department of Diagnostic and Interventional Radiology, Klinikum rechts der Isar, Technical University of Munich, Munich, Germany
e-mail: dimitrios.karampinos@tum.de

H. H. Hu
Department of Radiology, Nationwide Children's Hospital, Columbus, OH, USA

A. Pfeifer et al. (eds.), *Brown Adipose Tissue*,
Handbook of Experimental Pharmacology 251, https://doi.org/10.1007/164_2018_158

tissue (BAT). The work summarizes the features and mechanisms that allow MRI to differentiate BAT from white adipose tissue (WAT) by making use of their distinct morphological appearance and the functional characteristics of BAT. MR is a versatile imaging modality with multiple contrast mechanisms as potential candidates in the study of BAT, targeting properties of ^{1}H, ^{13}C, or ^{129}Xe nuclei. Techniques for assessing BAT morphometry based on fat fraction and markers of BAT microstructure, including intermolecular quantum coherence and diffusion imaging, are first described. Techniques for assessing BAT function based on the measurement of BAT metabolic activity, perfusion, oxygenation, and temperature are then presented. The application of the above methods in studies of BAT in animals and humans is described, and future directions in MR study of BAT are finally discussed.

Keywords
Activation · Brown adipose tissue · Magnetic resonance imaging · Magnetic resonance spectroscopy · Morphology · White adipose tissue

1 Introduction

Biomedical imaging has played a central role in the renaissance of scientific interest on brown adipose tissue (BAT). Increased metabolic activity in the cervical and supraclavicular regions of humans has been reported for a long time in the nuclear medicine literature (Cohade et al. 2003; Hany et al. 2002). However, it was not until 2009 when three simultaneous studies, published in the New England Journal of Medicine, reported that activated BAT can be observed in a significant percentage of the adult population using positron emission tomography (PET) in combination with adipose tissue biopsies and molecular biology analysis (Cypess et al. 2009; van Marken Lichtenbelt et al. 2009; Virtanen et al. 2009). Since then numerous imaging studies have investigated the identification of BAT using primarily fluorine 18 (^{18}F) fluorodeoxyglucose (FDG) PET (^{18}F-FDG-PET). PET is nowadays considered the reference modality in BAT imaging (Chen et al. 2016) in particular as it is depicting metabolic activity of BAT depots. However, the exposure to ionizing radiation remains a major concern when applying PET for studying BAT, especially in studies requiring multiple imaging sessions and pediatric populations. In addition, ^{18}F-FDG-PET likely underestimates BAT activity under physiologically meaningful conditions such as mild cold (Cypess et al. 2014).

Magnetic resonance imaging (MRI) and magnetic resonance spectroscopy (MRS) methods have been recently suggested for the assessment of BAT morphometry and function (Hu 2015). BAT is characterized by clear differences in both its morphometry and function compared to white adipose tissue (WAT): Specifically, BAT is specialized for non-shivering thermogenesis via mitochondrial uncoupling. BAT has a higher concentration of intracellular and extracellular water, contains more mitochondria, and is more densely vascularized than WAT. Furthermore, BAT exhibits a higher metabolic activity during thermogenesis than non-activated BAT.

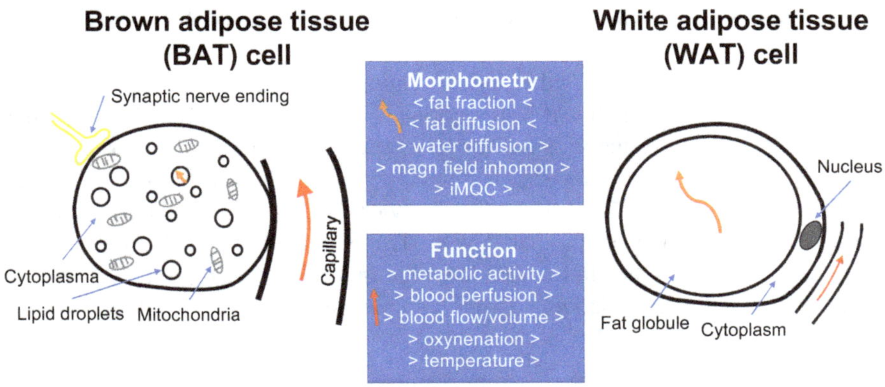

Fig. 1 Conceptual schematic relating the morphometry and function of BAT and WAT to the already proposed MR contrast mechanisms: BAT is characterized by lower fat fraction, more restricted fat diffusion (indicated by orange arrow), less restricted water diffusion, stronger microscopic magnetic field inhomogeneity (due to the abundance of mitochondria), a close proximity of water and fat molecules (generating intermolecular quantum coherence signals), higher metabolic activity, stronger blood perfusion and higher blood flow/volume (indicated by red arrow), higher oxygenation, and higher temperature (during thermogenesis) than WAT

WAT is specialized for energy storage and contains triglycerides in large lipid droplets. MRI and MRS methods have been thus proposed for differentiating BAT from WAT, relying on both the morphological and functional differences of the two tissues in both at rest and activation studies (Fig. 1).

2 Literature Search

MR techniques and applications of BAT have been in part summarized in some recent studies reviewing imaging modalities for BAT research (Bauwens et al. 2014; Borga et al. 2014; Chondronikola et al. 2018; Marzola et al. 2016; Sampath et al. 2016; Sun et al. 2017; van der Lans et al. 2014). Hu published in 2015 the only up-to-date review focusing entirely on MR of BAT (Hu 2015). The present study aims to provide a review of the state-of-the-art MR techniques and applications for BAT research. No differentiation is made between classical BAT and beige/brite adipose tissue due to the current lack of MR techniques to discriminate those two tissues.

Electronic searches in PubMed (http://www.ncbi.nlm.nih.gov/pubmed) were performed without a starting date up to April 2018 using as search terms the term "brown adipose tissue" and one of the following terms: "Magnetic Resonance Imaging, Magnetic Resonance Spectroscopy, fat fraction, diffusion MRI, perfusion MRI, metabolic MRI, temperature MRI." The search resulted in 237 entries and included studies both in rodents and in humans. The reference lists of relevant articles were also screened.

3 BAT-Containing Anatomical Regions Studied by MR

The majority of previous BAT MR studies in mice have focused on the interscapular adipose tissue depot, as a typical BAT region, and have compared its MR properties to the MR properties of the inguinal subcutaneous adipose tissue depot as a typical WAT region. In adult humans, six anatomical regions of activated BAT have been identified in ^{18}F-FDG-PET, including the cervical, supraclavicular, axillary, mediastinal, paraspinal, and abdominal depots (Leitner et al. 2017). However, most previous BAT MR studies in humans have focused on the supraclavicular adipose tissue depot, given its relatively larger volume and the easier definition of its anatomic borders.

4 BAT MRI in Radiological Reports

BAT and its appearance plays no role in the majority of radiological MRI reports. However, there are two clinical conditions where radiologists encounter BAT-like MRI appearance, and awareness of this appearance allows them for choosing the correct differential diagnosis.

The first condition is the so-called hibernoma. Hibernoma is a rare benign soft tissue tumor containing prominent brown adipocytes that resemble normal BAT. Intense uptake has been reported in ^{18}F-FDG-PET (Chatterton et al. 2002) of hibernoma, reflecting the high metabolic activity of the contained BAT. On MRI, the tumors are described to have high, thus, fat-resembling signal intensity on T_1-weighted imaging, however not the same signal intensity as subcutaneous fat. MR fat suppression techniques not always lead to suppression of the tumor. After injection of intravenous gadolinium, contrast enhancement could be found in most cases (Anderson et al. 2001), presumably due to the dense vascularization and the vascular leakage within the tumors.

The second condition is the so-called pheochromocytoma. Pheochromocytoma is a rare catecholamine-secreting tumor that arises from the adrenal glands. There is ample evidence that in this pathological condition of extremely high adrenergic activation, adult humans show a marked increase in thermogenically active BAT mass (Nedergaard et al. 2007). Early reports on the association of pheochromocytoma with BAT describe the angiographic appearance of BAT as hypervascular masses. In the last decade, there have been reports using ^{18}F-FDG-PET, where pheochromocytoma patients show extreme uptake in BAT with regard to both intensity and localization. However, one interesting MRI case report in a pheochromocytoma patient describes the perinephric soft tissue to be minimally hyperintense to muscle on T_1-weighted imaging and heterogeneously hyperintense on T_2-weighted imaging, with a signal drop in the opposed-phase images, but no suppression on chemical fat-suppressed T_2-weighted imaging, pointing to a type of adipose tissue being present that differs from the regular perinephric adipose tissue. Furthermore, the report describes prominent vessels within the tissue and mild to moderate contrast enhancement. The histopathological report confirmed the tissue to be composed of brown adipocytes (Dundamadappa et al. 2007).

5 MR Contrast Mechanisms in Characterizing BAT Morphometry

The MR contrast mechanisms proposed to characterize BAT morphometry and their use in differentiating BAT from WAT are first reviewed.

5.1 Water–Fat Environment and Quantitative MR

BAT is one of the very few tissues where both water and fat components can be equally present. The water–fat composition of BAT has two important implications when applying quantitative MR in the tissue. First, MRI and MRS can be used to perform quantitative measurements of the fat fraction and characterize BAT water–fat composition. Second, quantitative MRI measurements of individual properties of the BAT fat or water components require suppressing one component or accounting for the presence of the other component. Water and fat peaks show different T_2 decay, T_1 recovery, and diffusion attenuation depending on the underlying different T_2 relaxation times, T_1 relaxation times, and diffusion constants: the BAT water component has typically shorter T_2, longer T_1, and faster diffusion constant than the BAT fat component. The relaxation properties of the water and fat components in rodent WAT and BAT samples were assessed in a recent in vitro single-voxel MRS study by Hamilton et al. (2011), demonstrating in general small differences in relaxation parameters between the two tissues.

5.2 MR Measuring BAT Water–Fat Composition

MR has the ability to quantify the water and fat content of tissues. MR-based measurement of water–fat composition can be used in differentiating BAT and WAT, with BAT exhibiting smaller adipocytes containing a vast amount of intracellular water, iron-rich mitochondria, and less intracellular lipids and with WAT being composed of large adipocytes with limited intracellular water and large lipid droplets (Fig. 1).

5.2.1 Technical Considerations

MR can exploit the chemical shift difference between water and lipids for measuring water or fat fraction. Early adipose tissue fat fraction measurements relying on techniques exploiting chemical shift properties have been primarily signal-weighted (Sbarbati et al. 1997). Over the last few years, techniques have been emerging for measuring the proton-density fat fraction (PDFF), defined as the proportion of mobile proton density in tissue attributable to fat, primarily for fat quantification in abdominal organs (Reeder et al. 2011) and in bone marrow (Karampinos et al. 2018). PDFF is a standardized imaging biomarker of tissue fat content (Reeder et al. 2012) and can be measured using single-voxel proton MRS or chemical shift encoding-based water–fat imaging (Reeder et al. 2011).

Single-voxel proton MRS techniques can quantify fat fraction over a single voxel. Singe voxel MRS, in general, requires a good homogeneity of the magnetic field over the scanned voxel to maintain narrow water and fat peaks in the acquired spectrum. Single-voxel MRS is therefore challenged by the severe field inhomogeneity effects when scanning the head and neck region in humans for studying supraclavicular BAT. To measure PDFF in an MRS experiment, a long repetition time (TR) needs to be used (to minimize T_1-weighting effects), and an acquisition with multiple echo times (TEs) needs to be performed (to correct for T_2-weighting effects) (Fig. 2).

Chemical shift encoding-based water–fat imaging enables the separation of the MR signal into water and fat signals and therefore the reconstruction of spatially resolved 3D fat fraction maps. Its application in the adult head and neck region is also challenged by the presence of large magnetic field inhomogeneities which can result in the infamous water–fat swaps in the water–fat separation reconstruction. However, appropriate fieldmap estimation techniques can reduce such errors (Diefenbach et al. 2018; Hernando et al. 2010). Chemical shift encoding-based water–fat imaging determines PDFF after accounting for known confounding factors including the presence of multiple fat peaks, T_2* decay, T_1 bias, and phase error effects, as already shown in the liver (Reeder et al. 2011) and bone marrow (Karampinos et al. 2018). There is no report on the effect of confounding errors in PDFF measurements of BAT. However, the correction for T_2* decay effects should

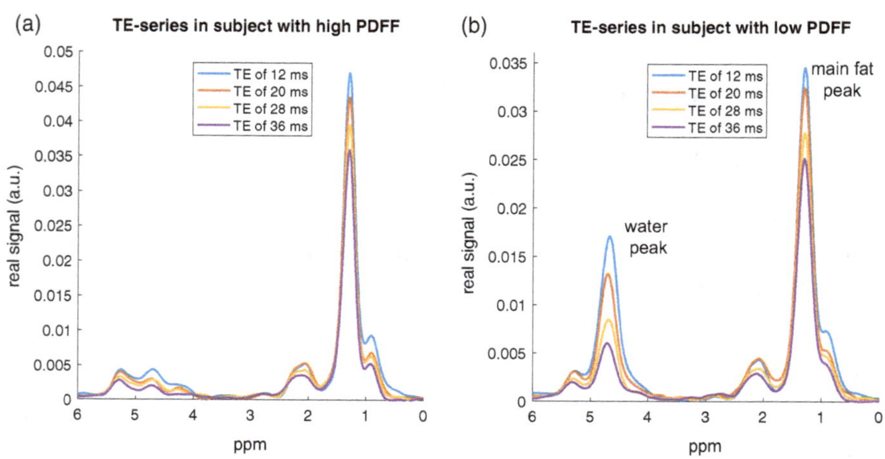

Fig. 2 Single voxel MRS of the human supraclavicular fossa: TE (echo time) series in (**a**) a subject with high PDFF and (**b**) a subject with low PDFF. Notice the faster T_2 decay of the water peak compared to the T_2 decay of the main fat peak [especially in the subject with low PDFF (**b**)], suggesting a shorter T_2 relaxation time for water compared to the main fat peak

be especially considered when applying chemical shift encoding-based water–fat imaging for differentiating BAT from WAT based on PDFF, as BAT has shorter T_2* than WAT due to its heterogeneous cellular composition.

Z-spectrum imaging was also recently proposed as an alternative technique based on direct saturation to measure the BAT fat fraction in mice and humans (Scotti et al. 2017).

5.2.2 Water–Fat Composition in BAT Studies at Rest

First investigations to examine BAT in rodents with MRS were conducted in the late 1980s (Osculati et al. 1989), followed by seminal studies showing the lower fat fraction of BAT compared to WAT in mice in combination with electron microscopy results (Osculati et al. 1991). Both water- and fat-selective imaging and single-voxel MRS techniques have been then applied to measure BAT water–fat composition in mice already since the 1990s (Lunati et al. 1999, 2001b; Sbarbati et al. 1997; Zancanaro et al. 1994). In parallel, multiple early MRS studies in the adipose tissue of mice showed a higher degree of poly-unsaturation in BAT compared to WAT depots (Lunati et al. 2001a; Strobel et al. 2008; Zancanaro et al. 1994).

Since the three seminal publications in the New England Journal of Medicine in 2009 rediscovered the presence of activated BAT in human adults (Cypess et al. 2009; van Marken Lichtenbelt et al. 2009; Virtanen et al. 2009), the interest of applying water–fat composition techniques for assessing the cellular composition differences between BAT and WAT has been rapidly growing.

Fat fraction measurements have been performed in ex vivo BAT samples from rodents (Hamilton et al. 2011; Rasmussen et al. 2013) and in vivo in the interscapular fat depot of mice (Hu et al. 2012a; Peng et al. 2013) and rats (Chen et al. 2012; Romu et al. 2015). Animal BAT fat fraction measurements have been validated qualitatively against hematoxylin/eosin (HE) staining (Peng et al. 2013; Romu et al. 2015) and immunofluorescence uncoupling protein 1 (UCP1) staining (Romu et al. 2015) and quantitatively against uncoupling protein 1 (UCP1) levels (Smith et al. 2013). Animal BAT fat fraction measurements have been also related to adipocyte size (Peng et al. 2013) and the Hounsfield units from dual-energy computed tomography (Romu et al. 2015). Fat fraction measurements even revealed seasonal growth of white and brown adipose tissue without cold exposure in a hibernator (MacCannell et al. 2017).

Fat fraction measurements have also been performed in multiple studies in humans. The supraclavicular fat depot shows typically significant spatial heterogeneity of the fat fraction and has an irregular geometric shape. Therefore, most human studies of BAT have applied chemical shift encoding-based water–fat imaging techniques for measuring BAT fat fraction. Although there is a great range of chemical shift encoding-based water–fat imaging technique variants applied in the study of human BAT, many of the most recent human BAT studies have implemented PDFF measurements in an effort to standardize the measured BAT fat fraction property (Franssens et al. 2016, 2017; Franz et al. 2018; McCallister et al. 2017).

Chemical shift encoding-based water–fat imaging technique have been applied in the supraclavicular fat regions of infants (Hu et al. 2012b, 2014; Lidell et al. 2013), children (Hu et al. 2013), adolescents (Kim et al. 2014), and adults (Chen et al. 2013; Franssens et al. 2016, 2017; Franz et al. 2018; Gifford et al. 2015, 2016; Koskensalo et al. 2017; McCallister et al. 2017; Romu et al. 2016). The comparison of the human supraclavicular MR fat measurements against histological results validating the presence of brown adipocytes has been performed only in a limited number of postmortem subjects (Hu et al. 2012b; Lidell et al. 2013) and in the single case of a living human adult (Reddy et al. 2014). Supraclavicular MR fat fraction mapping remains nowadays the most popular technique for detecting the presence of BAT in humans, thanks to its good sensitivity for detecting the presence of BAT, its wide availability in clinical MR scanners, and its excellent reproducibility (Franssens et al. 2016) and repeatability independent of BAT activation status (Franz et al. 2015; Holstila et al. 2013).

5.2.3 Water–Fat Composition in BAT Activation Studies

MR fat fraction measurements have been also applied in studies investigating BAT function. In mice, measurement of the interscapular BAT fat fraction has shown that fat fraction can also reflect the functional status of the tissue, reporting that a lower housing ambient temperature results in significantly reduced BAT fact fraction (Smith et al. 2013). Early microscopy studies have shown that activation of BAT can lead to nearly complete depletion of the intracellular lipids (Ito et al. 1991), presumably resulting in a fat fraction near to 0%. Grimpo et al. used single MRS to track the metabolic dynamics in murine BAT. They detected a major loss of free fatty acids (FFAs) from BAT after pharmacological stimulation, as FFAs serve as fuel for thermogenesis and activate UCP1 for uncoupling of oxidative phosphorylation during thermogenesis in BAT (Grimpo et al. 2014).

Activation studies in humans using a cooling/reheating protocol likewise showed a decrease in imaging fat fraction in BAT following cold exposure, indicating a combustion of lipid stores to produce heat in active BAT (Lundstrom et al. 2015; Stahl et al. 2017). Fat fraction changes under thermal challenges were also correlated with hypermetabolic BAT volume and with BAT activity as measured by PET/CT (Deng et al. 2018).

5.2.4 Limitations

Despite the popularity of MR-based fat fraction mapping for detecting BAT presence, a significant limitation of the method is its strong sensitivity to partial volume effects. The voxel size of a typical high-resolution acquisition in a clinical MR scanner is the order of 1 mm^3. First, based on the above spatial resolution, fat fraction cannot discriminate between intracellular water content and non-lipid tissue portions (e.g., from muscle, visceral organs, vessels, or connective tissue) within a voxel. Second, a differentiation between a cluster of brown adipocytes and a mixed cluster of white and brown adipocytes is not possible using fat fraction mapping, giving only average percentages of the fat fraction in volumes of interest and thus not directly reflecting mass or volume of pure BAT. Therefore, the reported percentage always needs to be seen in the anatomic context, as a certain range of fat fraction (50–80%) is sensitive for the detection of BAT but not specific for it (Hu 2015; Hu and Kan 2013).

5.3 MR Measuring BAT Microstructure

The aforementioned sensitivity of the water–fat composition methods to partial volume effects limits their applicability in quantifying BAT volume in regions where BAT and WAT are mixed or in detecting BAT situated around visceral organs. However, BAT and WAT have distinct microstructural differences (Fig. 1). Techniques like T_2*/R_2* mapping, intermolecular zero-quantum coherence (iZQC), intermolecular multiple (double)-quantum coherence (iMQC), and diffusion-weighted imaging (DWI) try to overcome the limitations of water–fat composition methods by measuring markers of BAT microstructure.

5.3.1 T_2* and R_2* Mapping at Rest

T_2* mapping can be a measure of microscopic field inhomogeneities in tissues enclosing phases with different magnetic susceptibility. Currently, T_2* mapping is most frequently used for measuring liver iron concentration, but it has been also proposed for measuring the concentration of paramagnetic contrast agent and trabecular bone density. T_2* mapping can be performed in combination with fat fraction mapping using a chemical shift encoding-based water–fat imaging technique. Multiple previous studies have shown that BAT has shorter T_2* and thus longer R_2* (R_2* is defined as the inverse of T_2*) than WAT in mice (Hu et al. 2012a) and in humans (Deng et al. 2015; Gifford et al. 2016; Holstila et al. 2017; Hu et al. 2013; Hui et al. 2017; Lundstrom et al. 2015). The shorter T_2* in BAT compared to WAT has been primarily attributed to the abundance of mitochondria and the increased iron content within BAT.

5.3.2 Intermolecular Quantum Coherence

Zero-quantum coherence and multiple-quantum coherence were known as measureable dipolar coupling mechanism within a molecule. Only in 1993, Warren et al. showed the effects also between distant molecules (Warren et al. 1993). Unlike the case of intramolecular scalar-coupled spins, the intermolecular dipolar interaction originates from spins at different spatial locations and with different Larmor frequencies. iZQC and iMQC measure the demagnetizing field arising from these long-range dipolar interactions. Due to the non-coherent molecular Brownian self-diffusion in liquids, the intermolecular dipolar interactions of nearby molecules are typically averaged to zero. However, when the spatial distance of the coupled spins is greater than a certain diffusion length, the signal coherence is conserved. This diffusion length can be experimentally manipulated to correlate the detected signal with tissue microstructure.

In BAT, the distance between intracellular water protons and lipid protons is also in a suitable range for observing intermolecular zero-quantum coherence (imZQC) or intermolecular multiple-quantum coherence (imMQC). Both rely on the spatial correlation between fat and water protons at a cellular level (in the range of about 100 μm) in BAT. As spatial correlation between water and fat spins is very different for WAT and BAT tissue, those techniques were proposed as an alternative for BAT detection even when the tissue is infiltrated into or mixed with other tissues, which is often the case in human BAT.

The water–methylene intermolecular zero-quantum coherence signal was shown to be characteristic of only BAT and not WAT in ex vivo scanning of rodent BAT and WAT samples (Branca and Warren 2011). Intermolecular zero-quantum coherence was also applied in vivo in rats, thereby detecting a mixture of BAT cells and WAT cells in an older rat that was undetectable by other noninvasive methods (Bao et al. 2013). Finally, Branca et al. tested the in vivo feasibility of intermolecular zero-quantum coherence method both in mice and in humans on a small number of subjects (Branca et al. 2013). Despite the immunity of the iZQC and iMQC methods to partial volume effects, their inherent low sensitivity remains a major technical challenge (Branca et al. 2013).

5.3.3 Diffusion-Weighted MR Measurements

DWI is a technique widely used in the clinical setting, especially in tumor assessment and neuroimaging. DWI measures the microscopic molecular mobility (diffusion) properties on the basis of incoherent motion of molecules in tissues, causing intravoxel signal-phase cancellation, resulting in MR signal decay with increasing diffusion weightings (b-values). For differentiating BAT from WAT, the diffusion properties of both the water molecules and the lipid molecules have been investigated (Deng et al. 2015; Verma et al. 2017).

The higher water content within BAT cells leads to increasing water mobility and thus results in a faster water DWI signal decay with increasing b-values. Furthermore, water diffusion is supposed to be increased in metabolically active BAT. The assumption behind this is an increase in water diffusion due to the depletion of intracellular lipids in active BAT. In a cohort of 28 children including normal-weight and obese children, the water diffusion of WAT cells in obese children was found to be lower in comparison to the normal-weight subjects (Deng et al. 2015). This was explained with the increased cell size in obese subjects and the resulting smaller water-containing extracellular space.

The microstructural characteristics of the lipids within BAT and WAT also differ from each other: the lipid content in the BAT cells is distributed in many small droplets, whereas the WAT cells contain one large connected lipid pool (Fig. 1). Assuming that the long diffusion time limit is met in a medium with restricted diffusion, Verma et al. measured BAT droplet size distribution and compared lipid diffusion properties of BAT versus WAT in ex vivo BAT samples of rats using diffusion-weighted MR spectroscopy (Verma et al. 2017).

Despite the above reports, BAT DWI remains technically challenging. DWI of the BAT water component requires the suppression of all lipid signals, as recently shown in the bone marrow (Dieckmeyer et al. 2017). In addition, in vivo examination of the diffusion properties of the BAT fat component requires strong diffusion weightings, since fat diffuses 100 times slower than water. It is thus especially prone to motion artifacts. In the human supraclavicular fossa, breathing and tissue deformation due to pulsation of nearby larger and smaller blood vessels are particularly severe and lead to intravoxel dephasing and a motion-biased loss of signal intensity at strong diffusion weightings.

6 MR Contrast Mechanisms in Characterizing BAT Function

Besides utilizing the typical morphological features of BAT in order to detect it with MRI, its function can be also of help for imaging purposes. Besides the commonly used ^{18}F-FDG-PET for detecting and measuring the activation status of BAT via its uptake of glucose, MRI also holds possibilities to depict BAT activation via its versatile contrast mechanisms. The MR contrast mechanisms proposed to characterize BAT function and their use in differentiating BAT from WAT are reviewed next.

6.1 MR Measuring Directly BAT Metabolic Activity

MR techniques measuring directly metabolic activity include the measurement of pyruvate to lactate conversion during glycolysis using ^{13}C MR spectroscopy techniques and the measurement of ATP synthesis rate during oxidative phosphorylation based on ^{31}P MR spectroscopy techniques. Lau et al. used hyperpolarized ^{13}C imaging to noninvasively identify pharmacologically activated depots of BAT in an in vivo rat model (Lau et al. 2014). Regions of activated BAT could be detected by an increased conversion of pre-polarized [1-^{13}C] pyruvate into its downstream products ^{13}C bicarbonate and [1-^{13}C] lactate. Grimpo et al. used ^{31}P MRS to measure energy-rich phosphate levels during BAT activation in mice (Grimpo et al. 2014). However, during cold-activated thermogenesis, the levels of all phosphates were reported of staying constant.

6.2 MR Measuring BAT Blood Flow and Perfusion Effects

Sufficient blood flow to BAT is crucial for its function. In order to transport nutrients to and the produced heat away from the BAT, blood capillaries capable of adjusting the blood flow to the activation status are essential. Consequently, direct or indirect MR-based sensitization of blood flow and perfusion of the tissue is one alternative method to assess BAT activation.

6.2.1 MR Measuring Directly BAT Perfusion

The seemingly easiest way to depict perfusion of a tissue is to perform a dynamic T_1-weighted contrast-enhanced MRI (DCE-MRI) using an intravenous gadolinium-based contrast agent. It is based on a technique broadly used in clinical protocols, for example, in stroke and tumor imaging. The acquired data can be used to compute parameters such as blood volume, blood flow, time to peak, and mean transit time of the relevant tissue. The activation of BAT using contrast-enhanced techniques was reported in rats in 2006 by Sbarbati et al. (2006). They showed that at rest, BAT exhibited a 2.3 times higher contrast enhancement than limb muscle in the same group of animals, suggesting a high blood perfusion in BAT. After pharmacological activation of the BAT depot, the response found in BAT was significantly higher compared to muscle. Another recent study showed increased uptake of contrast agent in activated interscapular BAT than thermoneutral interscapular BAT in rats (Yaligar et al. 2017).

Practical reasons have prevented the application of the technique in human studies, as repeated contrast injections of gadolinium-based contrast agent would be required, bringing up various concerns and obstacles such as the risk of gadolinium deposits in the brain after repeated injections. So far, the rich blood perfusion of BAT in humans was described as incidental finding within the scope of an MR angiography in a newborn (Hu 2015). Another example of a contrast enhancement within the supraclavicular fat in the neck region of a pediatric subject is shown in Fig. 3.

Another way to depict perfusion is by quantifying changes in the T_2*-weighted signal after the injection of supermagnetic contrast agents. Chen et al. also used monocrystalline iron oxide nanoparticles (MION), a superparamagnetic iron oxide MR contrast agent with long blood pool half-life for determining BAT perfusion. However, BAT volume estimation with MION was less accurate than with the other MR measurements used in the study (Chen et al. 2012). Jung et al. quantified BAT activity in mice after the injection of radioactively labeled superparamagnetic iron oxide nanoparticles (SPIOs) embedded into a lipoprotein layer (Jung et al. 2016).

An MRI technique to illustrate local tissue perfusion without using contrast material is the so-called arterial spin labeling. It allows quantitative measurements of regional tissue perfusion using magnetically labeled arterial water as a diffusible

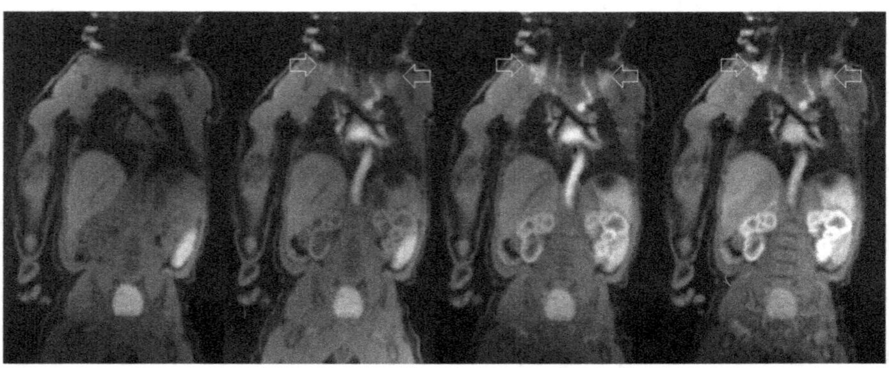

Fig. 3 Dynamic temporal frames from fat-suppressed 3D first-pass contrast-enhanced MR exam of a 5-day-old boy. Four frames are shown, approximately 7 s apart. Note intense gadolinium contrast uptake by supraclavicular BAT, indicative of the tissue's dense perfusion (arrows). Data acquired on a 3T Siemens Prisma platform, at Nationwide Children's Hospital

tracer, resulting in the possibility to quantify blood flow in units of blood flow per gram of tissue per unit time. Dai et al. used this technique in a small study with healthy volunteers, showing increased perfusion activity of 86% $\pm$ 32% in BAT after cold stimulation compared to the thermoneutral conditions (Dai et al. 2015). However, the authors mentioned that the perfusion images of BAT areas tended to be contaminated with large vessels and therefore the perfusion values within the BAT were not reliable.

6.2.2 MR Measuring BAT Oxygenation Effects

A popular method to depict the effect of tissue blood perfusion is by using the T_2*-weighted blood-oxygen-level-dependent (BOLD) effect. In BOLD, increases in oxygen consumption and blood flow produce a detectable change in the intensity of the MR signal via T_2* relaxation time. The BOLD MR signal change can then be used to sensitize changes in these physiological parameters. Khanna and Branca used BOLD-MRI for detection of BAT metabolic activity in mice (Khanna and Branca 2012). Multiple studies have performed T_2*-weighted imaging to measure BAT activation status in humans (Chen et al. 2013; Gifford et al. 2016; van Rooijen et al. 2013). For example, van Roojen et al. used BOLD-MRI in humans to depict BAT activation during cold stimulation, leading to modulations in the T_2*-weighted signal (van Rooijen et al. 2013). T_2*-weighted imaging however comes with the challenge of being sensitive to susceptibility artifacts and physiologically induced breathing artifacts due to the supraclavicular localization of BAT depots in humans.

Alternative methods for measuring tissue oxygenation include quantitative susceptibility mapping and T_1 mapping. A recent study used quantitative susceptibility mapping for monitoring BAT activation in mice showing significant changes in magnetic susceptibility of the interscapular fat during BAT activation (Simchick et al. 2017). Fat T_1 mapping was also recently shown to be able to depict BAT oxygenation changes by investigating interscapular BAT modifications during an air/carbogen challenge in rats (Franconi et al. 2018).

6.2.3 MR Measuring BAT Perfusion Effects Using Hyperpolarized Xenon

Branca et al. used hyperpolarized xenon MRI for detection of BAT and its thermogenic activity in mice (Branca et al. 2014). In this technique, xenon is being inhaled after prior hyperpolarization. The blood-dissolved lipophilic gas is then transported to distal organs, where it accumulates proportionally to the respective tissue's perfusion rate and its blood partition coefficient. Chemical shift changes enable a differentiation between xenon dissolved in blood and xenon dissolved in tissue or lipids. Due to the lipophilic nature of xenon and the increase in blood flow in BAT during activation, BAT can be detected, and its thermogenic activation can be quantified. However, in humans, this technique can be challenging as the amount of xenon being transported from lung to tissue is rather small and the increase in BAT perfusion during activation is smaller in humans compared to rodents.

6.2.4 MR Measuring BAT Perfusion Effects Using DWI

Perfusion effects can be also measured using an augmented version of DWI using the so-called intravoxel incoherent motion (IVIM) signal model. An IVIM experiment relies on breaking the DWI signal decay in each imaging voxel down to two components: the capillary perfusion and the molecular diffusion of water. The result is a quantitative capillary perfusion fraction map, reflecting the fraction of signal in each voxel originating from capillary perfusion versus the water diffusion in extra- and intracellular space. The technique using perfusion fraction estimates in BAT in normal-weight and obese children was also described by Deng et al., resulting in significant differences between separate prepubertal and pubertal groups (Deng et al. 2015). However, the reliable extraction of the IVIM signal using DWI remains challenging due to the need to suppress all lipid peak signals. Results of a preliminary study on the feasibility of using DW-MRS for measuring the IVIM signal in the human adult supraclavicular fossa were reported by Weidlich et al. (2017). However, the acquisition of diffusion measurements even at low b-values in the supraclavicular fossa requires the design of techniques that compensate for physiological motion effects (Fig. 4).

6.3 MR Measuring BAT Temperature

Temperature measurements during BAT activation studies in humans are used to assess thermogenesis and to evaluate BAT function. Body core temperature and surface skin temperature can be sampled with temperature probes (Stahl et al. 2017). Another method aiming for a larger measurement coverage is infrared thermography of the skin surface temperature. However, the isolating effect of the subcutaneous fat covering the supraclavicular area varies with the fat layer thickness and introduces a bias (Gatidis et al. 2016).

One widespread technique to measure relative temperature change with MRI is based on the linear proton resonance frequency shift (PRFS) of water with temperature (Ishihara et al. 1995). This method has the advantage of being independent from what type of aqueous tissue is imaged, but it is prone to phase-change confounders including motion, changes in perfusion, electrical conductivity, or any other source of non-temperature-induced magnetic field changes (Winter et al. 2016). Absolute MR thermometry on the other hand is possible using imZQC in water–fat mixed tissues (Galiana et al. 2008) by exploiting the PRFS of the water and the invariant frequency of the methylene signal. Recent efforts toward a higher temperature precision with imZQC have been made by implementing a different readout sequence (Davis et al. 2016).

Temperature values have been integrated into the chemical shift-encoded fat quantification model to enhance fat fraction calculations (Hernando et al. 2014). Inversely, this method was used in a BAT study to estimate absolute temperature at a known fat fraction in phantoms and humans (Gifford et al. 2014; Welch et al. 2014). In subjects with activated BAT during PET examinations, temperature changes within BAT after cold exposure were significantly higher compared to a subject

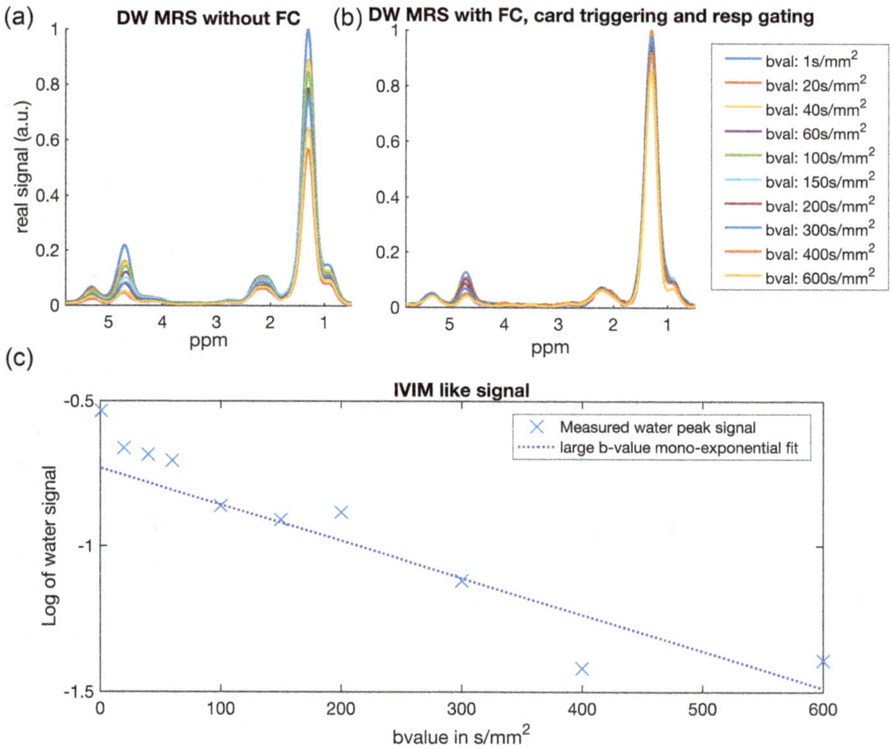

Fig. 4 DW-MRS in the supraclavicular fat of an adult healthy volunteer: (**a**) DW spectra without any flow compensation (FC) and (**b**) DW spectra with flow compensation, cardiac triggering, and respiratory gating. The fat peaks are expected to stay constant at the employed b-values (fat diffuses 100 times slower than water), but instead fat peaks show a significant reduction with increasing b-value in the DW spectra without flow compensation (**a**). The combination of flow compensation, cardiac triggering, and respiratory gating is thus necessary in order to reduce the effect of physiological motion on the measured data (**b**). (**c**) shows the plot of the logarithm of the experimentally measured DW water signal as a function of the b-value and the corresponding mono-exponential fit at b-values higher than 300 s/mm^2. Notice the deviation from the mono-exponential signal decay at b-values lower than 200 s/mm^2, suggesting an IVIM-like signal behavior and the presence of a significant microvascular signal pool

with no visibly active BAT during previous PET imaging. Other MR thermometry methods include T_1 and T_2 relaxation, diffusion, magnetization transfer, and proton-density measurements (Rieke and Butts Pauly 2008). In the context of BAT measurements, special attention needs to be paid to disentangle the change of these MR parameters related to temperature and the change related to the BAT activation.

7 Multimodal Imaging

The combination of MRI with other imaging modalities has been shown to provide additional benefits given the complementary nature of the properties of BAT measured with the different imaging modalities. MR has been previously combined with PET/CT performed in separate scans (Gifford et al. 2016). Simultaneous ^{18}F-FDG PET/MRI allows assessment of both metabolic activity and MR features in a single examination (McCallister et al. 2017; Rossato et al. 2016). PET enables visualization of metabolic processes in BAT with a high sensitivity by visualizing the regional glucose uptake in BAT, while MRI can be used to visualize perfusion and intracellular properties (lipid content, water content, or even mitochondrial occurrence). When information on both the composition or extent of BAT and its specific activation status are desired, the use of hybrid PET/MRI provides the required combined information. Thus, ^{18}F-FDG PET/MRI might become a promising tool for combined morphologic and functional noninvasive evaluation of BAT.

Other than PET and MRI, several studies during the last years have used optical techniques such as Cerenkov luminescence imaging, near-infrared fluorescence imaging, and multispectral optoacoustic imaging (MSOT) for BAT assessment (Dinish et al. 2017; Ganesan et al. 2016; Hartwig et al. 2017; Reber et al. 2018). The combination of those techniques with MRI again offers a greater range of diagnostic data. MSOT in the 700–970 nm spectral range allows noninvasive imaging of BAT activation by label-free real-time near-infrared optoacoustic sensing of hemoglobin (Reber et al. 2018). The possibility of combining MRI to identify BAT and in a next step the ability of MSOT to visualize BAT activation in real-time might also be a highly encouraging development for future studies.

8 MR Image Analysis Aspects

Typical MR imaging techniques for measuring BAT morphometry and function result in the generation of quantitative maps using one of the contrast mechanisms introduced above. The next important step in assessing BAT morphometry and function is the definition of an appropriate region of interest over which the measured property will be analyzed. The BAT depots in the human cervical-supraclavicular region are especially enclosed in three-dimensional regions with highly complex geometrical shape, making any manual segmentation step highly time consuming (Fig. 5). Despite the plethora of methods available for the automatic segmentation of the different types of white adipose tissue on MR images (Hu et al. 2016), automatic segmentation of BAT regions on MR images remains a challenging topic with only a limited number of studies available.

Neural network automatic segmentation techniques have been applied on fat fraction and T_2 relaxation maps of mice (Bhanu Prakash et al. 2016a) and on the fat fraction, T_2 and T_2* maps of rats (Bhanu Prakash et al. 2016b). An atlas-based automatic segmentation method has been applied for segmenting the human

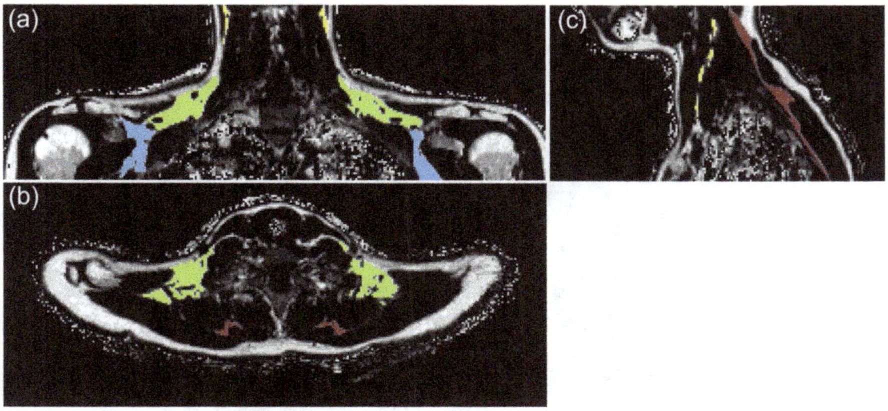

Fig. 5 Segmentation of the different fat depots in the cervical-supraclavicular region of an adult subject, in a coronal view (**a**), in an axial view (**b**), and in a sagittal view (**c**): masks indicate cervical fat (in yellow), paravertebral fat (in red), axillary fat (in blue), and supraclavicular fat (in green), superimposed on the grayscale PDFF map

cervical-supraclavicular adipose tissue with good agreement to reference masks (Lundstrom et al. 2017).

9 Applications of BAT MR

MR techniques evaluating the relationship between MR-based BAT biomarkers and other metabolic markers, including anthropometric markers and blood parameters, have been applied in studies of healthy subjects and in a limited number of patient studies.

9.1 MR BAT Markers Versus Anthropometric Metrics

Several studies related MRI markers to clinical, anthropometric, and laboratory markers. A study in humans showed that PDFF of supraclavicular adipose tissue depots closely correlated to the anthropometric obesity markers body mass index (BMI), waist circumference, and weight-to-height ratio, as well as to visceral and subcutaneous fat volumes (Franz et al. 2018) (Fig. 6). Another study resulted in the triglyceride content in supraclavicular fat depots measured by ^{1}H-MRS being correlated positively with BMI, waist circumference, subcutaneous and visceral fat masses, and 8-year diabetes risk based on the Framingham risk score. Furthermore, triglyceride content in supraclavicular fat depots correlated inversely with HDL-cholesterol and insulin sensitivity. Thus, the authors suggested the triglyceride content in supraclavicular fat depots as a marker of whole-body insulin sensitivity, independent of BAT metabolic activation (Raiko et al. 2015).

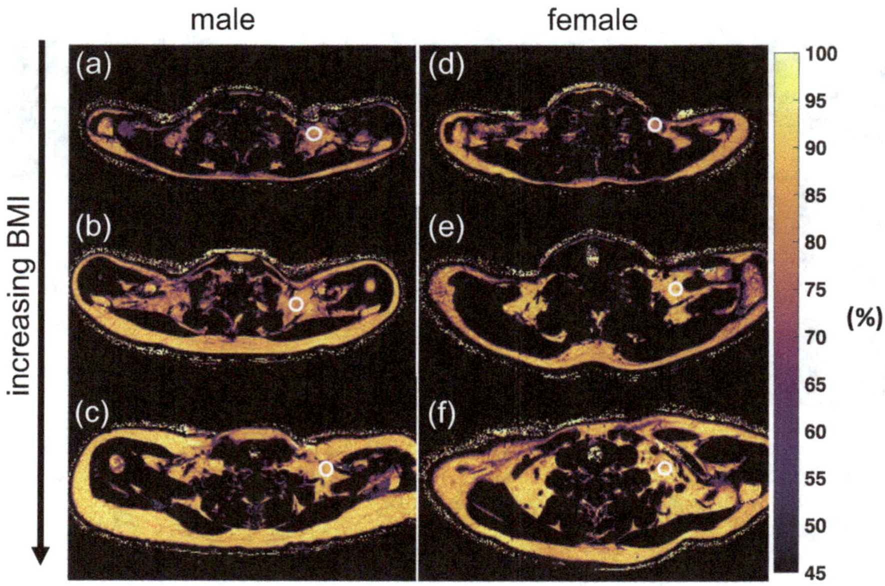

Fig. 6 PDFF maps in the supraclavicular region of six subjects with different BMI, showing a significant increase of supraclavicular PDFF with BMI: (**a**), (**b**), and (**c**) – female subjects with BMI of 17.2, 22, and 43.1 kg/m^2. The mean PDFF in the delineated circular ROI was 78.7 ± 4.3%, 82.0 ± 3.5%, and 85.9 ± 7.6%, respectively. (**d**), (**e**), and (**f**): male subjects with BMI of 19.3, 24.3, and 35.6 kg/m^2. The mean PDFF in the delineated circular ROI was 77.2 ± 7.5%, 87.2 ± 5.5%, and 88.1 ± 7.9%, respectively. Color bar of PDFF values in %

9.2 MRI of BAT in Disease

BAT activity has been shown to be correlated with total and resting energy expenditure. The inverse correlation between the presence of BAT and BMI is of great interest in the fight against obesity and its various comorbidities. MR techniques have been so far only applied in a significantly smaller number of BAT activation studies in disease compared to ^{18}F-FDG-PET.

With the use of SPIO-labeled triglyceride-rich lipoproteins, an MRI study in mice showed that cold-activated BAT controls triglyceride-rich lipoprotein metabolism, leading to an increased plasma clearance of triglycerides as a result of increased uptake into BAT (Bartelt et al. 2011), thus correcting hyperlipidemia and improving effects of insulin resistance. Another study in male humans demonstrated that obese subjects showed higher MRI-fat fraction R_2* than nonobese subjects under thermal challenges and fat fraction changes under thermal challenges correlated with hypermetabolic BAT volume and with BAT activity, as measured by PET/CT (Deng et al. 2018). In adult patients with a clinical manifestation of cardiovascular disease, BAT presence determined by MRI-fat fraction in supraclavicular adipose tissue was associated with less obesity and a more favorable metabolic profile (Franssens et al. 2017).

With regard to diabetes, a study in diabetic and prediabetic patients showed that patients with high insulin resistance had a higher triglyceride content in the supraclavicular fat depot, measured by ^{1}H-MRS, than patients with normal and lower insulin resistance (Koksharova et al. 2017). A study using T_2*-weighted imaging in mice also showed that heart failure results in the chronic activation of BAT, decreased BAT lipid stores, and decreased BAT volume (Panagia et al. 2016).

Aside from obesity, diabetes, and heart disease, diseases with energy deficits such as cachexia are relevant with regard to BAT research. One example for this approach is the Huntington disease, a neurodegenerative disease that leads not only to neuropsychiatric symptoms but also to a progressive energy deficit which is at least in part thought to be due to BAT dysfunction. In a mouse model of Huntington disease, the mice had significantly reduced BAT volume, leading to a highly cold-sensitive phenotype (Lindenberg et al. 2014).

10 Future Perspectives

Imaging studies have made significant contributions to our understanding of BAT function (Izzi-Engbeaya et al. 2015). Each imaging modality has some limitations. MR remains an attractive modality due to the lack of ionizing radiation and the versatility of its contrast mechanisms. Despite the aforementioned plethora of MR contrast mechanisms to assess BAT, most human BAT MR studies nowadays incorporate primarily techniques to measure the fat fraction. The MR-based fat fraction is a sensitive marker for the detection of BAT presence. However, it remains non-specific about the exact volume of BAT, and its application in previous studies has resulted in non-standardized metrics, which are sensitive to the employed experimental settings.

All other MR methods proposed for assessing human BAT morphometry remain experimental and can be in most cases performed at imaging sites with a strong technical expertise. The head and neck region, where the supraclavicular fossa is located, remains one of the most challenging regions in the human body to scan with MR: it is prone to magnetic field inhomogeneities and motion effects. Similarly, many of the MR techniques proposed for assessing BAT function require a strong technical expertise and remain challenged by technical aspects.

Refinements of existing MRI techniques and the development of novel MRI techniques are likely to improve the noninvasive study of BAT. This is especially important as in vivo studies will be required if modulation of BAT activity is to be employed therapeutically for improvements in metabolic health. Specifically, future research efforts should focus (1) on standardizing the supraclavicular fat fraction measurements using the PDFF metric in order to increase the comparability of data across studies, (2) on developing new solutions for overcoming the technical challenges in studying the BAT depot within the human supraclavicular region or other body parts using existing MR techniques, and (3) on developing new MR methods for improving the specificity of MR BAT biomarkers.

11 Conclusion

MR imaging and spectroscopy are versatile and powerful noninvasive imaging tools for assessing BAT morphometry and function with encouraging existing results in both animal and human studies. The MR-based fat fraction has been especially shown to be a sensitive marker of BAT presence and is already of wide use. The application of MR in larger cohort human BAT studies in the coming years could add important information in the imaging phenotyping of BAT depots at rest and complement the functional information of the currently considered imaging gold standard of ^{18}F-FDG-PET. The versatility of MR contrast mechanisms could make MRI the imaging modality for a comprehensive analysis of BAT morphometry and function in the future.

Acknowledgment The present work was supported by the European Research Council (grant agreement No 677661, ProFatMRI), the German Research Foundation (DFG-SFB824/A9), and Philips Healthcare.

References

Anderson SE, Schwab C, Stauffer E, Banic A, Steinbach LS (2001) Hibernoma: imaging characteristics of a rare benign soft tissue tumor. Skelet Radiol 30:590–595

Bao J, Cui X, Cai S, Zhong J, Cai C, Chen Z (2013) Brown adipose tissue mapping in rats with combined intermolecular double-quantum coherence and Dixon water–fat MRI. NMR Biomed 26:1663–1671

Bartelt A, Bruns OT, Reimer R, Hohenberg H, Ittrich H, Peldschus K, Kaul MG, Tromsdorf UI, Weller H, Waurisch C, Eychmuller A, Gordts PL, Rinninger F, Bruegelmann K, Freund B, Nielsen P, Merkel M, Heeren J (2011) Brown adipose tissue activity controls triglyceride clearance. Nat Med 17:200–205

Bauwens M, Wierts R, van Royen B, Bucerius J, Backes W, Mottaghy F, Brans B (2014) Molecular imaging of brown adipose tissue in health and disease. Eur J Nucl Med Mol Imaging 41:776–791

Bhanu Prakash KN, Srour H, Velan SS, Chuang KH (2016a) A method for the automatic segmentation of brown adipose tissue. MAGMA 29:287–299

Bhanu Prakash KN, Verma SK, Yaligar J, Goggi J, Gopalan V, Lee SS, Tian X, Sugii S, Leow MK, Bhakoo K, Velan SS (2016b) Segmentation and characterization of interscapular brown adipose tissue in rats by multi-parametric magnetic resonance imaging. MAGMA 29:277–286

Borga M, Virtanen KA, Romu T, Leinhard OD, Persson A, Nuutila P, Enerback S (2014) Brown adipose tissue in humans: detection and functional analysis using PET (positron emission tomography), MRI (magnetic resonance imaging), and DECT (dual energy computed tomography). Methods Enzymol 537:141–159

Branca RT, Warren WS (2011) In vivo brown adipose tissue detection and characterization using water-lipid intermolecular zero-quantum coherences. Magn Reson Med 65:313–319

Branca RT, Zhang L, Warren WS, Auerbach E, Khanna A, Degan S, Ugurbil K, Maronpot R (2013) In vivo noninvasive detection of Brown adipose tissue through intermolecular zero-quantum MRI. PLoS One 8:e74206

Branca RT, He T, Zhang L, Floyd CS, Freeman M, White C, Burant A (2014) Detection of brown adipose tissue and thermogenic activity in mice by hyperpolarized xenon MRI. Proc Natl Acad Sci USA 111:18001–18006

Chatterton BE, Mensforth D, Coventry BJ, Cohen P (2002) Hibernoma: intense uptake seen on Tc-99m tetrofosmin and FDG positron emission tomographic scanning. Clin Nucl Med 27:369–370

Chen YI, Cypess AM, Sass CA, Brownell AL, Jokivarsi KT, Kahn CR, Kwong KK (2012) Anatomical and functional assessment of brown adipose tissue by magnetic resonance imaging. Obesity (Silver Spring) 20:1519–1526

Chen YC, Cypess AM, Chen YC, Palmer M, Kolodny G, Kahn CR, Kwong KK (2013) Measurement of human brown adipose tissue volume and activity using anatomic MR imaging and functional MR imaging. J Nucl Med 54:1584–1587

Chen KY, Cypess AM, Laughlin MR, Haft CR, Hu HH, Bredella MA, Enerback S, Kinahan PE, Lichtenbelt W, Lin FI, Sunderland JJ, Virtanen KA, Wahl RL (2016) Brown adipose reporting criteria in imaging STudies (BARCIST 1.0): recommendations for standardized FDG-PET/CT experiments in humans. Cell Metab 24:210–222

Chondronikola M, Beeman SC, Wahl RL (2018) Non-invasive methods for the assessment of brown adipose tissue in humans. J Physiol 596:363–378

Cohade C, Mourtzikos KA, Wahl RL (2003) "USA-Fat": prevalence is related to ambient outdoor temperature-evaluation with 18F-FDG PET/CT. J Nucl Med 44:1267–1270

Cypess AM, Lehman S, Williams G, Tal I, Rodman D, Goldfine AB, Kuo FC, Palmer EL, Tseng YH, Doria A, Kolodny GM, Kahn CR (2009) Identification and importance of brown adipose tissue in adult humans. N Engl J Med 360:1509–1517

Cypess AM, Haft CR, Laughlin MR, Hu HH (2014) Brown fat in humans: consensus points and experimental guidelines. Cell Metab 20:408–415

Dai W, Weines L, Alsop D, Cypess A (2015) Feasibility and repeatability of brown adipose tissue volume and perfusion activity using MRI. In: Proceedings of 23rd international society for magnetic resonance in medicine, Toronto, p 539

Davis RM, Zhou Z, Chung H, Warren WS (2016) Multi-spin echo spatial encoding provides threefold improvement of temperature precision during intermolecular zero quantum thermometry. Magn Reson Med 75:1958–1966

Deng J, Schoeneman SE, Zhang H, Kwon S, Rigsby CK, Shore RM, Josefson JL (2015) MRI characterization of brown adipose tissue in obese and normal-weight children. Pediatr Radiol 45:1682–1689

Deng J, Neff LM, Rubert NC, Zhang B, Shore RM, Samet JD, Nelson PC, Landsberg L (2018) MRI characterization of brown adipose tissue under thermal challenges in normal weight, overweight, and obese young men. J Magn Reson Imaging 47:936–947

Dieckmeyer M, Ruschke S, Eggers H, Kooijman H, Rummeny EJ, Kirschke JS, Baum T, Karampinos DC (2017) ADC quantification of the vertebral bone marrow water component: removing the confounding effect of residual fat. Magn Reson Med 78:1432–1441

Diefenbach MN, Ruschke S, Eggers H, Meineke J, Rummeny EJ, Karampinos DC (2018) Improving chemical shift encoding-based water–fat separation based on a detailed consideration of magnetic field contributions. Magn Reson Med 80:990

Dinish US, Wong CL, Sriram S, Ong WK, Balasundaram G, Sugii S, Olivo M (2017) Diffuse optical spectroscopy and imaging to detect and quantify adipose tissue browning. Sci Rep 7:41357

Dundamadappa SK, Shankar S, Danrad R, Singh A, Vijayaraghavan G, Kim Y, Perugini R (2007) Imaging of brown fat associated with adrenal pheochromocytoma. Acta Radiol 48:468–472

Franconi F, Lemaire L, Saint-Jalmes H, Saulnier P (2018) Tissue oxygenation mapping by combined chemical shift and T1 magnetic resonance imaging. Magn Reson Med 79:1981–1991

Franssens BT, Eikendal AL, Leiner T, van der Graaf Y, Visseren FL, Hoogduin JM (2016) Reliability and agreement of adipose tissue fat fraction measurements with water–fat MRI in patients with manifest cardiovascular disease. NMR Biomed 29:48–56

Franssens BT, Hoogduin H, Leiner T, van der Graaf Y, Visseren FLJ (2017) Relation between brown adipose tissue and measures of obesity and metabolic dysfunction in patients with cardiovascular disease. J Magn Reson Imaging 46:497–504

Franz D, Karampinos DC, Rummeny EJ, Souvatzoglou M, Beer AJ, Nekolla SG, Schwaiger M, Eiber M (2015) Discrimination between brown and white adipose tissue using a 2-point dixon water–fat separation method in simultaneous PET/MRI. J Nucl Med 56:1742–1747

Franz D, Weidlich D, Freitag F, Holzapfel C, Drabsch T, Baum T, Eggers H, Witte A, Rummeny EJ, Hauner H, Karampinos DC (2018) Association of proton density fat fraction in adipose tissue with imaging-based and anthropometric obesity markers in adults. Int J Obes 42:175–182

Galiana G, Branca RT, Jenista ER, Warren WS (2008) Accurate temperature imaging based on intermolecular coherences in magnetic resonance. Science 322:421–424

Ganesan G, Warren RV, Leproux A, Compton M, Cutler K, Wittkopp S, Tran G, O'Sullivan T, Malik S, Galassetti PR, Tromberg BJ (2016) Diffuse optical spectroscopic imaging of subcutaneous adipose tissue metabolic changes during weight loss. Int J Obes 40:1292–1300

Gatidis S, Schmidt H, Pfannenberg CA, Nikolaou K, Schick F, Schwenzer NF (2016) Is it possible to detect activated brown adipose tissue in humans using single-time-point infrared thermography under thermoneutral conditions? Impact of BMI and subcutaneous adipose tissue thickness. PLoS One 11:e0151152

Gifford A, Towse T, Avison M, Welch E (2014) Temperature mapping in human brown adipose tissue using fat–water MRI with explicit fitting of water peak location. In: Proceedings of 22nd international society for magnetic resonance in medicine, Milan, p 2354

Gifford A, Towse TF, Walker RC, Avison MJ, Welch EB (2015) Human brown adipose tissue depots automatically segmented by positron emission tomography/computed tomography and registered magnetic resonance images. J Vis Exp 96:e52415–e52415

Gifford A, Towse TF, Walker RC, Avison MJ, Welch EB (2016) Characterizing active and inactive brown adipose tissue in adult humans using PET-CT and MR imaging. Am J Physiol Endocrinol Metab 311:E95–E104

Grimpo K, Volker MN, Heppe EN, Braun S, Heverhagen JT, Heldmaier G (2014) Brown adipose tissue dynamics in wild-type and UCP1-knockout mice: in vivo insights with magnetic resonance. J Lipid Res 55:398–409

Hamilton G, Smith DL Jr, Bydder M, Nayak KS, Hu HH (2011) MR properties of brown and white adipose tissues. J Magn Reson Imaging 34:468–473

Hany TF, Gharehpapagh E, Kamel EM, Buck A, Himms-Hagen J, von Schulthess GK (2002) Brown adipose tissue: a factor to consider in symmetrical tracer uptake in the neck and upper chest region. Eur J Nucl Med Mol Imaging 29:1393–1398

Hartwig V, Guiducci L, Marinelli M, Pistoia L, Tegrimi TM, Iervasi G, Quinones-Galvan A, L'Abbate A (2017) Multimodal imaging for the detection of brown adipose tissue activation in women: a pilot study using NIRS and infrared thermography. J Healthc Eng 2017:5986452

Hernando D, Kellman P, Haldar JP, Liang ZP (2010) Robust water/fat separation in the presence of large field inhomogeneities using a graph cut algorithm. Magn Reson Med 63:79–90

Hernando D, Sharma SD, Kramer H, Reeder SB (2014) On the confounding effect of temperature on chemical shift-encoded fat quantification. Magn Reson Med 72:464–470

Holstila M, Virtanen KA, Gronroos TJ, Laine J, Lepomaki V, Saunavaara J, Lisinen I, Komu M, Hannukainen JC, Nuutila P, Parkkola R, Borra RJ (2013) Measurement of brown adipose tissue mass using a novel dual-echo magnetic resonance imaging approach: a validation study. Metabolism 62:1189–1198

Holstila M, Pesola M, Saari T, Koskensalo K, Raiko J, Borra RJ, Nuutila P, Parkkola R, Virtanen KA (2017) MR signal-fat-fraction analysis and T2* weighted imaging measure BAT reliably on humans without cold exposure. Metabolism 70:23–30

Hu HH (2015) Magnetic resonance of brown adipose tissue: a review of current techniques. Crit Rev Biomed Eng 43:161–181

Hu HH, Kan HE (2013) Quantitative proton MR techniques for measuring fat. NMR Biomed 26:1609–1629

Hu HH, Hines CD, Smith DL Jr, Reeder SB (2012a) Variations in T2* and fat content of murine brown and white adipose tissues by chemical-shift MRI. Magn Reson Imaging 30:323–329

Hu HH, Tovar JP, Pavlova Z, Smith ML, Gilsanz V (2012b) Unequivocal identification of brown adipose tissue in a human infant. J Magn Reson Imaging 35:938–942

Hu HH, Yin L, Aggabao PC, Perkins TG, Chia JM, Gilsanz V (2013) Comparison of brown and white adipose tissues in infants and children with chemical-shift-encoded water–fat MRI. J Magn Reson Imaging 38:885–896

Hu HH, Wu TW, Yin L, Kim MS, Chia JM, Perkins TG, Gilsanz V (2014) MRI detection of brown adipose tissue with low fat content in newborns with hypothermia. Magn Reson Imaging 32:107–117

Hu HH, Chen J, Shen W (2016) Segmentation and quantification of adipose tissue by magnetic resonance imaging. MAGMA 29:259–276

Hui SCN, Ko JKL, Zhang T, Shi L, Yeung DKW, Wang D, Chan Q, Chu WCW (2017) Quantification of brown and white adipose tissue based on Gaussian mixture model using water–fat and T2* MRI in adolescents. J Magn Reson Imaging 46:758–768

Ishihara Y, Calderon A, Watanabe H, Okamoto K, Suzuki Y, Kuroda K, Suzuki Y (1995) A precise and fast temperature mapping using water proton chemical shift. Magn Reson Med 34:814–823

Ito T, Tanuma Y, Yamada M, Yamamoto M (1991) Morphological studies on brown adipose tissue in the bat and in humans of various ages. Arch Histol Cytol 54:1–39

Izzi-Engbeaya C, Salem V, Atkar RS, Dhillo WS (2015) Insights into brown adipose tissue physiology as revealed by imaging studies. Adipocytes 4:1–12

Jung CS, Heine M, Freund B, Reimer R, Koziolek EJ, Kaul MG, Kording F, Schumacher U, Weller H, Nielsen P, Adam G, Heeren J, Ittrich H (2016) Quantitative activity measurements of brown adipose tissue at 7 T magnetic resonance imaging after application of triglyceride-rich lipoprotein 59Fe-superparamagnetic Iron oxide nanoparticle: intravenous versus intraperitoneal approach. Investig Radiol 51:194–202

Karampinos DC, Ruschke S, Dieckmeyer M, Diefenbach M, Franz D, Gersing AS, Krug R, Baum T (2018) Quantitative MRI and spectroscopy of bone marrow. J Magn Reson Imaging 47:332–353

Khanna A, Branca RT (2012) Detecting brown adipose tissue activity with BOLD MRI in mice. Magn Reson Med 68:1285–1290

Kim MS, Hu HH, Aggabao PC, Geffner ME, Gilsanz V (2014) Presence of brown adipose tissue in an adolescent with severe primary hypothyroidism. J Clin Endocrinol Metab 99:E1686–E1690

Koksharova E, Ustyuzhanin D, Philippov Y, Mayorov A, Shestakova M, Shariya M, Ternovoy S, Dedov I (2017) The relationship between brown adipose tissue content in supraclavicular fat depots and insulin sensitivity in patients with type 2 diabetes mellitus and prediabetes. Diabetes Technol Ther 19:96–102

Koskensalo K, Raiko J, Saari T, Saunavaara V, Eskola O, Nuutila P, Saunavaara J, Parkkola R, Virtanen KA (2017) Human brown adipose tissue temperature and fat fraction are related to its metabolic activity. J Clin Endocrinol Metab 102:1200–1207

Lau AZ, Chen AP, Gu Y, Ladouceur-Wodzak M, Nayak KS, Cunningham CH (2014) Noninvasive identification and assessment of functional brown adipose tissue in rodents using hyperpolarized 13C imaging. Int J Obes 38:126–131

Leitner BP, Huang S, Brychta RJ, Duckworth CJ, Baskin AS, McGehee S, Tal I, Dieckmann W, Gupta G, Kolodny GM, Pacak K, Herscovitch P, Cypess AM, Chen KY (2017) Mapping of human brown adipose tissue in lean and obese young men. Proc Natl Acad Sci USA 114:8649–8654

Lidell ME, Betz MJ, Dahlqvist Leinhard O, Heglind M, Elander L, Slawik M, Mussack T, Nilsson D, Romu T, Nuutila P, Virtanen KA, Beuschlein F, Persson A, Borga M, Enerback S (2013) Evidence for two types of brown adipose tissue in humans. Nat Med 19:631–634

Lindenberg KS, Weydt P, Muller HP, Bornstedt A, Ludolph AC, Landwehrmeyer GB, Rottbauer W, Kassubek J, Rasche V (2014) Two-point magnitude MRI for rapid mapping of brown adipose tissue and its application to the R6/2 mouse model of Huntington disease. PLoS One 9:e105556

Lunati E, Marzola P, Nicolato E, Fedrigo M, Villa M, Sbarbati A (1999) In vivo quantitative lipidic map of brown adipose tissue by chemical shift imaging at 4.7 Tesla. J Lipid Res 40:1395–1400
Lunati E, Farace P, Nicolato E, Righetti C, Marzola P, Sbarbati A, Osculati F (2001a) Polyunsaturated fatty acids mapping by (1)H MR-chemical shift imaging. Magn Reson Med 46:879–883
Lunati E, Marzola P, Nicolato E, Sbarbati A (2001b) In-vivo quantitative hydrolipidic map of perirenal adipose tissue by chemical shift imaging at 4.7 Tesla. Int J Obes Relat Metab Disord 25:457–461
Lundstrom E, Strand R, Johansson L, Bergsten P, Ahlstrom H, Kullberg J (2015) Magnetic resonance imaging cooling-reheating protocol indicates decreased fat fraction via lipid consumption in suspected brown adipose tissue. PLoS One 10:e0126705
Lundstrom E, Strand R, Forslund A, Bergsten P, Weghuber D, Ahlstrom H, Kullberg J (2017) Automated segmentation of human cervical-supraclavicular adipose tissue in magnetic resonance images. Sci Rep 7:3064
MacCannell A, Sinclair K, Friesen-Waldner L, McKenzie CA, Staples JF (2017) Water–fat MRI in a hibernator reveals seasonal growth of white and brown adipose tissue without cold exposure. J Comp Physiol B 187:759–767
Marzola P, Boschi F, Moneta F, Sbarbati A, Zancanaro C (2016) Preclinical in vivo imaging for fat tissue identification, quantification, and functional characterization. Front Pharmacol 7:336
McCallister A, Zhang L, Burant A, Katz L, Branca RT (2017) A pilot study on the correlation between fat fraction values and glucose uptake values in supraclavicular fat by simultaneous PET/MRI. Magn Reson Med 78:1922–1932
Nedergaard J, Bengtsson T, Cannon B (2007) Unexpected evidence for active brown adipose tissue in adult humans. Am J Physiol Endocrinol Metab 293:E444–E452
Osculati F, Leclercq F, Sbarbati A, Zancanaro C, Cinti S, Antonakis K (1989) Morphological identification of brown adipose tissue by magnetic resonance imaging in the rat. Eur J Radiol 9:112–114
Osculati F, Sbarbati A, Leclercq F, Zancanaro C, Accordini C, Antonakis K, Boicelli A, Cinti S (1991) The correlation between magnetic resonance imaging and ultrastructural patterns of brown adipose tissue. J Submicrosc Cytol Pathol 23:167–174
Panagia M, Chen YC, Chen HH, Ernande L, Chen C, Chao W, Kwong K, Scherrer-Crosbie M, Sosnovik DE (2016) Functional and anatomical characterization of brown adipose tissue in heart failure with blood oxygen level dependent magnetic resonance. NMR Biomed 29:978–984
Peng XG, Ju S, Fang F, Wang Y, Fang K, Cui X, Liu G, Li P, Mao H, Teng GJ (2013) Comparison of brown and white adipose tissue fat fractions in ob, seipin, and Fsp27 gene knockout mice by chemical shift-selective imaging and (1)H-MR spectroscopy. Am J Physiol Endocrinol Metab 304:E160–E167
Raiko J, Holstila M, Virtanen KA, Orava J, Saunavaara V, Niemi T, Laine J, Taittonen M, Borra RJ, Nuutila P, Parkkola R (2015) Brown adipose tissue triglyceride content is associated with decreased insulin sensitivity, independently of age and obesity. Diabetes Obes Metab 17:516–519
Rasmussen JM, Entringer S, Nguyen A, van Erp TG, Burns J, Guijarro A, Oveisi F, Swanson JM, Piomelli D, Wadhwa PD, Buss C, Potkin SG (2013) Brown adipose tissue quantification in human neonates using water–fat separated MRI. PLoS One 8:e77907
Reber J, Willershauser M, Karlas A, Paul-Yuan K, Diot G, Franz D, Fromme T, Ovsepian SV, Beziere N, Dubikovskaya E, Karampinos DC, Holzapfel C, Hauner H, Klingenspor M, Ntziachristos V (2018) Non-invasive measurement of brown fat metabolism based on optoacoustic imaging of hemoglobin gradients. Cell Metab 27:689–701.e4
Reddy NL, Jones TA, Wayte SC, Adesanya O, Sankar S, Yeo YC, Tripathi G, McTernan PG, Randeva HS, Kumar S, Hutchinson CE, Barber TM (2014) Identification of brown adipose tissue using MR imaging in a human adult with histological and immunohistochemical confirmation. J Clin Endocrinol Metab 99:E117–E121
Reeder SB, Cruite I, Hamilton G, Sirlin CB (2011) Quantitative assessment of liver fat with magnetic resonance imaging and spectroscopy. J Magn Reson Imaging 34:729–749

Reeder SB, Hu HH, Sirlin CB (2012) Proton density fat-fraction: a standardized MR-based biomarker of tissue fat concentration. J Magn Reson Imaging 36:1011–1014

Rieke V, Butts Pauly K (2008) MR thermometry. J Magn Reson Imaging 27:376–390

Romu T, Elander L, Leinhard OD, Lidell ME, Betz MJ, Persson A, Enerback S, Borga M (2015) Characterization of brown adipose tissue by water–fat separated magnetic resonance imaging. J Magn Reson Imaging 42:1639–1645

Romu T, Vavruch C, Dahlqvist-Leinhard O, Tallberg J, Dahlstrom N, Persson A, Heglind M, Lidell ME, Enerback S, Borga M, Nystrom FH (2016) A randomized trial of cold-exposure on energy expenditure and supraclavicular brown adipose tissue volume in humans. Metabolism 65:926–934

Rossato M, Cecchin D, Vettor R (2016) Brown adipose tissue localization using (18)F-FDG PET/MRI in adult. Endocrine 54:562–563

Sampath SC, Sampath SC, Bredella MA, Cypess AM, Torriani M (2016) Imaging of brown adipose tissue: state of the art. Radiology 280:4–19

Sbarbati A, Guerrini U, Marzola P, Asperio R, Osculati F (1997) Chemical shift imaging at 4.7 tesla of brown adipose tissue. J Lipid Res 38:343–347

Sbarbati A, Cavallini I, Marzola P, Nicolato E, Osculati F (2006) Contrast-enhanced MRI of brown adipose tissue after pharmacological stimulation. Magn Reson Med 55:715–718

Scotti A, Tain RW, Li W, Gil V, Liew CW, Cai K (2017) Mapping brown adipose tissue based on fat water fraction provided by Z-spectral imaging. J Magn Reson Imaging 47:1527–33

Simchick G, Yin A, Yin H, Zhao Q (2017) Dynamic monitoring of brown adipose tissue activation and white adipose tissue beiging. In: Proceedings of 25th international society for magnetic resonance in medicine, Honolulu, p 5045

Smith DL Jr, Yang Y, Hu HH, Zhai G, Nagy TR (2013) Measurement of interscapular brown adipose tissue of mice in differentially housed temperatures by chemical-shift-encoded water–fat MRI. J Magn Reson Imaging 38:1425–1433

Stahl V, Maier F, Freitag MT, Floca RO, Berger MC, Umathum R, Berriel Diaz M, Herzig S, Weber MA, Dimitrakopoulou-Strauss A, Rink K, Bachert P, Ladd ME, Nagel AM (2017) In vivo assessment of cold stimulation effects on the fat fraction of brown adipose tissue using DIXON MRI. J Magn Reson Imaging 45:369–380

Strobel K, van den Hoff J, Pietzsch J (2008) Localized proton magnetic resonance spectroscopy of lipids in adipose tissue at high spatial resolution in mice in vivo. J Lipid Res 49:473–480

Sun L, Yan J, Sun L, Velan SS, Leow MKS (2017) A synopsis of brown adipose tissue imaging modalities for clinical research. Diabetes Metab 43:401–410

van der Lans AA, Wierts R, Vosselman MJ, Schrauwen P, Brans B, van Marken Lichtenbelt WD (2014) Cold-activated brown adipose tissue in human adults: methodological issues. Am J Physiol Regul Integr Comp Physiol 307:R103–R113

van Marken Lichtenbelt WD, Vanhommerig JW, Smulders NM, Drossaerts JM, Kemerink GJ, Bouvy ND, Schrauwen P, Teule GJ (2009) Cold-activated brown adipose tissue in healthy men. N Engl J Med 360:1500–1508

van Rooijen BD, van der Lans AA, Brans B, Wildberger JE, Mottaghy FM, Schrauwen P, Backes WH, van Marken Lichtenbelt WD (2013) Imaging cold-activated brown adipose tissue using dynamic T2*-weighted magnetic resonance imaging and 2-deoxy-2-[18F]fluoro-D-glucose positron emission tomography. Investig Radiol 48:708–714

Verma SK, Nagashima K, Yaligar J, Michael N, Lee SS, Xianfeng T, Gopalan V, Sadananthan SA, Anantharaj R, Velan SS (2017) Differentiating brown and white adipose tissues by high-resolution diffusion NMR spectroscopy. J Lipid Res 58:289–298

Virtanen KA, Lidell ME, Orava J, Heglind M, Westergren R, Niemi T, Taittonen M, Laine J, Savisto NJ, Enerback S, Nuutila P (2009) Functional brown adipose tissue in healthy adults. N Engl J Med 360:1518–1525

Warren WS, Richter W, Andreotti AH, Farmer BT 2nd (1993) Generation of impossible cross-peaks between bulk water and biomolecules in solution NMR. Science 262:2005–2009

Weidlich D, Hock A, Ruschke S, Franz D, Hauner H, Rummeny EJ, Karampinos DC (2017) Improving the quality of DW spectra in the supraclavicular fossa with a navigator-gated and cardiac-triggered flow-compensated diffusion-weighted STEAM MRS acquisition. In: Proceedings of 25th international society for magnetic resonance in medicine, Honolulu, p 5490

Welch E, Gifford A, Towse T (2014) Phantom validation of temperature mapping using fat–water MRI with explicit fitting of water peak location. In: Proceedings of 22nd international society for magnetic resonance in medicine, Milan, p 3673

Winter L, Oberacker E, Paul K, Ji Y, Oezerdem C, Ghadjar P, Thieme A, Budach V, Wust P, Niendorf T (2016) Magnetic resonance thermometry: methodology, pitfalls and practical solutions. Int J Hyperth 32:63–75

Yaligar J, Verma SK, Gopalan V, Rengaraj A, Xianfeng T, Velan SS (2017) Evaluation of the vascular perfusion in activated brown adipose tissue by dynamic contrast enhanced MR imaging. In: Proceedings of 25th international society for magnetic resonance in medicine, Honolulu, p 1067

Zancanaro C, Nano R, Marchioro C, Sbarbati A, Boicelli A, Osculati F (1994) Magnetic resonance spectroscopy investigations of brown adipose tissue and isolated brown adipocytes. J Lipid Res 35:2191–2199

Multispectral Optoacoustic Tomography of Brown Adipose Tissue

Angelos Karlas, Josefine Reber, Evangelos Liapis, Korbinian Paul-Yuan, and Vasilis Ntziachristos

Contents

1 The Multispectral Optoacoustic Tomography Principle 3
2 Sources of Contrast in Multispectral Optoacoustic Tomography 3
3 Label-Free Brown Fat Tissue Imaging Using Multispectral Optoacoustic Tomography 5
3.1 Preclinical Brown Fat Tissue Imaging Using Multispectral Optoacoustic Tomography 5
3.2 Clinical Brown Fat Tissue Imaging Using Multispectral Optoacoustic Tomography 6
4 Clinical Multispectral Optoacoustic Tomography Challenges and Perspectives for the Future 7
References 7

Abstract

MSOT has revolutionized biomedical imaging because it allows anatomical, functional, and molecular imaging of deep tissues in vivo in an entirely noninvasive, label-free, and real-time manner. This imaging modality works by pulsing light onto tissue, triggering the production of acoustic waves, which can be collected and reconstructed to provide high-resolution images of features as deep as several centimeters below the body surface. Advances in hardware and software continue to bring MSOT closer to clinical translation. Most recently, a clinical handheld MSOT system has been used to image brown fat tissue (BAT) and its metabolic activity by directly resolving the spectral signatures of hemoglobin and lipids. This opens up new possibilities for studying BAT physiology

---

Angelos Karlas and Josefine Reber contributed equally to this work.

A. Karlas · J. Reber · E. Liapis · K. Paul-Yuan · V. Ntziachristos (✉)
Chair of Biological Imaging, Technical University Munich, Munich, Germany

Institute of Biological and Medical Imaging (IBMI), Helmholtz Zentrum München, Neuherberg, Germany
e-mail: v.ntziachristos@tum.de

A. Pfeifer et al. (eds.), *Brown Adipose Tissue*,
Handbook of Experimental Pharmacology 251, https://doi.org/10.1007/164_2018_141

and its role in metabolic disease without the need to inject animals or humans with contrast agents. In this chapter, we overview how MSOT works and how it has been implemented in preclinical and clinical contexts. We focus on our recent work using MSOT to image BAT in resting and activated states both in mice and humans.

Keywords
Brown adipose tissue · Hemoglobin oxygenation · Metabolic imaging · MSOT · Optoacoustics · Photoacoustics · Spectral unmixing

Purely optical imaging techniques such as optical microscopy, endoscopy, and optical coherence tomography rely on light-tissue interactions for high-contrast imaging in vivo and ex vivo (Weissleder and Pittet 2008). However, light scattering and absorption degrade image resolution with increasing depth and gradually attenuate the available light energy, limiting the effective imaging depth to a few hundred microns (Ntziachristos 2010). Non-optical techniques are usually used to image beyond these depths, including X-ray computed tomography (CT), magnetic resonance imaging (MRI), ultrasound (US), positron emission tomography (PET), and single-photon emission tomography (SPECT). These non-optical techniques present disadvantages that limit their use in the clinic, including the need for ionizing radiation (CT, PET, and SPECT) or for expensive, bulky equipment (MRI, PET, and SPECT).

Optoacoustics (OA) overcomes the depth limitations of optical imaging techniques because, as a hybrid technique, it generates an image based not only on light but also on acoustic waves. In OA, the tissue is illuminated with pulsed laser light, which is absorbed and causes minimal local heating, which leads in turn to thermoelastic expansion (Ntziachristos and Razansky 2010). This expansion generates acoustic waves, usually within the range of ultrasound, which travel out of the tissue and are detected by ultrasound transducers. The detected waves are then reconstructed into planar images. Since acoustic waves are scattered much less strongly than light as they travel through tissue, OA can image down to depths of ~2–5 cm, compared to a maximum of only ~1 mm for optical imaging techniques (Ntziachristos 2010). While OA cannot yet achieve the penetration depth of US, it offers superior, optical contrast (Ntziachristos and Razansky 2010) while retaining the usability of US imaging systems.

Of the various types of OA developed so far, multispectral optoacoustic tomography (MSOT) has made the greatest progress toward clinical translation (Dean-Ben et al. 2017). MSOT has already demonstrated its usefulness in several fields, including vascular medicine, breast oncology, thyroid imaging, muscle hemodynamics, and white adipose tissue (WAT) imaging (Karlas et al. 2017; Taruttis et al. 2016; Diot et al. 2015, 2017; Dima and Ntziachristos 2016; Buehler et al. 2017). Most recently, our group has shown that MSOT can image brown adipose tissue (BAT) and its metabolic activity in mice and humans without the need for injecting potentially toxic contrast agents. This opens up new possibilities for noninvasive, longitudinal investigation of BAT composition and physiology as well as their changes in disease (Reber et al. 2018).

1 The Multispectral Optoacoustic Tomography Principle

In MSOT, tissue is repeatedly excited with sequential pulses of near-infrared (NIR) laser light covering, for example, such wavelength ranges as 680–980 nm in 10-nm steps. The resulting ultrasound responses are captured using an array of usually 256 or 512 piezoelectric sensors. MSOT works in single-pulse-per-frame (SPPF) mode: each ultrafast laser pulse (duration of ~10 ns) generates a broadband ultrasound response with energy in the frequency range of ~0.5–7 MHz (Fig. 1). Custom-developed reconstruction methods are used to generate a tomographic image from the recorded ultrasound response to each single-wavelength pulse (Ntziachristos and Razansky 2010). Modern MSOT systems achieve frame rates of up to 50 Hz, allowing a complete series of single-wavelength frames covering all user-selected wavelengths (known as a "multispectral stack") to be recorded in less than a second. The varying intensity of each pixel along a multispectral stack gives the absorption spectrum of the tissue at that pixel position. In this way, MSOT adds the fifth dimension of spectrum to four-dimensional spatiotemporal imaging. Finally, the absorption spectrum at each pixel can be decomposed into known absorption spectra of biomedically relevant chromophores, such as hemoglobin, lipids, and water. This step allows the recorded absorption spectra to be translated into contributions from the various chromophores through a process known as "spectral unmixing." Such a process can reveal the distribution of light-absorbing molecules in living tissue with picomolar sensitivity (Ntziachristos and Razansky 2010; Diot et al. 2017).

2 Sources of Contrast in Multispectral Optoacoustic Tomography

Like purely optical imaging techniques, MSOT can image the contrast produced by externally administered agents, such as fluorescent dyes, nanoparticles, and photosensitizers, provided they absorb in the NIR region and have low quantum yield (Gujrati et al. 2017). Indocyanine green (ICG), already in clinical use for more than half a century, is the NIR dye most often used in OA (Philip et al. 1996). Nevertheless, the real advantage of MSOT over all other purely optical as well as non-optical imaging techniques is its ability to simultaneously detect several endogenous chromophores, including hemoglobin, melanin, lipids, and water, without the need for injecting external agents such as ICG. This equips MSOT with the ability to measure a broad range of physiological and pathophysiological processes such as tissue oxygenation, vascularization, and atherosclerosis (Weber et al. 2016).

MSOT can image and quantify various endogenous tissue chromophores through its ability to recognize the spectral signatures of each chromophore within acoustic signals collected over a range of wavelengths (Weissleder 2001). The most abundant intrinsic chromophore is hemoglobin, the iron-containing protein inside the red blood cells of all vertebrates that delivers oxygen throughout the body. When oxygen binds to the heme group of hemoglobin, the protein undergoes structural and electronic changes that alter its absorption spectrum. MSOT can detect these spectral changes, allowing the discrimination between oxy- and deoxyhemoglobin,

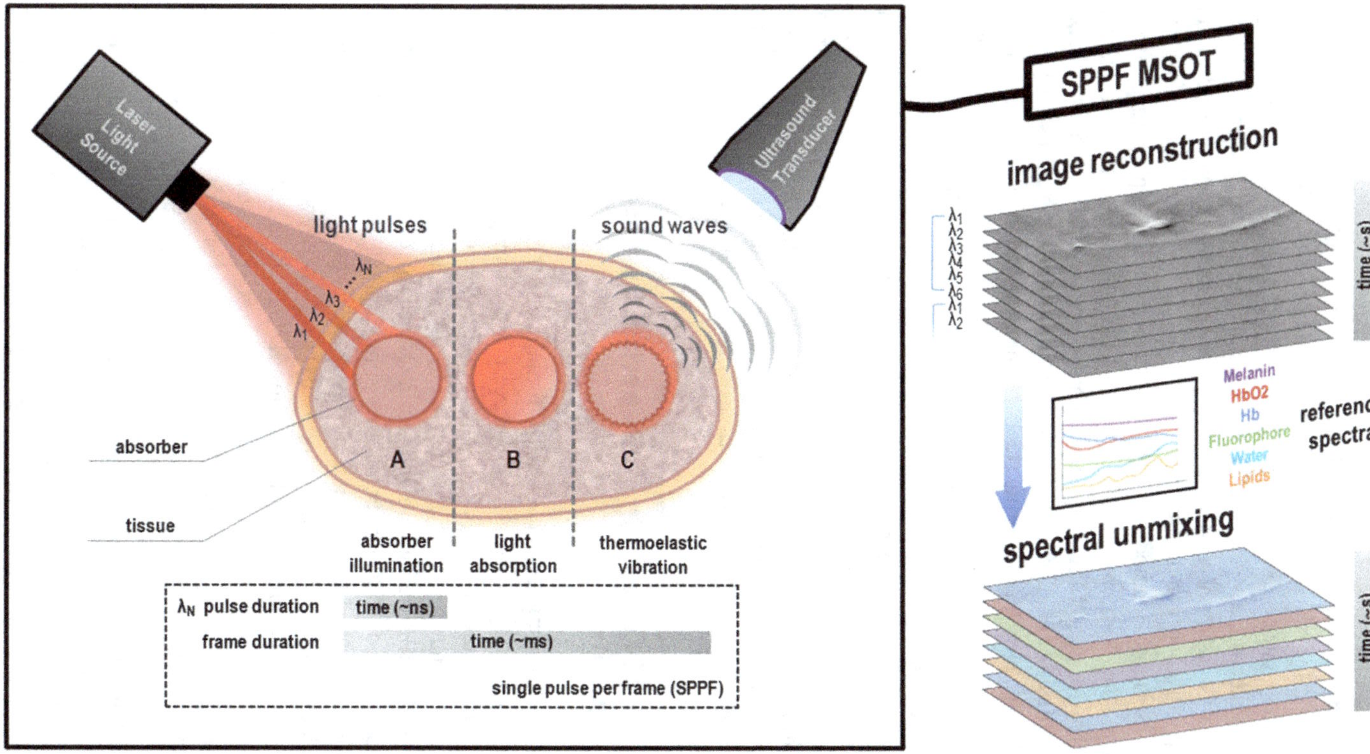

Fig. 1 The operating principle of MSOT. Light pulses with different NIR wavelengths (λ_1, λ_2, λ_3 ... λ_N) illuminate the tissue of interest, and chromophores within the tissue absorb the light, leading to transient local heating that triggers thermoelastic expansion. This expansion generates acoustic waves that propagate in all directions and are detected by an array of ultrasound detectors at the surface of the tissue. Multispectral stacks of single-wavelength tomographic images are produced using specific reconstruction methods. Decomposition of measured spectrum at each pixel across the *N* wavelengths in the stack (six in this example) based on the known reference spectra of biological chromophores such as melanin and hemoglobin allows "unmixing" of the total image into images of different chromophores. HbO_2 oxygenated hemoglobin, *Hb* deoxygenated hemoglobin

measurement of their respective concentrations, calculation of total hemoglobin concentration [also known as total blood volume (TBV)], and estimation of blood oxygen saturation (sO_2) (Laufer et al. 2012). Since MSOT illuminates tissue with light covering a broad range of NIR wavelengths, it can detect a similarly broad range of endogenous chromophores. The natural skin pigment melanin absorbs strongly in the visible and near-infrared ranges. Lipids absorb strongly around 930 nm, while water absorbs strongly around 970 nm. The NIR optical window of 680–980 nm is particularly useful because hemoglobin and water absorb much less in this window than at other wavelengths, allowing more sensitive detection of other chromophores even down to depths of several centimeters. For example, MSOT at these wavelengths can assess intra- and peritumoral vascularity and fat and water content in breast tumors in patients, greatly expanding on the information extracted by US (Diot et al. 2017).

Genetically encoded chromophores expand the contrast agents that MSOT can image noninvasively in preclinical animal studies (Weber et al. 2016). Reporter genes encoding OA-compatible proteins can be expressed in specific tissues at specific points in development, creating unique experimental opportunities. For example, green fluorescent protein (GFP) and its derivatives, which revolutionized anatomical and functional optical microscopy, can also be detected by MSOT (Razansky et al. 2009). However, none of the GFP variants described so far absorbs strongly in the NIR window of 680–980 nm (Razansky et al. 2009). Starting from phytochromes, which are photo-sensory receptors that absorb light when covalently bound to a linear tetrapyrrole such as biliverdin, researchers have recently developed fluorescent proteins that absorb light in the NIR range (Shu et al. 2009). For example, near-infrared fluorescent protein (iRFP) has been used for single-wavelength OA tomography in vivo, where it showed an absorption maximum at ~690 nm and good photodynamic stability (Filonov et al. 2012).

Another strategy when using genetically encoded chromophores is to express enzymes that generate OA-compatible small molecules. The prokaryotic *lacZ* gene can be expressed in mammalian tissues to generate the enzyme β-galactosidase, which can hydrolyze exogenously added X-gal to produce an intensely blue product readily detectable by OA imaging in the visible range (Cai et al. 2012). Another example is expressing the genes to endogenously produce violacein, which shows good photobleaching resistance similar to that of X-gal (Jiang et al. 2015). The tyrosinase gene, which encodes the key enzyme in melanin biosynthesis, can be expressed in otherwise non-melanogenic cells (Jathoul et al. 2015). Expression of tyrosinase allows creation of MSOT contrast without the need to administer an exogenous precursor.

Despite this range of potential chromophores, most MSOT studies have focused on the strong contrast provided by hemoglobin. The technique can provide noninvasive, longitudinal assessment of slow pathological processes such as angiogenesis and hypermetabolism (Omar et al. 2015; Herzog et al. 2012) as well as tumor hypoxia/oxygenation (Tzoumas et al. 2016). It can also monitor fast (sub-second) processes such as neural activity that alter hemodynamics and so can be detected as changes in blood oxygen saturation and total hemoglobin concentration (Gottschalk et al. 2015). In the next section, we discuss the recently demonstrated ability of MSOT to track several endogenous chromophores in vivo in order to characterize BAT and monitor its activation and changes related to disease.

3 Label-Free Brown Fat Tissue Imaging Using Multispectral Optoacoustic Tomography

Our group reasoned that MSOT should be able to differentiate BAT from WAT on the basis of their differences in hemoglobin, lipid, and water composition, which should translate to different spectral characteristics. If so, MSOT could turn out to be a powerful tool for studying BAT activation in a noninvasive, longitudinal manner. Our work suggests that, indeed, by measuring changes in local hemoglobin gradients over time, MSOT can quantify BAT activation in mice following pharmacological stimulation and BAT activation in humans following cold exposure (Fig. 2). Below we discuss, in turn, the preclinical and clinical evidence showing that MSOT can image BAT activation.

3.1 Preclinical Brown Fat Tissue Imaging Using Multispectral Optoacoustic Tomography

MSOT has been validated in mouse, fish, and other animal models of health and disease for being able to quantitatively analyze endogenous and exogenous chromophores (Razansky et al. 2007). High-quality MSOT imaging depends on homogeneous illumination and ultrasound detection around the sample. In state-of-the-art preclinical MSOT systems, ultrasound detectors cover approximately 270° (Fig. 2). The animal

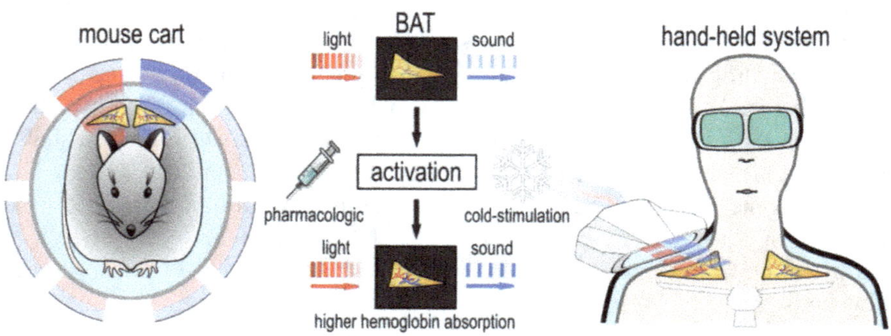

Fig. 2 Studying BAT activation in mice and humans using MSOT. The *image on the left* depicts the experimental setup for preclinical imaging, in which an anesthetized mouse is placed inside a cylindrical chamber within a larger measuring setup ("mouse cart"). Laser illumination of interscapular BAT deposits at various wavelengths generates acoustic waves, which are reconstructed into an image of BAT in the resting state. Then norepinephrine is injected intravenously to metabolically activate BAT, and the BAT deposits are imaged again. The *image on the right* depicts the experimental setup for clinical imaging of supraclavicular BAT activation with a handheld MSOT system. BAT was activated by cold stimulation using a cooling suit. Adapted with permission from Reber et al. (2018)

or excised tissue is placed in thin transparent foil (~100 μm) and then submerged in water at approximately 34°C. Typically, 45–60 min are needed to obtain a whole-body mouse scan in 300-μm steps along the *z*-axis.

3.1.1 Spectral Characterization of Mouse Adipose Tissue Ex Vivo

By designing and manufacturing appropriate biological imaging phantoms, the spectral signature of an excised tissue sample, such as BAT or WAT, can be accurately determined under tightly controlled experimental conditions. An ideal phantom should mimic the basic physical properties of living tissues such as optical absorption, scattering, and speed of sound. Phantoms can be used to analyze and optimize the imaging setup for subsequent in vivo or postmortem experiments. Excised tissues are typically examined within cylindrical phantoms with a diameter of ~2 cm and a composition of 1.3% (v/v) agar and 1.2% (v/v) fat emulsion. To investigate absorption spectra of mouse BAT and WAT, tissue samples were inserted into plastic tubes with a diameter of 3 mm, which were then inserted into the cylindrical phantom (Tzoumas et al. 2014). MSOT showed that OA signal intensity of BAT was more than two-fold higher than that of WAT over the entire NIR range of 700–900 nm (Reber et al. 2018). This may be because the high density of iron-rich mitochondria makes BAT dark brown (Enerback 2009), which may explain its greater light absorption. BAT is also more highly vascularized than WAT, and the higher hemoglobin content may contribute to the greater absorption. MSOT has shown promising ability to detect lipid-based differences among BAT, WAT, and beige adipose tissue. Beige adipose tissue is thought to have a composition intermediate between that of BAT and WAT (Cedikova et al. 2016) and a function closer to that of BAT (Giralt and Villarroya 2013). The lipid spectrum of beige adipose tissue showed greater intensity than the lipid spectrum of WAT in the NIR range from 700 to 900 nm, yet the beige spectrum retained the characteristic WAT peak at 930 nm.

3.1.2 Imaging Mouse Adipose Tissue In Vivo

MSOT can track the contrast of hemoglobin to analyze tissue pathophysiology hallmarks (Tzoumas and Ntziachristos 2017), and the same contrast can allow tracking of BAT activation. BAT activation is followed by a substantial increase in blood flow (Ernande et al. 2016). MSOT can image interscapular BAT (iBAT), including the underlying Sulzer vein (SV), which provides the main venous drainage (Fig. 3a, b) (Reber et al. 2018). Spectral unmixing allows quantification of oxy- and deoxyhemoglobin content throughout iBAT and surrounding tissues, first in the resting state (Fig. 3d, g) and then following metabolic activation with norepinephrine (Fig. 3c, e, h), which increases BAT perfusion and induces BAT thermogenesis via consumption of glucose and lipid (Cypess et al. 2015). In our mouse studies, we found that norepinephrine altered hemoglobin levels only in the iBAT, not in surrounding muscle or other soft tissues (Fig. 3f, i). These results suggest that MSOT can serve as a powerful method for characterizing iBAT activation in mice and that the extent of iBAT vascularization can be quantified based on hemoglobin contrast.

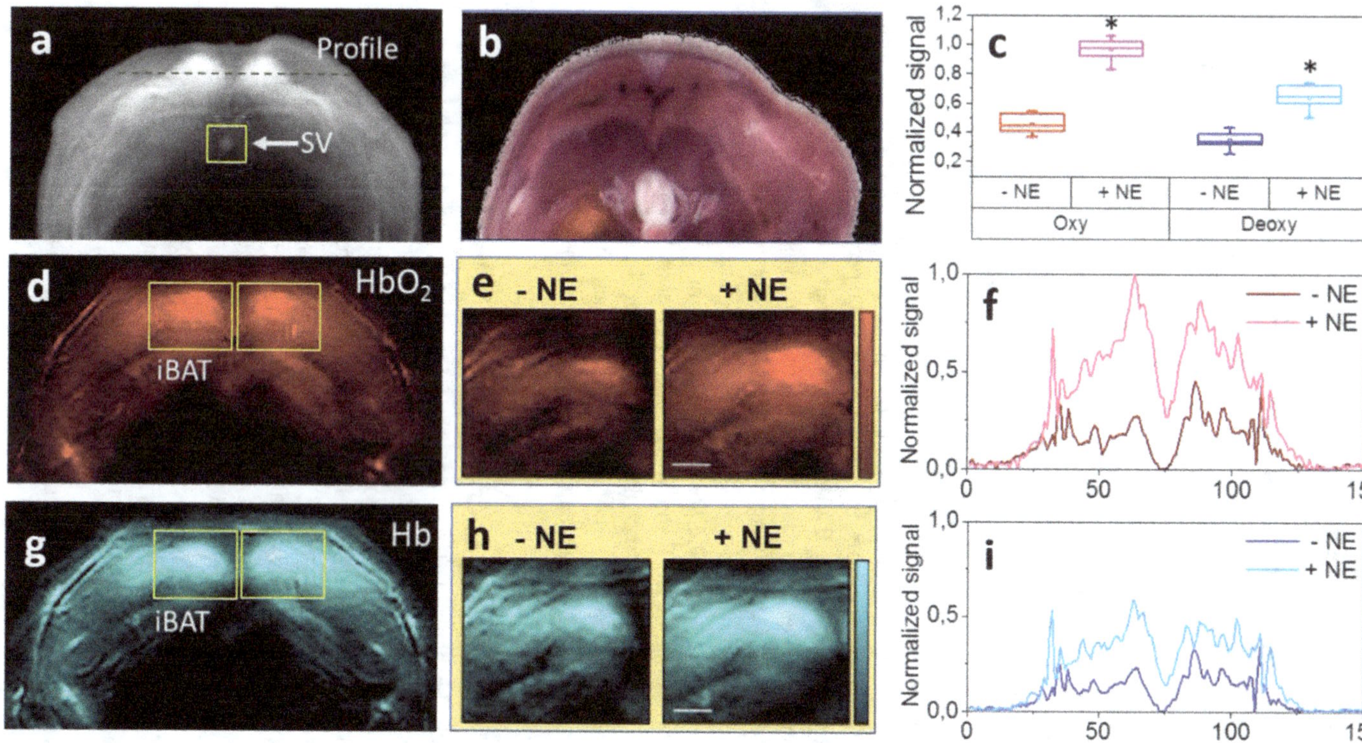

Fig. 3 In vivo imaging of interscapular BAT activation in mice using MSOT. (**a**) Reconstructed MSOT image (800 nm) showing interscapular BAT (iBAT) and the Sulzer vein (SV; yellow square). The black dashed line indicates the horizontal intensity profiles shown in panels (**f**) and (**i**). (**b**) Color image of a transverse cryoslice of the neck area, showing iBAT. (**c**) Boxplot of normalized signal intensities of oxyhemoglobin (HbO_2) and deoxyhemoglobin (Hb) in iBAT before and after activation with norepinephrine (NE). (**d–f**) Relative HbO_2 intensity in iBAT lobes before and after activation with NE. The close-up view in panel (**e**) shows the right lobe. The corresponding horizontal intensity profile is shown in panel (**f**). (**g–i**) Relative Hb intensity in iBAT lobes before and after activation with NE. The close-up view in panel (**h**) shows the right lobe. The corresponding horizontal intensity profile is shown in panel (**i**). Reproduced with permission from Reber et al. (Reber et al. 2018)

3.2 Clinical Brown Fat Tissue Imaging Using Multispectral Optoacoustic Tomography

One of the factors driving the application of MSOT to an expanding range of clinical problems is the ability to conduct high-resolution imaging with a handheld scanner that ensures patient comfort and flexibility for the clinician. The scanner can be used to analyze various parts of the body without extra equipment or special operator training. Handheld MSOT scanners emit near-infrared light, usually in the range of 700–980 nm; they carry ultrasound detectors operating at central frequencies of 4–11 MHz; and they record data at video rates of up to 50 Hz (Karlas et al. 2017). These portable clinical systems achieve penetration depths of 2–5 cm (depending on the central frequency and tissue type) and spatial resolution better than 100 μm.

Building on our studies of BAT activation in mice, we succeeded in imaging supraclavicular BAT in humans previously shown to have BAT deposits by PET or MRI (Fig. 4a–c) (Reber et al. 2018). By illuminating the tissue at 28 NIR wavelengths from 700 to 970 nm in 10-nm steps, we were able to differentiate BAT from WAT based on their spectral characteristics (Fig. 4d). We were also able to detect BAT activation in response to cold exposure, which led to significant increases in oxy- and deoxyhemoglobin OA signal and therefore to increases in TBV (Fig. 4e, f). These results suggest that MSOT has the capacity to track hemodynamic changes as a marker of BAT metabolic state, without the need for exogenous contrast agents. This may provide a unique opportunity for clinical application of MSOT, since the technique can provide rich, quantitative information about tissue physiology and function that is inaccessible to US, without requiring the extremely expensive infrastructure or radiation risks of other clinical imaging modalities such as MRI and PET.

4 Clinical Multispectral Optoacoustic Tomography: Challenges and Perspectives for the Future References

The time needed to acquire a full multispectral stack of images leaves today's handheld MSOT vulnerable to motion artifacts, which compromise spatial resolution and the accuracy of spectral unmixing. Such artifacts can be minimized as the clinician-operator becomes familiar with the setup and procedure. Other examples of motion artifacts which may threaten clinical MSOT imaging even when the scan head and patient remain still are the respiratory motion or the motion related to arterial pulsation. Recording data at higher frame rates or employing specialized motion correction algorithms usually suppress these and other types of motion artifacts (Taruttis et al. 2012). Another limitation of MSOT is that although it provides impressive imaging depth in the absence of exogenous contrast agents, the depth is currently insufficient for reliable localization and quantification of deeper BAT deposits such as in the retroperitoneum.

It is likely that future improvements in illumination schemes, ultrasound sensors, and analysis methods will miniaturize scanning probes, improve image quality, and decrease post-acquisition processing times. This will enable large clinical studies to

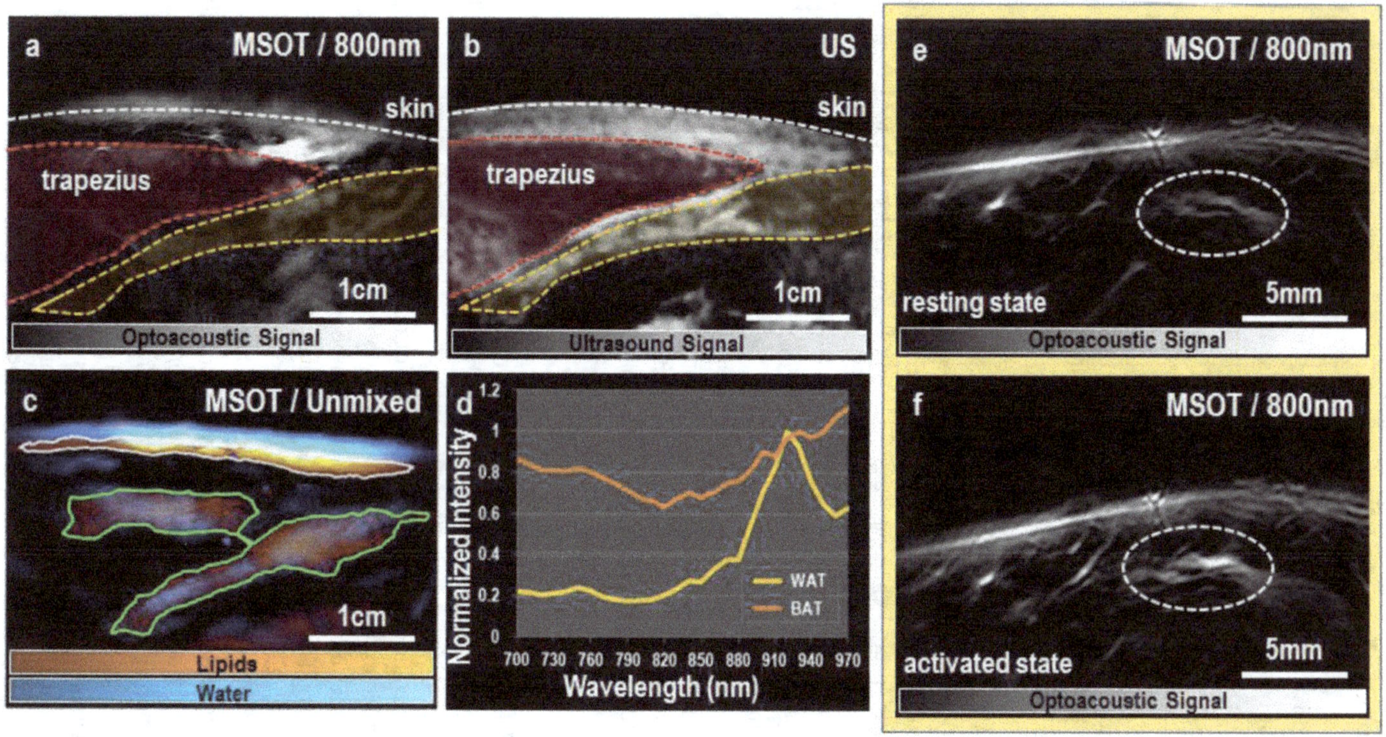

Fig. 4 In vivo imaging of supraclavicular BAT activation in humans using MSOT. Subjects were confirmed to have BAT deposits based on PET and MRI. (**a**) MSOT image (800 nm) showing the expected position of BAT (yellow region). The trapezius muscle is tinted in red. (**b**) US image corresponding to the field of view in panel (**a**). (**c**) MSOT image showing signal intensity attributed to lipid or water following spectral unmixing. Putative subcutaneous WAT is enclosed with a white line in the upper part of the image; putative trapezius muscle, with a green line in the middle of the image; and putative BAT, with a green line extending from the lower left to upper right of the image. (**d**) Mean spectral profiles of the WAT and BAT regions delineated in panel (**c**). (**e**, **f**) MSOT images of supraclavicular BAT (**e**) in the resting state and (**f**) after 20 min of cold exposure to induce BAT activation. The region bounded inside the white dashed line shows an increased optoacoustic signal after BAT activation due to an increase in hemoglobin. Wavelength of 800 nm corresponds to the isosbestic point of HbO_2 and Hb in the NIR. Adapted with permission from Reber et al. (2018)

validate and exploit the potential of MSOT for imaging tissue physiology and disease. In the case of BAT activation, further work should build on our findings so far (Reber et al. 2018) to establish the reproducibility of MSOT-based quantification and its correlation with the results of PET, MRI, and US. If MSOT can be validated, it can be applied in large trials to compare BAT mass and metabolic activity across patients with various metabolic disorders in the presence or absence of other comorbidities (e.g., cardiovascular).

References

Buehler A, Diot G, Volz T, Kohlmeyer J, Ntziachristos V (2017) Imaging of fatty tumors: appearance of subcutaneous lipomas in optoacoustic images. J Biophotonics 10:983–989

Cai X, Li L, Krumholz A, Guo Z, Erpelding TN, Zhang C, Zhang Y, Xia Y, Wang LV (2012) Multi-scale molecular photoacoustic tomography of gene expression. PLoS One 7:e43999

Cedikova M, Kripnerova M, Dvorakova J, Pitule P, Grundmanova M, Babuska V, Mullerova D, Kuncova J (2016) Mitochondria in white, brown, and beige adipocytes. Stem Cells Int 2016:6067349

Cypess AM, Weiner LS, Roberts-Toler C, Elía EF, Kessler SH, Kahn PA, English J, Chatman K, Trauger SA, Doria A, Kolodny GM (2015) Activation of human brown adipose tissue by a β3-adrenergic receptor agonist. Cell Metab 21:33–38

Dean-Ben XL, Gottschalk S, Mc Larney B, Shoham S, Razansky D (2017) Advanced optoacoustic methods for multiscale imaging of in vivo dynamics. Chem Soc Rev 46:2158–2198

Dima A, Ntziachristos V (2016) In-vivo handheld optoacoustic tomography of the human thyroid. Photoacoustics 4:65–69

Diot G, Dima A, Ntziachristos V (2015) Multispectral opto-acoustic tomography of exercised muscle oxygenation. Opt Lett 40:1496–1499

Diot G, Metz S, Noske A, Liapis E, Schroeder B, Ovsepian SV, Meier R, Rummeny E, Ntziachristos V (2017) Multispectral optoacoustic tomography (MSOT) of human breast cancer. Clin Cancer Res 23:6912–6922

Enerback S (2009) The origins of brown adipose tissue. N Engl J Med 360:2021–2023

Ernande L, Stanford KI, Thoonen R, Zhang H, Clerte M, Hirshman MF, Goodyear LJ, Bloch KD, Buys ES, Scherrer-Crosbie M (2016) Relationship of brown adipose tissue perfusion and function: a study through β2-adrenoreceptor stimulation. J Appl Physiol (1985) 120:825–832

Filonov GS, Krumholz A, Xia J, Yao J, Wang LV, Verkhusha VV (2012) Deep-tissue photoacoustic tomography of a genetically encoded near-infrared fluorescent probe. Angew Chem Int Ed Engl 51:1448–1451

Giralt M, Villarroya F (2013) White, brown, beige/brite: different adipose cells for different functions? Endocrinology 154:2992–3000

Gottschalk S, Felix Fehm T, Luís Deán-Ben X, Razansky D (2015) Noninvasive real-time visualization of multiple cerebral hemodynamic parameters in whole mouse brains using five-dimensional optoacoustic tomography. J Cereb Blood Flow Metab 35:531–535

Gujrati V, Mishra A, Ntziachristos V (2017) Molecular imaging probes for multi-spectral optoacoustic tomography. Chem Commun (Camb) 53:4653–4672

Herzog E, Taruttis A, Beziere N, Lutich AA, Razansky D, Ntziachristos V (2012) Optical imaging of cancer heterogeneity with multispectral optoacoustic tomography. Radiology 263:461–468

Jathoul AP, Laufer J, Ogunlade O, Treeby B, Cox B, Zhang E, Johnson P, Pizzey AR, Philip B, Marafioti T, Lythgoe MF, Pedley RB, Pule MA, Beard P (2015) Deep in vivo photoacoustic imaging of mammalian tissues using a tyrosinase-based genetic reporter. Nat Photonics 9:239–246

Jiang Y, Sigmund F, Reber J, Dean-Ben XL, Glasl S, Kneipp M, Estrada H, Razansky D, Ntziachristos V, Westmeyer GG (2015) Violacein as a genetically-controlled, enzymatically amplified and photobleaching-resistant chromophore for optoacoustic bacterial imaging. Sci Rep 5:11048

Karlas A, Reber J, Diot G, Bozhko D, Anastasopoulou M, Ibrahim T, Schwaiger M, Hyafil F, Ntziachristos V (2017) Flow-mediated dilatation test using optoacoustic imaging: a proof-of-concept. Biomed Opt Express 8:3395–3403

Laufer J, Johnson P, Zhang E, Treeby B, Cox B, Pedley B, Beard P (2012) In vivo preclinical photoacoustic imaging of tumor vasculature development and therapy. J Biomed Opt 17:056016

Ntziachristos V (2010) Going deeper than microscopy: the optical imaging frontier in biology. Nat Methods 7:603–614

Ntziachristos V, Razansky D (2010) Molecular imaging by means of multispectral optoacoustic tomography (MSOT). Chem Rev 110:2783–2794

Omar M, Schwarz M, Soliman D, Symvoulidis P, Ntziachristos V (2015) Pushing the optical imaging limits of cancer with multi-frequency-band raster-scan optoacoustic mesoscopy (RSOM). Neoplasia 17:208–214

Philip R, Penzkofer A, Bäumler W, Szeimies RM, Abels C (1996) Absorption and fluorescence spectroscopic investigation of indocyanine green. J Photochem Photobiol A Chem 96:137–148

Razansky D, Vinegoni C, Ntziachristos V (2007) Multispectral photoacoustic imaging of fluorochromes in small animals. Opt Lett 32:2891–2893

Razansky D, Distel M, Vinegoni C, Ma R, Perrimon N, Köster RW, Ntziachristos V (2009) Multispectral opto-acoustic tomography of deep-seated fluorescent proteins in vivo. Nat Photonics 3:412–417

Reber J, Willershäuser M, Karlas A, Paul-Yuan K, Diot G, Franz D, Fromme T, Ovsepian SV, Bézière N, Dubikovskaya E, Karampinos DC, Holzapfel C, Hauner H, Klingenspor M, Ntziachristos V (2018) Non-invasive measurement of brown fat metabolism based on optoacoustic imaging of hemoglobin gradients. Cell Metab 27:689–701.e684

Shu X, Royant A, Lin MZ, Aguilera TA, Lev-Ram V, Steinbach PA, Tsien RY (2009) Mammalian expression of infrared fluorescent proteins engineered from a bacterial phytochrome. Science (New York, NY) 324:804–807

Taruttis A, Claussen J, Razansky D, Ntziachristos V (2012) Motion clustering for deblurring multispectral optoacoustic tomography images of the mouse heart. J Biomed Opt 17:016009

Taruttis A, Timmermans AC, Wouters PC, Kacprowicz M, van Dam GM, Ntziachristos V (2016) Optoacoustic imaging of human vasculature: feasibility by using a handheld probe. Radiology 281:256–263

Tzoumas S, Ntziachristos V (2017) Spectral unmixing techniques for optoacoustic imaging of tissue pathophysiology. Philos Transact A Math Phys Eng Sci 375

Tzoumas S, Zaremba A, Klemm U, Nunes A, Schaefer K, Ntziachristos V (2014) Immune cell imaging using multi-spectral optoacoustic tomography. Opt Lett 39:3523–3526

Tzoumas S, Nunes A, Olefir I, Stangl S, Symvoulidis P, Glasl S, Bayer C, Multhoff G, Ntziachristos V (2016) Eigenspectra optoacoustic tomography achieves quantitative blood oxygenation imaging deep in tissues. Nat Commun 7:12121

Weber J, Beard PC, Bohndiek SE (2016) Contrast agents for molecular photoacoustic imaging. Nat Methods 13:639–650

Weissleder R (2001) A clearer vision for in vivo imaging. Nat Biotechnol 19:316–317

Weissleder R, Pittet MJ (2008) Imaging in the era of molecular oncology. Nature 452:580–589

BAT Exosomes: Metabolic Crosstalk with Other Organs and Biomarkers for BAT Activity

Deborah Goody and Alexander Pfeifer

Contents

1 Exosomes 338
2 Biological Function of Exosomes 339
  2.1 Exosomes and Brown Adipose Tissue 339
3 Diagnostic Potential of Exosomes 341
  3.1 Exosomes as Biomarkers of Brown Fat Activity 341
4 Conclusions 342
References 343

Abstract

In the last decade, exosomes have gained interest as a new type of intercellular communication between cells and tissues. Exosomes are circulating, cell-derived lipid vesicles smaller than 200 nm that contain proteins and nucleic acids, including microRNAs (miRNAs), and are able to modify cellular targets. Exosomal miRNAs function as signalling molecules that regulate the transcription of their target genes and can cause phenotypic transformation of recipient cells. Recent studies have shown that brown fat secretes exosomes as a form of communication with other metabolic organs such as the liver. Moreover, it has been shown that levels of miRNAs in BAT-derived exosomes change after BAT activation in vitro and in vivo. Thus, BAT-derived exosomes can be used as potential biomarkers of BAT activity. Here, we review the present knowledge about BAT-derived exosomes and their role in metabolism.

Keywords

Biomarker · Brown adipose tissue · Exosomes · MicroRNA

D. Goody · A. Pfeifer (✉)
Institute of Pharmacology and Toxicology, University Hospital Bonn, University of Bonn, Bonn, Germany
e-mail: alexander.pfeifer@uni-bonn.de

A. Pfeifer et al. (eds.), *Brown Adipose Tissue*,
Handbook of Experimental Pharmacology 251, https://doi.org/10.1007/164_2018_114

1 Exosomes

As part of the intercellular communication system, cells secrete a broad range of factors, including extracellular vesicles (EVs), into the extracellular space. EVs can modify target cells (Zhang et al. 2014; Nolte-'t Hoen and Wauben 2012). The term "exosome" was first used in 1981 by the group of Trams et al. to refer to exfoliated vesicles originated from cells (Trams et al. 1981). Six years later, Johnstone et al. proposed the name exosome to describe secreted, small membrane vesicles derived from multivesicular bodies (MVBs), which were first characterized in the transformation of mammalian reticulocyte to a mature erythrocyte (Johnstone et al. 1987). These vesicles were shown to be responsible for the removal of membranes and proteins (e.g. transferrin) during the reticulocyte maturation and were termed exosomes to describe the "reverse endocytosis" process. Exosomes are secreted from a wide variety of cells including immune cells, tumour cells and adipocytes (Lotvall and Valadi 2007; Roma-Rodrigues et al. 2014; Lin et al. 2013; Chen et al. 2016). Exosomes can be found in different body fluids such as serum, cerebrospinal fluid and urine (Rekker et al. 2014; Ho et al. 2014). Exosomes are 30–150 nm big and can be visualized by electron microscopy (Akers et al. 2013; Kowal et al. 2014). Despite growing knowledge on the biogenesis and characteristics of exosomes, the terminology used to describe and define the different groups of extracellular vesicles has still not been completely standardized.

The biogenesis of exosomes is distinct from other EVs. While exosomes are formed via endocytosis, another group of EVs named microvesicles arise by direct budding and fission of the plasma membrane. Microvesicles are larger in size (50–2,000 nm) and have also been described as "ectosomes" (Hess et al. 1999). The biogenesis of exosomes starts with endocytosis of the plasma membrane. On one hand, the endosomal network targets proteins and lipids for lysosomal degradation. On the other hand, it marks proteins/lipids for recycling and exocytosis. Many of these mechanisms are regulated by tetraspanins, a group of proteins enriched on the surface of exosomes (Sala-Valdes et al. 2012). Once endocytic vesicles are formed, they are transported to early endosomes. After various transformations, late endosomes are developed. Subsequently, membranes of late endosomes bud to form small intraluminal vesicles, also called multivesicular bodies (MVBs) (Akers et al. 2013; Keller et al. 2006; Stoorvogel et al. 1991). Finally, MVBs can fuse with the plasma membrane, and exosomes are released into the extracellular space (Johnstone et al. 1987; Glebov et al. 2015; Kajimoto et al. 2013). The endosomal sorting complex (ESCRT), which is made up of four protein complexes (ESCRT-0, I, II, III) and associated proteins (e.g. ALIX), is required for this process (Kowal et al. 2014; Colombo et al. 2013). Furthermore, tetraspanins play a role in the biogenesis of exosomes. These proteins are highly enriched on endosome membranes and are thought to support formation of MVBs through recruitment of other proteins (Stuffers et al. 2009; Jansen et al. 2009; Kosaka et al. 2010). Tetraspanins CD63, CD9 and CD81 are abundant in exosomes and serve as the most commonly used exosome markers (Roma-Rodrigues et al. 2014). Other proteins involved in exosome biogenesis (e.g. Alix, TSG101, Rab GTPases) have also been used for defining exosomes (Beninson et al. 2014; Mathew et al. 1995; Simpson et al. 2008). Previous studies have shown that various stress signals, e.g. oxidative stress, shear stress and hypoxia, induce the secretion of exosomes as well as microvesicles (Roma-Rodrigues et al. 2014). Hence, exosomes are not merely inert debris from cells. Once released, exosomes target neighbouring or remote recipient cells and alter cellular targets.

2 Biological Function of Exosomes

Exosomes are involved in many physiological and pathological processes as a novel mode of cell-to-cell communication and exchange of material between cells (Zhang et al. 2015). Over several years, multiple studies revealed their role in proliferation, differentiation, immunology and tumorigenesis (De Jong et al. 2014; Fernandez-Messina et al. 2015; Hoshino et al. 2015; Melo et al. 2014). In addition, exosomes can also be used as biomarkers to evaluate tissue function and in disease diagnosis (Tickner et al. 2014; Taylor and Gercel-Taylor 2014).

Exosomes comprise RNAs (mRNA and miRNAs), as well as proteins (e.g. Alix) and surface molecules (e.g. CD63). By transferring their cargo, exosomes act as carriers of information that may modify the function or fate of the target/recipient cells (Camussi et al. 2013). In this context, miRNAs are considered essential exosomal effectors and have gained interest as they directly regulate gene expression (Zhang et al. 2015). For example, in breast cancer, exosomal miR-105 decreases ZO-1 (tight junction protein 1) gene expression in endothelial cells and promotes metastases to the lung and brain (Zhou et al. 2014). MiRNAs are small, noncoding regulatory RNA molecules, which mediate specific post-transcriptional gene silencing of their target mRNA via the 3′-untranslated region (3′-UTR) and are involved in many biological processes (Bartel 2004). Some reports have demonstrated that specific miRNAs (e.g. miR-150, miR-451) are preferentially packaged into exosomes (Guduric-Fuchs et al. 2012). Although recent studies describe potential mechanisms for the sorting of miRNAs into exosomes, the definite underlying regulation remains unclear. One proposed mode of sorting is through the miRNA-induced silencing complex (miRISC) (Gibbings et al. 2009). MiRISC incorporates miRNA in order to identify target mRNA and, upon complementarity, silences the mRNA. This hypothesis is further supported by the co-localization of the main components of miRISC in exosomes (Gibbings et al. 2009).

2.1 Exosomes and Brown Adipose Tissue

There are mainly two types of adipose tissue: While white adipose tissue (WAT) represents the main energy storage in the body, brown adipose tissue (BAT) dissipates energy in form of heat (non-shivering thermogenesis) (Lowell and Flier 1997). In addition to WAT and BAT, inducible brown adipocytes also called beige or brite ("brown-in-white") adipocytes exist in subcutaneous WAT and can dissipate energy as heat upon cold induction, in a process also known as "browning" (Cousin et al. 1992; Harms and Seale 2013). Adipose tissue is composed of various types of cells such as preadipocytes, adipose stem cells, fibroblasts, endothelial cells and immune cells. The secretion of exosomes by adipocytes and adipose tissue has been reported by different groups (Deng et al. 2009; Ogawa et al. 2010). Exosomes from 3T3-L1 cells contain miRNAs that regulate adipogenesis such as miR-103 and miR-146b (Ogawa et al. 2010; Chen et al. 2014; Li et al. 2015). Recent studies have demonstrated that exosomes may function as intercellular signals within the different cells found in the adipose tissue as well as with other organs (Ying et al. 2017; Thomou et al. 2017; Chen et al. 2016). Ying et al. showed that adipose tissue macrophages (ATMs) in obese mice secrete exosomes, which contain high amounts of miR-155, an inhibitor of brown adipogenesis (Ying et al. 2017; Chen et al. 2013). MiR-155 was originally identified as a regulator of the immune system, but it was shown by Chen et al. that it impairs BAT function through inhibition of the central adipogenic transcription factor C/EBPβ (Chen et al. 2013). Interestingly, inhibition of miR-155 enhances brown adipocyte differentiation and induces the browning of white adipocytes (Chen et al. 2013). While miR-155 transgenic mice showed reduced BAT mass and UCP1 expression, mice lacking miR-155 had a significant increase in BAT function, and exhibit increased browning of WAT. Ying et al. later showed that administration of exosomes from obese mice induces glucose intolerance and insulin resistance in lean mice, supporting the concept of communication between ATM and adipocytes via exosomal miRNAs (Ying et al. 2017).

In order to understand the role of circulating miRNAs in adipose tissue, Thomou et al. generated mice specifically lacking the miRNA-processing enzyme Dicer in adipose tissue and analysed their serum-derived exosomes (Thomou et al. 2017). ADicerKO mice were characterized by reduced levels of circulating miRNAs, resulting in lipodystrophy (reduction of WAT), whitening of BAT as well as insulin resistance. Although the number of exosomes isolated from these mice was comparable, exosomes from ADicerKO mice exhibit significant alterations in miRNA content (Ortega et al. 2010). ADicerKO mice also showed a decrease in circulating miRNAs. Interestingly, ADicerKO mice presented an increase in circulating FGF21, which is mainly produced in the liver and upon release influences metabolism in multiple tissues (Potthoff et al. 2012; Badman et al. 2007). A database search for predicted miRNAs targeting *Fgf21* and luciferase reporter assays identified miR-99b as potential regulator of *Fgf21* (Thomou et al. 2017) (Fig. 1). Transplantation of wild-type BAT into ADicerKO mice led to a reduction of hepatic *Fgf21* as well as circulating FGF21, suggesting specific targeting of BAT-derived exosomes to the liver (Thomou et al. 2017). These results confirm the initial hypothesis that BAT exosomes and exosomal miRNAs can target other tissues, such as the liver. Thus, BAT might play also other roles in metabolism apart from non-shivering thermogenesis and energy expenditure.

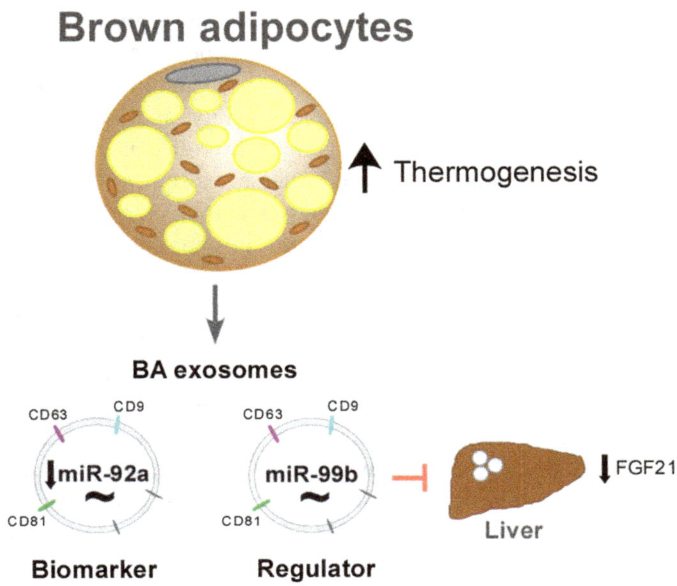

Fig. 1 Dual role of brown adipocyte-derived exosomes as biomarkers and regulators of metabolism. Thomou et al. identified the brown adipose tissue as an important source of circulating exosomal miR-99b, which regulates the expression and secretion of FGF21 from the liver (Thomou et al. 2017). Chen et al. showed an inverse correlation between BAT activity and exosomal miR-92a in mouse and human, leading to the potential of exosomal miR-92a as a serum biomarker of BAT acitivity (Chen et al. 2016)

3 Diagnostic Potential of Exosomes

The potential of exosomal miRNAs as diagnostic tools has been documented for cancer, such as colon and gastric cancer (Ogata-Kawata et al. 2014; Nedaeinia et al. 2017). These studies showed that the pattern of exosomal miRNAs changes in patients compared to healthy individuals, as well as during the progression of the disease. Furthermore, CD24 and EpCAM (epithelial cell adhesion molecule) have also been described as exosomal breast cancer markers (Rupp et al. 2011). Circulating miRNAs have been identified as potential biomarkers of metabolic disorder, including miR-152 and miR-17 (Wu et al. 2015). However, exosomal miRNAs were only recently described as indicators for BAT activity in the context of metabolism (Chen et al. 2016).

3.1 Exosomes as Biomarkers of Brown Fat Activity

BAT activity correlates with leanness in human adults, and the standard method for measuring this activity is ^{18}F-fluorodeoxyglucose (^{18}F-FDG) positron emission tomography coupled with computer tomography (PET/CT), exposing patients to cold and ionizing radiation (Virtanen et al. 2009). Despite growing concern about the rapid increase of obesity worldwide, novel diagnostic tools to asses BAT activity are still sparse. To determine if exosomes may be viable biomarkers of BAT metabolism, Chen et al. analysed exosome quantity and their miRNA cargo in BAT activation (Chen et al. 2016). Chen and colleagues were the first to demonstrate that brown adipocytes indeed release exosomes (Chen et al. 2016). Interestingly, the activation of brown adipocytes and BAT resulted in significant changes in both number and cargo of exosomes in vitro and in vivo (Chen et al. 2016). MiRNA array analysis of exosomes from activated murine brown adipocytes (treated with cAMP) and murine BAT (cold exposure or β_3-adrenoreceptor agonist) showed coherent changes in levels of miR-92a, miR-34c and miR-133a. Validation with quantitative real-time PCR (qPCR) revealed a significant down- and upregulation of miR-92a and miR-34c in activated brown adipocytes as well as BAT, respectively (Chen et al. 2016). In contrast, miR-133a abundance was not significantly altered in exosomes during activation of BAT in vivo and brown adipocytes in vitro. Furthermore, the authors investigated the exosomal miRNA content in human cohorts. Subjects in these cohorts had been previously studied by ^{18}F-FDG PET/CT regarding their BAT activity. In contrast to miR-92a and miR-133a, miR-34c could not be detected in exosomes derived from human serum samples (Chen et al. 2016). Moreover, miR-133a levels were similar between subjects with high BAT activity compared to subjects with low BAT activity. Therefore, the authors focused on miR-92a. Importantly, serum exosomal miR-92a was significantly lower in subjects with high BAT activity compared to those with low BAT activity. Taken together, these results indicate that exosomal miR-92a present in serum inversely correlates with murine as well as human BAT activity (Fig. 1) (Chen et al. 2016).

4 Conclusions

In the last few years, extracellular vesicles and, more importantly, exosomes have gained interest as important intercellular communicators and relevant biomarkers for various diseases, including metabolic disease (Chen et al. 2016; Ogawa et al. 2010). BAT has been linked to amelioration of overweight and to the development of therapeutic tools for the fight of obesity (Cannon and Nedergaard 2004). In addition, miRNA levels have been shown to be dysregulated in metabolic diseases, including obesity (McGregor and Choi 2011). MiRNAs can be packaged in extracellular vesicles, including exosomes, and transported in body fluids, e.g. blood and urine (Chevillet et al. 2014). Exosomes and its cargo have been shown to regulate the biological function of different cells (Zhang et al. 2016). Circulating exosome-derived miRNAs might play a key role in the metabolic crosstalk between adipose tissue and different organs. Recently, Chen et al. showed that BAT is an important source of circulating exosomes and that the activation of BAT resulted in significant changes in both the number and the miRNA content of exosomes (Chen et al. 2016). Exosomal miR-92a was shown to inversely correlate with BAT activity and energy expenditure and may serve as a serum biomarker of BAT (Fig. 1). Moreover, Thomou et al. identified hepatic FGF21 as a physiological target of BAT-derived exosomal miR-99b (Fig. 1) (Thomou et al. 2017).

However, further proteomic and miRNA analysis are essential in order to decipher the content of adipose tissue-derived exosomes. Despite growing interest and knowledge on the function of exosomes in fat, further studies are needed to fully reveal the potential of these small vesicles as biomarkers of metabolism as well as regulators of adipose tissue function and of crosstalk between multiple organs.

References

Akers JC, Gonda D, Kim R, Carter BS, Chen CC (2013) Biogenesis of extracellular vesicles (EV): exosomes, microvesicles, retrovirus-like vesicles, and apoptotic bodies. J Neurooncol 113:1–11

Badman MK, Pissios P, Kennedy AR, Koukos G, Flier JS, Maratos-Flier E (2007) Hepatic fibroblast growth factor 21 is regulated by PPARalpha and is a key mediator of hepatic lipid metabolism in ketotic states. Cell Metab 5:426–437

Bartel DP (2004) MicroRNAs: genomics, biogenesis, mechanism, and function. Cell 116:281–297

Beninson LA, Brown PN, Loughridge AB, Saludes JP, Maslanik T, Hills AK, Woodworth T, Craig W, Yin H, Fleshner M (2014) Acute stressor exposure modifies plasma exosome-associated heat shock protein 72 (Hsp72) and microRNA (miR-142-5p and miR-203). PLoS One 9:e108748

Camussi G, Deregibus MC, Cantaluppi V (2013) Role of stem-cell-derived microvesicles in the paracrine action of stem cells. Biochem Soc Trans 41:283–287

Cannon B, Nedergaard J (2004) Brown adipose tissue: function and physiological significance. Physiol Rev 84:277–359

Chen Y, Siegel F, Kipschull S, Haas B, Frohlich H, Meister G, Pfeifer A (2013) miR-155 regulates differentiation of brown and beige adipocytes via a bistable circuit. Nat Commun 4:1769

Chen L, Dai YM, Ji CB, Yang L, Shi CM, Xu GF, Pang LX, Huang FY, Zhang CM, Guo XR (2014) MiR-146b is a regulator of human visceral preadipocyte proliferation and differentiation and its expression is altered in human obesity. Mol Cell Endocrinol 393:65–74

Chen Y, Buyel JJ, Hanssen MJ, Siegel F, Pan R, Naumann J, Schell M, Van Der Lans A, Schlein C, Froehlich H, Heeren J, Virtanen KA, Van Marken Lichtenbelt W, Pfeifer A (2016) Exosomal microRNA miR-92a concentration in serum reflects human brown fat activity. Nat Commun 7:11420

Chevillet JR, Kang Q, Ruf IK, Briggs HA, Vojtech LN, Hughes SM, Cheng HH, Arroyo JD, Meredith EK, Gallichotte EN, Pogosova-Agadjanyan EL, Morrissey C, Stirewalt DL, Hladik F, Yu EY, Higano CS, Tewari M (2014) Quantitative and stoichiometric analysis of the microRNA content of exosomes. Proc Natl Acad Sci U S A 111:14888–14893

Colombo M, Moita C, Van Niel G, Kowal J, Vigneron J, Benaroch P, Manel N, Moita LF, Thery C, Raposo G (2013) Analysis of ESCRT functions in exosome biogenesis, composition and secretion highlights the heterogeneity of extracellular vesicles. J Cell Sci 126:5553–5565

Cousin B, Agou K, Leturque A, Ferre P, Girard J, Penicaud L (1992) Molecular and metabolic changes in white adipose tissue of the rat during development of ventromedial hypothalamic obesity. Eur J Biochem 207:377–382

De Jong OG, Van Balkom BW, Schiffelers RM, Bouten CV, Verhaar MC (2014) Extracellular vesicles: potential roles in regenerative medicine. Front Immunol 5:608

Deng ZB, Poliakov A, Hardy RW, Clements R, Liu C, Liu Y, Wang J, Xiang X, Zhang S, Zhuang X, Shah SV, Sun D, Michalek S, Grizzle WE, Garvey T, Mobley J, Zhang HG (2009) Adipose tissue exosome-like vesicles mediate activation of macrophage-induced insulin resistance. Diabetes 58:2498–2505

Fernandez-Messina L, Gutierrez-Vazquez C, Rivas-Garcia E, Sanchez-Madrid F, De La Fuente H (2015) Immunomodulatory role of microRNAs transferred by extracellular vesicles. Biol Cell 107:61–77

Gibbings DJ, Ciaudo C, Erhardt M, Voinnet O (2009) Multivesicular bodies associate with components of miRNA effector complexes and modulate miRNA activity. Nat Cell Biol 11(9):1143

Glebov K, Lochner M, Jabs R, Lau T, Merkel O, Schloss P, Steinhauser C, Walter J (2015) Serotonin stimulates secretion of exosomes from microglia cells. Glia 63:626–634

Guduric-Fuchs J, O'connor A, Camp B, O'neill CL, Medina RJ, Simpson DA (2012) Selective extracellular vesicle-mediated export of an overlapping set of microRNAs from multiple cell types. BMC Genomics 13:357

Harms M, Seale P (2013) Brown and beige fat: development, function and therapeutic potential. Nat Med 19:1252–1263

Hess C, Sadallah S, Hefti A, Landmann R, Schifferli JA (1999) Ectosomes released by human neutrophils are specialized functional units. J Immunol 163:4564–4573

Ho DH, Yi S, Seo H, Son I, Seol W (2014) Increased DJ-1 in urine exosome of Korean males with Parkinson's disease. Biomed Res Int 2014:704678

Hoshino A, Costa-Silva B, Shen TL, Rodrigues G, Hashimoto A, Tesic Mark M, Molina H, Kohsaka S, Di Giannatale A, Ceder S, Singh S, Williams C, Soplop N, Uryu K, Pharmer L, King T, Bojmar L, Davies AE, Ararso Y, Zhang T, Zhang H, Hernandez J, Weiss JM, Dumont-Cole VD, Kramer K, Wexler LH, Narendran A, Schwartz GK, Healey JH, Sandstrom P, Labori KJ, Kure EH, Grandgenett PM, Hollingsworth MA, De Sousa M, Kaur S, Jain M, Mallya K, Batra SK, Jarnagin WR, Brady MS, Fodstad O, Muller V, Pantel K, Minn AJ, Bissell MJ, Garcia BA, Kang Y, Rajasekhar VK, Ghajar CM, Matei I, Peinado H, Bromberg J, Lyden D (2015) Tumour exosome integrins determine organotropic metastasis. Nature 527:329–335

Jansen FH, Krijgsveld J, Van Rijswijk A, Van Den Bemd GJ, Van Den Berg MS, Van Weerden WM, Willemsen R, Dekker LJ, Luider TM, Jenster G (2009) Exosomal secretion of cytoplasmic prostate cancer xenograft-derived proteins. Mol Cell Proteomics 8:1192–1205

Johnstone RM, Adam M, Hammond JR, Orr L, Turbide C (1987) Vesicle formation during reticulocyte maturation. Association of plasma membrane activities with released vesicles (exosomes). J Biol Chem 262:9412–9420

Kajimoto T, Okada T, Miya S, Zhang L, Nakamura S (2013) Ongoing activation of sphingosine 1-phosphate receptors mediates maturation of exosomal multivesicular endosomes. Nat Commun 4:2712

Keller S, Sanderson MP, Stoeck A, Altevogt P (2006) Exosomes: from biogenesis and secretion to biological function. Immunol Lett 107:102–108

Kosaka N, Iguchi H, Yoshioka Y, Takeshita F, Matsuki Y, Ochiya T (2010) Secretory mechanisms and intercellular transfer of microRNAs in living cells. J Biol Chem 285:17442–17452

Kowal J, Tkach M, Thery C (2014) Biogenesis and secretion of exosomes. Curr Opin Cell Biol 29:116–125

Li M, Liu Z, Zhang Z, Liu G, Sun S, Sun C (2015) miR-103 promotes 3T3-L1 cell adipogenesis through AKT/mTOR signal pathway with its target being MEF2D. Biol Chem 396:235–244

Lin R, Wang S, Zhao RC (2013) Exosomes from human adipose-derived mesenchymal stem cells promote migration through Wnt signaling pathway in a breast cancer cell model. Mol Cell Biochem 383:13–20

Lotvall J, Valadi H (2007) Cell to cell signalling via exosomes through esRNA. Cell Adh Migr 1:156–158

Lowell BB, Flier JS (1997) Brown adipose tissue, beta 3-adrenergic receptors, and obesity. Annu Rev Med 48:307–316

Mathew A, Bell A, Johnstone RM (1995) Hsp-70 is closely associated with the transferrin receptor in exosomes from maturing reticulocytes. Biochem J 308(Pt 3):823–830

Mcgregor RA, Choi MS (2011) microRNAs in the regulation of adipogenesis and obesity. Curr Mol Med 11:304–316

Melo SA, Sugimoto H, O'connell JT, Kato N, Villanueva A, Vidal A, Qiu L, Vitkin E, Perelman LT, Melo CA, Lucci A, Ivan C, Calin GA, Kalluri R (2014) Cancer exosomes perform cell-independent microRNA biogenesis and promote tumorigenesis. Cancer Cell 26:707–721

Nedaeinia R, Manian M, Jazayeri MH, Ranjbar M, Salehi R, Sharifi M, Mohaghegh F, Goli M, Jahednia SH, Avan A, Ghayour-Mobarhan M (2017) Circulating exosomes and exosomal microRNAs as biomarkers in gastrointestinal cancer. Cancer Gene Ther 24:48–56

Nolte-'T Hoen EN, Wauben MH (2012) Immune cell-derived vesicles: modulators and mediators of inflammation. Curr Pharm Des 18:2357–2368

Ogata-Kawata H, Izumiya M, Kurioka D, Honma Y, Yamada Y, Furuta K, Gunji T, Ohta H, Okamoto H, Sonoda H, Watanabe M, Nakagama H, Yokota J, Kohno T, Tsuchiya N (2014) Circulating exosomal microRNAs as biomarkers of colon cancer. PLoS One 9:e92921

Ogawa R, Tanaka C, Sato M, Nagasaki H, Sugimura K, Okumura K, Nakagawa Y, Aoki N (2010) Adipocyte-derived microvesicles contain RNA that is transported into macrophages and might be secreted into blood circulation. Biochem Biophys Res Commun 398:723–729

Ortega FJ, Moreno-Navarrete JM, Pardo G, Sabater M, Hummel M, Ferrer A, Rodriguez-Hermosa JI, Ruiz B, Ricart W, Peral B, Fernandez-Real JM (2010) MiRNA expression profile of human subcutaneous adipose and during adipocyte differentiation. PLoS One 5:e9022

Potthoff MJ, Kliewer SA, Mangelsdorf DJ (2012) Endocrine fibroblast growth factors 15/19 and 21: from feast to famine. Genes Dev 26:312–324

Rekker K, Saare M, Roost AM, Kubo AL, Zarovni N, Chiesi A, Salumets A, Peters M (2014) Comparison of serum exosome isolation methods for microRNA profiling. Clin Biochem 47:135–138

Roma-Rodrigues C, Fernandes AR, Baptista PV (2014) Exosome in tumour microenvironment: overview of the crosstalk between normal and cancer cells. Biomed Res Int 2014:179486

Rupp AK, Rupp C, Keller S, Brase JC, Ehehalt R, Fogel M, Moldenhauer G, Marme F, Sultmann H, Altevogt P (2011) Loss of EpCAM expression in breast cancer derived serum exosomes: role of proteolytic cleavage. Gynecol Oncol 122:437–446

Sala-Valdes M, Ailane N, Greco C, Rubinstein E, Boucheix C (2012) Targeting tetraspanins in cancer. Expert Opin Ther Targets 16:985–997

Simpson RJ, Jensen SS, Lim JW (2008) Proteomic profiling of exosomes: current perspectives. Proteomics 8:4083–4099

Stoorvogel W, Strous GJ, Geuze HJ, Oorschot V, Schwartz AL (1991) Late endosomes derive from early endosomes by maturation. Cell 65:417–427

Stuffers S, Sem Wegner C, Stenmark H, Brech A (2009) Multivesicular endosome biogenesis in the absence of ESCRTs. Traffic 10:925–937

Taylor DD, Gercel-Taylor C (2014) Exosome platform for diagnosis and monitoring of traumatic brain injury. Philos Trans R Soc Lond Ser B Biol Sci 369(1652). pii: 20130503. doi: https://doi.org/10.1098/rstb.2013.0503

Thomou T, Mori MA, Dreyfuss JM, Konishi M, Sakaguchi M, Wolfrum C, Rao TN, Winnay JN, Garcia-Martin R, Grinspoon SK, Gorden P, Kahn CR (2017) Adipose-derived circulating miRNAs regulate gene expression in other tissues. Nature 542:450–455

Tickner JA, Urquhart AJ, Stephenson SA, Richard DJ, O'byrne KJ (2014) Functions and therapeutic roles of exosomes in cancer. Front Oncol 4:127

Trams EG, Lauter CJ, Salem N Jr, Heine U (1981) Exfoliation of membrane ecto-enzymes in the form of micro-vesicles. Biochim Biophys Acta 645:63–70

Virtanen KA, Lidell ME, Orava J, Heglind M, Westergren R, Niemi T, Taittonen M, Laine J, Savisto NJ, Enerback S, Nuutila P (2009) Functional brown adipose tissue in healthy adults. N Engl J Med 360:1518–1525

Wu L, Dai X, Zhan J, Zhang Y, Zhang H, Zeng S, Xi W (2015) Profiling peripheral microRNAs in obesity and type 2 diabetes mellitus. APMIS 123:580–585

Ying W, Riopel M, Bandyopadhyay G, Dong Y, Birmingham A, Seo JB, Ofrecio JM, Wollam J, Hernandez-Carretero A, Fu W, Li P, Olefsky JM (2017) Adipose tissue macrophage-derived exosomal miRNAs can modulate in vivo and in vitro insulin sensitivity. Cell 171:372–384.e12

Zhang B, Yin Y, Lai RC, Lim SK (2014) Immunotherapeutic potential of extracellular vesicles. Front Immunol 5:518

Zhang J, Li S, Li L, Li M, Guo C, Yao J, Mi S (2015) Exosome and exosomal microRNA: trafficking, sorting, and function. Genomics Proteomics Bioinformatics 13:17–24

Zhang Y, Yu M, Tian W (2016) Physiological and pathological impact of exosomes of adipose tissue. Cell Prolif 49:3–13

Zhou W, Fong MY, Min Y, Somlo G, Liu L, Palomares MR, Yu Y, Chow A, O'connor ST, Chin AR, Yen Y, Wang Y, Marcusson EG, Chu P, Wu J, Wu X, Li AX, Li Z, Gao H, Ren X, Boldin MP, Lin PC, Wang SE (2014) Cancer-secreted miR-105 destroys vascular endothelial barriers to promote metastasis. Cancer Cell 25:501–515

Part IV

Recruitment and Activation of Human BAT

Activation of Human Brown Adipose Tissue (BAT): Focus on Nutrition and Eating

Kirsi A. Virtanen

Contents

1 Cold-Induced Activation in BAT 350
2 Oxidative Metabolism in Activated Human BAT 350
3 Insulin-Stimulated Glucose Uptake in BAT 351
4 Meal-Induced Activation of BAT 353
5 Nutritional Status Affects BAT Activity 354
References 355

Abstract

Brown adipose tissue activation occurs most effectively by cold exposure. In the modern world, we do not spend long periods in cold environment, and eating and meals may be other activators of brown fat function. Short-term regulation of brown fat functional activity by eating involves most importantly insulin. Insulin is capable to increase glucose uptake in human brown adipose tissue fivefold to fasting conditions. Oxidative metabolism in brown fat is doubled both by cold and by a meal. Human brown adipose tissue is an insulin-sensitive tissue type, and insulin resistance impairs the function, as is found in obesity. Body weight reduction improves cold-induced activation of human brown adipose tissue.

Keywords

Cold · Human brown fat · Insulin · PET · Postprandial

K. A. Virtanen (✉)
Turku PET Centre, Turku University Hospital and University of Turku, Turku, Finland

Clinical Nutrition, Institute of Public Health and Clinical Nutrition, University of Eastern Finland (UEF), Kuopio, Finland
e-mail: kirsi.virtanen@utu.fi

A. Pfeifer et al. (eds.), *Brown Adipose Tissue*,
Handbook of Experimental Pharmacology 251, https://doi.org/10.1007/164_2018_136

1 Cold-Induced Activation in BAT

At fasting and in normal room temperature, BAT function is silent and comparable to white adipose tissue (WAT) metabolic activity (Virtanen et al. 2009; Orava et al. 2011). Cold is the most potent natural and physiological activator of human BAT (Orava et al. 2011). Along with acute cold exposure, skin temperature decreases (van Marken Lichtenbelt et al. 2009; Yoneshiro et al. 2011) but not on the skin area of the most prominent human BAT depot, supraclavicular region, in subjects with functionally active BAT, which indicates heat production by BAT (Yoneshiro et al. 2011). In addition, cold increases resting energy expenditure (Orava et al. 2011; van Marken Lichtenbelt et al. 2009; Saito et al. 2009; Ouellet et al. 2012) especially in the subjects with increased metabolic activity in BAT in cold (Yoneshiro et al. 2011).

In humans, tissue-specific BAT activation may be measured in vivo using non-invasive, combined PET/CT imaging. Several physiological and metabolic functions may be accurately measured with different tracers. Quantitative tissue-specific glucose uptake rate or semi-quantitative FDG uptake is most frequently used in BAT studies, and it has been regarded as an indirect measure of thermogenesis. The FDG (18F-fluoro-D-deoxyglucose) tracer is a glucose analog, and the uptake of the tracer gives an overall estimation of the metabolic activity of the tissue.

Increased uptake of FDG in supraclavicular and neck region has been described more often during cold season than during summer (Saito et al. 2009). In addition to seasonal variation, acute cold exposure may provoke BAT metabolic activity. Acutely induced glucose uptake rate by cold in BAT is more than ten times higher to measurement in the normal room temperature (Orava et al. 2011). Notably, only a few tissues are able to increase their metabolic rate so extensively during short stimulation. Cold-induced glucose uptake in BAT is comparable to cerebral glucose uptake (Orava et al. 2014), and in addition to BAT, skeletal may increase the metabolic rate several tenfolds during acute activation, i.e. exercise. Acute cold exposure increases the probability to detect metabolic activity in BAT region and in normal weighed subjects, the probability may be 60–90% (Orava et al. 2011; van Marken Lichtenbelt et al. 2009). If no cold exposure is performed, increased metabolic activity in BAT can be found in 0.6–25% of the patients (Cohade et al. 2003; Hany et al. 2002). Acclimation by repeated daily cold exposures increases metabolic activity of BAT (Blondin et al. 2014; van der Lans et al. 2013; Yoneshiro et al. 2013) resembling natural, seasonal acclimation by thermal winter.

2 Oxidative Metabolism in Activated Human BAT

BAT oxidative metabolism may be measured indirectly using ^{11}C-acetate-PET or radiowater, or directly measuring oxygen uptake in BAT with ^{15}O-O_2-PET (Ouellet et al. 2012; Orava et al. 2011; U Din et al. 2016). Measures of oxidative metabolism are more descriptive indicators of thermogenesis and mitochondrial substrate oxidation than substrate uptake per se, and cold exposure activates oxidative metabolism significantly (Ouellet et al. 2012). In addition, perfusion of BAT is elevated significantly during cold, approximately twofold (Orava et al. 2011) further supporting the increased oxidative role of BAT in cold. Oxygen consumption is 50% higher in the subjects with functionally active BAT – their BAT oxygen consumption is higher also in resting state when compared to controls with nonfunctional BAT (Muzik et al. 2013). In general, BAT oxygen uptake is doubled during acute cold exposure, along with doubled perfusion (U Din et al. 2016), and oxygen consumption and perfusion are strongly associated. Collectively, these indicate activated thermogenesis in human BAT during cold exposure.

Tissue-specific oxygen consumption may also be used for the estimation of BAT energy expenditure, which associates strongly with BAT fatty acid uptake, both in cold and in room temperature (U Din et al. 2016). Fatty acid uptake in BAT is measured using PET and ^{18}F-fluoro-thia-heptadecanoid-acid (^{18}F-FTHA), a palmitate analog, which may enter either intracellular lipid pool or directly to mitochondria. During cold, activated BAT utilizes both glucose and fatty acids, but if intracellular triglyceride lipolysis is inhibited by nicotinic acid during cold, oxidative metabolism in BAT is abolished, and shivering intensity is increased in muscles (Blondin et al. 2017b). The importance of intracellular lipolysis for BAT oxidative metabolism is further supported by the findings that BAT radiodensity is not changed if nicotinic acid is administered (Blondin et al. 2017b). Radiodensity of the tissue is measured with computed tomography (CT) providing Hounsfield units (HUs) for the tissue, as an indirect measure of triglyceride content. Cold increases HUs, and 3-h cold exposure may combust one third of the intracellular lipid pool (Ouellet et al. 2012). Mild cold also results in greater dietary fatty acid extraction by BAT than by other tissues, and it remains stable by cold acclimation (Blondin et al. 2017a). Additionally another modality, magnetic resonance imaging (MRI) and proton spectroscopy (^{1}H-MRS), may be used for the estimation of fat fraction and triglyceride content of the tissue, respectively. BAT triglyceride content measured with ^{1}H-MRS is significantly lower in subjects with functionally active BAT, compared to subjects with inactive BAT, and is independently related to whole-body insulin sensitivity (Raiko et al. 2015). Interestingly in whole-body level, during 2–3 h of acute cold exposure, fat oxidation is dominant (U Din et al. 2016), while prolonged cold exposure of 5–8 h increases also glucose oxidation in subjects with functionally active BAT along with increased insulin sensitivity (Chondronikola et al. 2014).

3 Insulin-Stimulated Glucose Uptake in BAT

While cold is highly potent activator and recruiter of BAT function, people of the modern world do not spend long times in cold environment. Both eating and exercise have been related to BAT activation (Rothwell and Stock 1979; Boström et al. 2012), and number of related hormonal and metabolic factors has been recognized. Their role in the regulation of BAT function is clarified gradually. As effects of exercise on BAT function were discussed in the previous chapter, the focus of this chapter is in eating and in postprandial activation of BAT.

Eating is a complex chain of reactions where the first signals to alter metabolism and prepare the body for the food intake and utilization of nutrients take place early before the initiation of eating or feeding. Cephalic phase of appetite and eating starts with the thoughts or smell of food, and saliva secretion is accelerated. During this phase number of hormonal signals is transferred, and among others, early release of insulin and peak in insulin concentration are recognized (Siegel et al. 1980). Interestingly in laboratory animals, body temperature is increased preceding eating, and this is accompanied with the increase of subscapular BAT temperature approximately 15 min prior eating (Blessing et al. 2012) – as if BAT would prepare mitochondria for substrate flow and initiate thermogenesis by utilizing intracellular lipids well before effective oxidation of nutritional substrates takes place.

Insulin has an obvious role also after eating, in postprandial state. After the first-phase boost of insulin, concentration is gradually increased to facilitate digestion. Typically, in healthy subjects, plasma fasting level of insulin is approximately 3–10 mU/l (20–60 pmol/l), and in postprandial state, the plasma level is increased up to 70–100 mU/l (420–600 pmol/l). Postprandial level is reflecting the fasting level; the higher fasting insulin concentration, the higher is the postprandial concentration. The knowledge of the postprandial insulin concentrations is utilized in experimental settings, and insulin stimulation produced by euglycemic hyperinsulinemic clamp technique aims to reach insulin concentrations of 70–100 mU/l, similar to postprandial levels. During this kind of insulin stimulation, tissue-specific substrate uptake can be measured with PET, and especially glucose uptake rate is increased (Nuutila et al. 1995). Partly, insulin stimulation produced by clamp technique may be regarded as mimicking postprandial state, at least in terms of plasma insulin concentration.

Similar to cold, insulin activates the sympathetic nervous system (SNS), and through activation of SNS, insulin may activate BAT thermogenesis. However, during steady state of hyperinsulinemic clamp, perfusion in BAT is not increased similarly to cold (Orava et al. 2011) suggesting that insulin might not have direct effect on BAT thermogenesis. It is noteworthy that steady state in hyperinsulinemic clamp is typically reached 45–60 min after initiation of insulin infusion, and the acute effect of insulin may have passed over.

Insulin stimulation increases BAT glucose uptake, approximately fivefold when compared to short-term fasting state after overnight fasting (Orava et al. 2011). The level of insulin-stimulated glucose uptake rate in BAT is comparable to glucose uptake rate in skeletal muscle, which is known to be highly insulin-sensitive tissue type (Orava et al. 2011; Nuutila et al. 1995). Thus, BAT may be regarded as an insulin-sensitive tissue type. Although BAT is a small tissue in size and the contribution of BAT glucose uptake to whole-body insulin sensitivity is small, insulin-stimulated glucose uptake rate in BAT correlates with M-value, the measure of whole-body insulin sensitivity (Orava et al. 2011).

The effects of cold and insulin stimulation on BAT metabolism are partly different. Both stimulations increase glucose uptake in BAT and energy expenditure in whole body (Orava et al. 2011), while plasma glucose concentration remains stable. However, otherwise the metabolic milieu is very different: during cold sympathetic activation results in high plasma fatty acid concentration which is, during insulin stimulation, suppressed to low concentrations of nonesterified fatty acids. Thus, during cold lipolysis from white adipose tissue dominates, and during insulin stimulation, it is suppressed by high concentration of insulin. This is related to high plasma noradrenaline (norepinephrine) levels during cold, but during insulin stimulation, such changes in noradrenaline concentration cannot be detected. Obviously plasma insulin concentration is high during hyperinsulinemic clamp, but during cold, insulin levels are decreased in all subjects, even in those who have higher fasting levels (obese and insulin-resistant subjects) (Orava et al. 2011). Thyroid hormones, thyroxine (T4) and especially triiodothyronine (T3), are decreased during cold in subjects with functionally active BAT (Orava et al. 2011). The thyroid hormone levels are not changed by insulin stimulation.

4 Meal-Induced Activation of BAT

Meal-induced thermogenesis, which is a different topic than diet-induced thermogenesis, refers to heat production taking place in response to eating. Thermogenesis reflects tissue respiration where mitochondria have a key role. Thereby, tissues with higher mitochondrial content and function have higher contribution to whole-body thermogenesis. In laboratory animals, meal- and diet-induced thermogenesis is well recognized, but in humans, it has been under debate for at least couple of decades. The role of human BAT including a high amount of mitochondria has been in the focus in this debate, and it has been questioned whether BAT thermogenesis has any role in energy balance, especially in obesity. In the 1980s the findings in rodents (Rothwell and Stock 1979) encouraged researchers to assume that diet- or meal-induced thermogenesis might explain why some people gain weight more easily than others (Himms-Hagen 1979). Still, the question whether meal-induced thermogenesis has a role in human metabolism is valid.

During diagnostic FDG-PET scan for tumour detection, the aim is to diminish accumulation of the tracer by other metabolically active tissues. Such tissues are among others skeletal muscle and BAT. Tracer (FDG) accumulation to BAT in diagnostic scans has been successfully diminished with beta-blockers (Parysow et al. 2007), but equally effective results may be achieved by keeping the patient warm before and during the scans. In addition to premedication and/or controlled ambient temperature during scanning, fatty meals have been used for the reduction of FDG uptake in BAT (Williams and Kolodny 2008). One group of patients ($n = 741$) prepared for the scan with high-fat, very-low-carbohydrate, protein-permitted meal and another group of patients ($n = 1{,}229$) by fasting. High-fat diet group had lower frequency of high FDG uptake in BAT (Williams and Kolodny 2008). Therefore, meal composition may effect on substrate preference in BAT, and Randle's cycle appears to function also in BAT, in addition to other tissues, such as myocardium and skeletal muscle (Nuutila et al. 1992).

If high-calorie, carbohydrate-rich meal is given for a healthy subject, postprandial FDG uptake is higher in BAT than in subcutaneous or visceral adipose tissue (Vosselman et al. 2013), but whether FDG uptake is increased postprandially compared to fasting state, preceding eating, is not known. Overall, postprandial FDG uptake remains lower than during acute cold exposure (Vosselman et al. 2013).

When oral glucose load is given after 3.5 h in an ambient temperature of 20°C or 25°C, the insulin response appears to be higher at 20°C, based on higher insulin to glucose ratio at 2 h of OGTT (Pathak et al. 2018). If the insulin concentration is lower in 20°C than 25°C and the lipolysis is activated by catecholamines, glucose load may induce a pronounced and compensatory release of insulin to favour glucose oxidation after mild cooling. However, the insulin concentrations in different temperatures were not shown, neither were the catecholamine or fatty acid levels measured, and thus the preceding speculation remains to be confirmed in the controlled setting.

Thus, substrate uptake by BAT in postprandial state may be affected by meal composition, although it may be expected that postprandial insulin levels are sufficient to increase BAT glucose uptake. As glucose uptake is not a perfect indicator of thermogenesis, postprandial oxidative metabolism could give better insight to meal-induced thermogenesis. Indeed, one normocaloric, carbohydrate-dominant meal increases both perfusion and oxygen uptake in BAT to the same degree than acute cold exposure (U Din et al. 2015). Therefore, meal-induced thermogenesis seems to exist in human adults.

5 Nutritional Status Affects BAT Activity

Long-term fasting for 54 h results in decreased cold-induced glucose uptake rate in BAT being approximately half of the rate measured in the baseline conditions (Hanssen et al. 2015). Also obesity decreases the probability to detect metabolically active BAT, and only 30% of the obese study subjects show significant increase in cold-induced glucose uptake in BAT (Orava et al. 2013). Metabolic activity of BAT in obesity is blunted, and also insulin-stimulated glucose uptake is less than half of the uptake measured in normal-weighed subjects (Orava et al. 2013). It may be that brown adipocytes of obese subjects are transdifferentiated to white adipocytes being entirely filled with triglyceride storages or that insulation by a thick subcutaneous adipose tissue in obesity is effective enough to prevent similar degree of activation in BAT than in lean subjects. On the other hand, some individuals may be prone to obesity due to poorly functioning BAT.

Recruitment of BAT (beiging) in other adipose depots, such as visceral white fat depot or perirenal white fat depot, could be beneficial in obesity. Morbidly obese subjects have lower UCP1 presence in the intraperitoneal adipose tissue than lean subjects (Oberkofler et al. 1997). Weight reduction preferentially targets intra-abdominal fat tissues, and thereby UCP1 expression and function could be promoted after weight loss in these depots. Further, cold-activated BAT FDG uptake is inversely related to visceral fat area (Saito et al. 2009).

Conventional weight reduction by dieting and exercise for 5 months results in 12% reduction of the original weight, and simultaneously, cold-induced BAT metabolic activity tends to be higher than before weight loss (Orava et al. 2013).

In morbidly obese subjects, bariatric surgery results in remarkable weight reduction (approximately 30% of the original weight), and 1 year after surgery, BAT metabolic activity is enhanced (Vijgen et al. 2012).

References

Blessing W, Mohammed M, Ootsuka Y (2012) Heating and eating: brown adipose tissue thermogenesis precedes food ingestion as part of the ultradian basic rest-activity cycle in rats. Physiol Behav 105(4):966–974. https://doi.org/10.1016/j.physbeh.2011.11.009

Blondin DP, Labbé SM, Tingelstad HC, Noll C, Kunach M, Phoenix S, Guérin B, Turcotte EE, Carpentier AC, Richard D, Haman F (2014) Increased brown adipose tissue oxidative capacity in cold-acclimated humans. J Clin Endocrinol Metab 99(3):E438–E446. https://doi.org/10.1210/jc.2013-3901

Blondin DP, Tingelstad HC, Noll C, Frisch F, Phoenix S, Guérin B, Turcotte ÉE, Richard D, Haman F, Carpentier AC (2017a) Dietary fatty acid metabolism of brown adipose tissue in cold-acclimated men. Nat Commun 8:14146. https://doi.org/10.1038/ncomms14146

Blondin DP, Frisch F, Phoenix S, Guérin B, Turcotte ÉE, Haman F, Richard D, Carpentier AC (2017b) Inhibition of intracellular triglyceride lipolysis suppresses cold-induced brown adipose tissue metabolism and increases shivering in humans. Cell Metab 25(2):438–447. https://doi.org/10.1016/j.cmet.2016.12.005

Boström P, Wu J, Jedrychowski MP, Korde A, Ye L, Lo JC, Rasbach KA, Boström EA, Choi JH, Long JZ, Kajimura S, Zingaretti MC, Vind BF, Tu H, Cinti S, Højlund K, Gygi SP, Spiegelman BM (2012) A PGC1-α-dependent myokine that drives brown-fat-like development of white fat and thermogenesis. Nature 481(7382):463–468. https://doi.org/10.1038/nature10777

Chondronikola M, Volpi E, Børsheim E, Porter C, Annamalai P, Enerbäck S, Lidell ME, Saraf MK, Labbe SM, Hurren NM, Yfanti C, Chao T, Andersen CR, Cesani F, Hawkins H, Sidossis LS (2014) Brown adipose tissue improves whole-body glucose homeostasis and insulin sensitivity in humans. Diabetes 63(12):4089–4099. https://doi.org/10.2337/db14-0746

Cohade C, Mourtzikos KA, Wahl RL (2003) "USA-Fat": prevalence is related to ambient outdoor temperature-evaluation with 18F-FDG PET/CT. J Nucl Med 44(8):1267–1270

van der Lans AA, Hoeks J, Brans B, Vijgen GH, Visser MG, Vosselman MJ, Hansen J, Jörgensen JA, Wu J, Mottaghy FM, Schrauwen P, van Marken Lichtenbelt WD (2013) Cold acclimation recruits human brown fat and increases nonshivering thermogenesis. J Clin Invest 123 (8):3395–3403. https://doi.org/10.1172/JCI68993

Hanssen MJ, Wierts R, Hoeks J, Gemmink A, Brans B, Mottaghy FM, Schrauwen P, van Marken Lichtenbelt WD (2015) Glucose uptake in human brown adipose tissue is impaired upon fasting-induced insulin resistance. Diabetologia 58(3):586–595. https://doi.org/10.1007/s00125-014-3465-8

Hany TF, Gharehpapagh E, Kamel EM, Buck A, Himms-Hagen J, von Schulthess GK (2002) Brown adipose tissue: a factor to consider in symmetrical tracer uptake in the neck and upper chest region. Eur J Nucl Med Mol Imaging 29(10):1393–1398

Himms-Hagen J (1979) Obesity may be due to a malfunctioning of brown fat. Can Med Assoc J 121 (10):1361–1364

van Marken Lichtenbelt WD, Vanhommerig JW, Smulders NM, Drossaerts JM, Kemerink GJ, Bouvy ND, Schrauwen P, Teule GJ (2009) Cold-activated brown adipose tissue in healthy men. N Engl J Med 360(15):1500–1508. Erratum in: N Engl J Med. 2009;360(18):1917

Muzik O, Mangner TJ, Leonard WR, Kumar A, Janisse J, Granneman JG (2013) 15O PET measurement of blood flow and oxygen consumption in cold-activated human brown fat. J Nucl Med 54(4):523–531. https://doi.org/10.2967/jnumed.112.111336

Nuutila P, Koivisto VA, Knuuti J, Ruotsalainen U, Teräs M, Haaparanta M, Bergman J, Solin O, Voipio-Pulkki LM, Wegelius U et al (1992) Glucose-free fatty acid cycle operates in human heart and skeletal muscle in vivo. J Clin Invest 89(6):1767–1774

Nuutila P, Knuuti MJ, Mäki M, Laine H, Ruotsalainen U, Teräs M, Haaparanta M, Solin O, Yki-Järvinen H (1995) Gender and insulin sensitivity in the heart and in skeletal muscles. Studies using positron emission tomography. Diabetes 44(1):31–36

Oberkofler H, Dallinger G, Liu YM, Hell E, Krempler F, Patsch W (1997) Uncoupling protein gene: quantification of expression levels in adipose tissues of obese and non-obese humans. J Lipid Res 38(10):2125–2133

Orava J, Nuutila P, Lidell ME, Oikonen V, Noponen T, Viljanen T, Scheinin M, Taittonen M, Niemi T, Enerbäck S, Virtanen KA (2011) Different metabolic responses of human brown adipose tissue to activation by cold and insulin. Cell Metab 14(2):272–279. https://doi.org/10.1016/j.cmet.2011.06.012

Orava J, Nuutila P, Noponen T, Parkkola R, Viljanen T, Enerbäck S, Rissanen A, Pietiläinen KH, Virtanen KA (2013) Blunted metabolic responses to cold and insulin stimulation in brown adipose tissue of obese humans. Obesity (Silver Spring) 21(11):2279–2287. https://doi.org/10.1002/oby.20456

Orava J, Nummenmaa L, Noponen T, Viljanen T, Parkkola R, Nuutila P, Virtanen KA (2014) Brown adipose tissue function is accompanied by cerebral activation in lean but not in obese humans. J Cereb Blood Flow Metab 34(6):1018–1023. https://doi.org/10.1038/jcbfm.2014.50

Ouellet V, Labbé SM, Blondin DP, Phoenix S, Guérin B, Haman F, Turcotte EE, Richard D, Carpentier AC (2012) Brown adipose tissue oxidative metabolism contributes to energy expenditure during acute cold exposure in humans. J Clin Invest 122(2):545–552

Parysow O, Mollerach AM, Jager V, Racioppi S, San Roman J, Gerbaudo VH (2007) Low-dose oral propranolol could reduce brown adipose tissue F-18 FDG uptake in patients undergoing PET scans. Clin Nucl Med 32(5):351–357

Pathak K, Woodman RJ, James AP, Soares MJ (2018) Fasting and glucose induced thermogenesis in response to three ambient temperatures: a randomized crossover trial in the metabolic syndrome. Eur J Clin Nutr. https://doi.org/10.1038/s41430-017-0058-x

Raiko J, Holstila M, Virtanen KA, Orava J, Saunavaara V, Niemi T, Laine J, Taittonen M, Borra RJ, Nuutila P, Parkkola R (2015) Brown adipose tissue triglyceride content is associated with decreased insulin sensitivity, independently of age and obesity. Diabetes Obes Metab 17(5):516–519. https://doi.org/10.1111/dom.12433

Rothwell NJ, Stock MJ (1979) A role for brown adipose tissue in diet-induced thermogenesis. Nature 281(5726):31–35

Saito M, Okamatsu-Ogura Y, Matsushita M, Watanabe K, Yoneshiro T, Nio-Kobayashi J, Iwanaga T, Miyagawa M, Kameya T, Nakada K, Kawai Y, Tsujisaki M (2009) High incidence of metabolically active brown adipose tissue in healthy adult humans: effects of cold exposure and adiposity. Diabetes 58(7):1526–1531

Siegel EG, Trimble ER, Renold AE, Berthoud HR (1980) Importance of preabsorptive insulin release on oral glucose tolerance: studies in pancreatic islet transplanted rats. Gut 21(11):1002–1009

U Din M, Raiko J, Saari T, Kudomi N, Parkkola R, Nuutila P, Virtanen KA (2015) Human brown adipose tissue oxygen consumption after meal is similar to cold activated consumption. Diabetologia 58(Suppl 1):S331, 691

U Din M, Raiko J, Saari T, Kudomi N, Tolvanen T, Oikonen V, Teuho J, Sipilä HT, Savisto N, Parkkola R, Nuutila P, Virtanen KA (2016) Human brown adipose tissue [(15)O]O2 PET imaging in the presence and absence of cold stimulus. Eur J Nucl Med Mol Imaging 43(10):1878–1886. https://doi.org/10.1007/s00259-016-3364-y

Vijgen GH, Bouvy ND, Teule GJ, Brans B, Hoeks J, Schrauwen P, van Marken Lichtenbelt WD (2012) Increase in brown adipose tissue activity after weight loss in morbidly obese subjects. J Clin Endocrinol Metab 97(7):E1229–E1233. https://doi.org/10.1210/jc.2012-1289

Virtanen KA, Lidell ME, Orava J, Heglind M, Westergren R, Niemi T, Taittonen M, Laine J, Savisto NJ, Enerbäck S, Nuutila P (2009) Functional brown adipose tissue in healthy adults. N Engl J Med 360(15):1518–1525. Erratum in: N Engl J Med. 2009;361(11):1123

Vosselman MJ, Brans B, van der Lans AA, Wierts R, van Baak MA, Mottaghy FM, Schrauwen P, van Marken Lichtenbelt WD (2013) Brown adipose tissue activity after a high-calorie meal in humans. Am J Clin Nutr 98(1):57–64. https://doi.org/10.3945/ajcn.113.059022

Williams G, Kolodny GM (2008) Method for decreasing uptake of 18F-FDG by hypermetabolic brown adipose tissue on PET. AJR Am J Roentgenol 190(5):1406–1409. https://doi.org/10.2214/AJR.07.3205

Yoneshiro T, Aita S, Matsushita M, Kameya T, Nakada K, Kawai Y, Saito M (2011) Brown adipose tissue, whole-body energy expenditure, and thermogenesis in healthy adult men. Obesity (Silver Spring) 19(1):13–16

Yoneshiro T, Aita S, Matsushita M, Kayahara T, Kameya T, Kawai Y, Iwanaga T, Saito M (2013) Recruited brown adipose tissue as an antiobesity agent in humans. J Clin Invest 123 (8):3404–3408. https://doi.org/10.1172/JCI67803

Translational Aspects of Brown Fat Activation by Food-Derived Stimulants

Takeshi Yoneshiro, Mami Matsushita, and Masayuki Saito

Contents

1 Introduction 360
2 Food-Derived Stimulants with Agonistic Activity on Transient Receptor Potential Channels 361
2.1 Capsaicin and Capsinoids as TRP Vanilloid 1 Agonists 363
2.2 Paradol as a TRPV1 and TRPA1 Agonist 365
2.3 DHA and EPA as TRPV1 Agonists 366
2.4 Catechins as TRPA1/V1 Agonists and Catechol-*O*-Methyltransferase Inhibitors 366
2.5 Menthol as a TRP Melastatin 8 Agonist 368
2.6 Allyl Isothiocyanate as a TRPA1 Agonist 368
2.7 Cinnamaldehyde as a TRPA1 Agonist 369
2.8 Other Ingredients with Agonistic Activity at Thermosensitive TRPs 369
3 Food-Derived Stimulants Acting via TRP-Independent Mechanisms 370
3.1 Retinoic Acid as an Activator of the Transcriptional Regulatory Enhancer of the Ucp1 Gene 370
3.2 Resveratrols as Sirt1 Activators 371
3.3 Other Ingredients with BAT-Activating Effects 371
4 Concluding Remarks 372
5 Acknowledgments 373
References 373

T. Yoneshiro (✉)
Diabetes Center, University of California, San Francisco, San Francisco, CA, USA
e-mail: takeshi.yoneshiro@ucsf.edu

M. Matsushita
Department of Nutrition, Tenshi College, Sapporo, Japan

M. Saito
Department of Biomedical Sciences, Graduate School of Veterinary Medicine, Hokkaido University, Sapporo, Japan

A. Pfeifer et al. (eds.), *Brown Adipose Tissue*,
Handbook of Experimental Pharmacology 251, https://doi.org/10.1007/164_2018_159

Abstract

Since the rediscovery of brown adipose tissue (BAT) in humans, its energy-dissipating ability has been well-recognized. The negative correlations of BAT activity with adiposity and insulin sensitivity provided an obvious rationale for discerning reliable and practical strategies for stimulating BAT. Though cold exposure or use of pharmacological adrenomimetics can activate BAT, they may have adverse effects. Therefore, determining alternative stimulants of BAT with lower risks such as commonly used food ingredients is highly desirable. Recent observations revealed that chemical activation of temperature-sensitive transient receptor potential (TRP) channels by food ingredients can recruit BAT in humans. Furthermore, animal studies have identified several food-derived stimulants of BAT acting through multiple mechanisms distinct from a TRP-mediated process. Dietary compounds acting as an activator of Sirtuin 1, a critical regulator of mitochondrial biogenesis and brown adipocyte differentiation, are one such class of promising food-derived BAT activators in humans. While the individual effects of various dietary factors are increasingly established in a laboratory setting, the potential synergistic effects of multiple stimulants on BAT remain to be tested in a clinical environment. These investigations may support the development of efficient, flexible dietary regimens capable of boosting BAT thermogenesis.

Keywords

Brown adipose tissue · Capsinoids · Catechins · Food ingredients · Obesity · Recruitment · SIRT1 · Thermogenesis · TRP channels

1 Introduction

The prevalence of obesity and related metabolic diseases such as type 2 diabetes is ever-increasing. Advanced strategies aimed at controlling appetite and promoting exercise have been attempted. However, sustained interventions are difficult to achieve owing to poor adherence (Moroshko et al. 2011). The rediscovery of metabolically active brown adipose tissue (BAT) in healthy adult humans (Cypess et al. 2009; Saito et al. 2009; van Marken Lichtenbelt et al. 2009; Virtanen et al. 2009) propelled the idea that BAT could be a promising target to boost energy expenditure (EE) and prevent obesity.

BAT thermogenesis is primarily dependent on uncoupling protein 1 (UCP1), a unique mitochondrial membrane protein, and is mainly regulated by the sympathetic nervous system (SNS) (Cannon and Nedergaard 2011). In small rodents, UCP1-mediated thermogenesis in BAT is activated during cold exposure and after hyperphagia to control body temperature and energy balance (Cannon and Nedergaard 2011). Over the past decade, the physiological role of BAT in humans has been increasingly elucidated. Radionuclide studies in healthy adults using fluorodeoxyglucose (FDG)-positron emission tomography (PET) and computed tomography (CT) revealed that BAT is activated by acute cold exposure (Saito et al. 2009) or administration of a β3-adrenergic receptor agonist (Cypess et al. 2015) and contributes to increasing whole-body EE (Cypess et al. 2015; Yoneshiro et al. 2011a) and fatty acid oxidation (Chondronikola et al. 2016). Two independent groups documented BAT activation after food intake (Vosselman et al. 2013) and its involvement in diet-induced thermogenesis (Hibi et al. 2016), as confirmed in mice (von Essen et al. 2017). However, the activity and prevalence of BAT are substantially decreased in older populations (Yoneshiro et al. 2011b). This age-related decrease in BAT is closely related to visceral fat accumulation and increased hemoglobin A1C, an index of average blood glucose (Matsushita et al. 2014). These findings suggest that reactivation and recruitment of BAT may be preventive against obesity, in particular in individuals with decreased BAT facing at high risks of obesity and diabetes. In fact, prolonged exposure to cold, which results in a successful recruitment of BAT, increases nonshivering thermogenesis (Blondin et al. 2014; van der Lans et al. 2013), decreases body fat content (Yoneshiro et al. 2013), and improves insulin sensitivity (Lee et al. 2014; Hanssen et al. 2015).

Despite these current advances in the therapeutic potentials, acceptable human interventions aimed at activating BAT have been limited due to potential unfavorable side effects. For example, chronic use of cold regimen or adrenomimetics could increase blood pressure (Cypess et al. 2015) and induce atherosclerosis (Dong et al. 2013). There have been substantial efforts in animals searching for chemical agents in foods that stimulate BAT thermogenesis (Saito et al. 2016). Food ingredients consumed traditionally by humans have low risks of adverse effects, some of which might be safe, practical, and effective agents to recruit BAT. In this chapter, we summarize natural compounds in foods that can stimulate BAT thermogenesis, together with the recent progresses in the underlying mechanisms and the effectiveness in humans.

2 Food-Derived Stimulants with Agonistic Activity on Transient Receptor Potential Channels

Cold-induced activation of BAT is initiated through the stimulation of temperature-sensitive transient receptor potential (TRP) channels located in peripheral tissues such as the skin. Stimulation of these receptors in turn causes the sympathetic efferent nerves innervating BAT to discharge, releasing norepinephrine (NE) and triggering β-adrenergic receptor-mediated intracellular signaling pathways in brown adipocytes, culminating in UCP1 activation and thermogenesis (Cannon and Nedergaard 2011; Nakamura 2011). This TRP-SNS-BAT signaling axis can also be activated by a number of environmental cues, including certain foods that act as TRP agonists (Fig. 1).

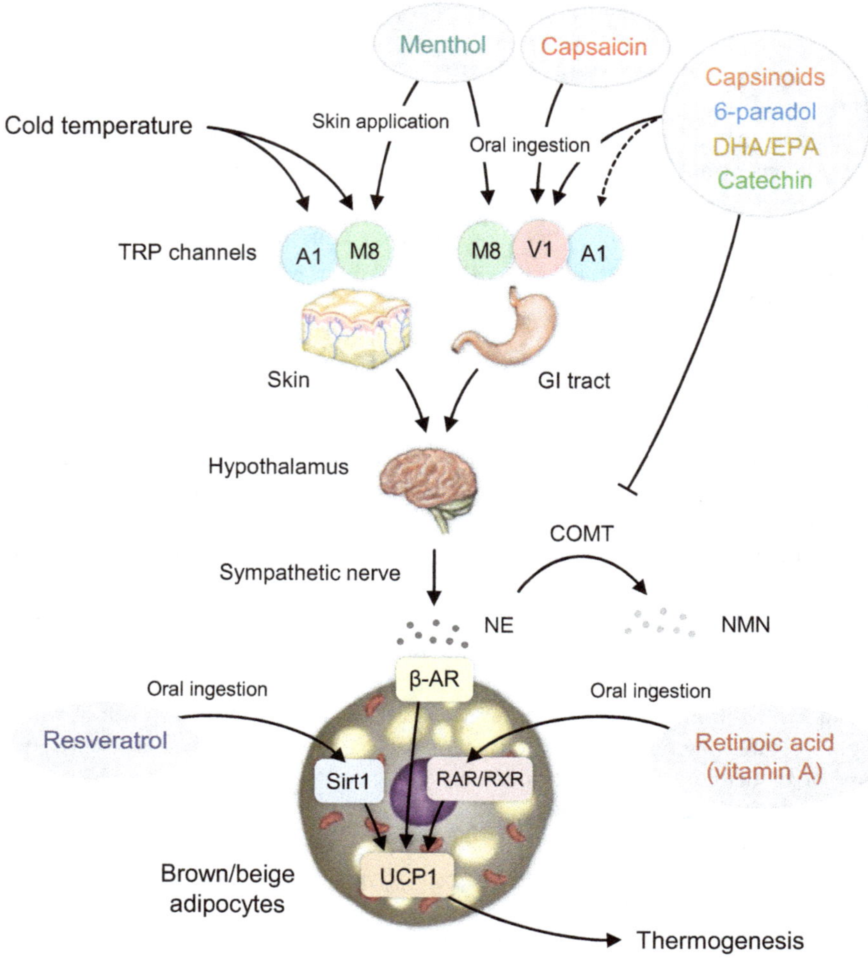

Fig. 1 Brown fat activation by food-derived stimulants with multiple mechanisms of action. Cold exposure activates temperature-sensitive TRP channels in the peripheral tissues such as the skin, triggering thermogenesis via β-adrenergic receptor-mediated pathway in UCP1-expressing thermogenic adipocytes. Orally ingested food ingredients with agonistic activity on TRP channels can mimic cold-induced nonshivering thermogenesis through the activation of gastrointestinal TRP-SNS-UCP1 axis. Retinoic acid and resveratrol activate RAR/RXR heterodimers and SIRT1, respectively, inducing *Ucp1* transcription, leading to enhanced thermogenic capacities of brown and/or beige fat. *COMT* catechol-*O*-methyltransferase, *DHA* docosahexaenoic acid, *EPA* eicosapentaenoic acid, *GI tract* gastrointestinal tract, *NE* norepinephrine, *NMN* normetanephrine, *RAR* retinoid acid receptors, *RXR* retinoid X receptors, *Sirt1* sirtuin 1, *SNS* sympathetic nervous system, *TRP* transient receptor potential, *UCP1* uncoupling protein 1, *β-AR* β-adrenergic receptors

2.1 Capsaicin and Capsinoids as TRP Vanilloid 1 Agonists

Capsaicin is the major component of chili peppers responsible for its pungent taste and is itself a potent activator of TRP vanilloid 1 (TRPV1), a thermoreceptor that senses temperatures greater than 42°C (Caterina et al. 1997). In addition, capsinoids, capsaicin-like compounds found in a nonpungent type of red pepper, are another class of known TRPV1 agonists. Capsaicin and capsinoids bind to TRPV1 with comparable affinities; however, pungency is much less defined in capsinoids than in capsaicin (1/1,000). The low pungency exhibited by capsinoids may be due to the high lipophilicity of capsinoids, which render these molecules unable to access the termini of trigeminal nerves in the oral cavity that is covered with epithelium (Uchida et al. 2017).

Animal studies have demonstrated that oral administration of capsaicin can activate TRPV1 expressed in sensory nerves within the gastrointestinal tract and increase sympathetic nerve activity innervating BAT, inducing a rapid increase in BAT temperature, and increasing whole-body EE (Ono et al. 2011). These physiological responses can be mimicked through administration of nonpungent capsinoids but are blunted by the administration of β-adrenergic blockers (Kawada et al. 1986) or through the denervation of vagal afferents and extrinsic nerves connected to the jejunum (Kawabata et al. 2009; Ono et al. 2011). Moreover, capsaicin (and probably capsinoids) evokes adrenal sympathetic nerve activity, inducing adrenaline secretion from the adrenal medulla, further stimulating BAT thermogenesis. This adrenalin secretion by capsaicin is largely diminished in TRPV1-KO mice (Uchida et al. 2017). It is possible that capsaicin additionally acts directly on TRPV1 expressed in BAT (Baskaran et al. 2017). By contrast, the direct action of capsinoids on TRPV1 in brown adipocytes is unlikely because orally ingested capsinoids are rapidly hydrolyzed and thus are usually undetectable in the general circulation. Most importantly, chronic administration of either capsaicin or capsinoids upregulates UCP1 expression in BAT and decreases body fat (Kawabata et al. 2009). The thermogenic and fat-reducing effects of capsinoids are abolished in UCP1-KO mice (Okamatsu-Ogura et al. 2015). Capsinoids, but not capsaicin, can also activate TRP ankyrin 1 (TRPA1), which senses temperatures below 17°C. However, the thermogenic effects of capsinoids are largely diminished in TRPV1-KO mice (Kawabata et al. 2009). These findings clearly demonstrate that oral administration of either pungent capsaicin or nonpungent capsinoids increases whole-body EE and prevents obesity through the activation of the TRPV1-SNS-BAT axis in small rodents (Fig. 1).

In humans, capsaicin is consumed worldwide as red pepper, and its anti-obesity effects have been documented experimentally (Ludy et al. 2012). However, despite its posited benefits, doses of capsaicin provided in the earlier studies (up to 10 g red pepper/meal) exceeded the amount preferred by the general population (0.3 g red pepper/meal). It is thus unlikely that most people could ingest sufficient amounts of capsaicin to replicate the desired effects seen in these studies. Given that capsinoids have much lower pungency but can exert metabolic effects comparable to capsaicin, capsinoids appear a more feasible tool for recruitment of BAT in humans. In fact, single oral ingestion of capsulated capsinoids increases whole-body EE in human individuals with metabolically active BAT, but not in those without (Yoneshiro et al. 2012). Furthermore, daily ingestion of capsinoids for 6 weeks augments nonshivering cold-induced thermogenesis (CIT) in individuals with low BAT activities (Yoneshiro et al. 2013). Since interindividual (Yoneshiro and Saito 2015) and intraindividual variations of CIT (Yoneshiro et al. 2016) are significantly related to BAT activity assessed by FDG-PET/CT, the capsinoid-induced increase in CIT reflects recruitment of BAT. Recently, this was directly confirmed by using FDG-PET/CT. Cold-induced BAT activity assessed by FDG uptake was significantly elevated after the capsinoid treatment for 6 weeks (Fig. 2) (Nirengi et al. 2016b). A similar result was reported in a study using near-infrared time-resolved spectroscopy (NIR_{TRS}), a novel method for evaluating BAT density in the region of interest (Nirengi et al. 2015). The BAT density assessed by $NIRS_{TRS}$ was increased after the 8-week capsinoid treatment; however, the increase was lost after 8 weeks from capsinoid withdrawal (Nirengi et al. 2016b). This indicates that a continuous supplementation is required for achieving metabolic benefits such as body fat reduction and improvement of insulin sensitivity. Indeed, relatively long treatment with capsinoids (for 12 weeks) generates a slight but significant decrease in abdominal fat by 1.1% in an obese population (Snitker et al. 2009).

The beneficial effects of capsinoids are greatly attenuated in individuals who carry a mutated (Val585Ile) TRPV1 (Snitker et al. 2009), consistent with the crucial role of TRPV1 in mice, as described earlier. Although the beneficial effects of capsinoids are similar to those of cold exposure, TRPV1 is intriguingly not a cold sensor, but rather a sensor of noxious hot temperatures and low pH (Tominaga 2007). It is therefore likely

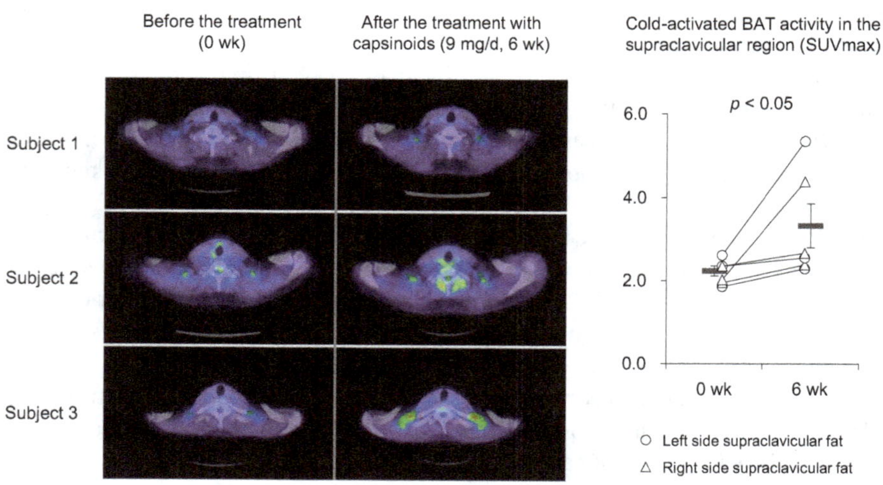

Fig. 2 BAT recruitment by daily ingestion of capsinoids. FDG-PET/CT examination combined with acute mild cold exposure (19°C, 2 h) was performed before and after the treatment with capsinoids (9 mg/day, 6 weeks). BAT activity was quantified as SUVmax of FDG. Cold-activated BAT activity was significantly increased after the treatment with capsinoids ($p < 0.05$) (Nirengi et al. 2016b). *BAT* brown adipose tissue. *FDG-PET/CT* fluorodeoxyglucose-positron emission tomography and computed tomography, *SUVmax* maximal standardized uptake value

that human BAT is activated by nociceptive stimuli, including TRPV1 activation. In agreement with this, chronic adrenergic stress induced by burn trauma results in browning of white adipose tissue (WAT) (Sidossis et al. 2015). Hence, it is conceivable that oral ingestion of capsinoids leads to the activation of BAT thermogenesis through the TRPV1-mediated pathway in humans, while cold exposure is more potent to induce BAT activation than capsinoid ingestion (Sun et al. 2018). Given a recent report that capsinoid treatment in mice potentiates cold-induced browning of WAT (Ohyama et al. 2016), a combination of capsinoid supplementation and mild cold exposure may be an effective strategy for recruitment of BAT in humans. It remains unknown if capsinoid treatment can improve insulin sensitivity, as cold exposure does in type 2 diabetic subjects (Hanssen et al. 2015).

2.2 Paradol as a TRPV1 and TRPA1 Agonist

There are various organic compounds that contain the vanilloid structure found in capsaicin and capsinoids. For example, gingerols, shogaol, zingerone, and 6-paradol are found in several types of ginger and known to act as agonists of TRPV1 (Yoneshiro and Saito 2015) (Fig. 1).

One of the common foods containing these compounds is "grains of paradise" [*Aframomum melegueta* (Rosco) K. Schum.] (GP), also called Guinea pepper or alligator pepper. GP seeds are used as a spice for flavoring food in West Africa and are very rich in 6-paradol, 6-gingerol, and 6-shogaol. Intragastric administration of an alcohol extract of GP seeds results in an enhanced discharge of efferent sympathetic nerves to BAT and a significant rise in BAT temperature in small rodents (Iwami et al. 2011). Among vanilloids in the GP extract, 6-paradol is the compound most likely responsible for the observed effects, judging by its high content and relatively high affinity to TRPV1 (Riera et al. 2009; Morera et al. 2012). Indeed, BAT thermogenesis induced by GP extract administration can be mimicked by intragastric administration of 6-paradol (Iwami et al. 2011). Like capsinoids, 6-paradol can activate both TRPV1 and TRPA1 (Riera et al. 2009), but no study has yet examined the TRP subtype responsible for 6-paradol-induced activation of BAT. It remains to be elucidated whether the UCP1-mediated thermogenesis in BAT and/or browned WAT is indispensable for the thermogenic and anti-obesity effects of 6-paradol.

In humans, single ingestion of capsulated GP extract increases whole-body EE in individuals with metabolically active BAT, but not in those without (Sugita et al. 2013). Moreover, daily ingestion of GP extract (40 mg/day, 3 weeks) results in an increase in CIT, an index of BAT activity in healthy young volunteers (Yoneshiro et al. 2018). This thermogenic effect of GP was coupled with a significant decrease in visceral fat area (−7.0%) (Sugita et al. 2014). As these responses are almost similar to those seen after capsinoid ingestion, it is likely that GP extract enhances EE as a result of BAT activation and recruitment. Thus, GP extract is a potential food ingredient for recruiting BAT and decreasing body fat content in humans. Whether the treatment with 6-paradol or the GP extract can improve insulin resistance in patients with type 2 diabetes is unknown.

2.3 DHA and EPA as TRPV1 Agonists

Dietary fatty acids are not only major nutrients but also key signaling molecules that regulate energy metabolism. Docosahexaenoic acid (DHA) and eicosapentaenoic acid (EPA) are *n*-3 polyunsaturated fatty acids enriched in fish oil. An earlier study has demonstrated that supplementation of DHA and EPA reduces the accumulation of body fat, which may be mediated by several mechanisms, including upregulation of UCP1 in BAT and WAT (Oudart et al. 1997). UCP1 induction by dietary EPA and DHA is blocked by either subdiaphragmatic vagotomy or treatment with a β-adrenergic blocker (Kim et al. 2015). EPA and DHA have agonistic activity on TRPV1. Accordingly, the thermogenic and anti-obesity effects of EPA and DHA are abolished in TRPV1-KO mice (Kim et al. 2015). Thus, the activation of the TRPV1-SNS-UCP1 pathway is one mechanism by which EPA and DHA activate BAT thermogenesis. A recent report proposed an additional mechanism. EPA is sensed by the membrane receptor free fatty acid receptor 4 in brown adipocytes, resulting in biogenesis of the microRNAs miR-30b and miR-378 and a rise of intracellular cAMP levels, both of which promotes transcriptional activation of brown adipogenesis, including Ucp1 induction (Kim et al. 2016). UCP1-inducing effects of ω3 fatty acids such as DHA and EPA are also reportedly mediated via inhibition of production of ω6-derived oxygenated metabolites, such as oxylipins, that can impair UCP1 activation and induction (Ghandour et al. 2018).

However, despite the abundance of evidence in rodents, thermogenic effects of EPA and DHA are poorly understood in humans. Previous studies suggest that *n*-3 fatty acids such as DHA and EPA can ameliorate symptoms of a variety of inflammatory disorders (Serhan 2014), coronary heart disease (Del Gobbo et al. 2016), and obesity (Okla et al. 2017). Further studies in healthy adults designed to examine potential effects of EPA/DHA on BAT activity and insulin sensitivity are vital to establish the effectiveness of *n*-3 fatty acids in combating metabolic disorders.

2.4 Catechins as TRPA1/V1 Agonists and Catechol-*O*-Methyltransferase Inhibitors

Catechin, catechin gallate, gallocatechin, gallocatechin gallate, epicatechin, epicatechin gallate, epigallocatechin, and epigallocatechin gallate (EGCG) are low-molecular-weight polyphenols collectively referred to as catechins. Catechins are enriched in green tea and, especially in the case of EGCG, have various benefits including thermogenic and anti-obesity effects in humans (Dulloo et al. 1999; Thavanesan 2011). The thermogenic response to catechins has been proposed to be mediated by the inhibitory effect of catechins against the noradrenaline-degrading enzyme catechol-*O*-methyltransferase (COMT), stimulating hormone-sensitive lipase to degrade triglycerides into fatty acids and glycerol in the liver and adipose tissue (Ferreira et al. 2016; Westerterp-Plantenga 2010). However, this is still debated because a study in humans suggested that COMT activity is not impaired by oral ingestion of high doses of EGCG in vivo (Lorenz et al. 2014). This may be explained by the fact that the circulating levels of catechins after a single ingestion of catechin-rich tea (~0.1 μM at maximum) (Takahashi et al. 2014) are much lower than the half-maximal inhibitory concentration of catechins for the COMT activity (~14 μM) (Kadowaki et al. 2005). Thus, the role of COMT inhibition as a primary target of the catechin action on triglyceride degradation and thermogenesis remains controversial. Furthermore, it has not been fully understood how the energy produced by catechin-induced substrate oxidation, such as proton electrochemical gradient (H^+) or ATP, is harnessed and dissipated as heat.

To this end, we suggest the TRP-mediated activation of BAT as a potential mechanism underlying catechin-induced thermogenesis because UCP1 in BAT is capable of dissipating proton gradient in the mitochondria, thus generating heat. EGCG and its auto-oxidation products are able to activate TRPA1 and TRPV1 of intestinal enteroendocrine cells with a concentration of 100–200 μM (Kurogi et al. 2012, 2015). Given that the concentration of catechins in the catechin-rich tea used in the previous clinical studies was approximately 700 μM (Takahashi et al. 2014), TRPA1 and/or TRPV1 located in the sensory neuron of the gastrointestinal tract could be activated by ingested catechins, thus triggering SNS-mediated BAT thermogenesis. Consistent with this hypothesis, recent studies in small rodents demonstrated that dietary supplementation of catechins upregulates UCP1 expression in BAT and WAT (Nomura et al. 2008; Sae-Tan et al. 2015). Studies using TRP and UCP1 knockout mice seem highly likely to yield further understanding for potential roles of TRP-BAT axis in beneficial effects of catechins. Meanwhile, a recent study in healthy humans reported that oral ingestion of tea containing 615 mg catechins in 350 mL (~700 μM) significantly increases EE in hours in subjects with detectable BAT activity, but not in those with undetectable BAT activity (Yoneshiro et al. 2017b). The thermogenic effect of catechins was significantly and positively correlated with BAT activity, suggesting the involvement of BAT in catechin-induced thermogenesis. Moreover, a 5-week daily ingestion of catechin-rich tea (1,230 mg/day) resulted in a significant increase in CIT, a marker for BAT activity. Although the active and placebo beverages contained a moderate amount of caffeine (~80 mg per bottle), the placebo ingestion did not produce any change in EE and CIT. Thus, it is highly likely that observed thermogenesis is due to the catechin action, rather than caffeine action. This does not exclude, however, a possible synergistic action of catechins and caffeine (Ferreira et al. 2016; Westerterp-Plantenga 2010). The chronic effects of catechins on BAT were also confirmed by using the $NIRS_{TRS}$ technique (Nirengi et al. 2016a). Collectively, thermogenic and fat-reducing effects of catechins would be attributable to the activation of BAT (Fig. 1).

Protective effects of catechins against body fat accumulation and type 2 diabetes have been documented (Ferreira et al. 2016; Yang et al. 2014). The association between these beneficial effects and the potential recruitment of BAT is to be examined in obese and diabetic patients. It also remains to be investigated whether BAT-mediated thermogenesis is responsible for the catechin effects on body fat content and insulin sensitivity, by using genetic mouse models such as UCP1-KO mice.

2.5 Menthol as a TRP Melastatin 8 Agonist

Menthol is a cooling and flavorful compound enriched in mint and is a potent activator of TRP melastatin 8 (TRPM8) (Yoneshiro and Saito 2013). TRPM8 is a sensor of cold temperatures lower than 26°C and expressed in sensory nerves in the skin, involving in BAT activation during cold exposure (Nakamura 2011). The topical application of menthol increases oxygen consumption and core body temperature in mice (Tajino et al. 2007). Although no study reported a BAT-stimulating effect of menthol in human in vivo, a single skin menthol administration induces a slight but significant elevation of metabolic rate in young healthy humans (Valente et al. 2015). Moreover, in diet-induced obese mice, oral administration of menthol activates UCP1-mediated thermogenesis in a TRPM8-dependent manner, thereby preventing body fat accumulation (Ma et al. 2012). While TRPM8 expression in gastrointestinal sensory nerves has been shown (de Jong et al. 2015), further studies are required to elucidate a potential involvement of gastrointestinal TRP-SNS-BAT axis in thermogenic effect of menthol. It may also be possible that menthol acts directly on TRPM8 expressed in brown adipocytes, because both oral administration and skin application of menthol can increase the circulating level of menthol (Valente et al. 2015). In fact, a study on human white adipocytes cultured with menthol showed an increase in UCP1 expression and basal- and insulin-stimulated glucose uptakes (Rossato et al. 2014).

2.6 Allyl Isothiocyanate as a TRPA1 Agonist

Allyl isothiocyanate (AITC), which is enriched in wasabi (Japanese horse radish) and mustard, is a typical TRPA1 agonist (Yoneshiro and Saito 2013). Dietary supplementation of AITC is reported to improve mitochondrial function and ameliorate insulin resistance in mice (Ahn et al. 2014), but the involvement of BAT or UCP1 in the beneficial metabolic effects of AITC is poorly understood.

In humans, the thermogenic effect of AITC remains elusive due to the limited availability of studies. A recent report in human participants was unable to identify any metabolic effects of AITC (Langeveld et al. 2017). It should be noted, however, that BAT activity is attenuated by aging and fluctuates seasonally, being low or negligible in the elderly or in the summer season (Au-Yong et al. 2009; Yoneshiro et al. 2011b). This was the rationale to perform most of our previous studies with young participants in winter when testing BAT activation by food (Sugita et al. 2013; Yoneshiro et al. 2012, 2013, 2017b). Therefore, there is still a need for the age- and season-matched studies to minimize the inter- and intraindividual variations of BAT activity, facilitating to investigate the effects of AITC on human BAT and EE.

2.7 Cinnamaldehyde as a TRPA1 Agonist

Cinnamaldehyde, found in cinnamon, is another typical TRPA1 agonist. Dietary cinnamaldehyde protects mice against diet-induced obesity (Tamura et al. 2012; Zuo et al. 2017). The body fat-reducing effect has been associated with UCP1 upregulation in BAT (Tamura et al. 2012) and WAT (Zuo et al. 2017). This was also confirmed in murine adipocytes. However, the UCP1-inducing effect of cinnamaldehyde can be seen in both TRPA1-KO and wild-type adipocytes (Jiang et al. 2017). It is thus likely that the direct action of cinnamaldehyde on adipocytes may be mediated by a TRPA1-independent pathway, while an indirect action via gastrointestinal TRPA1 might be involved in the thermogenic activation. To date, little information is available on whether beneficial metabolic effects of cinnamaldehyde are direct consequences of UCP1 thermogenic activation. In addition, whether TRPA1 expressed in the gastrointestinal sensory nerves is indispensable for the potential activation of BAT by cinnamaldehyde is not known.

In humans, cinnamon consumption correlates with lower fasting blood glucose levels (Allen et al. 2013). Importantly, the UCP1-inducing effect of cinnamaldehyde was demonstrated in human adipocytes differentiated from human adipose stem cells, regardless of donor's age, ethnicity, or BMI (Jiang et al. 2017). While the effects of cinnamaldehyde on human BAT activity in vivo remain to be tested, single ingestion of this substance can increase metabolic rate in humans (Michlig et al. 2016). Given the wide use of cinnamon as a popular food additive, cinnamaldehyde (or cinnamon) may be a useful solution for developing therapeutic strategies that are much better adhered to by participants.

2.8 Other Ingredients with Agonistic Activity at Thermosensitive TRPs

Based on animal studies, many other food ingredients that can activate TRPs have been proposed as potential agents to boost BAT thermogenesis. For example, royal jelly was recently reported to prevent body fat accumulation, insulin resistance, and hepatic steatosis as a result of BAT thermogenic activation in diet-induced obese mice (Yoneshiro et al. 2017a). Royal jelly contains the unique fatty acids hydroxydecenoic acid (HDEA) and hydroxydecanoic acid (HDAA), which are able to activate TRPA1 (Terada et al. 2011). It seems thus conceivable that the beneficial metabolic effect of royal jelly could be mediated by the activation of the TRPA1-SNS-BAT axis. Furthermore, allicin in garlic and onion, eugenol in clove, piperine in black pepper, and eucalyptol in eucalyptus are also known as agonists of temperature-sensitive TRPs such as TRPV1, TRPM8, and TRPA1 (Uchida et al. 2017). All of these compounds may serve as BAT-activating agents in humans.

3 Food-Derived Stimulants Acting via TRP-Independent Mechanisms

In addition to TRP-activating substances, there are several anti-obesity food ingredients that stimulate BAT via multiple mechanisms of action distinct from TRP-mediated processes. Here we describe these substances and their impacts on systemic energy homeostasis.

3.1 Retinoic Acid as an Activator of the Transcriptional Regulatory Enhancer of the Ucp1 Gene

Retinoic acid is a metabolite of vitamin A (retinol) and an important regulator of transcription of over 500 genes through binding to the retinoic acid response elements in the regulatory regions of target genes. The major sources of retinoic acid are dietary retinol, retinyl esters, and provitamin A carotenoids (β-carotene) that are absorbed in the small intestine and delivered to peripheral tissues via multiple mechanisms such as binding to retinol-binding protein and packaging into chylomicrons (Li et al. 2014). In the peripheral cells, retinols are enzymatically oxidized to retinoic acid and transported to the nucleus. Once inside the nucleus, retinoic acid binds to retinoid acid receptors (RAR), which results in the dimerization of RAR with retinoid X receptors (RXR) (Berry and Noy 2009). In mice, retinoic acid treatment appears to induce UCP1 expression in BAT and WAT and to improve insulin sensitivity (Bargut et al. 2017; Flajollet et al. 2013). The stimulatory effect of retinoic acid on Ucp1 expression is independent of adrenergic pathways but mediated by the activation of the heterodimers RAR and RXR, which bind to specific regulatory regions in the UCP1 enhancer (Okla et al. 2017) (Fig. 1). Conversely, consuming a vitamin A-deficient diet increases body fat mass and reduces BAT thermogenic capacity in mice (Bonet et al. 2000). The novel adipokine lipocalin 2 (Lcn2), one of the target genes of retinoic acid, is indispensable for retinoid-induced activation of UCP1 expression and thermogenesis (Guo et al. 2016). These findings, coupled with the fact that mice lacking Lcn2 are prone to develop obesity (Guo et al. 2010), suggest retinoic acid as a potent dietary factor to regulate BAT activity and energy homeostasis.

Nevertheless, a study reported that the Ucp1-inducing effect of retinoic acid is only marginal in primary cultured human white adipocytes (Murholm et al. 2013), suggesting that retinoic acid does not have an effect on the recruitment of thermogenic adipocytes. However, it is well-known that the ability of primary stromal vascular cells (SVC) to differentiate into thermogenic adipocytes is highly region-dependent: high in primary SVC isolated from the supraclavicular BAT depot but much low in those isolated from the subcutaneous WAT depot (Lee et al. 2011; Shinoda et al. 2015). Thus, the effectiveness of retinoic acid supplementation needs to be retested in primary or clonal human brown/beige adipocytes (Shinoda et al. 2015) and especially in human in vivo.

3.2 Resveratrols as Sirt1 Activators

Resveratrols, which include *trans*-resveratrol, gnetin C, gnemonoside A, and gnemonoside D, are natural polyphenolic compounds found in the skin of grapes. Dietary supplementation of resveratrols is capable of preventing obesity: consumption of resveratrols prevents body fat accumulation, adipose tissue inflammation, and insulin resistance in mice fed with a high-fat diet (Lagouge et al. 2006). In cultured adipocytes, resveratrol treatment downregulates adipogenesis-related genes and fatty acid synthesis-related genes (Rayalam et al. 2008). Dietary resveratrol also increases thermogenesis-related genes in BAT and WAT in mice. Mechanistically, resveratrols can activate sirtuin 1 (Sirt1), which is a critical regulator of mitochondrial biogenesis (Chang and Guarente 2014), thus increasing mitochondrial content in BAT and WAT (Andrade et al. 2014; Wang et al. 2015). Moreover, SIRT1 deacetylates peroxisome proliferator-activated receptor γ (PPARγ), thus allowing recruitment of PR domain-containing 16 (PRDM16) (Chang and Guarente 2014), a master regulator of brown adipocyte differentiation (Kajimura et al. 2009; Ohno et al. 2013). It is thus assumed that resveratrols activate adaptive adipose tissue thermogenesis through Sirt1-mediated mitochondrial biogenesis and UCP1 transcription in BAT and WAT (Fig. 1). However, it remains to be determined if the UCP1-mediated thermogenesis is indispensable for the body fat-reducing and insulin-sensitizing effects of resveratrols.

In addition to purified resveratrols, there are several foods enriched in resveratrols and traditionally consumed by humans. One of these is melinjo (*Gnetum gnemon* L.), an arboreal dioecious plant widely cultivated in Southeast Asia. Seeds and fruits of melinjo are eaten as an ordinary vegetable. Consistent with the findings on purified resveratrols, dietary supplementation of the lyophilized powder of melinjo seed extract (MSE) containing >20% resveratrol derivatives protects mice against diet-induced obesity (Ikuta et al. 2015). Moreover, dietary MSE consumption is sufficient to increase UCP1 expression and mitochondrial content in BAT, thus protecting mice against adipose inflammation, insulin resistance, and hepatic steatosis (Yoneshiro et al. 2018). In humans, effects of consumption of either purified resveratrols or MSE on BAT thermogenesis are scarce, and clear anti-obesity action has not been revealed so far (Fernández-Quintela et al. 2017).

3.3 Other Ingredients with BAT-Activating Effects

Animal studies demonstrate that dietary nitric oxide (NO), which is synthesized from non-caloric inorganic nitrate rich in green leafy vegetables, is able to recruit BAT (Roberts et al. 2015). Dietary nitrate upregulates UCP1 and other thermogenic genes in BAT and WAT and increases oxygen consumption in adipocytes in mice.

Moreover, dietary supplementation with amino acids, particularly arginine, protects against diet-induced obesity (Fu et al. 2005; Kohli et al. 2004). As arginine is the major source of NO production, it may be able to stimulate BAT through the NO-related mechanisms (Jobgen et al. 2009; Kohli et al. 2004).

Rutin, a natural compound found in mulberry, is a novel small molecule that activates BAT. Rutin treatment results in a parallel increase in whole-body EE and UCP1 expression in BAT and WAT and a concomitant decrease in body fat content in obese mice (Yuan et al. 2017). Such rutin effects seem to be due to its ability to stabilize SIRT1, leading to hypoacetylation of PPARγ, mitochondrial biogenesis, and UCP1 transcription.

Quercetin, also known as hydrolysate of rutin, is rich in onion peel and has various biological functions, including antioxidant and anti-obesity effects (Kim and Yim 2015; Moon et al. 2013). Dietary onion peel extract induces UCP1 expression and browning of WAT through AMP-activated protein kinase (AMPK)-SIRT1 signaling (Ahn et al. 2008; Lee et al. 2017).

Curcumin, a naturally occurring curcuminoid of turmeric, which is a member of the ginger family (Zingiberaceae), upregulates thermogenesis-related genes in cultured cells by activating the phosphorylation of AMPK and SIRT1 (Lone et al. 2016).

4 Concluding Remarks

Since the rediscovery of metabolically active BAT in healthy humans in 2009, a growing body of literature has revealed the energy-dissipating abilities of human brown adipocytes both in vivo and in vitro (Boon and van Marken Lichtenbelt 2016). Important observations that reduced BAT activity during aging is closely associated with visceral fat accumulation and development of insulin resistance (Matsushita et al. 2014) provided an impetus for determining reliable and practical strategies for recruiting BAT. The most significant merits of the use of food ingredients, which were traditionally consumed by humans, are low risks of adverse effects and better likelihood of adherence in subjects, as compared to cold regimen or the use of pharmacological adrenomimetics.

This chapter summarized the recent updates on dietary compounds that could be included as candidates in a dietary regimen for activating BAT-mediated thermogenesis. It was also highlighted that numerous remaining challenges still lie ahead. Presently, our knowledge of the mechanisms underlying the food ingredient-induced activation of BAT is limited. Particularly, further studies using genetically modified mouse models such as UCP1-, TRP-, and SIRT1-KO mice are highly desirable for unveiling the responsible thermogenic molecules and their regulatory pathways. Since recent investigations uncovered UCP1-independent thermogenesis in brown-like, beige/brite adipocytes and myocytes through activation of creatine cycling (Kazak et al. 2015) or calcium cycling (Bal et al. 2017; Ikeda et al. 2017), the possible involvements of these noncanonical thermogenic processes in the anti-obesity effects of food ingredients should also be considered. These investigations will not only promote our scientific curiosity but further fuel translational research in humans.

Compared with studies in rodents, available evidence in humans for the thermogenic function of food substances, as well as their relation to BAT, is highly limited. One reason may be the few options available to quantify BAT activity in humans.

Although FDG-PET/CT is currently the standard tool, this technique has several limitations, including limited accessibility of the device, cost, and the unpleasantness of acute cold exposure. Particularly, inevitable radiation exposure by the administration of FDG and CT scanning makes it difficult to quantify BAT repeatedly in the same subjects; there is an immense need for developing innovative but noninvasive technologies that detect human BAT with high sensitivity. A realizable option may be $NIRS_{TRS}$. This $NIRS_{TRS}$ technique can quantify optical properties of tissue and provide total hemoglobin concentration, an indicator of the vascularity. As BAT has abundant capillaries compared to other tissues such as white fat, the indices of vascular density assessed by $NIRS_{TRS}$ is closely correlated with BAT activity assessed by FDG-PET/CT, thus reflecting BAT density (Nirengi et al. 2015). While further validations are needed, $NIRS_{TRS}$ technique has already been used to confirm the stimulatory effects of dietary factors for BAT and has provided us reasonable results, comparable to those by FDG-PET/CT (Nirengi et al. 2016a, b). The investigation with novel noninvasive techniques will bridge the disproportionate gaps between accumulating evidence in animal models and limited confirmation in humans. In contrast, little is known about the food constituents that could inhibit potential BAT activation by thermogenic compounds. In addition to the function of individual dietary factors, future studies should also focus on the combination effects of the multiple substances. These investigations will allow us to devise efficient but flexible eating plans to boost BAT thermogenesis, helping to develop strategies targeting BAT for obesity prevention.

5 Acknowledgments

This work was supported in part by a Grant-in-Aid for Scientific Research from the Ministry of Education, Culture, Sports, Science, and Technology of Japan from the Japan Society for the Promotion of Science. We thank Zachary Brown for editing a draft of this manuscript.

References

Ahn J, Lee H, Kim S, Park J, Ha T (2008) The anti-obesity effect of quercetin is mediated by the AMPK and MAPK signaling pathways. Biochem Biophys Res Commun 373:545–549

Ahn J, Lee H, Im SW, Jung CH, Ha TY (2014) Allyl isothiocyanate ameliorates insulin resistance through the regulation of mitochondrial function. J Nutr Biochem 25:1026–1034

Allen RW, Schwartzman E, Baker WL, Coleman CI, Phung OJ (2013) Cinnamon use in type 2 diabetes: an updated systematic review and meta-analysis. Ann Fam Med 11:452–459

Andrade JM, Frade AC, Guimarães JB, Freitas KM, Lopes MT, Guimarães AL et al (2014) Resveratrol increases brown adipose tissue thermogenesis markers by increasing SIRT1 and energy expenditure and decreasing fat accumulation in adipose tissue of mice fed a standard diet. Eur J Nutr 53:1503–1510

Au-Yong IT, Thorn N, Ganatra R, Perkins AC, Symonds ME (2009) Brown adipose tissue and seasonal variation in humans. Diabetes 58:2583–2587

Bal NC, Singh S, Reis FCG, Maurya SK, Pani S, Rowland LA et al (2017) Both brown adipose tissue and skeletal muscle thermogenesis processes are activated during mild to severe cold adaptation in mice. J Biol Chem 292:16616–16625

Bargut TCL, Souza-Mello V, Aguila MB, Mandarim-de-Lacerda CA (2017) Browning of white adipose tissue: lessons from experimental models. Horm Mol Biol Clin Investig 31(1). doi: https://doi.org/10.1515/hmbci-2016-0051. Review

Baskaran P, Krishnan V, Fettel K, Gao P, Zhu Z, Ren J et al (2017) TRPV1 activation counters diet-induced obesity through sirtuin-1 activation and PRDM-16 deacetylation in brown adipose tissue. Int J Obes (Lond) 41:739–749

Berry DC, Noy N (2009) All-trans-retinoic acid represses obesity and insulin resistance by activating both peroxisome proliferation-activated receptor beta/delta and retinoic acid receptor. Mol Cell Biol 29:3286–3296

Blondin DP, Labbé SM, Tingelstad HC, Noll C, Kunach M, Phoenix S et al (2014) Increased brown adipose tissue oxidative capacity in cold-acclimated humans. J Clin Endocrinol Metab 99: E438–E446

Bonet ML, Oliver J, Pico C, Felipe F, Ribot J, Cinti S et al (2000) Opposite effects of feeding a vitamin A-deficient diet and retinoic acid treatment on brown adipose tissue uncoupling protein 1 (UCP1), UCP2 and leptin expression. J Endocrinol 166:511–517

Boon MR, van Marken Lichtenbelt WD (2016) Brown adipose tissue: a human perspective. Handb Exp Pharmacol 233:301–319

Cannon B, Nedergaard J (2011) Nonshivering thermogenesis and its adequate measurement in metabolic studies. J Exp Biol 214:242–253

Caterina MJ, Schumacher MA, Tominaga M, Rosen TA, Levine JD, Julius D (1997) The capsaicin receptor: a heat-activated ion channel in the pain pathway. Nature 389:816–824

Chang HC, Guarente L (2014) SIRT1 and other sirtuins in metabolism. Trends Endocrinol Metab 25:138–145

Chondronikola M, Volpi E, Børsheim E, Porter C, Saraf MK, Annamalai P et al (2016) Brown adipose tissue activation is linked to distinct systemic effects on lipid metabolism in humans. Cell Metab 23:1200–1206

Cypess AM, Lehman S, Williams G, Tal I, Rodman D, Goldfine AB et al (2009) Identification and importance of brown adipose tissue in adult humans. N Engl J Med 360:1509–1517

Cypess AM, Weiner LS, Roberts-Toler C, Franquet Elía E, Kessler SH, Kahn PA et al (2015) Activation of human brown adipose tissue by a β3-adrenergic receptor agonist. Cell Metab 21:33–38

Del Gobbo LC, Imamura F, Aslibekyan S, Marklund M, Virtanen JK, Wennberg M et al (2016) ω-3 polyunsaturated fatty acid biomarkers and coronary heart disease: pooling project of 19 cohort studies. JAMA Intern Med 176:1155–1166

de Jong PR, Takahashi N, Peiris M, Bertin S, Lee J, Gareau MG et al (2015) TRPM8 on mucosal sensory nerves regulates colitogenic responses by innate immune cells via CGRP. Mucosal Immunol 8:491–504

Dong M, Yang X, Lim S, Cao Z, Honek J, Lu H et al (2013) Cold exposure promotes atherosclerotic plaque growth and instability via UCP1-dependent lipolysis. Cell Metab 18:118–129

Dulloo AG, Duret C, Rohrer D, Girardier L, Mensi N, Fathi M et al (1999) Efficacy of a green tea extract rich in catechin polyphenols and caffeine in increasing 24-h energy expenditure and fat oxidation in humans. Am J Clin Nutr 70:1040–1045

Fernández-Quintela A, Milton-Laskibar I, González M, Portillo MP (2017) Antiobesity effects of resveratrol: which tissues are involved? Ann N Y Acad Sci 1403:118–131

Ferreira MA, Silva DM, de Morais AC Jr, Mota JF, Botelho PB (2016) Therapeutic potential of green tea on risk factors for type 2 diabetes in obese adults – a review. Obes Rev 17:1316–1328

Flajollet S, Staels B, Lefebvre P (2013) Retinoids and nuclear retinoid receptors in white and brown adipose tissues: physiopathologic aspects. Horm Mol Biol Clin Investig 14:75–86

Fu WJ, Haynes TE, Kohli R, Hu J, Shi W, Spencer TE et al (2005) Dietary L-arginine supplementation reduces fat mass in Zucker diabetic fatty rats. J Nutr 135:714–721

Ghandour RA, Colson C, Giroud M, Maurer S, Rekima S, Ailhaud G et al (2018) Impact of dietary ω3 polyunsaturated fatty acid supplementation on brown and brite adipocyte function. J Lipid Res 59:452–461

Guo H, Jin D, Zhang Y, Wright W, Bazuine M, Brockman DA et al (2010) Lipocalin-2 deficiency impairs thermogenesis and potentiates diet-induced insulin resistance in mice. Diabetes 59:1376–1385

Guo H, Foncea R, O'Byrne SM, Jiang H, Zhang Y, Deis JA et al (2016) Lipocalin 2, a regulator of retinoid homeostasis and retinoid-mediated thermogenic activation in adipose tissue. J Biol Chem 291:11216–11229

Hanssen MJ, Hoeks J, Brans B, van der Lans AA, Schaart G, van den Driessche JJ et al (2015) Short-term cold acclimation improves insulin sensitivity in patients with type 2 diabetes mellitus. Nat Med 21:863–865

Hibi M, Oishi S, Matsushita M, Yoneshiro T, Yamaguchi T, Usui C et al (2016) Brown adipose tissue is involved in diet-induced thermogenesis and whole-body fat utilization in healthy humans. Int J Obes (Lond) 40:1655–1661

Ikeda K, Kang Q, Yoneshiro T, Camporez JP, Maki H, Homma M et al (2017) UCP1-independent signaling involving SERCA2b-mediated calcium cycling regulates beige fat thermogenesis and systemic glucose homeostasis. Nat Med 23:1454–1465

Ikuta T, Saito S, Tani H, Tatefuji T, Hashimoto K (2015) Resveratrol derivative-rich melinjo (Gnetum gnemon L.) seed extract improves obesity and survival of C57BL/6 mice fed a high-fat diet. Biosci Biotechnol Biochem 79:2044–2049

Iwami M, Mahmoud FA, Shiina T, Hirayama H, Shima T, Sugita J et al (2011) Extract of grains of paradise and its active principle 6-paradol trigger thermogenesis of brown adipose tissue in rats. Auton Neurosci 161:63–67

Jiang J, Emont MP, Jun H, Qiao X, Liao J, Kim DI et al (2017) Cinnamaldehyde induces fat cell-autonomous thermogenesis and metabolic reprogramming. Metabolism 77:58–64

Jobgen W, Meininger CJ, Jobgen SC, Li P, Lee MJ, Smith SB et al (2009) Dietary L-arginine supplementation reduces white fat gain and enhances skeletal muscle and brown fat masses in diet-induced obese rats. J Nutr 139:230–237

Kadowaki M, Ootani E, Sugihara N, Furuno K (2005) Inhibitory effects of catechin gallates on o-methyltranslation of protocatechuic acid in rat liver cytosolic preparations and cultured hepatocytes. Biol Pharm Bull 28:1509–1513

Kajimura S, Seale P, Kubota K, Lunsford E, Frangioni JV, Gygi SP et al (2009) Initiation of myoblast to brown fat switch by a PRDM16-C/EBP-beta transcriptional complex. Nature 460:1154–1158

Kawabata F, Inoue N, Masamoto Y, Matsumura S, Kimura W, Kadowaki M et al (2009) Non-pungent capsaicin analogs (capsinoids) increase metabolic rate and enhance thermogenesis via gastrointestinal TRPV1 in mice. Biosci Biotechnol Biochem 73:2690–2697

Kawada T, Watanabe T, Takaishi T, Tanaka T, Iwai K (1986) Capsaicin-induced beta-adrenergic action on energy metabolism in rats: influence of capsaicin on oxygen consumption, the respiratory quotient, and substrate utilization. Proc Soc Exp Biol Med 183:250–256

Kazak L, Chouchani ET, Jedrychowski MP, Erickson BK, Shinoda K, Cohen P et al (2015) A creatine-driven substrate cycle enhances energy expenditure and thermogenesis in beige fat. Cell 163:643–655

Kim KA, Yim JE (2015) Antioxidative activity of onion peel extract in obese women: a randomized, double-blind, placebo controlled study. J Cancer Prev 20:202–207

Kim M, Goto T, Yu R, Uchida K, Tominaga M, Kano Y et al (2015) Fish oil intake induces UCP1 upregulation in brown and white adipose tissue via the sympathetic nervous system. Sci Rep 5:18013

Kim J, Okla M, Erickson A, Carr T, Natarajan SK, Chung S (2016) Eicosapentaenoic acid potentiates brown ahermogenesis through FFAR4-dependent up-regulation of miR-30b and miR-378. J Biol Chem 291:20551–20562

Kohli R, Meininger CJ, Haynes TE, Yan W, Self JT, Wu G (2004) Dietary L-arginine supplementation enhances endothelial nitric oxide synthesis in streptozotocin-induced diabetic rats. J Nutr 134:600–608

Kurogi M, Miyashita M, Emoto Y, Kubo Y, Saitoh O (2012) Green tea polyphenol epigallocatechin gallate activates TRPA1 in an intestinal enteroendocrine cell line, STC-1. Chem Senses 37:167–177

Kurogi M, Kawai Y, Nagatomo K, Tateyama M, Kubo Y, Saitoh O (2015) Auto-oxidation products of epigallocatechin gallate activate TRPA1 and TRPV1 in sensory neurons. Chem Senses 40:27–46

Lagouge M, Argmann C, Gerhart-Hines Z, Meziane H, Lerin C, Daussin F et al (2006) Resveratrol improves mitochondrial function and protects against metabolic disease by activating SIRT1 and PGC-1alpha. Cell 127:1109–1122

Langeveld M, Tan CY, Soeters MR, Virtue S, Watson LP, Murgatroyd PR et al (2017) No metabolic effects of mustard allyl-isothiocyanate compared with placebo in men. Am J Clin Nutr 106:1197–1205

Lee P, Swarbrick MM, Zhao JT, Ho KK (2011) Inducible brown adipogenesis of supraclavicular fat in adult humans. Endocrinology 152:3597–3602

Lee P, Smith S, Linderman J, Courville AB, Brychta RJ, Dieckmann W et al (2014) Temperature-acclimated brown adipose tissue modulates insulin sensitivity in humans. Diabetes 63:3686–3698

Lee SG, Parks JS, Kang HW (2017) Quercetin, a functional compound of onion peel, remodels white adipocytes to brown-like adipocytes. J Nutr Biochem 42:62–71

Li Y, Wongsiriroj N, Blaner WS (2014) The multifaceted nature of retinoid transport and metabolism. Hepatobiliary Surg Nutr 3:126–139

Lone J, Choi JH, Kim SW, Yun JW (2016) Curcumin induces brown fat-like phenotype in 3T3-L1 and primary white adipocytes. J Nutr Biochem 27:193–202

Lorenz M, Paul F, Moobed M, Baumann G, Zimmermann BF, Stangl K et al (2014) The activity of catechol-O-methyltransferase (COMT) is not impaired by high doses of epigallocatechin-3-gallate (EGCG) in vivo. Eur J Pharmacol 740:645–651

Ludy MJ, Moore GE, Mattes RD (2012) The effects of capsaicin and capsiate on energy balance: critical review and meta-analyses of studies in humans. Chem Senses 37:103–121

Ma S, Yu H, Zhao Z, Luo Z, Chen J, Ni Y et al (2012) Activation of the cold-sensing TRPM8 channel triggers UCP1-dependent thermogenesis and prevents obesity. J Mol Cell Biol 4:88–96

Matsushita M, Yoneshiro T, Aita S, Kameya T, Sugie H, Saito M (2014) Impact of brown adipose tissue on body fatness and glucose metabolism in healthy humans. Int J Obes (Lond) 38:812–817

Michlig S, Merlini JM, Beaumont M, Ledda M, Tavenard A, Mukherjee R et al (2016) Effects of TRP channel agonist ingestion on metabolism and autonomic nervous system in a randomized clinical trial of healthy subjects. Sci Rep 6:20795

Moon J, Do HJ, Kim OY, Shin MJ (2013) Antiobesity effects of quercetin-rich onion peel extract on the differentiation of 3T3-L1 preadipocytes and the adipogenesis in high fat-fed rats. Food Chem Toxicol 58:347–354

Morera E, De Petrocellis L, Morera L, Moriello AS, Nalli M, Di Marzo V et al (2012) Synthesis and biological evaluation of [6]-gingerol analogues as transient receptor potential channel TRPV1 and TRPA1 modulators. Bioorg Med Chem Lett 22:1674–1677

Moroshko I, Brennan L, O'Brien P (2011) Predictors of dropout in weight loss interventions: a systematic review of the literature. Obes Rev 12:912–934

Murholm M, Isidor MS, Basse AL, Winther S, Sørensen C, Skovgaard-Petersen J et al (2013) Retinoic acid has different effects on UCP1 expression in mouse and human adipocytes. BMC Cell Biol 14:41

Nakamura K (2011) Central circuitries for body temperature regulation and fever. Am J Physiol Regul Integr Comp Physiol 301:R1207–R1228

Nirengi S, Yoneshiro T, Sugie H, Saito M, Hamaoka T (2015) Human brown adipose tissue assessed by simple, noninvasive near-infrared time-resolved spectroscopy. Obesity (Silver Spring) 23:973–980

Nirengi S, Amagasa S, Homma T, Yoneshiro T, Matsumiya S, Kurosawa Y et al (2016a) Daily ingestion of catechin-rich beverage increases brown adipose tissue density and decreases extramyocellular lipids in healthy young women. Springerplus 5:1363

Nirengi S, Homma T, Inoue N, Sato H, Yoneshiro T, Matsushita M et al (2016b) Assessment of human brown adipose tissue density during daily ingestion of thermogenic capsinoids using near-infrared time-resolved spectroscopy. J Biomed Opt 21:091305

Nomura S, Ichinose T, Jinde M, Kawashima Y, Tachiyashiki K, Imaizumi K (2008) Tea catechins enhance the mRNA expression of uncoupling protein 1 in rat brown adipose tissue. J Nutr Biochemv 19:840–847

Ohno H, Shinoda K, Ohyama K, Sharp LZ, Kajimura S (2013) EHMT1 controls brown adipose cell fate and thermogenesis through the PRDM16 complex. Nature 504:163–167

Ohyama K, Nogusa Y, Shinoda K, Suzuki K, Bannai M, Kajimura S (2016) A synergistic antiobesity effect by a combination of capsinoids and cold temperature through promoting beige adipocyte biogenesis. Diabetes 65:1410–1423

Okamatsu-Ogura Y, Tsubota A, Ohyama K, Nogusa Y, Saito M, Kimura K (2015) Capsinoids suppress diet-induced obesity through uncoupling protein 1-dependent mechanism in mice. J Funct Food 19:1–9

Okla M, Kim J, Koehler K, Chung S (2017) Dietary factors promoting brown and beige fat development and thermogenesis. Adv Nutr 8:473–483

Ono K, Tsukamoto-Yasui M, Hara-Kimura Y, Inoue N, Nogusa Y, Okabe Y et al (2011) Intragastric administration of capsiate, a transient receptor potential channel agonist, triggers thermogenic sympathetic responses. J Appl Physiol (1985) 110:789–798

Oudart H, Groscolas R, Calgari C, Nibbelink M, Leray C, Le Maho Y et al (1997) Brown fat thermogenesis in rats fed high-fat diets enriched with n-3 polyunsaturated fatty acids. Int J Obes Relat Metab Disord 21:955–962

Rayalam S, Yang JY, Ambati S, Della-Fera MA, Baile CA (2008) Resveratrol induces apoptosis and inhibits adipogenesis in 3T3-L1 adipocytes. Phytother Res 22:1367–1371

Riera CE, Menozzi-Smarrito C, Affolter M, Michlig S, Munari C, Robert F et al (2009) Compounds from Sichuan and Melegueta peppers activate, covalently and non-covalently, TRPA1 and TRPV1 channels. Br J Pharmacol 157:1398–1409

Roberts LD, Ashmore T, Kotwica AO, Murfitt SA, Fernandez BO, Feelisch M et al (2015) Inorganic nitrate promotes the browning of white adipose tissue through the nitrate–nitrite–nitric oxide pathway. Diabetes 64:471–484

Rossato M, Granzotto M, Macchi V, Porzionato A, Petrelli L, Calcagno A et al (2014) Human white adipocytes express the cold receptor TRPM8 which activation induces UCP1 expression, mitochondrial activation and heat production. Mol Cell Endocrinol 383:137–146

Sae-Tan S, Rogers CJ, Lambert JD (2015) Decaffeinated green tea and voluntary exercise induce gene changes related to beige adipocyte formation in high fat-fed obese mice. J Funct Foods 14:210–214

Saito M, Okamatsu-Ogura Y, Matsushita M, Watanabe K, Yoneshiro T, Nio-Kobayashi J et al (2009) High incidence of metabolically active brown adipose tissue in healthy adult humans: effects of cold exposure and adiposity. Diabetes 58:1526–1531

Saito M, Yoneshiro T, Matsushita M (2016) Activation and recruitment of brown adipose tissue by cold exposure and food ingredients in humans. Best Pract Res Clin Endocrinol Metab 30:537–547

Serhan CN (2014) Pro-resolving lipid mediators are leads for resolution physiology. Nature 510:92–101

Shinoda K, Luijten IH, Hasegawa Y, Hong H, Sonne SB, Kim M et al (2015) Genetic and functional characterization of clonally derived adult human brown adipocytes. Nat Med 21:389–394

Sidossis LS, Porter C, Saraf MK, Børsheim E, Radhakrishnan RS, Chao T et al (2015) Browning of subcutaneous white adipose tissue in humans after severe adrenergic stress. Cell Metab 22:219–227

Snitker S, Fujishima Y, Shen H, Ott S, Pi-Sunyer X, Furuhata Y et al (2009) Effects of novel capsinoid treatment on fatness and energy metabolism in humans: possible pharmacogenetic implications. Am J Clin Nutr 89:45–50

Sugita J, Yoneshiro T, Hatano T, Aita S, Ikemoto T, Uchiwa H et al (2013) Grains of paradise (Aframomum melegueta) extract activates brown adipose tissue and increases whole-body energy expenditure in men. Br J Nutr 110:733–738

Sugita J, Yoneshiro T, Sugishima Y, Ikemoto T, Uchiwa H, Suzuki I et al (2014) Daily ingestion of grains of paradise (Aframomum melegueta) extract increases whole-body energy expenditure and decreases visceral fat in humans. J Nutr Sci Vitaminol (Tokyo) 60:22–27

Sun L, Camps SG, Goh HJ, Govindharajulu P, Schaefferkoetter JD, Townsend DW et al (2018) Capsinoids activate brown adipose tissue (BAT) with increased energy expenditure associated with subthreshold 18-fluorine fluorodeoxyglucose uptake in BAT-positive humans confirmed by positron emission tomography scan. Am J Clin Nutr 107:62–70

Tajino K, Matsumura K, Kosada K, Shibakusa T, Inoue K, Fushiki T et al (2007) Application of menthol to the skin of whole trunk in mice induces autonomic and behavioral heat-gain responses. Am J Physiol Regul Integr Comp Physiol 293:R2128–R2135

Takahashi M, Miyashita M, Suzuki K, Bae SR, Kim HK, Wakisaka T et al (2014) Acute ingestion of catechin–rich green tea improves postprandial glucose status and increases serum thioredoxin concentrations in postmenopausal women. Br J Nutr 112:1542–1550

Tamura Y, Iwasaki Y, Narukawa M, Watanabe T (2012) Ingestion of cinnamaldehyde, a TRPA1 agonist, reduces visceral fats in mice fed a high-fat and high-sucrose diet. J Nutr Sci Vitaminol (Tokyo) 58:9–13

Terada Y, Narukawa M, Watanabe T (2011) Specific hydroxy fatty acids in royal jelly activate TRPA1. J Agric Food Chem 59:2627–2635

Thavanesan N (2011) The putative effects of green tea on body fat: an evaluation of the evidence and a review of the potential mechanisms. Br J Nutr 106:1297e309

Tominaga M (2007) Nociception and TRP channels. Handb Exp Pharmacol 179:489–505

Uchida K, Dezaki K, Yoneshiro T, Watanabe T, Yamazaki J, Saito M et al (2017) Involvement of thermosensitive TRP channels in energy metabolism. J Physiol Sci 67:549–560

Valente A, Carrillo AE, Tzatzarakis MN, Vakonaki E, Tsatsakis AM, Kenny GP et al (2015) The absorption and metabolism of a single L-menthol oral versus skin administration: Effects on thermogenesis and metabolic rate. Food Chem Toxicol 86:262–273

van der Lans AA, Hoeks J, Brans B, Vijgen GH, Visser MG, Vosselman MJ et al (2013) Cold acclimation recruits human brown fat and increases nonshivering thermogenesis. J Clin Invest 123:3395–3403

van Marken Lichtenbelt WD, Vanhommerig JW, Smulders NM, Drossaerts JM, Kemerink GJ, Bouvy ND et al (2009) Cold-activated brown adipose tissue in healthy men. N Engl J Med 360:1500–1508

Virtanen KA, Lidell ME, Orava J, Heglind M, Westergren R, Niemi T et al (2009) Functional brown adipose tissue in healthy adults. N Engl J Med 360:1518–1525

von Essen G, Lindsund E, Cannon B, Nedergaard J (2017) Adaptive facultative diet-induced thermogenesis in wild-type but not in UCP1-ablated mice. Am J Physiol Endocrinol Metab 313:E515–E527

Vosselman MJ, Brans B, van der Lans AA, Wierts R, van Baak MA, Mottaghy FM et al (2013) Brown adipose tissue activity after a high-calorie meal in humans. Am J Clin Nutr 98:57–64

Wang S, Liang X, Yang Q, Fu X, Rogers CJ, Zhu M et al (2015) Resveratrol induces brown-like adipocyte formation in white fat through activation of AMP-activated protein kinase (AMPK) α1. Int J Obes (Lond) 39:967–976

Westerterp-Plantenga MS (2010) Green tea catechins, caffeine and body-weight regulation. Physiol Behav 100:42–46

Yang WS, Wang WY, Fan WY, Deng Q, Wang X (2014) Tea consumption and risk of type 2 diabetes: a meta-analysis of cohort studies. Br J Nutr 111:1329–1339

Yoneshiro T, Saito M (2013) Transient receptor potential activated brown fat thermogenesis as a target of food ingredients for obesity management. Curr Opin Clin Nutr Metab Care 16:625–631

Yoneshiro T, Saito M (2015) Activation and recruitment of brown adipose tissue as anti-obesity regimens in humans. Ann Med 47:133–141

Yoneshiro T, Aita S, Matsushita M, Kameya T, Nakada K, Kawai Y et al (2011a) Brown adipose tissue, whole-body energy expenditure, and thermogenesis in healthy adult men. Obesity (Silver Spring) 19:13–16

Yoneshiro T, Aita S, Matsushita M, Okamatsu-Ogura Y, Kameya T, Kawai Y et al (2011b) Age-related decrease in cold-activated brown adipose tissue and accumulation of body fat in healthy humans. Obesity (Silver Spring) 19:1755–1760

Yoneshiro T, Aita S, Kawai Y, Iwanaga T, Saito M (2012) Nonpungent capsaicin analogs (capsinoids) increase energy expenditure through the activation of brown adipose tissue in humans. Am J Clin Nutr 95:845–850

Yoneshiro T, Aita S, Matsushita M, Kayahara T, Kameya T, Kawai Y et al (2013) Recruited brown adipose tissue as an antiobesity agent in humans. J Clin Invest 123:3404–3408

Yoneshiro T, Matsushita M, Nakae S, Kameya T, Sugie H, Tanaka S et al (2016) Brown adipose tissue is involved in the seasonal variation of cold-induced thermogenesis in humans. Am J Physiol Regul Integr Comp Physiol. doi: https://doi.org/10.1152/ajpregu.00057.2015

Yoneshiro T, Kaede R, Nagaya K, Aoyama J, Saito M, Okamatsu-Ogura Y et al (2017a) Royal jelly ameliorates diet-induced obesity and glucose intolerance by promoting brown adipose tissue thermogenesis in mice. Obes Res Clin Pract 12:127–137

Yoneshiro T, Matsushita M, Hibi M, Tone H, Takeshita M, Yasunaga K et al (2017b) Tea catechin and caffeine activate brown adipose tissue and increase cold-induced thermogenic capacity in humans. Am J Clin Nutr 105:873–881

Yoneshiro T, Kaede R, Nagaya K, Saito M, Aoyama J, Elfeky M et al (2018) Melinjo (Gnetum gnemon L.) seed extract induces uncoupling protein 1 expression in brown fat and protects mice against diet-induced obesity, inflammation, and insulin resistance. Nutr Res 58:17–25

Yuan X, Wei G, You Y, Huang Y, Lee HJ, Dong M et al (2017) Rutin ameliorates obesity through brown fat activation. FASEB J 31:333–345

Zuo J, Zhao D, Yu N, Fang X, Mu Q, Ma Y et al (2017) Cinnamaldehyde ameliorates diet-induced obesity in mice by inducing browning of white adipose tissue. Cell Physiol Biochem 42:1514–1525

Translational Pharmacology and Physiology of Brown Adipose Tissue in Human Disease and Treatment

Christopher J. Larson

Contents

1 Introduction 383
2 Basic Biology 384
3 Energy Balance 385
4 Cellular Energetics 386
5 Physiology of Energetics and Adaptive Thermogenesis 387
6 BAT and Human Adaptive Thermogenesis 388
7 Human BAT and Adaptive Thermogenic Response to Cold 389
8 Human BAT and Transition from Physiology to Pharmacology 390
9 Human BAT and Adrenergic Pharmacology 391
10 Human BAT and Selective β3-Adrenergic Agonism 395
11 Human BAT and Other Molecules and Nutritional Approaches 401
12 Human BAT and Poorly Modeled Mechanisms 403
13 Future Directions 405
14 Conclusions 407
References 407

Abstract

Human brown adipose tissue (BAT) is experimentally modeled to better understand the biology of this important metabolic tissue, and also to enable the potential discovery and development of novel therapeutics for obesity and sequelae resulting from the persistent positive energy balance. This chapter focuses on translation into humans of findings and hypotheses generated in nonhuman models of BAT pharmacology. Given the demonstrated challenges of sustainably reducing caloric intake in modern humans, potential solutions to obesity likely lie in increasing energy expenditure. The energy-transforming

C. J. Larson (✉)
Development, Aging and Regeneration Program, Sanford Burnham Prebys Medical Discovery Institute, La Jolla, CA, USA
e-mail: clarson@SBPdiscovery.org

A. Pfeifer et al. (eds.), *Brown Adipose Tissue*,
Handbook of Experimental Pharmacology 251, https://doi.org/10.1007/164_2018_184

activities of a single cell in any given tissue can be conceptualized as a flow of chemical energy from energy-rich substrate molecules into energy-expending, endergonic biological work processes through oxidative degradation of organic molecules ingested as nutrients. Despite the relatively tight coupling between metabolic reactions and products, some expended energy is incidentally lost as heat, and in this manner a significant fraction of the energy originally captured from the environment nonproductively transforms into heat rather than into biological work. In human and other mammalian cells, some processes are even completely uncoupled, and therefore purely energy consuming. These molecular and cellular actions sum up at the physiological level to adaptive thermogenesis, the endogenous physiology in which energy is nonproductively released as heat through uncoupling of mitochondria in brown fat and potentially skeletal muscle. Adaptive thermogenesis in mammals occurs in three forms, mostly in skeletal muscle and brown fat: shivering thermogenesis in skeletal muscle, non-shivering thermogenesis in brown fat, and diet-induced thermogenesis in brown fat. At the cellular level, the greatest energy transformations in humans and other eukaryotes occur in the mitochondria, where creating energetic inefficiency by uncoupling the conversion of energy-rich substrate molecules into ATP usable by all three major forms of biological work occurs by two primary means. Basal uncoupling occurs as a passive, general, nonspecific leak down the proton concentration gradient across the membrane in all mitochondria in the human body, a gradient driving a key step in ATP synthesis. Inducible uncoupling, which is the active conduction of protons across gradients through processes catalyzed by proteins, occurs only in select cell types including BAT. Experiments in rodents revealed UCP1 as the primary mammalian molecule accounting for the regulated, inducible uncoupling of BAT, and responsive to both cold and pharmacological stimulation. Cold stimulation of BAT has convincingly translated into humans, and older clinical observations with nonselective 2,4-DNP validate that human BAT's participation in pharmacologically mediated, though nonselective, mitochondrial membrane decoupling can provide increased energy expenditure and corresponding body weight loss. In recent times, however, neither beta-adrenergic antagonism nor unselective sympathomimetic agonism by ephedrine and sibutramine provide convincing evidence that more BAT-selective mechanisms can impact energy balance and subsequently body weight. Although BAT activity correlates with leanness, hypothesis-driven selective β3-adrenergic agonism to activate BAT in humans has only provided robust proof of pharmacologic activation of β-adrenergic receptor signaling, limited proof of the mechanism of increased adaptive thermogenesis, and no convincing evidence that body weight loss through negative energy balance upon BAT activation can be accomplished outside of rodents. None of the five demonstrably β3 selective molecules with sufficient clinical experience to merit review provided significant weight loss in clinical trials (BRL 26830A, TAK 677, L-796568, CL 316,243, and BRL 35135). Broader conclusions regarding the human BAT therapeutic hypothesis are limited by the absence of data from most studies demonstrating specific activation of BAT thermogenesis in most studies. Additionally, more limited data sets with older or less selective β3 agonists also did not provide strong evidence of body weight effects. Encouragingly, β3-adrenergic agonists, catechins, capsinoids, and nutritional extracts, even without robust negative energy balance outcomes, all demonstrated increased total energy expenditure that in some cases could be associated with concomitant activation of BAT, though the absence of body weight loss indicates that in no cases did the magnitude of negative energy balance reach sufficient levels. Glucocorticoid receptor agonists, PPARg agonists, and thyroid hormone receptor agonists all possess defined molecular and cellular pharmacology that preclinical models predicted to be efficacious for negative energy balance and body weight loss, yet their effects on human BAT thermogenesis upon translation were inconsistent with predictions and disappointing. A few new mechanisms are nearing the stage of clinical trials and may yet provide a more quantitatively robust translation from preclinical to human experience with BAT. In conclusion, translation into humans has been demonstrated with BAT molecular pharmacology and cell biology, as well as with physiological response to cold. However, despite pharmacologically mediated, statistically significant elevation in total energy expenditure, translation into biologically meaningful negative energy balance was not achieved, as indicated by the absence of measurable loss of body weight over the duration of a clinical study.

Keywords
Adaptive thermogenesis · Adrenergic · Energetics · Pharmacology · Physiology

1 Introduction

This volume reviews the origins of brown and beige adipocytes, the molecular mechanisms of BAT function, in vivo detection of BAT, and recruitment and activation of BAT. The current chapter will focus on the field's experience with the translation into humans of findings and hypotheses generated in nonhuman models of BAT pharmacology. Much basic research into BAT biology and pharmacology implicitly aims to model human biology, rather than to more deeply understand the BAT biology of other species for their own intrinsic purposes. Indeed, in most domains of the BAT field, the purpose is explicitly medical pharmacology, where pharmacology is defined as the study of substances that interact with living systems through chemical processes, particularly by binding to regulatory molecules and activating or inhibiting normal bodily processes, and medical pharmacology is defined as the interventional science of such substances used to prevent, diagnose, or treat disease.

A primary objective of medical pharmacology for BAT modulation is the treatment of obesity, which derives from a persistent positive energy balance of calories taken in over calories expended and results in numerous metabolic sequelae. The ideal anti-obesity drug should result in sustained weight loss with few side effects (Rodgers et al. 2012). The earliest anti-obesity drugs clinically tested established that pharmacological intervention to generate negative energy balance is possible. The pancreatic lipase inhibitor orlistat has been available globally for many years, but is not widely used both because of limited efficacy and because of occasional, significant tolerability issues. Other medicines approved in some global regions include the 5HT2C agonist lorcaserin, the combination of phentermine and topiramate, and the combination of naltrexone and bupropion (Arch 2015; Clapham and Arch 2007). Most recently, efforts have been made to approximate the anti-obesity benefits of bariatric surgery by pharmacologically modulating or mimicking incretins and other gut hormones. Improved understanding of hunger and satiety signaling from the gut by cholecystokinin (CCK), peptide YY (PYY), ghrelin, and glucagon-like peptide-1 (GLP-1), and regulatory mechanisms linked to both GLP-1 and to leptin and related signaling in the hypothalamus, offered additional opportunities for intervention. Although some molecules associated with these mechanisms have entered clinical development, it is unclear whether they possess sufficient magnitude and duration of efficacy, or sufficient safety and tolerability, for chronic use. The most recent example of a weight loss agent approved in some regions of the world that represents a clear exception to this skepticism is GLP-1 receptor agonism, which has succeeded broadly in diabetes treatment and to a lesser extent at body weight-lowering action in humans. Numerous mechanisms and molecules that target pathways in metabolic tissues, such as the adipocytes, liver, and skeletal muscle, have shown potential in preclinical studies, but few have reached clinical development, and none has shown efficacy.

This chapter will review only those pharmacological mechanisms for which sufficient human experience exists to enable discussion. The majority of discussion will focus on modulation of adrenergic pharmacology for physiological benefit, though experience with catechins, caffeine, capsinoids, and nutritional extracts will also be discussed. A number of mechanisms that behaved in an inconsistent manner in humans versus nonclinical models, including PPAR, thyroid hormone, and glucocorticoid receptor ligands, will also be discussed. The chapter will end with a brief discussion of the emerging translational experiences of cGMP, adenosine, and FGF21.

2 Basic Biology

To address translation of preclinical pharmacology research of metabolism and obesity, it may help to start with bioenergetics or energy transformations in living organisms. The three overarching classes of biological energy transformation are photosynthesis, respiration, and performance of biological work. There are three types of biological work conducted by living organisms: the chemical work of synthesis and degradation, concentration work via transport, and mechanical work (Lehninger 1971). Each time a constituent chemical or physical process occurs during various types of biology work, physical inefficiencies result in an incomplete conversion of one kind of energy into another. In this manner, a significant fraction of the energy originally captured from the environment transforms into heat rather than into biological work. The energy-transforming activities of a cell can be conceptualized as a flow of chemical energy from energy-rich substrate molecules into energy-requiring, endergonic biological work processes through oxidative degradation of organic molecules ingested as nutrients. The original chemical energy of ingested organic molecule nutrients is conserved as chemical energy inside cells through reduction-oxidation transformation of such electron donors into the compound adenosine triphosphate (ATP), or else the ingested chemical energy would be nonproductively lost as heat. Mitochondria are the cellular sites where enzymes catalyze oxidation of nutrient organic molecules by molecular oxygen, and this oxidative chemical energy is captured as ATP.

3 Energy Balance

Obesity at a simple level is an arithmetic pathology of positive energy balance created by a persistent excess of caloric intake over caloric expenditure and egress (Tseng et al. 2010; Rolfe and Brown 1997; Lowell and Spiegelman 2000). The most obvious remedy of this imbalance would be reduction or elimination of any excess caloric intake while holding caloric expenditure and egress unaltered. Such thinking is supported by the observation that individuals experiencing forced starvation universally lose body mass, starting with fat mass. However, both theoretical and pragmatic challenges emerge. As demonstrated by clinical experience with the endocannabinoid system, satiety centers can be complexly intermingled with other centers and circuits in the brain, causing significant safety issues (Padwal and Majumdar 2007). Moreover, the human body appears to possess homeostasis mechanisms that engage upon weight loss to produce increase caloric efficiency, causing a reduced basal metabolic rate, which opposes further weight loss (Redman et al. 2009; Leibel et al. 1995).

If reducing caloric intake is unlikely to succeed for the prior reasons, then increasing energy expenditure, either productively through increased muscular mechanical work (exercise) or nonproductively though increased thermogenesis, may provide a viable alternative (Chadwick and Cardew 2008; Bouchard et al. 1990; Levine et al. 1999; Ravussin et al. 1988). In addition to the calories consumed to fuel the mechanical work of muscular exercise, a major benefit of exercise is to increase the resting metabolic rate and thus total energy expenditure by an amount greater than that resulting directly from exercise (Poehlman 1989). Thus medical pharmacology interventions that increase energy expenditure may not only mimic the transient increases in energy expenditure corresponding to bouts of muscular work but may potentially reset an overweight individual's set point for body weight to a lower range through permanently increasing their basal metabolic rate, raising the possibility that pharmacological intervention may not be chronic.

4 Cellular Energetics

Despite the relatively tight coupling between metabolic reactions and products, some energy is incidentally lost as heat. More interesting for the subject of this chapter, in human and other mammalian cells, some processes are completely uncoupled, or futile, and therefore purely energy consuming. As prior chapters of this volume have described, active brown adipose tissue exists in adult humans, suggesting this highly energy-consuming tissue may be an appropriate therapeutic target to increase basal energy expenditure through manipulation of cellular energetic processes such as uncoupling (Clapham and Arch 2007; Cypess et al. 2009; van Marken Lichtenbelt et al. 2009; Virtanen et al. 2009; Saito et al. 2009; Zingaretti et al. 2009; Harper et al. 2001; Wolfe et al. 1987, 1990; Klein and Wolfe 1990; Newsholme and Crabtree 1976; Randle et al. 1963; Mazzucotelli et al. 2007). The specific mechanisms of proton coupling and uncoupling in cell types, including brown adipose, have been described in previous chapters. The flow of electrons from organic substrate molecules to oxygen is coupled with a proton electrochemical gradient across the mitochondrial inner membrane to the synthesis of ATP from adenosine diphosphate (ADP) and inorganic phosphate in a process named oxidative phosphorylation. There are two primary means of proton uncoupling at the mitochondria. Baseline (basal) uncoupling occurs as a passive general nonspecific leak down the concentration gradient across the membrane in all mitochondria in the human body, and as would be predicted from its nonspecific nature, there are no regulatory mechanisms associated with this fairly fundamental energetic process. Inducible uncoupling, which is active conduction of protons across gradients through processes catalyzed by proteins, occurs only in select cell types – e.g., brown adipose cells – and is subject to increasingly understood regulatory mechanisms. Since mitochondrial proton cycling through uncoupling represents a significant proportion of basal or standard metabolic rate (Rolfe et al. 1994, 1999; Rolfe and Brand 1996), further uncoupling of mitochondria may further increase energy expenditure and thus offer a plausible medical pharmacology target for the treatment of obesity and its sequelae.

The discovery of UCP1 as the mediator of uncoupling originated from the finding of greater uncoupling of BAT mitochondria than of liver and heart mitochondria (Nicholls 1977, 1979), and numerous subsequent studies generated the conclusion that UCP1 is the primary molecule accounting for the regulated uncoupling of BAT (Enerbäck et al. 1997; Nicholls 1976, 1977; Heaton et al. 1978; Lin et al. 1980; Bathgate et al. 1992; Jezek et al. 1990; Katiyar and Shrago 1989; Jessen et al. 1980; Strieleman et al. 1985; Harper and Himms-Hagen 2001; Melnyk et al. 1997; Golozoubova et al. 2001; Kontani et al. 2005; Leonardsson et al. 2004; Nedergaard et al. 2001; Nicholls and Locke 1984). As would be predicted from its molecular function as a facilitator of mitochondrial membrane proton electrochemical gradient leak, UCP1 is capable of functioning outside of BAT if it were to be present there. Studies utilizing specific, engineered expression of UCP1 in either murine muscle (Li et al. 2000) or white adipose tissue (Kopecky et al. 1995) revealed amelioration of obesity and improved comorbid conditions, indicating increased uncoupling activities in these tissues (Li et al. 2000; Ricquier and Bouillaud 2000; Harper et al. 2002, 2008; Pecqueur et al. 2000).

5 Physiology of Energetics and Adaptive Thermogenesis

While currently available obesity therapies including bariatric surgery focus on reducing energy intake, findings reviewed earlier in this chapter and elsewhere suggest that increasing cellular energy expenditure may be a successful complementary approach (Tseng et al. 2010), with pharmacological agents that elevate metabolic rate through amplified mitochondrial uncoupling potentially mimicking an exercise effect on basal metabolic rate.

At the physiological level, this approach would seek to modulate adaptive thermogenesis, the endogenous physiology in which energy is nonproductively released as heat through uncoupling of mitochondria in brown fat and potentially skeletal muscle. Adaptive thermogenesis in mammals occurs in three forms, mostly in skeletal muscle and brown fat: shivering thermogenesis in skeletal muscle, non-shivering thermogenesis in brown fat, and diet-induced thermogenesis in brown fat, with different mechanisms active depending on the needs of the organism (Himms-Hagen 2004; Lean et al. 1986). Non-shivering thermogenesis appears to primarily occur in brown fat, although an absence of technologies distinguishing direct measurement of muscle non-shivering thermogenesis from other muscle thermogenic processes limits the strength of this conclusion (Stowell 2008; Wijers et al. 2008).

Experiments with rodents have revealed a critical role of the molecule UCP1 in BAT in causing both non-shivering and diet-induced thermogenesis. UCP1-deficient mice are more susceptible to cold temperatures and use shivering for thermoregulation (Enerbäck et al. 1997). At thermoneutrality (30°C for mice), UCP1-knockout mice lack diet-induced thermogenesis and develop obesity (Feldmann et al. 2009; Richard and Picard 2011), indicating that diet-induced thermogenesis is fully dependent on UCP1.

Diet-induced thermogenesis in humans was inferred in the early twentieth century as a physiological mechanism that must exist (Neumann 1902), and later diet-induced thermogenesis was discovered to be tightly linked with the recruitment of BAT by increased adrenergic activity (Rothwell and Stock 1983). The large differences in weight gain between individuals upon overfeeding have been proposed to be determined by differences in diet-induced thermogenesis, which therefore may participate in the onset and progression of obesity in humans (Rothwell and Stock 1979a) and in rodents (Feldmann et al. 2009; Rothwell and Stock 1979a; Nedergaard et al. 2010; Trayhurn et al. 1977, 1982).

Since adaptive thermogenesis nonproductively dissipates calories ingested in order to generate heat, it would be expected to contribute to positive, negative, and neutral energy balance, and various studies support this hypothesis (Clapham and Arch 2007; Bouchard et al. 1990; Levine et al. 1999; Maes et al. 1997; Joint 2001; Christiansen et al. 2005). To successfully modulate adaptive thermogenesis for the purposes of medical pharmacology, the process and its regulation would need to be understood. Both diet-induced thermogenesis and non-shivering thermogenesis processes happen in BAT, and both are controlled by the nervous system and by endocrine systems. Based on the evidence that cold exposure and diet increase sympathetic nerve activity (Landsberg et al. 1984), that exogenous administration of norepinephrine and epinephrine increases energy expenditure in both in vivo and in isolated tissue preparations, that BAT is highly vascularized and richly innervated by terminal fibers of the postganglionic neurons of the sympathetic nervous system (Cannon and Nedergaard 2004; Richard and Picard 2011; Doyon et al. 2006), and that thermogenic activity in BAT is dependent upon intact sympathetic stimulation (Himms-Hagen 1989), the sympathetic nervous system (SNS) is concluded to be the efferent pathway through which the nervous system regulates adaptive thermogenesis (Cannon and Nedergaard 2004; Hosaka et al. 2004; Nakamura and Morrison 2007; Morrison et al. 2008; Matsumoto et al. 2001; Wijers et al. 2007). Adaptive thermogenesis is also modulated by endocrine systems and hormones, including thyroid hormone (Golozoubova et al. 2004; Silva 2006) leptin and adenosine at least in rodents (Mistry et al. 1997; Commins et al. 1999; Elmquist et al. 1998; Harris 2000; Rosenbaum et al. 2005; Ruan et al. 2018, pp. 476–489; Gnad et al. 2014, pp. 395–399), insulin (Ferrannini et al. 1999), and glucagon, adrenaline, and glucocorticoids, though these three may function permissively to increase substrate availability to BAT rather than to exert regulatory control (Silva 2006).

6 BAT and Human Adaptive Thermogenesis

BAT is clearly critical for the thermogenic response and energy balance in small mammals (Ghorbani et al. 1997; Guerra et al. 1998; Lowell et al. 1993), but the translational relevance of this body of work is unclear for the larger mammal human. The thermogenic capacity of BAT in rodents is large relative to total body weight: a cold-acclimated rat weighing 350–400 g possesses 3 g of BAT consuming oxygen at a rate approximately twice the BMR (Foster and Frydman 1979). Humans have 200× greater body mass than rats, only 10× greater BAT mass, and 7× lower BMR per gram of body weight (Cypess et al. 2009; Foster and Frydman 1979; Mifflin et al. 1990), implying that in humans as little as 50 g of BAT (less than 0.1% of body weight) could utilize up to 20% of basal calorific needs if maximally stimulated (Rothwell and Stock 1983).

The existence of physiologically relevant human BAT has solidified after a period of controversy. The earlier phase of BAT research exploited multiple lines of histological data (Virtanen et al. 2009; Atit et al. 2006; Timmons et al. 2007; Seale et al. 2008; Kajimura et al. 2009; Waldén 2010) to induce that BAT is present in adult humans throughout life (Heaton 1972). However, concomitant efforts to confirm its activity (Astrup 1986) or exploit its presumed thermogenic capacity (Weyer et al. 1998; Larsen et al. 2002) yielded mixed results, which reasonably yielded the impression that no functional BAT existed in normal adult humans (Cunningham et al. 1985; Nedergaard et al. 2007). Combined positron emission tomography (PET) and computed tomography (CT) imaging work has provided additional support for the existence of functional adult human BAT (Cypess et al. 2009; Virtanen et al. 2009; Saito et al. 2009; Zingaretti et al. 2009; Cunningham et al. 1985; Nedergaard et al. 2007; Schöder et al. 2004; Hany et al. 2002; Cohade et al. 2003a), suggested an inverse correlation between adiposity and active BAT in adult humans (Cypess et al. 2009; van Marken Lichtenbelt et al. 2009; Virtanen et al. 2009; Saito et al. 2009; Zingaretti et al. 2009; Lean et al. 1986; Vijgen et al. 2011; Lee et al. 2011a). Moreover, imaging studies support BAT contribution to energy expenditure (Cypess et al. 2009; Virtanen et al. 2009; Saito et al. 2009; Zingaretti et al. 2009; Richard and Picard 2011; Hany et al. 2002; Ouellet et al. 2011; Kim et al. 2008; Garcia et al. 2006; Cohade et al. 2003b; Cannon and Nedergaard 2010; Enerbäck 2010), though earlier estimates of prevalence and magnitude based on retrospective studies (Cypess et al. 2009; van Marken Lichtenbelt et al. 2009; Saito et al. 2009; Hany et al. 2002; Cohade et al. 2003b) were updated with data from more recent, prospective studies (Cypess et al. 2009; van Marken Lichtenbelt et al. 2009; Virtanen et al. 2009; Saito et al. 2009; Zingaretti et al. 2009; Yoneshiro et al. 2011a) implying a relationship between decreased BAT activity and age-related obesity.

7 Human BAT and Adaptive Thermogenic Response to Cold

The confirmation of image-active BAT in adult humans suggests it may contribute to normal adult human physiology (Nedergaard et al. 2007; Ravussin and Kozak 2009), with human BAT demonstrated to respond detectably to cold, even in adult populations in whom functional BAT was considered absent. However, the transition from a qualitative to a quantitative conclusion is less robust, with the biomedical literature yielding numbers for not only detectable but also physiologically relevant prevalence of adult human BAT from 2% to close to 100% of the population. This range includes earlier retrospective studies (Inokuma et al. 2005; Marette and Bukowiecki 1991; Vallerand et al. 1987; Lee et al. 2010a; Rousseau et al. 2006) that documented activation of BAT upon cold exposure (van Marken Lichtenbelt et al. 2009; Virtanen et al. 2009; Saito et al. 2009) and increased energy expenditure in humans upon cold exposure (Wijers et al. 2007; van Marken Lichtenbelt et al. 2002; Dauncey 1981) and later studies that established relationships between cold-induced thermogenesis (Wijers et al. 2010; Ooijen et al. 2006; Contaldo et al. 1986) and detection of BAT (Cypess et al. 2009; van Marken Lichtenbelt et al. 2009) and body weight. Additional studies demonstrated BAT adaptive thermogenic differences between lean (Wijers et al. 2010) and obese (Wijers et al. 2010; Ooijen et al. 2006) subjects, with calculated differences between BAT adaptive thermogenic responses of roughly 4× between healthy lean and healthy overweight men (van Marken Lichtenbelt et al. 2009), though it is currently not possible to exclude a contribution from skeletal muscle (Tseng et al. 2010; Wijers et al. 2008). More recent cold exposure studies have included broader metabolic measures beyond the thermogenic response, have derived from multiple independent laboratories using a variety of intermittent cold exposure protocols (Saito et al. 2009; van der Lans et al. 2013; Blondin et al. 2014; Yoneshiro et al. 2013; Iwen et al. 2017; Ouellet et al. 2012; Peterson et al. 2016), and yield prevalence numbers in the high double digits. Overall, functionally avid BAT can be detected in the majority of adult humans over a large number of studies (van Marken Lichtenbelt et al. 2009; Virtanen et al. 2009; Saito et al. 2009; van der Lans et al. 2013; Yoneshiro et al. 2011b, 2013; Ouellet et al. 2012; Orava et al. 2011; Nakamura 2011).

8 Human BAT and Transition from Physiology to Pharmacology

As described in the previous section, the consensus over a number of studies to date is that cold-stimulated human BAT adaptive thermogenesis occurs, and moreover this activity is common in the adult human population, with a true prevalence of 30–100% depending on the sample cohorts examined in more recent, prospective studies. Thus we can conclude that cold-challenge *physiology* in rodents has convincingly translated into humans. The next question is: Do the *pharmacology* findings in rodents translate into humans? And is human BAT usefully available as a target of medical pharmacology?

Decreasing ambient temperature in human dwellings, or planned cold exposure, should activate cold-induced thermogenesis by increasing BAT activity, which might decrease body fat (Yoneshiro and Saito 2015). However, while health benefits may accrue from repeated cold exposure (van Marken Lichtenbelt et al. 2014), significant, prolonged cold exposure seems inconsistent with patient compliance. Moreover, the centrally derived signals that increase BAT activity and function upon cold exposure also are known to generate increased appetite and food intake in an attempt to maintain energy balance, and such cold-induced hyperphagia would counteract any increase energy expenditure via cold-induced thermogenesis (Ravussin et al. 2014; sun Yoo et al. 2014; Rothwell and Stock 1979b), making it unclear that correctly executed cold exposure would result in persistently negative energy balance in overweight humans absent enforced caloric restriction.

Pharmacological agents that increase metabolic rate by promoting uncoupling of mitochondrial proton gradients from ATP synthesis may provide an alternative route to achieving transient negative energy balance and a persistently lowered body weight. Pharmacological agents increasing uncoupled mitochondrial proton cycling in BAT without inducing central hyperphagic signals would be attractive (Carey and Kingwell 2013). Proof of concept in humans that mitochondrial uncoupling can elevate energy expenditure without triggering compensatory increased food intake that completely counteracts the increased energy expenditure has been established: the mitochondrial uncoupler 2,4-dinitrophenol (DNP) generated meaningful weight loss in humans in the mid-twentieth century (Parascandola 1974; Tainter et al. 1934). Thermogenic effects of DNP derivatives were reported as early as 1885, when scientists observed thermogenic effects of martius yellow (a dinitro-alpha-napthol), a substance used in the nineteenth century to give the impression that food was rich in eggs (Cazeneuve and Lépine 1885). French munitions workers during World War I, regularly exposed to a blend of 40% 2,4-dinitrophenol and 60% trinitrophenol for their munitions, reported body weight loss effects. DNP was introduced as a drug in the 1930s and demonstrated benefits on body weight, though cataracts and deaths from overdose caused the nascent FDA to ban its usage in 1938.

2,4-DNP crosses mitochondrial membranes in its protonated state, deprotonates and crosses back as the base, and then reprotonates to the acid and repeats, thereby uncoupling the mitochondrial membrane by increasing basal proton leak via this shuttling mechanism. 2,4-DNP stimulates cellular respiration and a rise in body temperature in rodents, and death at proximally higher doses (Magne et al. 1932; Cutting et al. 1933). Despite this narrow therapeutic index and history of deaths in humans, in recent years a series of controlled trials in obese patients were conducted, with efficacy equaling or surpassing that of current treatments (Sjöström et al. 1998; James et al. 2000; Bray et al. 1999). Importantly, unlike the older use of thyroid extract for obesity, 2,4-DNP in these trials did not result in nitrogen excretion, suggesting that the observed weight loss spared muscle and derived from a specific loss of fat (Cutting and Tainter 1933).

Ultimately, DNP's narrow therapeutic index between anti-obesity efficacy and risk of sudden death, as well as cataracts and other side effects related to its nonspecific uncoupling action, indicates that DNP is not a credible anti-obesity drug. Deploying pharmacological agents to uncouple all mitochondria throughout the body may be a high-risk treatment, particularly risking energy homeostasis in critical tissues such as the heart and brain, though highly respiring tissues such as the heart and brain may be less susceptible to mild uncoupling than less active ones like resting muscle or resting BAT since proton conductance has much less control over respiration rate in active mitochondria (Hafner et al. 1990). However, 2,4-DNP's historical and recent clinical efficacy generates the hypothesis that a related mechanism that is tissue selective may be sufficiently safe to uncouple mitochondria particularly in BAT, thereby increasing thermogenesis and basal metabolic rate, thereby causing body weight loss by liberating as fuel fat from adipose depots. In support of this, beyond the proof of concept in humans using 2,4-dinitrophenol (DNP) (Parascandola 1974), activating BAT UCP1 in animals (Weyer et al. 1999a) and overexpressing of UCP3 (uncoupling protein 3) in muscle (Clapham et al. 2000) show that selective uncoupling of brown adipose tissue mitochondria through UCP1 or of muscle mitochondria through UCP3 may be a viable approach to the development of drugs for the treatment of obesity.

9 Human BAT and Adrenergic Pharmacology

Adult humans, including overweight and obese, possess detectable BAT that is an active participant in normal adaptive thermogenesis physiology, including cold (Cypess et al. 2009; van Marken Lichtenbelt et al. 2009; Virtanen et al. 2009; Saito et al. 2009; Zingaretti et al. 2009; Nedergaard et al. 2007). As in rodents, UCP1 in human BAT presents as the ideal target for a strategy of BAT-specific promotion of nonproductive energy expenditure through increased mitochondrial uncoupling, and a large body of data supports intervention in the adrenergic system as the best approach to cause such action.

Adrenergic nervous transmission occurs through the actions of endogenous molecules norepinephrine (NE), the primary transmitter of most sympathetic postganglionic fibers and some tracts in the CNS; dopamine (DA), the major transmitter of the mammalian extrapyramidal system and of several mesocortical and mesolimbic neuronal pathways; and epinephrine, the predominant hormone of the adrenal medulla. Together, these three amines are called catecholamines, which bind to α and β classes of adrenergic receptors to generate excitatory or inhibitory responses, depending on the tissue (Brunton et al. 2011; Giese 1996). β receptors are further classified into three subtypes: β1 (myocardium), β2 (smooth muscle and most other sites), and β3 (adipose). Epinephrine and norepinephrine are equipotent on β1 in myocardium, but epinephrine is 10–50× more potent than norepinephrine on β2 smooth muscle and elsewhere, and β1 and β2 receptor selective antagonists were developed and are commonly used clinical agents. The β3 adrenergic receptor exists in both humans and rodents and is 10× more sensitive to norepinephrine than to epinephrine, with older antagonists displaying reduced potency. While β3 adrenergic receptors are primarily expressed in adipose tissue (Grujic et al. 1997, pp. 17686–17693; Krief et al. 1993, pp. 344–349), all three β adrenergic receptors can be detected in both white adipose tissue and brown adipose tissue (Krief et al. 1993, pp. 344–349).

In rodents, norepinephrine induces thermogenesis via the three known beta-adrenergic receptors, and an overfeeding challenge to mice lacking beta-adrenergic receptors results in obesity that does not occur in wild-type mice similarly challenged (Bachman et al. 2002). Associations of polymorphisms in the β3 receptor gene to risk of obesity or type 2 diabetes in certain populations (Arner and Hoffstedt 1999; Hoffstedt et al. 1999) lead to the hypothesis that β3 receptor – selective agonists – may be useful for the treatment of such conditions (Weyer et al. 1999b). Moreover, rodents treated with β3 receptor agonists display a robust thermogenic response, including lipolytic activity (Robidoux et al. 2004). Additional beta-3 selective ligands displayed significant anti-obesity actions in mice and rats (Arch et al. 1984a; Strosberg and Pietri-Rouxel 1996).

Blockade of the adrenergic system with nonselective antagonists has substantial clinical experience, but no evidence exists in the biomedical literature for antagonism of the beta-adrenergic system in human BAT contributing to energy balance and overweight. The nonselective beta-adrenergic antagonist propranolol inhibits FDG uptake by human BAT (George et al. 2017; Söderlund et al. 2007; Uslu et al. 2015; Parysow et al. 2007; Agrawal et al. 2009; Tatsumi et al. 2004), supporting the belief that human BAT can be pharmacologically modulated by beta-adrenergic ligands. Although the adrenergic system underlies the cold-adaptive thermogenic process mediated by UCP1 in BAT, Wijers et al. (2011) observed that beta-adrenergic receptor blockade does not inhibit cold-induced thermogenesis in humans, though a potentially compensatory rise in endogenous catecholamines may have confounded this particular study. As well, publications exist describing an association between positive energy balance and chronic exposure to pharmacological antagonism of the beta-adrenergic system (Leslie et al. 2007). Therefore, as there is no direct study linking beta-adrenergic antagonism in human BAT to overweight, on its own this limited set of conflicting studies permits no clear conclusions regarding pharmacological modulation of beta-adrenergic signaling in human BAT for the purposes of energy balance.

Is there evidence from agonistic pharmacology of the adrenergic system in human BAT contributing to energy balance and weight loss? Because of the critical role of the adrenergic system for sympathetic nervous system function, activators of adrenergic signaling are termed sympathomimetics. Catecholamines and sympathomimetic drugs can be typed as direct-acting, indirect-acting, or mixed-acting sympathomimetics. Direct-acting sympathomimetic drugs operate directly on one or more of the adrenergic receptors. They may have substantial selectivity for a specific receptor subtype, or little to no selectivity and activate multiple receptor types. Indirect-acting drugs increase the levels of norepinephrine (NE) or epinephrine available to activate adrenergic receptors. Mechanisms of indirect activation include displacing NE from sympathetic nerve varicosities, inhibiting transport of NE into sympathetic neurons, or inhibiting the metabolism of endogenous catecholamines. Agents indirectly release NE, and also directly activated receptors are named mixed-acting sympathomimetic drugs.

Ephedrine is mixed-acting sympathomimetic that is an agonist at both α and β receptors while simultaneously promoting release of another agonist – norepinephrine – from sympathetic neurons. Prior chapters in this volume and other reviews (Cannon and Nedergaard 2004, pp. 277–359) describe how β adrenergic receptor agonism activates BAT by promoting cAMP/PKA/hormone sensitive lipase action to release fatty acids, which activates UCP1 resulting in the release of heat at BAT thermogenesis.

Unlike physiological BAT activation upon cold exposure, ephedrine increases norepinephrine peripherally in postsynaptic peripheral nerve terminals and does not act centrally to stimulate food intake (Dulloo et al. 1991), making it an attractive if imperfect surrogate for adrenergic receptor agonism effects in BAT on energy balance. Despite this profile, the effects of ephedrine on human BAT biology, energy expenditure, energy balance, and overweight remain mixed. A meta-analysis of several dozen trials concluded that ephedrine stimulated short-term weight loss, though long-term data are lacking from this set of studies, and could only be collected under carefully designed and executed protocols due to ephedrine's combined increase in sympathetic activation and decrease in parasympathetic activation in the heart (Shekelle et al. 2003). Proof of concept could be considered to arise from reports that ephedrine can activate BAT in humans, though less in obese than lean individuals (Carey et al. 2013), to contribute to BAT-mediated thermogenesis in some studies (Astrup et al. 1985a) but not in others (Cypess et al. 2012; Vosselman et al. 2012). Repeat administration of ephedrine enhances thermogenesis in overweight (Astrup et al. 1985b), but both body weight loss effects (Malchow-Møller et al. 1981) or no effects on body weight despite side effects evidencing pharmacology have been observed (Pasquali et al. 1985).

Moreover, some effects were observed that contradict expectations. In one study, chronic ephedrine reduced BAT glucose disposal, implying that chronic ephedrine treatment decreased, rather than increased, BAT adaptive thermogenesis activity (Carey et al. 2015).

The combination of ephedrine and caffeine generates significant weight loss over short periods of time (Boozer et al. 2002), which has been proposed to be mediated by β3 adrenergic receptors (de Matteis et al. 2002). Though caffeine is prevalent in human populations (Magkos and Kavouras 2005), its mechanism of action is generally accepted to be inhibition of adenosine A2A receptors (Huang et al. 2005), and it has recently been demonstrated ex vivo in mouse and human adipocytes that adenosine A2A agonism increases BAT activation (Gnad et al. 2014, pp. 395–399). Thus it is unclear if caffeine would sufficiently stimulate the SNS to activate BAT (Dhar et al. 2005) despite opposing effects through adenosine A2A receptor in BAT.

Overall, human clinical experience with ephedrine is less clearly supportive than that in rodents. BAT activation in rodents and increased energy expenditure when challenged with sympathomimetic drugs are quantitatively much larger than in humans. As well, BAT activation generally is observed in all study participants in rodents, but human studies display a significant number of nonresponders, with the proportion rising with obesity. For example, in human studies cold exposure, ephedrine ingestion, and noradrenaline infusion elevate energy expenditure 5–30% (van Marken Lichtenbelt et al. 2009; Jessen et al. 1980; Astrup et al. 1985a; Sjostrom et al. 1983; Shannon et al. 1999; Joy 1963), with BAT calculated to provide 1–5% of the change in energy expenditure after cold (van Marken Lichtenbelt and Schrauwen 2011; Muzik et al. 2012). In rodents, however, cold exposure or noradrenaline administration elevated energy expenditure by 100% (Golozoubova et al. 2006), and priming rodents with chronic adaptive cold thermogenesis further increased rodent noradrenaline-stimulated energy expenditure approximately by an additional 100% (Golozoubova et al. 2006), but in humans the additional increase was 15% (Joy 1963). Finally, sympathomimetic-induced BAT activity appears to be more impaired in obese humans than in obese rodents, with the clearest evidence arising from cold exposure studies, with lower proportions of obese individuals exhibiting BAT activation (van Marken Lichtenbelt et al. 2009; Saito et al. 2009; Vijgen et al. 2011).

A second, partial sympathomimetic with human clinical experience is sibutramine, a monoamine reuptake inhibitor that inhibits reuptake of norepinephrine by ~73%, serotonin by ~54%, and dopamine by ~16%, thus increasing the levels of these substances in synaptic clefts. Sibutramine promotes energy expenditure by increasing thermogenesis in BAT (Luque and Rey 2002; Pagotto et al. 2008), with minimal weight loss of 4–5 kg resulting from a dual action of not only increased thermogenesis but also decreased food intake were observed in five studies (James et al. 2010). And as would be both predicted from known actions of its adrenergic receptors and by analogy with the observed clinical experience of ephedrine, sibutramine caused detectable elevations in pulse rate and blood pressure, rendering it inappropriate for use as a weight loss agent.

Catechins are indirect-acting sympathomimetics that increase the levels of norepinephrine (NE) or epinephrine available to activate adrenergic receptors by inhibiting catechol-O-methyl transferase, which degrades catecholamines including norepinephrine. As loss of norepinephrine and epinephrine is reduced, the resulting increased stimulation of adrenergic receptors promotes energy expenditure and fat oxidation. A combination of catechins and caffeine has been shown to activate brown adipose tissue and increase cold-induced thermogenic capacity in humans (Yoneshiro et al. 2017). Acute increases in energy expenditure associated with increased BAT activity were observed under one dosing protocol, and under a second chronic dosing protocol, cold-induced thermogenesis was promoted, suggesting recruitment of BAT.

In conclusion, the overall clinical experience of nonselective sympathomimetic adrenergic activation with ephedrine, sibutramine, and catechins provides qualitative proof of concept that negative energy balance sufficient to cause detectable weight loss can be achieved by presumably activating BAT adaptive thermogenesis and energy expenditure, though they simultaneously raise the possibility that the quantitative horsepower of this physiology in humans may be substantially less than in rodents.

10 Human BAT and Selective β3-Adrenergic Agonism

Although BAT activity correlates with leanness (Rothwell and Stock 1979a; Himms-Hagen 1979), selective β3-adrenergic agonism to activate BAT in humans has essentially only provided robust proof of pharmacologic activation of β adrenergic receptor signaling, limited proof of the mechanism of increased adaptive thermogenesis, and no convincing evidence that body weight loss through negative energy balance can be accomplished outside of rodents.

There has been considerable interest and effort in the drug discovery field in creating specific β3-adrenoceptor agonists that would selectively lead to the activation of uncoupling through UCP1 in BAT. The ability to selectively uncouple would avoid many of the side effects that might occur with DNP, and selectively promoting lipolysis in both white and brown fat leading to a subsequent activation of UCP1-mediated lipolysis in brown fat should promote a number of medically desirable consequences (Ricquier and Bouillaud 2000; Strosberg and Pietri-Rouxel 1996; Nicholls and Rial 1999). The β3-adrenergic receptor in rodents is essentially restricted to brown adipocytes, and many reports in rodents have described β3-adrenoceptor agonists generating significant improvements in insulin sensitivity while causing large weight loss due to selective fat loss (Harper et al. 2008; Weyer et al. 1999b; Arch and Wilson 1996). The body of work in rodents generated a compelling preclinical therapeutic hypothesis that promoting mitochondrial uncoupling selectively in brown fat could safely generate biologically effective nonproductive energy consumption, whole body negative energy balance, and corresponding weight loss.

The human experience to date with selective β3-adrenergic agonism to activate BAT, while qualitatively consistent with the experience in rodents in terms of mechanisms, has quantitatively disappointed in terms of weight loss (Clapham and Arch 2007), with a number of factors contributing to the overall lack of success. Generating β3-adrenoceptor agonists lacking activity against β1- and β2-adrenoceptors has proven challenging (Weyer et al. 1999b; Arch and Wilson 1996). While this technical problem may be soluble over time, relatively low levels of brown fat in adult humans compared to rodents may represent an insurmountable intrinsic refutation to this hypothesis and confound interpretation of clinical trials reported thus far. There are theoretical escapes from this essential obstacle, such as reactivating unidentified and dormant brown fat or converting plentiful white to brown adipose that would allow a future suitably selective agent to demonstrate safe efficacy, but robust evidence in humans for these scenarios is currently lacking. Later β3-adrenergic receptor agonists with improved selectivity had low oral bioavailability, unfavorable pharmacokinetics, and overall poor drug-like properties (Arch 2008). Complicating evaluation of the therapeutic hypothesis, several of these trials were performed without the use of PET and CT scanning for measuring adult human BAT function and mass, so for some of these integrated, overall failures, it is unclear at which specific step the experimental hypothesis fails. With the current ability to quantify human BAT activity, testing β3 adrenergic receptor agonists for BAT activation, thermogenesis, and weight loss could be conducted more scientifically. An ideal study would deliver a sufficiently potent and selective agonist to the target in human BAT at high, sustained concentrations, use PET/CT and calorimetry to confirm that overall energy expenditure increases were driven at least in part by increases in BAT thermogenesis, and then calculate whether the amount and duration of negative energy balance created merited further study in large weight loss trials.

Detection of the β3-adrenoceptor not only in intact human adipocytes but also in ventricular myocardium may preclude its usage (de Matteis et al. 2002), due to negative inotropic effects. The potential for body temperature rise to dangerous levels as seen with DNP (Colman 2007) should also be investigated early in clinical development under conditions in which the magnitude and duration of full pathway engagement in BAT was understood.

BRL 26830A, *TAK-677*, *L-796568*, *CL 316,243*, and *BRL 35135* are demonstrably β3 selective molecules for which enough clinical experience exists to merit discussion (Table 1). Unfortunately, conclusions regarding the therapeutic hypothesis are limited by the absence of any data demonstrating specific activation of BAT thermogenesis.

BRL 26830A has been evaluated in man with conflicting results and has been widely studied preclinically. BRL 26830A causes dose-dependent body weight loss in obese rats and mice without causing weight loss in lean counterparts. Adiposity is reduced in both overweight and lean rodents through increase energy expenditure with no caloric intake effects, but lean mass is preserved (Arch et al. 1984a, b).

The thermogenic benefit of BRL 26830A is increased by repeat dosing in rodent at and below thermoneutrality, with greater effects observed in obese than in lean animals. BRL 26830A stimulates BAT thermogenesis and depletes BAT lipid stores, and inhibition of fatty acid oxidation eliminates these effects (Wilson et al. 1987). While BRL 26830A increased energy expenditure, concomitantly increased food intake minimized negative energy balance and weight loss (Yen et al. 1984; Shaw et al. 1981; Rothwell and Stock 1987; Arch et al. 1984b; Arch and Ainsworth 1983). Skeletal muscle glucose oxidation was stimulated by BRL 26830 (Challiss et al. 1988), raising the question of whether BAT is BRL 26830's major target tissue as BAT contributes little to whole body glucose disposal since fatty acids rather than

Table 1 Clinical experience with selective β3-adrenergic agonism of BAT

Drug	Results	Adverse events	References
BRL 26830A	↑ Placebo-adjusted energy expenditure and ↓ body weight reduction after 6-week calorie restriction. Two other studies (−) body weight ↑ insulin sensitivity assessed by clamp	↑ Tremor	Abraham et al. (1987), Chapman et al. (1985, 1988), Connacher et al. (1988, 1990, 1992a, b), Munro et al. (1987), and Smith et al. (1987)
TAK-677	↑ Energy expenditure (−) body weight or other metabolic parameters	↑ Heart rate	Redman et al. (2006)
L-796568	(−) Body weight or energy expenditure upon chronic dosing. Acute ↑ energy expenditure and lipolysis	Slight ↑ systolic blood pressure	Larsen et al. (2002), Yoshitomi et al. (1998), Manara et al. (2000), de Ponti et al. (1999), Fletcher et al. (1998), and van Baak et al. (2002)
CL 316,243	(−) Body weight or total energy expenditure upon chronic dosing. ↓ Free fatty acids, ↑ fat oxidation, and ↑ insulin-stimulated glucose disposal	(−) Cardiovascular or tremor events	Weyer et al. (1998), Dolan et al. (1994), Himms-Hagen et al. (1994), Yoshida et al. (1994), and Arbeeny et al. (1995)
BRL 35135	(−) Body weight despite ↑ energy expenditure. ↑/↓ Glucose metabolism	↑ Heart rate and tremor	Wheeldon et al. (1993, 1994), Smith et al. (1989), Cawthorne et al. (1992, pp. 252S–257S), Mitchell et al. (1989)
Mirabegron	↑ FDG update ↑ Energy expenditure	↑ Heart rate and ↑ systolic blood pressure	Roberts-Toler et al. (2015, pp. 1765–1770), Baskin et al. (2018, pp. 2113–2125); Carey et al. (2013), Cypess et al. (2012, 2015, pp. 33–38), Vosselman et al. (2012), Villarroya et al. (2013), Stanford et al. (2012), Hall et al. (2011), and Broeders et al. (2016)
RO 16-8714	(−) Energy expenditure (−) Body weight	(−) Heart rate	Ferré et al. (1992), Assimacopoulos-Jeannet et al. (1992), Isler et al. (1987), and Jequier et al. (1992)
ICI D-7114	(−) Energy expenditure (−) Body weight		Toubro et al. (1993)

glucose are the major oxidative substrate in BAT (Isler et al. 1987). Yet BRL 26830 in other preclinical studies displayed increased glucose utilization specifically in BAT (Young et al. 1984, 1985), supporting BAT as a key target tissue for metabolic effects of this compound.

BRL 26830A was tested in at least four human studies, with unimpressive results on body weight overall. The most encouraging results were reported by Chapman et al. (1988) from a study in which placebo-adjusted increased energy expenditure, moderate body weight reduction, and preservation of protein mass (Chapman et al. 1988; Abraham et al. 1987) in obese patients after 6 weeks treatment with BRL 26830A under conditions of sufficient food restriction that placebo subjects also lost weight despite reduced energy expenditure on study. Other studies were less impressive. Chapman et al. (1985) reported a statistically insignificant difference in body weight change upon BRL 26830A administration after a calorie-restricted run-in phase, yet side effects were present (Chapman et al. 1988; Connacher et al. 1990, 1992a; Munro et al. 1987). Smith et al. reported that BRL26830A improved insulin sensitivity as assessed by clamp (Smith et al. 1987), which was replicated in a larger study in which no changes in weight loss or oxidative substrate preference were observed (Connacher et al. 1988, 1992a, b).

The β3-adrenoreceptor agonist *TAK-677* has been shown in humans to generate a slight increase in total energy expenditure that cannot be attributed to BAT, but no changes in substrate utilization or body weight (Redman et al. 2006). In 65 obese men and women, TAK-677 slightly increased total energy expenditure at the highest dose given, but resulted in no change in respiratory quotient, fat oxidation, total body weight, adipose mass, or fat-free mass, nor changes in circulating glucose, insulin, or free fatty acids. Large interindividual variability in plasma concentrations of TAK-677 suggest insufficient drug-like behaviors of this molecule, yet along with energy expenditure increases, the heart rate elevations observed upon administration suggest sufficient SNS promotion to trigger these physiological effects.

L-796568 is selective (>600× over human 1 and 2 receptors), potent (EC50 3.6 nmol/L), and fully efficacious human β3-adrenergic agonist (Mathvink et al. 2000) that has been evaluated in several clinical trials. L-796568 demonstrated acute increases in total energy expenditure and lipolysis, though sustained changes in body weight or energy expenditure are absent upon chronic dosing of humans.

L-796568 displayed anorectic effects in animal studies as well as chronically decreasing triglycerides in obese animals (Yoshitomi et al. 1998), which may be caused by reduced gastric emptying resulting from its effects on gastric motility (Manara et al. 2000; de Ponti et al. 1999; Fletcher et al. 1998), but little consistent effects on glucose metabolism.

In overweight individuals, L-796568 acutely elevated lipolysis and energy expenditure (van Baak et al. 2002), though after 28 days treatment, no such consequences of intervention were observed (Larsen et al. 2002). Almost 10% increase in energy expenditure could be measured after the single high dose to fasted subjects (van Baak et al. 2002), with increased circulating glycerol and free fatty acids implying the expected increase in lipolysis. Cardiovascular effects were mild, with elevated systolic blood pressure and no changes in diastolic pressure, heart rate, ear temperature, potassium, or circulating catecholamines. A follow-on study (Larsen et al. 2002) under less controlled, non-rested, and non-fasted conditions reported results from 28-day dosing of a dose of L-796568 achieving plasma concentrations equal to those associated with acute energy expenditure increases. While effects on circulating fatty acids were observed again and minor effects on plasma glucose may have been present, no changes were observed in body composition, energy expenditure, respiratory quotient, glucose tolerance, heart rate, or blood pressure.

CL 316,243 is a partial (60% activity compared with isoprenaline) but selective (>1,500-fold selectivity over 1 and 2) human beta(3)-adrenergic agonist (Dolan et al. 1994) but is only weakly potent ($EC_{50} = 3{,}700$ nmol/L) compared to L-796568 (1,000× weaker). CL 316,243's properties are sufficient to generate significant thermogenic effects in rodents (Himms-Hagen et al. 1994; Yoshida et al. 1994) and upon chronic administration generate significant weight loss and improvements in diabetic measures (Yoshida et al. 1994; Arbeeny et al. 1995).

CL 316,243 results in humans, however, were disappointing, with no changes in energy expenditure or body weight, though increased insulin action and fat oxidation were observed (Weyer et al. 1998) and though with this particular molecule there may have been a dosing issue, as plasma concentrations in the study were low compared with its EC50 value. No changes in body weight or total energy expenditure were observed after 8 weeks administration to lean subjects, though fasting free fatty acids, fat oxidation, and insulin-stimulated glucose disposal were improved as predicted from preclinical experience. No cardiovascular or tremor effects were observed.

BRL 35135 is a weakly selective (20-fold selective β(3)- relative to β(2)) adrenoceptor agonist that has been examined in several clinical studies, with no weight loss or blood glucose effects despite increased energy expenditure and elevations of heart rate. BRL35135 produced thermogenic effects that were only partially inhibited by a nonselective beta receptor antagonist (Wheeldon et al. 1993, 1994). In another study, a single dose stimulated energy expenditure and glucose-induced thermogenesis in healthy subjects, though no effects on blood glucose levels could be detected (Smith et al. 1989). Despite these moderate effects, minor elevations in heart rate were measured in both trials (Cawthorne et al. 1992). In a third study, 10 days treatment of obese patients with BRL 35135 did not result in detectable weight loss, (Wheeldon et al. 1994) though glucose tolerance through better glucose storage was observed (Mitchell et al. 1989). In obese patients measured under clamp conditions, 10 days BRL 35135 treatment again improved insulin-stimulated glucose disposal (Cawthorne et al. 1992, p. 763), though changes in glucose oxidation and any consequent glucose-stimulated thermogenesis were not detected. As mild tremor was detected in some patients in more than one trial (Mitchell et al. 1989), on balance the clinical experience with BRL 35135 was discouraging.

In conclusion, the overall clinical experience with β3 adrenergic selective molecules BRL 26830A, TAK-677, L-796568, CL 316,243, and BRL 35135 supports a statistically significant elevation in total energy expenditure, but when examined this did not translate into a biologically meaningful negative energy balance, for which convincing evidence would have been a measurable loss of body weight over the duration of the study.

More limited date sets with older or less selective β3 agonists will be reviewed next.

Mirabegron (Myrbetriq, Astellas Pharma, Inc.), a β3-AR agonist (Cypess et al. 2015, pp. 33–38; Roberts-Toler et al. 2015, pp. 1765–1770; Baskin et al. 2018, pp. 2113–2125) approved to treat overactive bladder through inhibiting urinary outflow, has been examined for its potential to acutely activate BAT (Cypess et al. 2015) in a study that did not look at body weight changes but did examine BAT physiology using limited methods (Carey et al. 2013; Cypess et al. 2012; Vosselman et al. 2012). 200 mg mirabegron, a significantly higher dose than the 25 and 50 mg doses indicated for overactive bladder, led to higher BAT metabolic activity as measured via (18)F fluorodeoxyglucose ((18)F-FDG) using positron emission tomography (PET) combined with computed tomography (CT) in all 12 healthy male subjects ($p = 0.001$), with all 12 subjects displaying glucose uptake significantly increased over placebo primarily at cervical-supraclavicular-axillary adipose tissue depots but also occasionally in the paraspinal, periaortic, perihepatic, perirenal, and perisplenic regions. Mirabegron increased resting metabolic rate (RMR) by 203 ± 40 kcal/day (+13%; $p = 0.001$), and BAT metabolic activity was also a significant predictor of the changes in RMR ($p = 0.006$). The reported maximal increase in energy expenditure, of 200 kcal/day, in the unlikely event that it were maintained close to peak through the day, would lead to an eventual weight loss of 5 kg in the first year and 10 kg by the end of 3 years and improved glucose disposal prior to significant weight loss (Villarroya et al. 2013; Stanford et al. 2012; Hall et al. 2011), though real-world losses due to not dosing continuously to peak exposure would almost certainly be lower. Importantly, heart rate and systolic but not diastolic blood pressure also increased, though the magnitude of cardiovascular changes were lower than is seen with sympathomimetics for this amount of RMR change (Carey et al. 2013; Cypess et al. 2012; Broeders et al. 2016).

β3-AR agonist *RO 16-8714* robustly stimulates thermogenesis and total energy expenditure in multiple rodent species (Ferré et al. 1992, pp. 180–183). RO 16-8714 is not specific to BAT, as glucose utilization was increased in oxidative muscles by RO 16 8714 (Ferré et al. 1992), but evidence for RO 16-8714 activity in BAT exists (Assimacopoulos-Jeannet et al. 1992) despite the generally low contribution of BAT to whole body glucose homeostasis (Ferré et al. 1992) even though glucose is a major energy substrate for BAT (Isler et al. 1987). However, in limited clinical trials, Ro 40-2148 acutely had no significant effect on either resting energy expenditure or heart rate, and body weight was unchanged (Jequier et al. 1992).

While significant preclinical data exists with *LY 104 119*, no clinical data have been published, and the mechanism of this molecule now appears mixed, though lipolytic activity particularly in brown fat is clear. LY 104 119 elevated norepinephrine turnover in interscapular brown adipose tissue of lean mice to a higher rate than in AVY/a mice (Yen et al. 1988). LY 104 119 promoted similar biochemical changes in lean and obese animals, including increased lipolysis and thermogenesis, though animals countered the energy loss by elevating food consumption and thus experienced no weight loss (Yen et al. 1984; Shaw et al. 1981; Rothwell and Stock 1987; Arch et al. 1984b). LY 104 119 promoted a varied relationship between calorie intake and energy expenditure. Minimally overweight mice experienced weight loss when dosed with LY104119 in spite of overeating (Yen et al. 1984).

Lean mice that were not allowed to overeat also experienced weight reduction when treated with LY104119 (Yen et al. 1984). Chronic treatment of obese animals with molecules of this class may (Yen et al. 1984; Shaw et al. 1981) or may not (Yen et al. 1984; Yen 1984) promote overeating, depending on the study considered. In the obese mice, LY 104 119 reduced body weight and adiposity with no changes in food intake, though alteration of a body weight set point was not achieved, as overweight returned upon cessation of drug treatment, whereas treatment of lean animals resulted in reduced adiposity with no overall weight loss if allowed to overeat (Yen et al. 1984).

The selectivity of *ICI D-7114* is unclear. A 14-day, double-blind, randomized trial was performed with 50 and 100 mg TID (Toubro et al. 1993), but no changes in body composition and weight, energy expenditure, or substrate oxidation were detected.

11 Human BAT and Other Molecules and Nutritional Approaches

Other molecules with enough clinical experience to merit discussion include *capsinoids* and *nutritional extracts*, though their individual mechanisms of action are either disputed or frankly unknown.

Capsinoids (capsiate, dihydrocapsiate, and nordihydrocapsiate) are nonpungent analogs of capsaicin (Yazawa et al. 1989; Kobata et al. 1998). The parent molecule capsaicin is known to activate the adrenosympathetic nervous system, increase energy expenditure and fat oxidation, and thereby lower body fat in rodents (Luo et al. 2011; Diepvens et al. 2007; Kawada et al. 1986a, b). Capsinoids also have been reported to increase energy expenditure in rodents and humans, potentially through activation of human BAT thermogenesis. The physiological effects of capsinoids are clearer than the pharmacological actions potentially causing them. Capsaicinoids are considered to activate transient receptor potential vanilloid 1 receptors, which are proposed to increase energy expenditure and catabolic processes in adipose tissue. Cold-adaptive thermogenesis in BAT is regulated through activation of SNS, triggered by somatic stimulation of transient receptor potential (TRP) channels in sensory neurons (Nakamura 2011; Tajino et al. 2011), a pathway also stimulated by capsaicin and nonpungent capsaicin analogs capsinoids (Ono et al. 2010; Shintaku et al. 2012). Overall, in rodents capsinoids appear to activate the TRPV1 pathway, which activates the SNS, which activates BAT, which stimulates energy expenditure.

SNS activation promotes BAT thermogenesis under cold exposure (Lowell and Spiegelman 2000; Cannon and Nedergaard 2004), and several studies have ascribed the thermogenic effect of capsaicin and capsinoids in rodents to action in BAT. Intragastric administration of capsinoids elevates BAT and rectal temperature in mice, which is reduced in mice lacking TRPV1 (Kawabata et al. 2009). Repeat administration of capsinoids stimulates sympathetic nerve activity and elevates UCP1 (Ono et al. 2010; Masuda et al. 2003), total energy expenditure and thermogenesis, and reduced adiposity concomitant with increased fat oxidation in rodents (Ohnuki et al. 2001a).

BAT has been broadly detected in adult humans (van Marken Lichtenbelt et al. 2009; Virtanen et al. 2009; Saito et al. 2009) and has been associated with cold-induced adaptive thermogenesis (Yoneshiro et al. 2011a, b). Thermogenic effects of capsinoids have not only been reported in rodents but also in humans (Yoshioka et al. 1998). Analogous to cold triggering temperature sensitive TRP channels, which excites the SNS, which activates BAT, administration of capsinoids has been shown to activate TRPV1, which promotes efferent discharge of SNS in BAT (Yoneshiro and Saito 2013), which stimulates BAT thermogenesis (Yoneshiro et al. 2012). Other studies have demonstrated that capsinoid administration to humans excites SNS and promotes BAT thermogenesis, elevating total energy expenditure and fat oxidation, which reduces fat mass (Yoneshiro et al. 2013; Ono et al. 2010; Lee et al. 2010b; Galgani and Ravussin 2010; Snitker et al. 2008; Inoue et al. 2007; Ohnuki et al. 2001b). These effects can be observed even in humans with low detectable BAT function (Yoneshiro et al. 2013), where both cold exposure increased energy expenditure and BAT activity and decreased fat mass, and capsinoid treatment in low-BAT subjects still increased cold-induced thermogenesis.

Nutritional extracts from *Kaempferia parviflora* (KP) and grains of paradise both have been reported to result in increased energy expenditure in humans. *Kaempferia parviflora* is a plant in the Zingiberaceae family native to Southeast Asia, where it is historically used to increase vitality. KP extract has been shown to have anti-gastric ulcer and anti-inflammatory benefits in humans (Yenjai et al. 2004; Kusirisin et al. 2009; Rujjanawate et al. 2005). KP extract is known to reduce body weight in normal mice and to reduce body weight, adiposity, and glucose intolerance in obese diabetic mice (Yoshino et al. 2014; Shimada et al. 2011; Akase et al. 2011), and the absence of caloric intake effects implies an energy expenditure mechanism. KP also elevates urinary excretion of noradrenaline, UCP1 expression, and energy expenditure in mice (Yoshino et al. 2014), indicating an effect on the sympathetic nervous system.

The effect of KP extract on BAT and on energy expenditure has been studied in humans (Matsushita et al. 2015). A single oral administration elevated energy expenditure, and based on FDG-PET, the energy expenditure increase was proportional to amounts of avid BAT.

Grains of paradise (*Aframomum melegueta* [Rosco] K. Schum.) (GP), or Guinea pepper or alligator pepper, is a member of the Zingiberaceae family indigenous to West Africa. GP seeds have a long ethnobotany history and are deployed against a variety of ailments. Some of the constituent molecular components, including 6-paradol, 6-gingerol, and 6-shogaol, possess the same vanillinoid moiety as do capsinoids, suggesting that they may activate the TRPV1 pathway and thus the SNS, BAT, and adaptive thermogenesis (Morera et al. 2012; Riera et al. 2009). Administration of GP extract and 6-paradol to rats promoted SNS efferent synaptic discharge in BAT and a significant rise in BAT temperature (Iwami et al. 2011).

Controlled studies in humans provide support for GP effects on BAT. A single dose of GP extract increased BAT activation and energy expenditure in men (Sugita et al. 2013). Repeated administration over 4 weeks resulted in a significant reduction of visceral fat in humans (Sugita et al. 2014). Whole body energy expenditure was increased, though no changes in BMI, body weight, total body fat, or subcutaneous body were observed. It is unclear why fat reduction was restricted to visceral depots, with total and subcutaneous fat depots unaffected, though a similar effect of capsinoids (Kawabata et al. 2006, pp. 2824–2835; Snitker et al. 2008, pp. 45–50) implies that visceral fat may be intrinsically more responsive to hormonal or nutritional perturbations than is subcutaneous fat.

12 Human BAT and Poorly Modeled Mechanisms

β3-adrenergic agonists, catechins, capsinoids, and nutritional extracts, even without robust negative energy balance outcomes, have all demonstrated increased total energy expenditure that in some cases can be associated with concomitant activation of BAT. Unlike catechins, capsinoids, and nutritional extracts, *glucocorticoid* receptor agonists, *PPARg* agonists, and *thyroid* hormone receptor agonists all possess defined molecular and cellular pharmacology, yet their effects on human BAT thermogenesis were inconsistent with predictions.

Glucocorticoids are known to suppress BAT activation in rodents, and chronic glucocorticoid excess causes obesity in humans, thus generating expectations that glucocorticoid antagonism would promote BAT activation and energy expenditure, at least in humans. Glucocorticoid receptor antagonist prednisolone administered under cold conditions promotes FDG uptake in BAT of lean healthy men and increased energy expenditure under cold but not thermoneutral conditions (Ramage et al. 2016). In vitro, glucocorticoid agonist cortisol acutely elevated isoprenaline-stimulated respiration and UCP1 in human primary brown adipocytes (Ramage et al. 2016) but had the opposite effect in primary murine brown and beige adipocytes. In a second human study (Scotney et al. 2017), however, the glucocorticoid receptor agonist hydrocortisone acutely increased BAT temperature in eight healthy males in basal and cold-induced conditions. Thus both glucocorticoid agonism and antagonism yielded increases in human BAT activity measures, and at least acute effects of glucocorticoid receptor modulation were also inconsistent between humans and model species.

Rodent models with *PPARγ* deletion in adipose display impaired BAT development (Imai et al. 2004), and mice treated with PPARγ agonists have significant expansion of BAT mass (Tai et al. 1996). PPARγ agonists induce browning in human adipocyte cells (Digby et al. 1998), and PPARγ agonist rosiglitazone is widely reported to promote brown differentiation and browning of human adipocyte cells (Digby et al. 1998; Jespersen et al. 2013; Elabd et al. 2009; Carey et al. 2014; Bogacka et al. 2005). A related TZD PPARγ agonist pioglitazone was tested ex vivo and in vivo in humans (Loh et al. 2018). Pioglitazone increased in vitro browning and adipogenesis of cultured human primary subacromioclavicular-derived adipocytes. In patients, however, pioglitazone decreased cold-induced

FDG uptake by BAT. These clinical trial results are inconsistent with experience in rodent models and in human cell models.

Cold remains the stimulation of human BAT thermogenesis with the best translational experience (Saito et al. 2009; van der Lans et al. 2013; Orava et al. 2011), and *thyroid* hormones are known regulators of basal metabolism that elevate basal metabolic rate (BMR) in humans (Goglia et al. 2002; Tata et al. 1962) and are linked to heat generation during both shivering and non-shivering thermogenesis. It is known that thyroid hormone levels can affect energy expenditure and thermogenesis during activity and at rest (Al-Adsani et al. 1997) and that in clinical hypothyroidism or hyperthyroidism, energy expenditure can decrease or increase (Bianco and Silva 1988). Thyroid hormone receptors can be detected in BAT, and thyroid hormone and norepinephrine are combined to elevate expression of UCP1 in animals.

In rodents models, thyroid hormones activate BAT directly (de Jesus et al. 2001), increase brown adipogenesis (Guerra et al. 1996, pp. 2076–2081), and may also induce BAT indirectly by stimulating the hypothalamic pathway (López et al. 2010). Systemic hyperthyroidism or central administration of thyroid hormone to rodents causes sympathetic activation and BAT induction (Andersen et al. 2012). Cold exposure in rodents promotes BAT activity by increasing SNS signaling and by modulating tissue thyroid hormone metabolism (Silva 2006; Bianco and Silva 1988). Both catecholamines and local thyroid hormone act together to elevate UCP1 expression and thermogenesis in BAT in rodents (Silva and Bianco 2008).

Thyroid hormone thus is clearly linked to energy expenditure in humans, and reasonably linked to BAT activation in rodents, but clinical data linking thyroid hormones specifically to BAT activation and BAT-mediated energy expenditure in humans is mixed (Kim et al. 2014; Zhang et al. 2014; Lahesmaa et al. 2014; Skarulis et al. 2010). Thyroid hormone stimulates the development of UCP1-positive cells in white preadipocytes (Lee et al. 2011b). Thyroid hormone increases UCP1 fatty acid oxidation genes, along with oxygen consumption, in multipotent adipose-derived stem cells (Lee et al. 2011b). Human studies have reported mixed results for BAT volume and activity in hypothyroid or thyrotoxic subjects (Kim et al. 2014; Zhang et al. 2014; Lahesmaa et al. 2014; Skarulis et al. 2010). A thyroid carcinoma patient with extreme insulin resistance and thyroid cancer was treated with high doses of thyroid hormone and demonstrated BAT activation at room temperature by PET-CT (Skarulis et al. 2010). As well, hyperthyroid patients display 3× greater BAT glucose uptake over healthy subjects (Skarulis et al. 2010). Thyrotoxicosis generates increased energy expenditure in rodents and humans (Møller et al. 1996). In patients with well-differentiated thyroid carcinoma eligible for surgical treatment and follow-on radioactive iodine ablation therapy, examination of the hypothyroid state after thyroidectomy, and the subsequent hyperthyroid state during TSH suppression, higher levels of thyroid hormone were associated with elevated cold-activated BAT (Broeders et al. 2016), including increased basal metabolic rate and cold-induced energy expenditure.

Other clinical investigations provide more discouraging evidence. In ten patients with overt hyperthyroidism and in eight healthy participants, hyperthyroidism had no effect on BAT metabolism under conditions where skeletal muscle metabolism, total fat oxidation, and total energy expenditure were increased (Lahesmaa et al. 2014). In newly diagnosted, previously untreated Graves' disease-caused hyperthyroid patients (Zhang et al. 2014), hyperthyroid patients had no detectably active BAT. In another protocol evaluating BAT mass and activity in adult humans in the thyrotoxic versus hypothyroid states (Gavrila et al. 2017), it was observed that cold-induced BAT activation was not inhibited by hypothyroidism present for several weeks, and the administration of thyroid hormone did not promote cold-induced thermogenesis. And in a juvenile severe Hashimoto's case with substantial supraclavicular BAT volume, 8 weeks of thyroid hormone treatment reduced BAT activity from pretreatment levels (Kim et al. 2014).

13 Future Directions

A few new mechanisms are nearing the stage of clinical trials and may yet provide a more quantitatively robust translation from preclinical to human experience with BAT.

Obese and type 2 diabetic humans have elevated levels of serum circulating *FGF21* levels (Chavez et al. 2009; Cuevas-Ramos et al. 2010; Gallego-Escuredo et al. 2015; Zhang et al. 2008), which may reflect an FGF21 compensatory response to metabolic disruptions, including reduced functional BAT. While the presence of pathology despite elevations of endogenous FGF21 in disease implies a limited therapeutic benefit of further elevations upon clinical administration, analogies to insulin and GLP1 suggest that benefits on glucose, lipids, and BAT energy expenditure could be realized by further exposure to exogenous FGF21.

Several different approaches have optimized production of FGF21 analogs with greater solubility, stability, and pharmacokinetics (Zhang and Li 2014, pp. 579–589). PEGylated FGF21 (Ye et al. 2015), FGF21-antibody conjugates (Talukdar et al. 2016), and antibody-based activation of the FGFR/β-Klotho complex (Reitman 2013) have been reported, with FGF21 analogs LY2405319 and PF05231023 reporting clinically positive and negative results in humans (Dong et al. 2015a; Gaich et al. 2013). PF-05231023 was administered in multiple studies, including intravenously in patients with type 2 diabetes. LY2405319 was administered subcutaneously once daily for 4 weeks in patients with diabetes and obesity. Both drugs showed improvement of glucose levels, lipid profile, and weight loss (Gaich et al. 2013; Dong et al. 2015b). PF-05231023 dosing of obese cynomolgus monkeys or type 2 diabetic humans caused measurable body weight loss, improved lipoprotein profile, and elevated adiponectin levels, though no clinically significant changes in glucose (Talukdar et al. 2016).

Likewise, LY2405319 (Gaich et al. 2013) improved dyslipidemia, body weight, fasting insulin, and adiponectin, though not circulating glucose. Other FGF21 analogs may soon enter clinical development (Kharitonenkov and DiMarchi 2015, pp. 606–617; Gimeno and Moller 2014, pp. 303–311; Bailey et al. 2016, pp. 350–359).

Release of FGF21 from BAT was recently shown to be promoted by *adenosine* A2A receptor (A2AR) signaling in BAT, which prevented hypertension-related cardiac damage in a mouse model (Ruan et al. 2018, pp. 476–489). Moreover, adenosine has been demonstrated to activate lipolysis and the thermogenic program in brown and white human and murine adipocytes and recruit beige adipocytes ex vivo via A2A receptors (Gnad et al. 2014, pp. 395–399). Also, in vivo pharmacological or genetic inhibition of adenosine A2A receptor signaling lowered BAT thermogenesis, whereas agonism increased thermogenesis and browned white adipose tissue and preserved lean phenotype and glucose tolerance in high-fat diet-fed mice (Gnad et al. 2014, pp. 395–399). This work generates the working hypothesis that clinical chronic adenosine receptor agonism could induce persistent negative energy balance through human BAT, though the proven robust effects of adenosine and analogs in the cardiovascular and CNS organ systems impose a significant subtype and tissue selectivity challenge.

cGMP signaling in BAT biology has been reviewed earlier in this volume, as has the observation that a subset of cells may exist in white adipose depots that is capable of transition to a brown-like phenotype, which may provide energy balance and other metabolic benefits (Petrovic et al. 2010; Boström et al. 2012; Barbatelli et al. 2010). Human preadipocytes have been stimulated to undergo this phenotype transition in vitro (Gustafson et al. 2015), and results obtained with human subcutaneous white adipose tissue (sWAT) undergoing such a transition are consistent with potential benefit in insulin responsiveness (Yang et al. 2003). As has been reviewed earlier, elevated adrenergic tone promotes more active adaptive thermogenic processes by BAT, so it is interesting that the prolonged adrenergic stress of burn trauma associates both with the browning of human sWAT and with an increased overall metabolic rate (Kern et al. 2014; Sidossis et al. 2015). Exogenous pharmacological interventions to more deliberately test this hypothesis are limited. Phosphodiesterase 5 (PDE5) is widely expressed in the body and modulates cyclic guanosine monophosphate (cGMP) by catalyzing hydrolytic degradation. The PDE5 inhibitor sildenafil has been demonstrated to stimulate UCP1 expression and thus browning of sWAT (Mitschke et al. 2013), generating the therapeutic hypothesis that elevating cGMP in humans may functionally increase by transdifferentiating abundant WAT toward a BAT-like phenotype. A small randomized, double-blinded, placebo-controlled clinical trial was executed to test this hypothesis, with 16 overweight males treated with sildenafil (Li et al. 2018). Sildenafil treatment in this study elevated cGMP in plasma, as well as circulating levels of catecholamines, though classical BAT itself was not activated concomitant with the latter. However, sWAT displayed elevated UCP1 at both mRNA and protein levels, increased respiratory capacity leak, and decreased adipocyte size, consistent with a browning transition. Browning of abundant WAT, rather than activation of a more sparse and potentially less energetic BAT compartment, may be a promising therapeutic strategy for the metabolic derangements discussed in this review.

14 Conclusions

The intent of this review, while not exhaustive, is to accurately reflect that state of translation of preclinical BAT pharmacology and physiology into humans. Specific results have been described earlier, and from that overview at least several general conclusions may be reached.

First, nonclinical BAT molecular pharmacology mostly translates, with consequent cellular behaviors broadly consistent between human and prior model species. Second, with the few exceptions noted in the text, the qualitative elements of physiology also translate, with activation mechanisms for adaptive thermogenesis and pharmacological triggers from nonclinical models apparently recapitulated in humans. Third, and perhaps most important for considerations of medical pharmacology, it is the quantitative elements of BAT physiology that have not yet translated well into humans. Specifically, there is relatively less functional BAT in adult humans than in adult rodents, the proportion of BAT that can be activated in obese humans is relatively less than in obese rodents, the magnitude of total energy expenditure increase that can be achieved by activating human BAT is significantly less than in rodents, and the threshold result from all of these trends is that, while activating BAT in rodents can tilt overall energy balance negative in rodents and promote weight loss, in humans this tilt into negative energy balance has not been clearly documented. Lastly, while comprehensive physiological characterization of BAT modulation preclinically and clinically is achievable, clinical translation suffers from a lack of acute biomarkers that connect pharmacology to physiology, and continued research into promising candidate biomarkers is urgently needed (Chen et al. 2016, p. 11420).

Beyond considerations specific to BAT biology, additional general points can be made. First the field of BAT biology, and in particular BAT human biology, is a couple decades old at best. Thus we are presumably only at the beginning of our maturity in both basic and translational research knowledge about BAT. Second, the typical asymmetry between preclinical and clinical data robustness exists. Most human studies of BAT biology represent the $N = 1$ situation, whereas conclusions from biochemical, cellular, and laboratory animal models derive from numerous to many to several iterations. Accordingly, it is possible that further iterations of one or more mechanism translating into humans will show a strong negative energy balance from BAT-promoted total energy expenditure increases. Finally, even if pharmacological BAT activation does not achieve large reductions in body weight, it could be a novel approach to treat metabolic sequelae of obesity, particularly diabetes.

References

Abraham R, Zed C, Mitchell T, Parr J, Wynn V (1987) The effect of a novel beta-agonist BRL-26830A on weight and protein loss in obese patients. Int J Obes 11(3):A306

Agrawal A, Nair N, Baghel N (2009) A novel approach for reduction of brown fat uptake on FDG PET. Br J Radiol 82(980):626–631

Akase T, Shimada T, Terabayashi S, Ikeya Y, Sanada H, Aburada M (2011) Antiobesity effects of Kaempferia parviflora in spontaneously obese type II diabetic mice. J Nat Med 65(1):73–80
Al-Adsani H, Hoffer LJ, Silva JE (1997) Resting energy expenditure is sensitive to small dose changes in patients on chronic thyroid hormone replacement. J Clin Endocrinol Metabol 82 (4):1118–1125
Andersen S, Kleinschmidt K, Hvingel B, Laurberg P (2012) Thyroid hyperactivity with high thyroglobulin in serum despite sufficient iodine intake in chronic cold adaptation in an Arctic Inuit hunter population. Eur J Endocrinol 166(3):433–440
Arbeeny CM, Meyers DS, Hillyer DE, Bergquist KE (1995) Metabolic alterations associated with the antidiabetic effect of beta 3-adrenergic receptor agonists in obese mice. Am J Physiol Endocrinol Metab 268(4):E678–E684
Arch JR (2008) The discovery of drugs for obesity, the metabolic effects of leptin and variable receptor pharmacology: perspectives from β 3-adrenoceptor agonists. Naunyn Schmiedeberg's Arch Pharmacol 378(2):225
Arch JR (2015) Horizons in the pharmacotherapy of obesity. Curr Obes Rep 4(4):451–459
Arch J, Ainsworth A (1983) Thermogenic and antiobesity activity of a novel β-adrenoceptor agonist (BRL 26830A) in mice and rats. Am J Clin Nutr 38(4):549–558
Arch J, Wilson S (1996) Prospects for beta 3-adrenoceptor agonists in the treatment of obesity and diabetes. Int J Obes Relat Metab Disord 20(3):191–199
Arch J, Ainsworth A, Cawthorne M, Piercy V, Sennitt M, Thody V, Wilson C, Wilson S (1984a) Atypical β-adrenoceptor on brown adipocytes as target for anti-obesity drugs. Nature 309 (5964):163
Arch JR, Ainsworth AT, Ellis RD, Piercy V, Thody VE, Thurlby PL, Wilson C, Wilson S, Young P (1984b) Treatment of obesity with thermogenic beta-adrenoceptor agonists: studies on BRL 26830A in rodents. Int J Obes 8(Suppl 1):1–11
Arner P, Hoffstedt J (1999) Adrenoceptor genes in human obesity. J Intern Med 245(6):667–672
Assimacopoulos-Jeannet F, Greco-Perotto R, Terrettaz J, Meier M, Jeanrenaud B (1992) Effect of a β-adrenergic agonist on glucose transport and insulin-responsive glucose transporters (GLUT4) in brown adipose tissue of control and obese fa/fa rats. Pflugers Arch 421(1):52–58
Astrup A (1986) Thermogenesis in human brown adipose tissue and skeletal muscle induced by sympathomimetic stimulation. Acta Endocrinol 112(3 Suppl):S9–S32
Astrup A, Bulow J, Madsen J, Christensen N (1985a) Contribution of BAT and skeletal muscle to thermogenesis induced by ephedrine in man. Am J Physiol Endocrinol Metab 248(5):E507–E515
Astrup A, Lundsgaard C, Madsen J, Christensen NJ (1985b) Enhanced thermogenic responsiveness during chronic ephedrine treatment in man. Am J Clin Nutr 42(1):83–94
Atit R, Sgaier SK, Mohamed OA, Taketo MM, Dufort D, Joyner AL, Niswander L, Conlon RA (2006) β-catenin activation is necessary and sufficient to specify the dorsal dermal fate in the mouse. Dev Biol 296(1):164–176
Bachman ES, Dhillon H, Zhang C-Y, Cinti S, Bianco AC, Kobilka BK, Lowell BB (2002) βAR signaling required for diet-induced thermogenesis and obesity resistance. Science 297 (5582):843–845
Bailey CJ, Tahrani AA, Barnett AH (2016) Future glucose-lowering drugs for type 2 diabetes. Lancet Diabetes Endocrinol 4(4):350–359
Barbatelli G, Murano I, Madsen L, Hao Q, Jimenez M, Kristiansen K, Giacobino J, de Matteis R, Cinti S (2010) The emergence of cold-induced brown adipocytes in mouse white fat depots is determined predominantly by white to brown adipocyte transdifferentiation. Am J Physiol Endocrinol Metab 298(6):E1244–E1253
Baskin AS, Linderman JD, Brychta RJ, McGehee S, Anflick-Chames E, Cero C, Johnson JW, O'Mara AE, Fletcher LA, Leitner BP, Duckworth CJ, Huang S, Cai H, Garraffo HM, Millo CM, Dieckmann W, Tolstikov V, Chen EY, Gao F, Narain NR, Kiebish MA, Walter PJ, Herscovitch P, Chen KY, Cypess AM (2018) Regulation of human adipose tissue activation, gallbladder size, and bile acid metabolism by a β3-adrenergic receptor agonist. Diabetes 67(10):2113–2125

Bathgate B, Freebairn E, Greenland A, Reid G (1992) Functional expression of the rat brown adipose tissue uncoupling protein in Saccharomyces cerevisiae. Mol Microbiol 6(3):363–370
Bianco AC, Silva JE (1988) Cold exposure rapidly induces virtual saturation of brown adipose tissue nuclear T3 receptors. Am J Physiol Endocrinol Metab 255(4):E496–E503
Blondin DP, Labbé SM, Tingelstad HC, Noll C, Kunach M, Phoenix S, Guérin B, Turcotte ÉE, Carpentier AC, Richard D (2014) Increased brown adipose tissue oxidative capacity in cold-acclimated humans. J Clin Endocrinol Metabol 99(3):E438–E446
Bogacka I, Xie H, Bray GA, Smith SR (2005) Pioglitazone induces mitochondrial biogenesis in human subcutaneous adipose tissue in vivo. Diabetes 54(5):1392–1399
Boozer C, Daly P, Homel P, Solomon J, Blanchard D, Nasser J, Strauss R, Meredith T (2002) Herbal ephedra/caffeine for weight loss: a 6-month randomized safety and efficacy trial. Int J Obes 26(5):593
Boström P, Wu J, Jedrychowski MP, Korde A, Ye L, Lo JC, Rasbach KA, Boström EA, Choi JH, Long JZ (2012) A PGC1-α-dependent myokine that drives brown-fat-like development of white fat and thermogenesis. Nature 481(7382):463
Bouchard C, Tremblay A, Després J-P, Nadeau A, Lupien PJ, Thériault G, Dussault J, Moorjani S, Pinault S, Fournier G (1990) The response to long-term overfeeding in identical twins. N Engl J Med 322(21):1477–1482
Bray GA, Blackburn GL, Ferguson JM, Greenway FL, Jain AK, Mendel CM, Mendels J, Ryan DH, Schwartz SL, Scheinbaum ML (1999) Sibutramine produces dose-related weight loss. Obes Res 7(2):189–198
Broeders EP, Vijgen GH, Havekes B, Bouvy ND, Mottaghy FM, Kars M, Schaper NC, Schrauwen P, Brans B, van Marken Lichtenbelt WD (2016) Thyroid hormone activates brown adipose tissue and increases non-shivering thermogenesis-a cohort study in a group of thyroid carcinoma patients. PLoS One 11(1):e0145049
Brunton L, Chabner B, Knollman B (eds) (2011) Goodman & Gilman's the pharmacological basis of therapeutics, 12th edn. McGraw Hill, New York
Cannon B, Nedergaard J (2004) Brown adipose tissue: function and physiological significance. Physiol Rev 84(1):277–359
Cannon B, Nedergaard J (2010) Metabolic consequences of the presence or absence of the thermogenic capacity of brown adipose tissue in mice (and probably in humans). Int J Obes 34(S1):S7
Carey AL, Kingwell BA (2013) Brown adipose tissue in humans: therapeutic potential to combat obesity. Pharmacol Ther 140(1):26–33
Carey AL, Formosa MF, van Every B, Bertovic D, Eikelis N, Lambert GW, Kalff V, Duffy SJ, Cherk MH, Kingwell BA (2013) Ephedrine activates brown adipose tissue in lean but not obese humans. Diabetologia 56(1):147–155
Carey AL, Vorlander C, Reddy-Luthmoodoo M, Natoli AK, Formosa MF, Bertovic DA, Anderson MJ, Duffy SJ, Kingwell BA (2014) Reduced UCP-1 content in in vitro differentiated beige/brite adipocytes derived from preadipocytes of human subcutaneous white adipose tissues in obesity. PLoS One 9(3):e91997
Carey AL, Pajtak R, Formosa MF, van Every B, Bertovic DA, Anderson MJ, Eikelis N, Lambert GW, Kalff V, Duffy SJ (2015) Chronic ephedrine administration decreases brown adipose tissue activity in a randomised controlled human trial: implications for obesity. Diabetologia 58 (5):1045–1054
Cawthorne MA, Sennitt MV, Arch JR, Smith SA (1992) BRL 35135, a potent and selective atypical beta-adrenoceptor agonist. Am J Clin Nutr 55(1 Suppl):252S–257S
Cazeneuve P, Lépine R (1885) Sur les effets produits par l'ingestion et l'infusion intra-veineuse de trois colorants jaunes, derives de la houille. Compt Rend Soc Biol 101:1167–1169
Chadwick DJ, Cardew G (2008) The origins and consequences of obesity. Wiley, Hoboken
Challiss RJ, Leighton B, Wllson S, Thurlby PL, Arch JR (1988) An investigation of the β-adrenoceptor that mediates metabolic responses to the novel agonist BRL28410 in rat soleus muscle. Biochem Pharmacol 37(5):947–950

Chapman B, Farquhar D, Galloway S, Simpson G, Munro J (1985) The effects of BRL-26830A, a new beta-adrenoceptor agonist in refractory obesity. Int J Obes 9:230

Chapman B, Farquahar D, Galloway S, Simpson G, Munro J (1988) The effects of a new beta-adrenoceptor agonist BRL 26830A in refractory obesity. Int J Obes 12(2):119–123

Chavez AO, Molina-Carrion M, Abdul-Ghani MA, Folli F, DeFronzo RA, Tripathy D (2009) Circulating fibroblast growth factor-21 (FGF-21) is elevated in impaired glucose tolerance and type 2 diabetes and correlates with muscle and hepatic insulin resistance. Diabetes Care 32:1542–1546

Chen Y, Buyel JJ, Hanssen MJ, Siegel F, Pan R, Naumann J, Schell M, van der Lans A, Schlein C, Froehlich H, Heeren J, Virtanen KA, van Marken Lichtenbelt W, Pfeifer A (2016) Exosomal microRNA miR-92a concentration in serum reflects human brown fat activity. Nat Commun 7:11420

Christiansen E, Garby L, Sørensen TI (2005) Quantitative analysis of the energy requirements for development of obesity. J Theor Biol 234(1):99–106

Clapham J, Arch J (2007) Thermogenic and metabolic antiobesity drugs: rationale and opportunities. Diabetes Obes Metab 9(3):259–275

Clapham JC, Arch JR, Chapman H, Haynes A, Lister C, Moore GB, Piercy V, Carter SA, Lehner I, Smith SA, Beeley LJ, Godden RJ, Herrity N, Skehel M, Changani KK, Hockings PD, Reid DG, Squires SM, Hatcher J, Trail B, Latcham J, Rastan S, Harper AJ, Cadenas S, Buckingham JA, Brand MD, Abuin A (2000) Mice overexpressing human uncoupling protein-3 in skeletal muscle are hyperphagic and lean. Nature 406(6794):415–418. https://doi.org/10.1038/35019082

Cohade C, Osman M, Pannu HK, Wahl RL (2003a) Uptake in supraclavicular area fat ("USA-Fat"): description on (18)F-FDG PET/CT. J Nucl Med 44(2):170

Cohade C, Mourtzikos KA, Wahl RL (2003b) "USA-Fat": prevalence is related to ambient outdoor temperature-evaluation with 18F-FDG PET/CT. J Nucl Med 44(8):1267–1270

Colman E (2007) Dinitrophenol and obesity: an early twentieth-century regulatory dilemma. Regul Toxicol Pharmacol 48(2):115–117

Commins SP, Marsh DJ, Thomas SA, Watson PM, Padgett MA, Palmiter R, Gettys TW (1999) Norepinephrine is required for leptin effects on gene expression in brown and white adipose tissue. Endocrinology 140(10):4772–4778

Connacher A, Jung R, Mitchell P (1988) Weight loss in obese subjects on a restricted diet given BRL 26830A, a new atypical β adrenoceptor agonist. Br Med J (Clin Res Ed) 296(6631):1217

Connacher A, Lakie M, Powers N, Elton R, Walsh E, Jung R (1990) Tremor and the anti-obesity drug BRL 26830A. Br J Clin Pharmacol 30(4):613–615

Connacher A, Bennet W, Jung R, Rennie M (1992a) Metabolic effects of three weeks administration of the beta-adrenoceptor agonist BRL 26830A. Int J Obes Relat Metab Disord 16 (9):685–694

Connacher AA, Bennet WM, Jung RT (1992b) Clinical studies with the β-adrenoceptor agonist BRL 26830A. Oxford University Press, Oxford

Contaldo F, Scalfi L, Coltorti A, Lanzilli A (1986) Reduced cold-induced thermogenesis in familial human obesity. Klin Wochenschr 64(4):177–180

Cuevas-Ramos D, Almeda-Valdes P, Gómez-Pérez FJ, Meza-Arana CE, Cruz-Bautista I, Arellano-Campos O, Navarrete-López M, Aguilar-Salinas CA (2010) Daily physical activity, fasting glucose, uric acid, and body mass index are independent factors associated with serum fibroblast growth factor 21 levels. Eur J Endocrinol 163(3):469–477

Cunningham S, Leslie P, Hopwood D, Illingworth P, Jung R, Nicholls D, Peden N, Rafael J, Rial E (1985) The characterization and energetic potential of brown adipose tissue in man. Clin Sci 69 (3):343–348

Cutting W, Tainter M (1933) Metabolic actions of dinitrophenol: with the use of balanced and unbalanced diets. J Am Med Assoc 101(27):2099–2102

Cutting W, Mehrtens H, Tainter M (1933) Actions and uses of dinitrophenol: promising metabolic applications. J Am Med Assoc 101(3):193–195

Cypess AM, Lehman S, Williams G, Tal I, Rodman D, Goldfine AB, Kuo FC, Palmer EL, Tseng Y-H, Doria A (2009) Identification and importance of brown adipose tissue in adult humans. N Engl J Med 360(15):1509–1517

Cypess AM, Chen Y-C, Sze C, Wang K, English J, Chan O, Holman AR, Tal I, Palmer MR, Kolodny GM (2012) Cold but not sympathomimetics activates human brown adipose tissue in vivo. Proc Natl Acad Sci 109(25):10001–10005

Cypess AM, Weiner LS, Roberts-Toler C, Franquet Elía E, Kessler SH, Kahn PA, English J, Chatman K, Trauger SA, Doria A, Kolodny GM (2015) Activation of human brown adipose tissue by a β3-adrenergic receptor agonist. Cell Metab 21(1):33–38

Dauncey M (1981) Influence of mild cold on 24 h energy expenditure, resting metabolism and diet-induced thermogenesis. Br J Nutr 45(2):257–267

de Jesus LA, Carvalho SD, Ribeiro MO, Schneider M, Kim S-W, Harney JW, Larsen PR, Bianco AC (2001) The type 2 iodothyronine deiodinase is essential for adaptive thermogenesis in brown adipose tissue. J Clin Invest 108(9):1379–1385

de Matteis R, Arch J, Petroni M, Ferrari D, Cinti S, Stock M (2002) Immunohistochemical identification of the β 3-adrenoceptor in intact human adipocytes and ventricular myocardium: effect of obesity and treatment with ephedrine and caffeine. Int J Obes 26(11):1442

de Ponti F, Modini C, Gibelli G, Crema F, Frigo G (1999) Atypical β-adrenoceptors mediating relaxation in the human colon: functional evidence for β3-rather than β4-adrenoceptors. Pharmacol Res 39(5):345–348

Dhar R, Stout CW, Link MS, Homoud MK, Weinstock J, Estes NM III (2005) Cardiovascular toxicities of performance-enhancing substances in sports. Mayo Clin Proc 80:1307–1315

Diepvens K, Westerterp KR, Westerterp-Plantenga MS (2007) Obesity and thermogenesis related to the consumption of caffeine, ephedrine, capsaicin, and green tea. Am J Phys Regul Integr Comp Phys 292(1):R77–R85

Digby JE, Montague CT, Sewter CP, Sanders L, Wilkison WO, O'Rahilly S, Prins JB (1998) Thiazolidinedione exposure increases the expression of uncoupling protein 1 in cultured human preadipocytes. Diabetes 47(1):138–141

Dolan JA, Muenkel HA, Burns MG, Pellegrino SM, Fraser CM, Pietri F, Strosberg AD, Largis EE, Dutia MD, Bloom JD (1994) Beta-3 adrenoceptor selectivity of the dioxolane dicarboxylate phenethanolamines. J Pharmacol Exp Ther 269(3):1000–1006

Dong JQ, Rossulek M, Somayaji VR, Baltrukonis D, Liang Y, Hudson K, Hernandez-Illas M, Calle RA (2015a) Pharmacokinetics and pharmacodynamics of PF-05231023, a novel long-acting FGF21 mimetic, in a first-in-human study. Br J Clin Pharmacol 80(5):1051–1063

Dong JQ, Rossulek M, Somayaji VR, Baltrukonis D, Liang Y, Hudson K, Hernandez-Illas M, Calle RA (2015b) Pharmacokinetics and pharmacodynamics of PF-05231023, a novel long-acting FGF21 mimetic, in a first-in-human study. Br J Clin Pharmacol 80(5):1051–1063. https://doi.org/10.1111/bcp.12676

Doyon C, Denis RG, Baraboi ED, Samson P, Lalonde J, Deshaies Y, Richard D (2006) Effects of rimonabant (SR141716) on fasting-induced hypothalamic-pituitary-adrenal axis and neuronal activation in lean and obese Zucker rats. Diabetes 55(12):3403–3410. https://doi.org/10.2337/db06-0504

Dulloo A, Seydoux J, Girardier L (1991) Peripheral mechanisms of thermogenesis induced by ephedrine and caffeine in brown adipose tissue. Int J Obes 15(5):317–326

Elabd C, Chiellini C, Carmona M, Galitzky J, Cochet O, Petersen R, Pénicaud L, Kristiansen K, Bouloumié A, Casteilla L (2009) Human multipotent adipose-derived stem cells differentiate into functional brown adipocytes. Stem Cells 27(11):2753–2760

Elmquist JK, Maratos-Flier E, Saper CB, Flier JS (1998) Unraveling the central nervous system pathways underlying responses to leptin. Nat Neurosci 1(6):445

Enerbäck S (2010) Human brown adipose tissue. Cell Metab 11(4):248–252

Enerbäck S, Jacobsson A, Simpson EM, Guerra C, Yamashita H, Harper M-E, Kozak LP (1997) Mice lacking mitochondrial uncoupling protein are cold-sensitive but not obese. Nature 387 (6628):90

Feldmann HM, Golozoubova V, Cannon B, Nedergaard J (2009) UCP1 ablation induces obesity and abolishes diet-induced thermogenesis in mice exempt from thermal stress by living at thermoneutrality. Cell Metab 9(2):203–209

Ferrannini E, Galvan A, Gastaldelli A, Camastra S, Sironi A, Toschi E, Baldi S, Frascerra S, Monzani F, Antonelli A (1999) Insulin: new roles for an ancient hormone. Eur J Clin Investig 29 (10):842–852

Ferré P, Pénicaud L, Hitier Y, Meier M, Girard J (1992) Hypoglycemic effects of a beta-agonist, Ro 16-8714, in streptozotocin-diabetic rats: decreased hepatic glucose production and increased glucose utilization in oxidative muscles. Metabolism 41(2):180–183

Fletcher DS, Candelore MR, Grujic D, Lowell BB, Luell S, Susulic VS, Macintyre DE (1998) Beta-3 adrenergic receptor agonists cause an increase in gastrointestinal transit time in wild-type mice, but not in mice lacking the beta-3 adrenergic receptor. J Pharmacol Exp Ther 287 (2):720–724

Foster DO, Frydman ML (1979) Tissue distribution of cold-induced thermogenesis in conscious warm-or cold-acclimated rats reevaluated from changes in tissue blood flow: the dominant role of brown adipose tissue in the replacement of shivering by nonshivering thermogenesis. Can J Physiol Pharmacol 57(3):257–270

Gaich G, Chien JY, Fu H, Glass LC, Deeg MA, Holland WL, Kharitonenkov A, Bumol T, Schilske HK, Moller DE (2013) The effects of LY2405319, an FGF21 analog, in obese human subjects with type 2 diabetes. Cell Metab 18(3):333–340

Galgani JE, Ravussin E (2010) Effect of dihydrocapsiate on resting metabolic rate in humans. Am J Clin Nutr 92(5):1089–1093

Gallego-Escuredo J, Gomez-Ambrosi J, Catalan V, Domingo P, Giralt M, Frühbeck G, Villarroya F (2015) Opposite alterations in FGF21 and FGF19 levels and disturbed expression of the receptor machinery for endocrine FGFs in obese patients. Int J Obes 39(1):121

Garcia CA, van Nostrand D, Atkins F, Acio E, Butler C, Esposito G, Kulkarni K, Majd M (2006) Reduction of brown fat 2-deoxy-2-[F-18] fluoro-D-glucose uptake by controlling environmental temperature prior to positron emission tomography scan. Mol Imaging Biol 8(1):24–29

Gavrila A, Hasselgren P-O, Glasgow A, Doyle AN, Lee AJ, Fox P, Gautam S, Hennessey JV, Kolodny GM, Cypess AM (2017) Variable cold-induced brown adipose tissue response to thyroid hormone status. Thyroid 27(1):1–10

George A, Sinha P, Conrad G, Memon AA, Dressler EV, Wagner LM (2017) Pilot study of propranolol premedication to reduce FDG uptake in brown adipose tissue on PET scans of adolescent and young adult oncology patients. Pediatr Hematol Oncol 34(3):136–143

Ghorbani M, Claus TH, Himms-Hagen J (1997) Hypertrophy of brown adipocytes in brown and white adipose tissues and reversal of diet-induced obesity in rats treated with a β3-adrenoceptor agonist. Biochem Pharmacol 54(1):121–131

Giese J (1996) Olestra: properties, regulatory concerns, and applications. Food Technol 50(3):86

Gimeno RE, Moller DE (2014) FGF21-based pharmacotherapy–potential utility for metabolic disorders. Trends Endocrinol Metab 25(6):303–311

Gnad T, Scheibler S, von Kügelgen I, Scheele C, Kilić A, Glöde A, Hoffmann LS, Reverte-Salisa L, Horn P, Mutlu S, El-Tayeb A, Kranz M, Deuther-Conrad W, Brust P, Lidell ME, Betz MJ, Enerbäck S, Schrader J, Yegutkin GG, Müller CE, Pfeifer A (2014) Adenosine activates brown adipose tissue and recruits beige adipocytes via A2A receptors. Nature 516(7531):395–399

Goglia F, Silvestri E, Lanni A (2002) Thyroid hormones and mitochondria. Biosci Rep 22(1):17–32

Golozoubova V, Hohtola E, Matthias A, Jacobsson A, Cannon B, Nedergaard J (2001) Only UCP1 can mediate adaptive nonshivering thermogenesis in the cold. FASEB J 15(11):2048–2050

Golozoubova V, Gullberg H, Matthias A, Cannon B, Vennström B, Nedergaard J (2004) Depressed thermogenesis but competent brown adipose tissue recruitment in mice devoid of all hormone-binding thyroid hormone receptors. Mol Endocrinol 18(2):384–401

Golozoubova V, Cannon B, Nedergaard J (2006) UCP1 is essential for adaptive adrenergic nonshivering thermogenesis. Am J Physiol Endocrinol Metab 291(2):E350–E357

Grujic D, Susulic VS, Harper ME, Himms-Hagen J, Cunningham BA, Corkey BE, Lowell BB (1997) Beta3-adrenergic receptors on white and brown adipocytes mediate beta3-selective agonist-induced effects on energy expenditure, insulin secretion, and food intake. A study using transgenic and gene knockout mice. J Biol Chem 272(28):17686–17693

Guerra C, Roncero C, Porras A, Fernández M, Benito M (1996) Triiodothyronine induces the transcription of the uncoupling protein gene and stabilizes its mRNA in fetal rat brown adipocyte primary cultures. J Biol Chem 271(4):2076–2081

Guerra C, Koza RA, Yamashita H, Walsh K, Kozak LP (1998) Emergence of brown adipocytes in white fat in mice is under genetic control. Effects on body weight and adiposity. J Clin Invest 102(2):412–420

Gustafson B, Hammarstedt A, Hedjazifar S, Hoffmann JM, Svensson P-A, Grimsby J, Rondinone C, Smith U (2015) BMP4 and BMP antagonists regulate human white and beige adipogenesis. Diabetes 64(5):1670–1681

Hafner RP, Brown GC, Brand MD (1990) Thyroid-hormone control of state-3 respiration in isolated rat liver mitochondria. Biochem J 265(3):731–734

Hall KD, Sacks G, Chandramohan D, Chow CC, Wang YC, Gortmaker SL, Swinburn BA (2011) Quantification of the effect of energy imbalance on bodyweight. Lancet 378(9793):826–837

Hany TF, Gharehpapagh E, Kamel EM, Buck A, Himms-Hagen J, von Schulthess GK (2002) Brown adipose tissue: a factor to consider in symmetrical tracer uptake in the neck and upper chest region. Eur J Nucl Med Mol Imaging 29(10):1393–1398

Harper M-E, Himms-Hagen J (2001) Mitochondrial efficiency: lessons learned from transgenic mice. Biochim Biophys Acta 1504(1):159–172

Harper J, Dickinson K, Brand M (2001) Mitochondrial uncoupling as a target for drug development for the treatment of obesity. Obes Rev 2(4):255–265

Harper JA, Stuart JA, Jekabsons MB, Roussel D, Brindle KM, Dickinson K, Jones RB, Brand MD (2002) Artifactual uncoupling by uncoupling protein 3 in yeast mitochondria at the concentrations found in mouse and rat skeletal-muscle mitochondria. Biochem J 361(1):49–56

Harper M-E, Green K, Brand MD (2008) The efficiency of cellular energy transduction and its implications for obesity. Annu Rev Nutr 28:13–33

Harris RB (2000) Leptin – much more than a satiety signal. Annu Rev Nutr 20(1):45–75

Heaton JM (1972) The distribution of brown adipose tissue in the human. J Anat 112(Pt 1):35

Heaton GM, Wagenvoord RJ, Kemp A Jr, Nicholls DG (1978) Brown-adipose-tissue mitochondria: photoaffinity labelling of the regulatory site of energy dissipation. Eur J Biochem 82(2):515–521

Himms-Hagen J (1979) Obesity may be due to a malfunctioning of brown fat. Can Med Assoc J 121 (10):1361–1364

Himms-Hagen J (1989) Brown adipose tissue thermogenesis and obesity. Prog Lipid Res 28 (2):67–115

Himms-Hagen J (2004) Exercise in a pill: feasibility of energy expenditure targets. Curr Drug Targets CNS Neurol Disord 3(5):389–409

Himms-Hagen J, Cui J, Danforth E Jr, Taatjes D, Lang S, Waters B, Claus T (1994) Effect of CL-316,243, a thermogenic beta 3-agonist, on energy balance and brown and white adipose tissues in rats. Am J Phys Regul Integr Comp Phys 266(4):R1371–R1382

Hoffstedt J, Poirier O, Thorne A, Lonnqvist F, Herrmann SM, Cambien F, Arner P (1999) Polymorphism of the human [[Beta]. sub. 3]-adrenoceptor gene forms a well-conserved haplotype that is associated with moderate obesity and altered receptor function. Diabetes 48 (1):203–204

Hosaka T, Biggs WH, Tieu D, Boyer AD, Varki NM, Cavenee WK, Arden KC (2004) Disruption of forkhead transcription factor (FOXO) family members in mice reveals their functional diversification. Proc Natl Acad Sci 101(9):2975–2980

Huang Z-L, Qu W-M, Eguchi N, Chen J-F, Schwarzschild MA, Fredholm BB, Urade Y, Hayaishi O (2005) Adenosine A 2A, but not A 1, receptors mediate the arousal effect of caffeine. Nat Neurosci 8(7):858

Imai T, Takakuwa R, Marchand S, Dentz E, Bornert J-M, Messaddeq N, Wendling O, Mark M, Desvergne B, Wahli W (2004) Peroxisome proliferator-activated receptor γ is required in mature white and brown adipocytes for their survival in the mouse. Proc Natl Acad Sci 101 (13):4543–4547

Inokuma K, Ogura-Okamatsu Y, Toda C, Kimura K, Yamashita H, Saito M (2005) Uncoupling protein 1 is necessary for norepinephrine-induced glucose utilization in brown adipose tissue. Diabetes 54(5):1385–1391

Inoue N, Matsunaga Y, Satoh H, Takahashi M (2007) Enhanced energy expenditure and fat oxidation in humans with high BMI scores by the ingestion of novel and non-pungent capsaicin analogues (capsinoids). Biosci Biotechnol Biochem 71(2):380–389

Isler D, Hill H-P, Meier MK (1987) Glucose metabolism in isolated brown adipocytes under β-adrenergic stimulation. Quantitative contribution of glucose to total thermogenesis. Biochem J 245(3):789–793

Iwami M, Mahmoud FA, Shiina T, Hirayama H, Shima T, Sugita J, Shimizu Y (2011) Extract of grains of paradise and its active principle 6-paradol trigger thermogenesis of brown adipose tissue in rats. Auton Neurosci 161(1–2):63–67

Iwen KA, Backhaus J, Cassens M, Waltl M, Hedesan OC, Merkel M, Heeren J, Sina C, Rademacher L, Windjäger A (2017) Cold-induced brown adipose tissue activity alters plasma fatty acids and improves glucose metabolism in men. J Clin Endocrinol Metabol 102 (11):4226–4234

James WPT, Astrup A, Finer N, Hilsted J, Kopelman P, Rössner S, Saris WH, van Gaal LF, Group SS (2000) Effect of sibutramine on weight maintenance after weight loss: a randomised trial. Lancet 356(9248):2119–2125

James WPT, Caterson ID, Coutinho W, Finer N, van Gaal LF, Maggioni AP, Torp-Pedersen C, Sharma AM, Shepherd GM, Rode RA (2010) Effect of sibutramine on cardiovascular outcomes in overweight and obese subjects. N Engl J Med 363(10):905–917

Jequier E, Munger R, Felber J (1992) Thermogenic effects of various β-adrenoceptor agonists in humans: their potential usefulness in the treatment of obesity. Oxford University Press, Oxford

Jespersen NZ, Larsen TJ, Peijs L, Daugaard S, Homøe P, Loft A, de Jong J, Mathur N, Cannon B, Nedergaard J (2013) A classical brown adipose tissue mRNA signature partly overlaps with brite in the supraclavicular region of adult humans. Cell Metab 17(5):798–805

Jessen K, Rabøl A, Winkler K (1980) Total body and splanchnic thermogenesis in curarized man during a short exposure to cold. Acta Anaesthesiol Scand 24(4):339–344

Jezek P, Orosz DE, Garlid K (1990) Reconstitution of the uncoupling protein of brown adipose tissue mitochondria. Demonstration of GDP-sensitive halide anion uniport. J Biol Chem 265 (31):19296–19302

Joint F (2001) Human energy requirements. Report of a Joint FAO/WHO/UNU Expert Consultation, Rome, 17–24 Oct 2001–2004

Joy RJ (1963) Responses of cold-acclimatized men to infused norepinephrine. J Appl Physiol 18 (6):1209–1212

Kajimura S, Seale P, Kubota K, Lunsford E, Frangioni JV, Gygi SP, Spiegelman BM (2009) Initiation of myoblast to brown fat switch by a PRDM16–C/EBP-β transcriptional complex. Nature 460(7259):1154

Katiyar SS, Shrago E (1989) Reconstitution of purified brown adipose tissue mitochondria uncoupling protein: demonstration of separate identity of nucleotide binding and proton translocation sites by chemical probes. Proc Natl Acad Sci 86(8):2559–2562

Kawabata F, Inoue N, Yazawa S, Kawada T, Inoue K, Fushiki T (2006) Effects of CH-19 sweet, a non-pungent cultivar of red pepper, in decreasing the body weight and suppressing body fat accumulation by sympathetic nerve activation in humans. Biosci Biotechnol Biochem 70 (12):2824–2835

Kawabata F, Inoue N, Masamoto Y, Matsumura S, Kimura W, Kadowaki M, Higashi T, Tominaga M, Inoue K, Fushiki T (2009) Non-pungent capsaicin analogs (capsinoids) increase metabolic rate and enhance thermogenesis via gastrointestinal TRPV1 in mice. Biosci Biotechnol Biochem 73(12):2690–2697

Kawada T, Hagihara K-I, Iwai K (1986a) Effects of capsaicin on lipid metabolism in rats fed a high fat diet. J Nutr 116(7):1272–1278

Kawada T, Watanabe T, Takaishi T, Tanaka T, Iwai K (1986b) Capsaicin-induced β-adrenergic action on energy metabolism in rats: influence of capsaicin on oxygen consumption, the respiratory quotient, and substrate utilization. Proc Soc Exp Biol Med 183(2):250–256

Kern PA, Finlin BS, Zhu B, Rasouli N, McGehee RE Jr, Westgate PM, Dupont-Versteegden EE (2014) The effects of temperature and seasons on subcutaneous white adipose tissue in humans: evidence for thermogenic gene induction. J Clin Endocrinol Metabol 99(12):E2772–E2779

Kharitonenkov A, DiMarchi R (2015) FGF21 Revolutions: recent advances illuminating FGF21 biology and medicinal properties. Trends Endocrinol Metab 26(11):608–617

Kim S, Krynyckyi BR, Machac J, Kim CK (2008) Temporal relation between temperature change and FDG uptake in brown adipose tissue. Eur J Nucl Med Mol Imaging 35(5):984–989

Kim MS, Hu HH, Aggabao PC, Geffner ME, Gilsanz V (2014) Presence of brown adipose tissue in an adolescent with severe primary hypothyroidism. J Clin Endocrinol Metabol 99(9):E1686–E1690

Klein S, Wolfe R (1990) Whole-body lipolysis and triglyceride-fatty acid cycling in cachectic patients with esophageal cancer. J Clin Invest 86(5):1403–1408

Kobata K, Todo T, Yazawa S, Iwai K, Watanabe T (1998) Novel capsaicinoid-like substances, capsiate and dihydrocapsiate, from the fruits of a nonpungent cultivar, CH-19 Sweet, of pepper (Capsicum annuum L.). J Agric Food Chem 46(5):1695–1697

Kontani Y, Wang Y, Kimura K, Inokuma KI, Saito M, Suzuki-Miura T, Wang Z, Sato Y, Mori N, Yamashita H (2005) UCP1 deficiency increases susceptibility to diet-induced obesity with age. Aging Cell 4(3):147–155

Kopecky J, Clarke G, Enerbäck S, Spiegelman B, Kozak LP (1995) Expression of the mitochondrial uncoupling protein gene from the aP2 gene promoter prevents genetic obesity. J Clin Invest 96 (6):2914–2923

Krief S, Lönnqvist F, Raimbault S, Baude B, Van Spronsen A, Arner P, Strosberg AD, Ricquier D, Emorine LJ (1993) Tissue distribution of beta 3-adrenergic receptor mRNA in man. J Clin Invest 91(1):344–349

Kusirisin W, Srichairatanakool S, Lerttrakarnnon P, Lailerd N, Suttajit M, Jaikang C, Chaiyasut C (2009) Antioxidative activity, polyphenolic content and anti-glycation effect of some Thai medicinal plants traditionally used in diabetic patients. Med Chem 5(2):139–147

Lahesmaa M, Orava J, Schalin-Jäntti C, Soinio M, Hannukainen JC, Noponen T, Kirjavainen A, Iida H, Kudomi N, Enerbäck S (2014) Hyperthyroidism increases brown fat metabolism in humans. J Clin Endocrinol Metabol 99(1):E28–E35

Landsberg L, Saville ME, Young JB (1984) Sympathoadrenal system and regulation of thermogenesis. Am J Physiol Endocrinol Metab 247(2):E181–E189

Larsen TM, Toubro S, van Baak MA, Gottesdiener KM, Larson P, Saris WH, Astrup A (2002) Effect of a 28-d treatment with L-796568, a novel β3-adrenergic receptor agonist, on energy expenditure and body composition in obese men. Am J Clin Nutr 76(4):780–788

Lean M, James W, Jennings G, Trayhurn P (1986) Brown adipose tissue uncoupling protein content in human infants, children and adults. Clin Sci (Lond) 71(3):291–297

Lee P, Greenfield JR, Ho KK, Fulham MJ (2010a) A critical appraisal of the prevalence and metabolic significance of brown adipose tissue in adult humans. Am J Physiol Endocrinol Metab 299(4):E601–E606

Lee TA, Li Z, Zerlin A, Heber D (2010b) Effects of dihydrocapsiate on adaptive and diet-induced thermogenesis with a high protein very low calorie diet: a randomized control trial. Nutr Metab 7(1):78

Lee P, Zhao JT, Swarbrick MM, Gracie G, Bova R, Greenfield JR, Freund J, Ho KK (2011a) High prevalence of brown adipose tissue in adult humans. J Clin Endocrinol Metabol 96 (8):2450–2455

Lee J-Y, Takahashi N, Yasubuchi M, Kim Y-I, Hashizaki H, Kim M-J, Sakamoto T, Goto T, Kawada T (2011b) Triiodothyronine induces UCP-1 expression and mitochondrial biogenesis in human adipocytes. Am J Phys Cell Phys 302(2):C463–C472

Lehninger AH (1971) Bioenergetics. Benjamin Cummings Publishing Group, San Francisco

Leibel RL, Rosenbaum M, Hirsch J (1995) Changes in energy expenditure resulting from altered body weight. N Engl J Med 332(10):621–628

Leonardsson G, Steel JH, Christian M, Pocock V, Milligan S, Bell J, So P-W, Medina-Gomez G, Vidal-Puig A, White R (2004) Nuclear receptor corepressor RIP140 regulates fat accumulation. Proc Natl Acad Sci 101(22):8437–8442

Leslie WS, Hankey CR, Lean ME (2007) Weight gain as an adverse effect of some commonly prescribed drugs: a systematic review. Int J Med 100(7):395–404

Levine JA, Eberhardt NL, Jensen MD (1999) Role of nonexercise activity thermogenesis in resistance to fat gain in humans. Science 283(5399):212–214

Li B, Nolte LA, Ju J-S, Han DH, Coleman T, Holloszy JO, Semenkovich CF (2000) Skeletal muscle respiratory uncoupling prevents diet-induced obesity and insulin resistance in mice. Nat Med 6 (10):1115

Li S, Li Y, Xiang L, Dong J, Liu M, Xiang G (2018) Sildenafil induces browning of subcutaneous white adipose tissue in overweight adults. Metabolism 78:106–117

Lin C, Hackenberg H, Klingenberg E (1980) The uncoupling protein from brown adipose tissue mitochondria is a dimer. A hydrodynamic study. FEBS Lett 113(2):304–306

Loh RKC, Formosa MF, Eikelis N, Bertovic DA, Anderson MJ, Barwood SA, Nanayakkara S, Cohen ND, La Gerche A, Reutens AT, Yap KS, Barber TW, Lambert GW, Cherk MH, Duffy SJ, Kingwell BA, Carey AL (2018) Pioglitazone reduces cold-induced brown fat glucose uptake despite induction of browning in cultured human adipocytes: a randomised, controlled trial in humans. Diabetologia 61(1):220–230

López M, Varela L, Vázquez MJ, Rodríguez-Cuenca S, González CR, Velagapudi VR, Morgan DA, Schoenmakers E, Agassandian K, Lage R (2010) Hypothalamic AMPK and fatty acid metabolism mediate thyroid regulation of energy balance. Nat Med 16(9):1001

Lowell BB, Spiegelman BM (2000) Towards a molecular understanding of adaptive thermogenesis. Nature 404(6778):652

Lowell BB, Vedrana S, Hamann A, Lawitts JA, Himms-Hagen J, Boyer BB, Kozak LP, Flier JS (1993) Development of obesity in transgenic mice after genetic ablation of brown adipose tissue. Nature 366(6457):740

Luo X-J, Peng J, Li Y-J (2011) Recent advances in the study on capsaicinoids and capsinoids. Eur J Pharmacol 650(1):1–7

Luque CA, Rey JA (2002) The discovery and status of sibutramine as an anti-obesity drug. Eur J Pharmacol 440(2–3):119–128

Maes HH, Neale MC, Eaves LJ (1997) Genetic and environmental factors in relative body weight and human adiposity. Behav Genet 27(4):325–351

Magkos F, Kavouras SA (2005) Caffeine use in sports, pharmacokinetics in man, and cellular mechanisms of action. Crit Rev Food Sci Nutr 45(7–8):535–562

Magne H, Mayer A, Plantefol L (1932) Etudes sur l'action du dinitrophenol 1-2-4 (thermol). Ann de Physiol 8:70

Malchow-Møller A, Larsen S, Hey H, Stokholm KH, Juhl E, Quaade F (1981) Ephedrine as an anorectic: the story of the 'Elsinore pill'. Int J Obes 5(2):183–187

Manara L, Croci T, Aureggi G, Guagnini F, Maffrand J, Le Fur G, Mukenge S, Ferla G (2000) Functional assessment of β adrenoceptor subtypes in human colonic circular and longitudinal (taenia coli) smooth muscle. Gut 47(3):337–342

Marette A, Bukowiecki LJ (1991) Noradrenaline stimulates glucose transport in rat brown adipocytes by activating thermogenesis. Evidence that fatty acid activation of mitochondrial respiration enhances glucose transport. Biochem J 277(1):119–124

Masuda Y, Haramizu S, Oki K, Ohnuki K, Watanabe T, Yazawa S, Kawada T, Hashizume S-i, Fushiki T (2003) Upregulation of uncoupling proteins by oral administration of capsiate, a nonpungent capsaicin analog. J Appl Physiol 95(6):2408–2415

Mathvink RJ, Tolman JS, Chitty D, Candelore MR, Cascieri MA, Colwell LF, Deng L, Feeney WP, Forrest MJ, Hom GJ (2000) Discovery of a potent, orally bioavailable β3 adrenergic receptor agonist,(R)-N-[4-[2-[[2-hydroxy-2-(3-pyridinyl) ethyl] amino] ethyl] phenyl]-4-[4-[4-(trifluoromethyl) phenyl] thiazol-2-yl] benzenesulfonamide. J Med Chem 43 (21):3832–3836

Matsumoto T, Miyawaki C, Ue H, Kanda T, Yoshitake Y, Moritani T (2001) Comparison of thermogenic sympathetic response to food intake between obese and non-obese young women. Obes Res 9(2):78–85. https://doi.org/10.1038/oby.2001.10

Matsushita M, Yoneshiro T, Aita S, Kamiya T, Kusaba N, Yamaguchi K, Takagaki K, Kameya T, Sugie H, Saito M (2015) Kaempferia parviflora extract increases whole-body energy expenditure in humans: roles of brown adipose tissue. J Nutr Sci Vitaminol 61(1):79–83

Mazzucotelli A, Viguerie N, Tiraby C, Annicotte J-S, Mairal A, Klimcakova E, Lepin E, Delmar P, Dejean S, Tavernier G (2007) The transcriptional coactivator peroxisome proliferator–activated receptor (PPAR) γ coactivator-1α and the nuclear receptor PPARα control the expression of glycerol kinase and metabolism genes independently of PPARγ activation in human white adipocytes. Diabetes 56(10):2467–2475

Melnyk A, Harper M, Himms-Hagen J (1997) Raising at thermoneutrality prevents obesity and hyperphagia in BAT-ablated transgenic mice. Am J Phys Regul Integr Comp Phys 272(4): R1088–R1093

Mifflin MD, St Jeor ST, Hill LA, Scott BJ, Daugherty SA, Koh YO (1990) A new predictive equation for resting energy expenditure in healthy individuals. Am J Clin Nutr 51(2):241–247

Mistry AM, Swick AG, Romsos DR (1997) Leptin rapidly lowers food intake and elevates metabolic rates in lean and Ob/Ob mice. J Nutr 127(10):2065–2072

Mitchell T, Ellis R, Smith S, Robb G, Cawthorne M (1989) Effects of BRL 35135, a beta-adrenoceptor agonist with novel selectivity, on glucose tolerance and insulin sensitivity in obese subjects. Int J Obes 13(6):757–766

Mitschke MM, Hoffmann LS, Gnad T, Scholz D, Kruithoff K, Mayer P, Haas B, Sassmann A, Pfeifer A, Kilić A (2013) Increased cGMP promotes healthy expansion and browning of white adipose tissue. FASEB J 27(4):1621–1630

Møller N, Nielsen S, Nyholm B, Pørksen N, George K, Alberti M, Weeke J (1996) Glucose turnover, fuel oxidation and forearm substrate exchange in patients with thyrotoxicosis before and after medical treatment. Clin Endocrinol 44(4):453–459

Morera E, de Petrocellis L, Morera L, Moriello AS, Nalli M, di Marzo V, Ortar G (2012) Synthesis and biological evaluation of [6]-gingerol analogues as transient receptor potential channel TRPV1 and TRPA1 modulators. Bioorg Med Chem Lett 22(4):1674–1677

Morrison SF, Nakamura K, Madden CJ (2008) Central control of thermogenesis in mammals. Exp Physiol 93(7):773–797

Munro J, Chapman B, Robb G, Zed C (1987) Clinical studies with thermogenic drugs. Recent Adv Obesity Res

Muzik O, Mangner TJ, Granneman JG (2012) Assessment of oxidative metabolism in brown fat using PET imaging. Front Endocrinol 3:15

Nakamura K (2011) Central circuitries for body temperature regulation and fever. Am J Phys Regul Integr Comp Phys 301(5):R1207–R1R28

Nakamura K, Morrison SF (2007) Central efferent pathways mediating skin cooling-evoked sympathetic thermogenesis in brown adipose tissue. Am J Phys Regul Integr Comp Phys 292 (1):R127–R136

Nedergaard J, Golozoubova V, Matthias A, Asadi A, Jacobsson A, Cannon B (2001) UCP1: the only protein able to mediate adaptive non-shivering thermogenesis and metabolic inefficiency. Biochim Biophys Acta 1504(1):82–106

Nedergaard J, Bengtsson T, Cannon B (2007) Unexpected evidence for active brown adipose tissue in adult humans. Am J Physiol Endocrinol Metab 293(2):E444–E452
Nedergaard J, Feldmann H, Cannon B (2010) Brown adipose tissue is essential for diet-induced thermogenesis: the absence of Ucp1 makes the obesity-resistant 129sv mouse obesity-prone, due to lack of adaptive adrenergic thermogenesis: T1: po. 06. Obes Rev 11:92
Neumann RO (1902) Experimentelle beiträge zur lehre von dem täglichen nahrungsbedarf des menschen unter besonderer berück-sichtigung der notwendigen eiweissmenge. Arch Hyg (Berl) 45:1
Newsholme E, Crabtree B (1976) Substrate cycles in metabolic regulation and in heat generation. Biochem Soc Symp 41:61–109
Nicholls DG (1976) Hamster brown-adipose-tissue mitochondria: purine nucleotide control of the ion conductance of the inner membrane, the nature of the nucleotide binding site. Eur J Biochem 62(2):223–228
Nicholls DG (1977) The effective proton conductance of the inner membrane of mitochondria from brown adipose tissue: dependency on proton electrochemical potential gradient. Eur J Biochem 77(2):349–356
Nicholls DG (1979) Brown adipose tissue mitochondria. Biochim Biophys Acta 549(1):1–29
Nicholls DG, Locke RM (1984) Thermogenic mechanisms in brown fat. Physiol Rev 64(1):1–64
Nicholls DG, Rial E (1999) A history of the first uncoupling protein, UCP1. J Bioenerg Biomembr 31(5):399–406
Ohnuki K, Haramizu S, Oki K, Watanabe T, Yazawa S, Fushiki T (2001a) Administration of capsiate, a non-pungent capsaicin analog, promotes energy metabolism and suppresses body fat accumulation in mice. Biosci Biotechnol Biochem 65(12):2735–2740
Ohnuki K, Niwa S, Maeda S, Inoue N, Yazawa S, Fushiki T (2001b) CH-19 sweet, a non-pungent cultivar of red pepper, increased body temperature and oxygen consumption in humans. Biosci Biotechnol Biochem 65(9):2033–2036
Ono K, Tsukamoto-Yasui M, Hara-Kimura Y, Inoue N, Nogusa Y, Okabe Y, Nagashima K, Kato F (2010) Intragastric administration of capsiate, a transient receptor potential channel agonist, triggers thermogenic sympathetic responses. J Appl Physiol 110(3):789–798
Ooijen AMC, Westerterp KR, Wouters L, Schoffelen PF, van Steenhoven AA, van Lichtenbelt WDM (2006) Heat production and body temperature during cooling and rewarming in overweight and lean men. Obesity 14(11):1914–1920
Orava J, Nuutila P, Lidell ME, Oikonen V, Noponen T, Viljanen T, Scheinin M, Taittonen M, Niemi T, Enerbäck S (2011) Different metabolic responses of human brown adipose tissue to activation by cold and insulin. Cell Metab 14(2):272–279
Ouellet V, Routhier-Labadie A, Bellemare W, Lakhal-Chaieb L, Turcotte E, Carpentier AC, Richard D (2011) Outdoor temperature, age, sex, body mass index, and diabetic status determine the prevalence, mass, and glucose-uptake activity of ^{18F}-FDG-detected BAT in humans. J Clin Endocrinol Metabol 96(1):192–199
Ouellet V, Labbé SM, Blondin DP, Phoenix S, Guérin B, Haman F, Turcotte EE, Richard D, Carpentier AC (2012) Brown adipose tissue oxidative metabolism contributes to energy expenditure during acute cold exposure in humans. J Clin Invest 122(2):545–552
Padwal RS, Majumdar SR (2007) Drug treatments for obesity: orlistat, sibutramine, and rimonabant. Lancet 369(9555):71–77
Pagotto U, Vanuzzo D, Vicennati V, Pasquali R (2008) Pharmacological therapy of obesity. G Ital Cardiol (Rome) 9(4 Suppl 1):83s–93s
Parascandola J (1974) Dinitrophenol and bioenergetics: an historical perspective. Mol Cell Biochem 5(1–2):69–77
Parysow O, Mollerach AM, Jager V, Racioppi S, San Roman J, Gerbaudo VH (2007) Low-dose oral propranolol could reduce brown adipose tissue F-18 FDG uptake in patients undergoing PET scans. Clin Nucl Med 32(5):351–357
Pasquali R, Baraldi G, Cesari M, Melchionda N, Zamboni M, Stefanini C, Raitano A (1985) A controlled trial using ephedrine in the treatment of obesity. Int J Obes 9(2):93–98

Pecqueur C, Alves-Guerra M-C, Gelly C, Lévi-Meyrueis C, Couplan E, Collins S, Ricquier D, Bouillaud F, Miroux B (2000) Uncoupling protein 2: in vivo distribution, induction upon oxidative stress and evidence for translational regulation. J Biol Chem 276:8705–8712

Peterson CM, Lecoultre V, Frost EA, Simmons J, Redman LM, Ravussin E (2016) The thermogenic responses to overfeeding and cold are differentially regulated. Obesity (Silver Spring) 24 (1):96–101. https://doi.org/10.1002/oby.21233

Petrovic N, Walden TB, Shabalina IG, Timmons JA, Cannon B, Nedergaard J (2010) Chronic peroxisome proliferator-activated receptor γ (PPARγ) activation of epididymally derived white adipocyte cultures reveals a population of thermogenically competent, UCP1-containing adipocytes molecularly distinct from classic brown adipocytes. J Biol Chem 285(10):7153–7164

Poehlman ET (1989) A review: exercise and its influence on resting energy metabolism in man. Med Sci Sports Exerc 21(5):515–525

Ramage LE, Akyol M, Fletcher AM, Forsythe J, Nixon M, Carter RN, van Beek EJ, Morton NM, Walker BR, Stimson RH (2016) Glucocorticoids acutely increase brown adipose tissue activity in humans, revealing species-specific differences in UCP-1 regulation. Cell Metab 24 (1):130–141

Randle P, Garland P, Hales C, Newsholme E (1963) The glucose fatty-acid cycle its role in insulin sensitivity and the metabolic disturbances of diabetes mellitus. Lancet 281(7285):785–789

Ravussin E, Kozak L (2009) Have we entered the brown adipose tissue renaissance? Obes Rev 10 (3):265–268

Ravussin E, Lillioja S, Knowler WC, Christin L, Freymond D, Abbott WG, Boyce V, Howard BV, Bogardus C (1988) Reduced rate of energy expenditure as a risk factor for body-weight gain. N Engl J Med 318(8):467–472

Ravussin Y, Xiao C, Gavrilova O, Reitman ML (2014) Effect of intermittent cold exposure on brown fat activation, obesity, and energy homeostasis in mice. PLoS One 9(1):e85876

Redman LM, de Jonge L, Fang X, Gamlin B, Recker D, Greenway FL, Smith SR, Ravussin E (2006) Lack of an effect of a novel β3-adrenoceptor agonist, TAK-677, on energy metabolism in obese individuals: a double-blind, placebo-controlled randomized study. J Clin Endocrinol Metabol 92(2):527–531

Redman LM, Heilbronn LK, Martin CK, de Jonge L, Williamson DA, Delany JP, Ravussin E (2009) Metabolic and behavioral compensations in response to caloric restriction: implications for the maintenance of weight loss. PLoS One 4(2):e4377

Reitman ML (2013) FGF21 mimetic shows therapeutic promise. Cell Metab 18(3):307–309

Richard D, Picard F (2011) Brown fat biology and thermogenesis. Front Biosci 16:1233–1260

Ricquier D, Bouillaud F (2000) The uncoupling protein homologues: UCP1, UCP2, UCP3, StUCP and AtUCP. Biochem J 345(2):161–179

Riera C, Menozzi-Smarrito C, Affolter M, Michlig S, Munari C, Robert F, Vogel H, Simon S, Le Coutre J (2009) Compounds from Sichuan and Melegueta peppers activate, covalently and non-covalently, TRPA1 and TRPV1 channels. Br J Pharmacol 157(8):1398–1409

Roberts-Toler C, O'Neill BT, Cypess AM (2015) Diet-induced obesity causes insulin resistance in mouse brown adipose tissue. Obesity (Silver Spring) 23(9):1765–1770

Robidoux J, Martin TL, Collins S (2004) β-Adrenergic receptors and regulation of energy expenditure: a family affair. Annu Rev Pharmacol Toxicol 44:297–323

Rodgers RJ, Tschöp MH, Wilding JP (2012) Anti-obesity drugs: past, present and future. Dis Model Mech 5(5):621–626

Rolfe D, Brand MD (1996) Contribution of mitochondrial proton leak to skeletal muscle respiration and to standard metabolic rate. Am J Physiol Cell Physiol 271(4):C1380–C1389

Rolfe D, Brown GC (1997) Cellular energy utilization and molecular origin of standard metabolic rate in mammals. Physiol Rev 77(3):731–758

Rolfe DF, Hulbert A, Brand MD (1994) Characteristics of mitochondrial proton leak and control of oxidative phosphorylation in the major oxygen-consuming tissues of the rat. Biochim Biophys Acta 1188(3):405–416

Rolfe DF, Newman JM, Buckingham JA, Clark MG, Brand MD (1999) Contribution of mitochondrial proton leak to respiration rate in working skeletal muscle and liver and to SMR. Am J Physiol Cell Physiol 276(3):C692–C699

Rosenbaum M, Goldsmith R, Bloomfield D, Magnano A, Weimer L, Heymsfield S, Gallagher D, Mayer L, Murphy E, Leibel RL (2005) Low-dose leptin reverses skeletal muscle, autonomic, and neuroendocrine adaptations to maintenance of reduced weight. J Clin Invest 115 (12):3579–3586

Rothwell NJ, Stock MJ (1979a) A role for brown adipose tissue in diet-induced thermogenesis. Nature 281(5726):31

Rothwell NJ, Stock MJ (1979b) Regulation of energy balance in two models of reversible obesity in the rat. J Comp Physiol Psychol 93(6):1024

Rothwell N, Stock M (1983) Luxuskonsumption, diet-induced thermogenesis and brown fat: the case in favour. Clin Sci 64(1):19–23

Rothwell N, Stock M (1987) Influence of clenbuterol on energy balance, thermogenesis and body composition in lean and genetically obese Zucker rats. Int J Obes 11(6):641–647

Rousseau C, Bourbouloux E, Campion L, Fleury N, Bridji B, Chatal J, Resche I, Campone M (2006) Brown fat in breast cancer patients: analysis of serial 18F-FDG PET/CT scans. Eur J Nucl Med Mol Imaging 33(7):785–791

Ruan CC, Kong LR, Chen XH, Ma Y, Pan XX, Zhang ZB, Gao PJ (2018) A2A receptor activation attenuates hypertensive cardiac remodeling via promoting brown adipose tissue-derived FGF21. Cell Metab 28(3):476–489

Rujjanawate C, Kanjanapothi D, Amornlerdpison D, Pojanagaroon S (2005) Anti-gastric ulcer effect of Kaempferia parviflora. J Ethnopharmacol 102(1):120–122

Saito M, Okamatsu-Ogura Y, Matsushita M, Watanabe K, Yoneshiro T, Nio-Kobayashi J, Iwanaga T, Miyagawa M, Kameya T, Nakada K (2009) High incidence of metabolically active brown adipose tissue in healthy adult humans: effects of cold exposure and adiposity. Diabetes 58:1526–1531. https://doi.org/10.2337/db09-0530

Schöder H, Larson SM, Yeung HW (2004) PET/CT in oncology: integration into clinical management of lymphoma, melanoma, and gastrointestinal malignancies. J Nucl Med 45(1 suppl):72S–81S

Scotney H, Symonds ME, Law J, Budge H, Sharkey D, Manolopoulos KN (2017) Glucocorticoids modulate human brown adipose tissue thermogenesis in vivo. Metabolism 70:125–132

Seale P, Bjork B, Yang W, Kajimura S, Chin S, Kuang S, Scime A, Devarakonda S, Conroe HM, Erdjument-Bromage H (2008) PRDM16 controls a brown fat/skeletal muscle switch. Nature 454(7207):961

Shannon JR, Gottesdiener K, Jordan J, Kong C, Flattery S, Larson PJ, Candelore MR, Gertz B, Robertson D, Ming S (1999) Acute effect of ephedrine on 24-h energy balance. Clin Sci 96 (5):483–491

Shaw WN, Schmiegel KK, Yen TT, Toomey RE, Meyers DB, Mills J (1981) LY79771: a novel compound for weight control. Life Sci 29(20):2091–2101

Shekelle PG, Hardy ML, Morton SC, Maglione M, Mojica WA, Suttorp MJ, Rhodes SL, Jungvig L, Gagné J (2003) Efficacy and safety of ephedra and ephedrine for weight loss and athletic performance: a meta-analysis. JAMA 289(12):1537–1545

Shimada T, Horikawa T, Ikeya Y, Matsuo H, Kinoshita K, Taguchi T, Ichinose K, Takahashi K, Aburada M (2011) Preventive effect of Kaempferia parviflora ethyl acetate extract and its major components polymethoxyflavonoid on metabolic diseases. Fitoterapia 82(8):1272–1278

Shintaku K, Uchida K, Suzuki Y, Zhou Y, Fushiki T, Watanabe T, Yazawa S, Tominaga M (2012) Activation of transient receptor potential A1 by a non-pungent capsaicin-like compound, capsiate. Br J Pharmacol 165(5):1476–1486

Sidossis LS, Porter C, Saraf MK, Børsheim E, Radhakrishnan RS, Chao T, Ali A, Chondronikola M, Mlcak R, Finnerty CC (2015) Browning of subcutaneous white adipose tissue in humans after severe adrenergic stress. Cell Metab 22(2):219–227

Silva JE (2006) Thermogenic mechanisms and their hormonal regulation. Physiol Rev 86 (2):435–464

Silva JE, Bianco SD (2008) Thyroid–adrenergic interactions: physiological and clinical implications. Thyroid 18(2):157–165

Sjostrom L, Schutz Y, Gudinchet F, Hegnell L, Pittet P, Jequier E (1983) Epinephrine sensitivity with respect to metabolic rate and other variables in women. Am J Physiol Endocrinol Metab 245(5):E431–E442

Sjöström L, Rissanen A, Andersen T, Boldrin M, Golay A, Koppeschaar HP, Krempf M, Group EMOS (1998) Randomised placebo-controlled trial of orlistat for weight loss and prevention of weight regain in obese patients. Lancet 352(9123):167–172

Skarulis MC, Celi FS, Mueller E, Zemskova M, Malek R, Hugendubler L, Cochran C, Solomon J, Chen C, Gorden P (2010) Thyroid hormone induced brown adipose tissue and amelioration of diabetes in a patient with extreme insulin resistance. J Clin Endocrinol Metabol 95(1):256–262

Smith S, Cawthorne M, Fay L, McCullough D, Mitchell T (1987) Effect of a novel-adrenoceptor agonist on insulin sensitivity in lean healthy male volunteers. Diabetes 36(Suppl 1):15A

Smith S, Zed C, McCullough D, Harris G, Cawthorne M (1989) Thermogenic activity in man of BRL 35135: a potent and selective atypical β-adrenoceptor agonist. Int J Obes 13(suppl 1):33

Snitker S, Fujishima Y, Shen H, Ott S, Pi-Sunyer X, Furuhata Y, Sato H, Takahashi M (2008) Effects of novel capsinoid treatment on fatness and energy metabolism in humans: possible pharmacogenetic implications. Am J Clin Nutr 89(1):45–50

Söderlund V, Larsson SA, Jacobsson H (2007) Reduction of FDG uptake in brown adipose tissue in clinical patients by a single dose of propranolol. Eur J Nucl Med Mol Imaging 34(7):1018–1022

Stanford KI, Middelbeek RJ, Townsend KL, An D, Nygaard EB, Hitchcox KM, Markan KR, Nakano K, Hirshman MF, Tseng Y-H (2012) Brown adipose tissue regulates glucose homeostasis and insulin sensitivity. J Clin Invest 123(1):215–223

Stowell KM (2008) Malignant hyperthermia: a pharmacogenetic disorder. Pharmacogenomics 9:1657–1672

Strieleman PJ, Schalinske KL, Shrago E (1985) Fatty acid activation of the reconstituted brown adipose tissue mitochondria uncoupling protein. J Biol Chem 260(25):13402–13405

Strosberg AD, Pietri-Rouxel F (1996) Function and regulation of the β3-adrenoceptor. Trends Pharmacol Sci 17(10):373–381

Sugita J, Yoneshiro T, Hatano T, Aita S, Ikemoto T, Uchiwa H, Iwanaga T, Kameya T, Kawai Y, Saito M (2013) Grains of paradise (Aframomum melegueta) extract activates brown adipose tissue and increases whole-body energy expenditure in men. Br J Nutr 110(4):733–738

Sugita J, Yoneshiro T, Sugishima Y, Ikemoto T, Uchiwa H, Suzuki I, Saito M (2014) Daily ingestion of grains of paradise (Aframomum melegueta) extract increases whole-body energy expenditure and decreases visceral fat in humans. J Nutr Sci Vitaminol 60(1):22–27

Yoo HS, Qiao L, Bosco C, Leong L-H, Lytle N, Feng G-S, Chi N-W, Shao J (2014) Intermittent cold exposure enhances fat accumulation in mice. PLoS One 9(5):e96432

Tai T-AC, Jennermann C, Brown KK, Oliver BB, MacGinnitie MA, Wilkison WO, Brown HR, Lehmann JM, Kliewer SA, Morris DC (1996) Activation of the nuclear receptor peroxisome proliferator-activated receptor γ promotes brown adipocyte differentiation. J Biol Chem 271 (47):29909–29914

Tainter ML, Cutting WC, Stockton A (1934) Use of dinitrophenol in nutritional disorders: a critical survey of clinical results. Am J Public Health Nat Health 24(10):1045–1053

Tajino K, Hosokawa H, Maegawa S, Matsumura K, Dhaka A, Kobayashi S (2011) Cooling-sensitive TRPM8 is thermostat of skin temperature against cooling. PLoS One 6(3):e17504

Talukdar S, Zhou Y, Li D, Rossulek M, Dong J, Somayaji V, Weng Y, Clark R, Lanba A, Owen BM, Brenner MB, Trimmer JK, Gropp KE, Chabot JR, Erion DM, Rolph TP, Goodwin B, Calle RA (2016) A long-acting FGF21 molecule, PF-05231023, decreases body weight and improves lipid profile in non-human primates and type 2 diabetic subjects. Cell Metab 23(3):427–440. https://doi.org/10.1016/j.cmet.2016.02.001

Tata J, Ernster L, Lindberg O (1962) Control of basal metabolic rate by thyroid hormones and cellular function. Nature 193(4820):1058–1060

Tatsumi M, Engles JM, Ishimori T, Cohade C, Wahl RL (2004) Intense ^{18F}-FDG uptake in brown fat can be reduced pharmacologically. J Nucl Med 45(7):1189–1193

Timmons JA, Wennmalm K, Larsson O, Walden TB, Lassmann T, Petrovic N, Hamilton DL, Gimeno RE, Wahlestedt C, Baar K (2007) Myogenic gene expression signature establishes that brown and white adipocytes originate from distinct cell lineages. Proc Natl Acad Sci 104 (11):4401–4406

Toubro S, Astrup A, Hardmann M (1993) A double-blind randomized 14 day trial of the effect of the β-3 agonist ICI D-7114 on 24 h energy expenditure and substrate oxidation in adipose patients. Int J Obes 17:S73

Trayhurn P, Thurlby P, James W (1977) Thermogenic defect in pre-obese Ob/Ob mice. Nature 266 (5597):60

Trayhurn P, Goodbody AE, James WP (1982) A role for brown adipose tissue in the genesis of obesity? Studies on experimental animals. Proc Nutr Soc 41(2):127–131

Tseng Y-H, Cypess AM, Kahn CR (2010) Cellular bioenergetics as a target for obesity therapy. Nat Rev Drug Discov 9(6):465

Uslu L, Donig J, Link M, Rosenberg J, Quon A, Daldrup-Link HE (2015) Value of ^{18F}-FDG PET and PET/CT for evaluation of pediatric malignancies. J Nucl Med 56:274–286

Vallerand AL, Perusse F, Bukowiecki LJ (1987) Cold exposure potentiates the effect of insulin on in vivo glucose uptake. Am J Physiol Endocrinol Metab 253(2):E179–E186

van Baak MA, Hul GB, Toubro S, Astrup A, Gottesdiener KM, DeSmet M, Saris WH (2002) Acute effect of L-796568, a novel β3-adrenergic receptor agonist, on energy expenditure in obese men. Clin Pharmacol Ther 71(4):272–279

van der Lans AA, Hoeks J, Brans B, Vijgen GH, Visser MG, Vosselman MJ, Hansen J, Jörgensen JA, Wu J, Mottaghy FM (2013) Cold acclimation recruits human brown fat and increases nonshivering thermogenesis. J Clin Invest 123(8):3395–3403

van Marken Lichtenbelt WD, Schrauwen P (2011) Implications of nonshivering thermogenesis for energy balance regulation in humans. Am J Phys Regul Integr Comp Phys 301(2):R285–R296

van Marken Lichtenbelt WD, Schrauwen P, van de Kerckhove S, Westerterp-Plantenga MS (2002) Individual variation in body temperature and energy expenditure in response to mild cold. Am J Physiol Endocrinol Metab 282(5):E1077–E1083

van Marken Lichtenbelt WD, Vanhommerig JW, Smulders NM, Drossaerts JM, Kemerink GJ, Bouvy ND, Schrauwen P, Teule GJ (2009) Cold-activated brown adipose tissue in healthy men. N Engl J Med 360(15):1500–1508

van Marken Lichtenbelt W, Kingma B, van der Lans A, Schellen L (2014) Cold exposure – an approach to increasing energy expenditure in humans. Trends Endocrinol Metab 25(4):165–167

Vijgen GH, Bouvy ND, Teule GJ, Brans B, Schrauwen P, van Marken Lichtenbelt WD (2011) Brown adipose tissue in morbidly obese subjects. PLoS One 6(2):e17247

Villarroya J, Cereijo R, Villarroya F (2013) An endocrine role for brown adipose tissue? Am J Physiol Endocrinol Metab 305(5):E567–E572

Virtanen KA, Lidell ME, Orava J, Heglind M, Westergren R, Niemi T, Taittonen M, Laine J, Savisto N-J, Enerbäck S (2009) Functional brown adipose tissue in healthy adults. N Engl J Med 360(15):1518–1525

Vosselman MJ, van der Lans AA, Brans B, Wierts R, van Baak MA, Schrauwen P, van Marken Lichtenbelt WD (2012) Systemic β-adrenergic stimulation of thermogenesis is not accompanied by brown adipose tissue activity in humans. Diabetes 61:3106–3113

Waldén TB (2010) Regulatory factors that reveal three distinct adipocytes: the brown, the white and the brite. The Wenner-Gren Institute, Stockholm University, Stockholm

Weyer C, Tataranni PA, Snitker S, Danforth E, Ravussin E (1998) Increase in insulin action and fat oxidation after treatment with CL 316,243, a highly selective beta3-adrenoceptor agonist in humans. Diabetes 47(10):1555–1561

Weyer C, Bogardus C, Mott DM, Pratley RE (1999a) The natural history of insulin secretory dysfunction and insulin resistance in the pathogenesis of type 2 diabetes mellitus. J Clin Invest 104(6):787–794

Weyer C, Gautier J, Danforth E Jr (1999b) Development of beta 3-adrenoceptor agonists for the treatment of obesity and diabetes an update. Diabetes Metab 25:11–21

Wheeldon N, McDevitt D, Lipworth B (1993) Do β3-adrenoceptors mediate metabolic responses to isoprenaline. Int J Med 86(9):595–600

Wheeldon N, McDevitt D, McFarlane L, Lipworth B (1994) β-Adrenoceptor subtypes mediating the metabolic effects of BRL 35135 in man. Clin Sci 86(3):331–337

Wijers SL, Saris WH, van Marken Lichtenbelt WD (2007) Individual thermogenic responses to mild cold and overfeeding are closely related. J Clin Endocrinol Metabol 92(11):4299–4305

Wijers SL, Schrauwen P, Saris WH, van Marken Lichtenbelt WD (2008) Human skeletal muscle mitochondrial uncoupling is associated with cold induced adaptive thermogenesis. PLoS One 3 (3):e1777

Wijers SL, Saris WH, Lichtenbelt WDM (2010) Cold-induced adaptive thermogenesis in lean and obese. Obesity 18(6):1092–1099

Wijers SL, Schrauwen P, van Baak MA, Saris WH, van Marken Lichtenbelt WD (2011) β-Adrenergic receptor blockade does not inhibit cold-induced thermogenesis in humans: possible involvement of brown adipose tissue. J Clin Endocrinol Metabol 96(4):E598–E605

Wilson S, Thurlby PL, Arch JR (1987) Substrate supply for thermogenesis induced by the beta-adrenoceptor agonist BRL 26830A. Can J Physiol Pharmacol 65(2):113–119

Wolfe RR, Herndon DN, Jahoor F, Miyoshi H, Wolfe M (1987) Effect of severe burn injury on substrate cycling by glucose and fatty acids. N Engl J Med 317(7):403–408

Wolfe RR, Klein S, Carraro F, Weber J-M (1990) Role of triglyceride-fatty acid cycle in controlling fat metabolism in humans during and after exercise. Am J Physiol Endocrinol Metab 258(2): E382–E389

Yang X, Enerbäck S, Smith U (2003) Reduced expression of FOXC2 and brown adipogenic genes in human subjects with insulin resistance. Obes Res 11(10):1182–1191

Yazawa S, Suetom N, Okamoto K, Namiki T (1989) Content of capsaicinoids and capsaicinoid-like substances in fruit of pepper (Capsicum annuum L.) hybrids made with 'CH-19 sweet' as a parent. J Jpn Soc Hortic Sci 58(3):601–607

Ye X, Qi J, Ren G, Xu P, Wu Y, Zhu S, Yu D, Li S, Wu Q, Muhi RL (2015) Long-lasting anti-diabetic efficacy of PEGylated FGF-21 and liraglutide in treatment of type 2 diabetic mice. Endocrine 49(3):683–692

Yen T (1984) The antiobesity and metabolic activities of LY79771 in obese and normal mice. Int J Obes 8(1):69–78

Yen T, McKee M, Stamm N (1984) Thermogenesis and weight control. Int J Obes 8:65–78

Yen T, Fuller R, Hemrick-Luecke S, Dininger N (1988) Effects of LY104119, a thermogenic weight-reducing compound, on norepinephrine concentrations and turnover in obese and lean mice. Int J Obes 12(1):59–67

Yenjai C, Prasanphen K, Daodee S, Wongpanich V, Kittakoop P (2004) Bioactive flavonoids from Kaempferia parviflora. Fitoterapia 75(1):89–92

Yoneshiro T, Saito M (2013) Transient receptor potential activated brown fat thermogenesis as a target of food ingredients for obesity management. Curr Opin Clin Nutr Metab Care 16 (6):625–631

Yoneshiro T, Saito M (2015) Activation and recruitment of brown adipose tissue as anti-obesity regimens in humans. Ann Med 47(2):133–141

Yoneshiro T, Aita S, Matsushita M, Okamatsu-Ogura Y, Kameya T, Kawai Y, Miyagawa M, Tsujisaki M, Saito M (2011a) Age-related decrease in cold-activated brown adipose tissue and accumulation of body fat in healthy humans. Obesity 19(9):1755–1760

Yoneshiro T, Aita S, Matsushita M, Kameya T, Nakada K, Kawai Y, Saito M (2011b) Brown adipose tissue, whole-body energy expenditure, and thermogenesis in healthy adult men. Obesity 19(1):13–16

Yoneshiro T, Aita S, Kawai Y, Iwanaga T, Saito M (2012) Nonpungent capsaicin analogs (capsinoids) increase energy expenditure through the activation of brown adipose tissue in humans. Am J Clin Nutr 95(4):845–850

Yoneshiro T, Aita S, Matsushita M, Kayahara T, Kameya T, Kawai Y, Iwanaga T, Saito M (2013) Recruited brown adipose tissue as an antiobesity agent in humans. J Clin Invest 123 (8):3404–3408

Yoneshiro T, Matsushita M, Hibi M, Tone H, Takeshita M, Yasunaga K, Katsuragi Y, Kameya T, Sugie H, Saito M (2017) Tea catechin and caffeine activate brown adipose tissue and increase cold-induced thermogenic capacity in humans, 2. Am J Clin Nutr 105(4):873–881

Yoshida T, Sakane N, Wakabayashi Y, Umekawa T, Kondo M (1994) Anti-obesity and anti-diabetic effects of CL 316, 243, a highly specific β3-adrenoceptor agonist, in yellow KK mice. Life Sci 54(7):491–498

Yoshino S, Kim M, Awa R, Kuwahara H, Kano Y, Kawada T (2014) Kaempferia parviflora extract increases energy consumption through activation of BAT in mice. Food Sci Nutr 2(6):634–637

Yoshioka M, St-Pierre S, Suzuki M, Tremblay A (1998) Effects of red pepper added to high-fat and high-carbohydrate meals on energy metabolism and substrate utilization in Japanese women. Br J Nutr 80(6):503–510

Yoshitomi H, Yamazaki K, Abe S, Tanaka I (1998) Differential regulation of mouse uncoupling proteins among brown adipose tissue, white adipose tissue, and skeletal muscle in chronic β3Adrenergic receptor agonist treatment. Biochem Biophys Res Commun 253(1):85–91

Young P, Cawthorne M, Levy AL, Wilson K (1984) Reduced maximum capacity of glycolysis in brown adipose tissue of genetically obese, diabetic (db/db) mice and its restoration following treatment with a thermogenic β-adrenoceptor agonist. FEBS Lett 176(1):16–20

Young P, Cawthorne M, Smith S (1985) Brown adipose tissue is a major site of glucose utilisation in C57B16obob mice treated with a thermogenic β-adrenoceptor agonist. Biochem Biophys Res Commun 130(1):241–248

Zhang J, Li Y (2014) Fibroblast growth factor 21, the endocrine FGF pathway and novel treatments for metabolic syndrome. Drug Discov Today 19(5):579–589

Zhang X, Yeung DC, Karpisek M, Stejskal D, Zhou Z-G, Liu F, Wong RL, Chow W-S, Tso AW, Lam KS (2008) Serum FGF21 levels are increased in obesity and are independently associated with the metabolic syndrome in humans. Diabetes 57:1246–1253

Zhang Q, Miao Q, Ye H, Zhang Z, Zuo C, Hua F, Guan Y, Li Y (2014) The effects of thyroid hormones on brown adipose tissue in humans: a PET-CT study. Diabetes Metab Res Rev 30 (6):513–520

Zingaretti MC, Crosta F, Vitali A, Guerrieri M, Frontini A, Cannon B, Nedergaard J, Cinti S (2009) The presence of UCP1 demonstrates that metabolically active adipose tissue in the neck of adult humans truly represents brown adipose tissue. FASEB J 23(9):3113–3120